SPIDER MITES
THEIR BIOLOGY,
NATURAL ENEMIES
AND CONTROL

World Crop Pests

Editor-in-Chief
W. Helle
University of Amsterdam
Laboratory of Experimental Entomology
Kruislaan 302
1098 SM Amsterdam
The Netherlands

Volumes in the Series

Spider Mites. Their Biology, Natural Enemies and Control. A
(ISBN 0-444-42372-9)
Spider Mites. Their Biology, Natural Enemies and Control. B
(ISBN 0-444-42374-5)
Aphids. Their Biology, Natural Enemies and Control (2 volumes)

SPIDER MITES THEIR BIOLOGY, NATURAL ENEMIES AND CONTROL

Volume 1A

Edited by

W. HELLE

*University of Amsterdam, Laboratory of Experimental Entomology,
Amsterdam, The Netherlands*

M.W. SABELIS

*University of Leiden, Department of Population Biology,
Leiden, The Netherlands*

ELSEVIER

Amsterdam — Oxford — New York — Tokyo   1985

ELSEVIER SCIENCE PUBLISHERS B.V.
Sara Burgerhartstraat 25
P.O. Box 211, 1000 AE Amsterdam, The Netherlands

Distributors for the United States and Canada:

ELSEVIER SCIENCE PUBLISHING COMPANY INC.
52, Vanderbilt Avenue
New York, NY 10017

ISBN 0-444-42372-9 (Vol. 1A)
ISBN 0-444-42373-7 (Series)

Printed in The Netherlands

Preface

The study of spider mites has made steady progress since World War II and has experienced a meteoric growth in the past decade. The impetus of many of these studies derives from the agricultural importance of spider mites as plant parasites of many crops all over the world. But, basic studies on spider mites have also contributed significantly to our knowledge, broadening our understanding of their biology and opening new vistas for further exploration. Unfortunately, reports on basic and applied aspects are scattered and it is impossible to keep oneself up-to-date in all facets of spider mite biology. Jeppson, Keifer and Baker (1975) attempted to compile the world's knowledge on mites injurious to economic plants. Their book is tremendously important in that it covers all acarine families containing economically important phytophagous mites, but the comprehensiveness of their objective is inevitably at the expense of profundity. Hence, we feel that there is not only a need for an up-to-date text, but also for bringing together facts and views of acarologists specialized in various aspects of the biology of spider mites. Certainly, the need for such a treatment of scientific progress and recommended topics for future research exists among students commencing in the study of acarology and plant protection, as well as among those engaged in acarological research and teaching. To fulfil this need the book on 'Spider Mites: Their Biology, Natural Enemies and Control' has been written by more than 50 specialists from all over the world.

Like all volumes in Elsevier's series on 'World Crop Pests', this book is divided into three parts:

1. a detailed treatise of the pest organism;
2. an analysis of its interaction with possible agents for biological control;
3. a discussion of the status and future trends of pest control in various crops.

Part 1 gives an account of today's knowledge on anatomy, phylogeny and systematics (Chapter 1), reproduction and development (Chapter 2), physiology and genetics (Chapter 3) and ecology (Chapter 4). There is an additional chapter on a number of acarological techniques.

Part 2 deals with the natural enemies of spider mites and focuses on phytoseiid predators, the most successful agents of biological control. Apart from phytoseiid mites, other predaceous acarines, predaceous insects and pathogens are considered. A section on techniques is also included in this part.

Part 3 not only gives an elaborate account of biological control and integrated control of spider mites, but also highlights topics of general interest to crop protection, such as damage assessment, host-plant resistance, pesticide resistance and genetic control. Moreover, there is a section on weed control, using spider mites as natural enemies of weeds.

Although the book contains a fairly full exposition of the matters treated, some topics are excluded. Clearly, striving for completeness would soon lead to an unmanageable edifice. Limiting the scope of the book, we have decided in favour of reviewing evolutionary aspects of spider mites in some depth, this at the expense of certain other aspects (e.g. population dynamics). Several authors in Part 1 and 2 have put their presentation in an evolutionary perspective. Accordingly, a special section is devoted to the adaptive consequences of male haploidy.

We trust that this book will serve several useful functions — provide a synthesis of much of the knowledge on basic and applied aspects of the biology of spider mites and their natural enemies; stimulate students to analyse critically the views propounded by the authors of this book; and instigate research into environmentally safe and cost effective means of pest control.

W. Helle   M.W. Sabelis

Plate I. Diapause induced in biparental sex forms of the two-spotted spider mite *Tetrany-chus urticae* Koch (see chapter 1.2.3, p. 136 and chapter 1.3.2, p. 190). The female at the left shows the orange coloration, typical of the diapause form, throughout the entire body. The intersex at the right exhibits the orange coloration in the anterior (= female) part of the body, but not in the posterior (= male) part. The intersex ('giant male') in the middle is predominantly male in appearance and lacks the orange coloration.

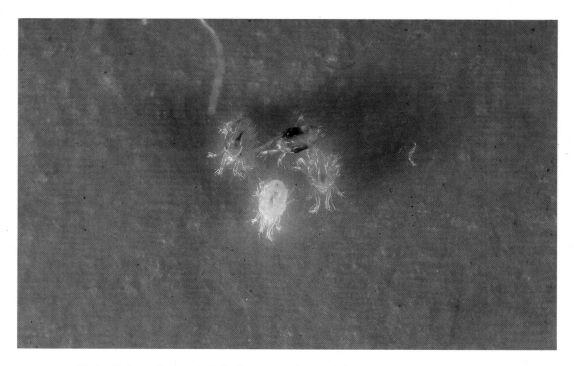

Plate II (see chapter 1.4.6). Summer (above, right) and diapausing (above, left) females of the wild type and diapausing females of albino (below) and lemon (below, right) mutants of *Tetranychus urticae*. (By courtesy of W. Helle.)

Contents

Contributors to this Volume

G. ALBERTI
Zoologisches Institut, Universität Heidelberg, D-6900 Heidelberg
R. DE BOER
Laboratory of Experimental Entomology, University of Amsterdam, 1098
SM Amsterdam
W.W. CONE
Washington State University, Irrigated Agriculture Research and Extension
Center, Prosser, Washington 99350-0030
A. CROOKER
University of Washington, Center for Bioengineering, WD-12, Seattle, WA
98195
R.H. CROZIER
The University of New South Wales, Kensington, NSW
L.J. DRENTH-DIEPHUIS
Dept. of Genetics, University of Groningen, 9751 NN Haren (GN)
C.C.M. FEIERTAG-KOPPEN
University of Groningen, 9752 VZ Haren (GN)
M.A. FERWERDA
Dept. of Genetics, University of Groningen, 9751 NN Haren (GN)
L.P.S. VAN DER GEEST
Laboratory of Experimental Entomology, University of Amsterdam, 1098
SM Amsterdam
U. GERSON
The Hebrew University of Jerusalem, Faculty of Agriculture, Rehovot
76-100
J. GUTIERREZ
Centre O.R.S.T.O.M., c/o I.A.M., Montpellier
W. HELLE
Laboratory of Experimental Entomology, University of Amsterdam, 1098
SM Amsterdam
G.G. KENNEDY
North Carolina State University, School of Agriculture and Life Sciences,
Dept of Entomology, Raleigh, NC 27695-7630
D. KROPCZYNSKA
Agricultural University of Warsaw, Dept. of Applied Entomology, 02-766
Warszawa
E.E. LINDQUIST
Biosystematics Research Institute, Ottawa, Ontario, K1A 0C6

W.P.J. OVERMEER
Laboratory of Experimental Entomology, University of Amsterdam, 1098
  SM Amsterdam
L.P. PIJNACKER
Dept. of Genetics, University of Groningen, 9751 NN Haren (GN)
M.W. SABELIS
Dept. of Population Biology, University of Leiden, 2300 RA Leiden
Y. SAITÔ
Institute of Applied Zoology, Faculty of Agriculture, Hokkaido University,
  060 Sapporo
D.R. SMITLEY
North Carolina State University, School of Agriculture and Life Sciences,
  Dept. of Entomology, Raleigh, NC 27695-7630
M.J. TEMPELAAR
Dept. of Genetics, University of Groningen, 9751 NN Haren (GN)
A. TOMCZYK
Agricultural University of Warsaw, Dept. of Applied Entomology, 02-766
  Warszawa
A. VEERMAN
Laboratory of Experimental Entomology, University of Amsterdam, 1098
  SM Amsterdam
F. WEYDA
Institute of Entomology, 37005 Ceské Budejovice
D.L. WRENSCH
Ohio State University, Dept. of Genetics, Columbus, Ohio 43210-1120

PART 1

THE TETRANYCHIDAE

Chapter 1.1 Anatomy, Phylogeny and Systematics

1.1.1 External Anatomy

E.E. LINDQUIST

GNATHOSOMA

In all spider mites and their relatives in the superfamily Tetranychoidea, the cheliceral bases are consolidated to form a stylophore (Figs. 1.1.1.1 and 1.1.1.2). The stylophore is deeply retractable and extrudable, independently of the infracapitulum, in a manner that is unique to this superfamily. The dorsal surface of the stylophore lacks the dorsal setae of the cheliceral bases. The cheliceral stylets are inserted within the stylophore. These greatly elongated modifications of the movable cheliceral digits are strongly recurved basally, thus allowing for a greater degree of retraction and protraction than in mites of other superfamilies also having a stylophore (Raphignathoidea, Cheyletoidea). The stylophore retains a fenestrated median vertical septum which derives from the coalesced median walls of the fused cheliceral bases and separates one basal cheliceral chamber from the other (Fig. 1.1.1.2). Within these chambers, each stylet attaches basally to a cheliceral lever which is oriented along the longitudinal and vertical axes of the gnathosoma, with its *base* inserted near the anterodorsal extremity of the stylophore (Figs. 1.1.1.2 and 1.1.1.3). The levers thus activate or pivot the stylets along a vertical plane. The levers' mechanism of forward thrust of the reverse-curved stylets has been clarified recently by André and Remacle (1984). The margins of the stylet shafts are smooth and of uniform size in cross-section, except for the few (5—10) microns nearest the apex, beyond a minute notch on either side, where they are slenderer and pointed for piercing and cutting into plant tissue (Fig. 1.1.1.2). When protruded, the stylets are juxtaposed to form a hollow tube; transmission electron microscopic (TEM) sections and scanning electron microscopic (SEM) micrographs reveal that the shafts may interlock by a 'tongue and groove' arrangement for this purpose (Summers et al., 1973; Hislop and Jeppson, 1976). The formation of a single hollow probe by the stylets when protruded for feeding, as shown in various SEM micrographs (Fig. 1.1.1.52; see also figures in Summers et al., 1973; Hislop and Jeppson, 1976; Tanigoshi and Davis, 1978; André and Remacle, 1984), apparently is a unique adaptation for feeding on vascular plants that is characteristic for all families of Tetranychoidea. Associated with the basal half of each stylet is a hyaline pointed process which may represent the vestige of the fixed cheliceral digit (Figs. 1.1.1.2 and 1.1.1.3). This structure appears to ensheath the stylet dorsally, and nearly to the extremity of the stylet when the latter is fully retracted. The base of the stylophore has a deep invagination medially, in which a pair of stigmata are located. From these, a pair of conspicuous peritremes arises which continues onto, and is embedded in, the membranous anterior surface of

Chapter 1.1.1. references, p. 26

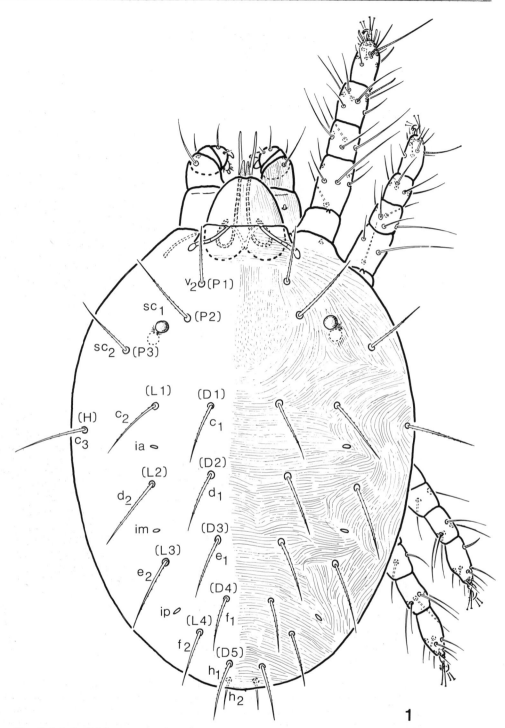

Fig. 1.1.1.1. *Schizotetranychus* sp., dorsal view of deutonymph, with equivalent notations for dorsal body setae as used in this chapter (primarily after Grandjean, 1939) and, in parentheses, as used in Chapter 1.1.4 (primarily after Pritchard and Baker, 1955).

the prodorsum. Since the peritremes are embedded in part of the membranous cuticle that ensheaths the stylophore, the degree of protraction or retraction of the stylophore affects the position and degree of exposure of the peritremes to the outer atmosphere (this is well shown by a series of illustrations, figs. 18—20, in Blauvelt, 1945). The peritremes are not closed tubes but open via a continuous slit along their outer wall from their origin

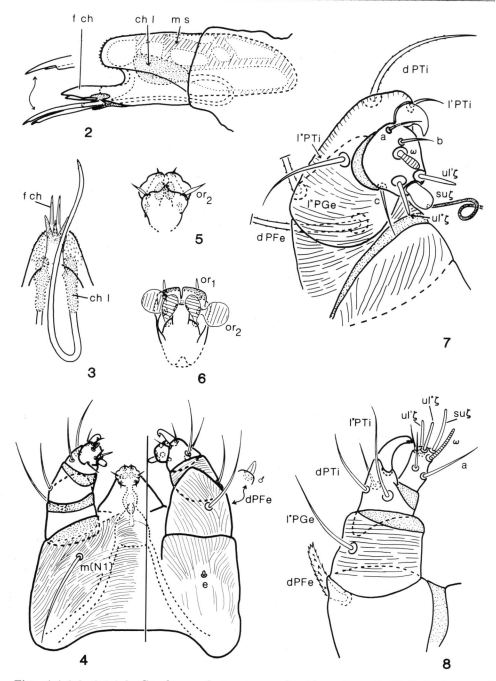

Figs. 1.1.1.2—1.1.1.8. Gnathosomal structures of spider mites. (2, 3) Stylophore and cheliceral stylets of *Lindquistiella* sp: (2) lateral aspect, with enlarged view of apex of stylet, (3) ventral aspect. (4) Infracapitulum and palpi of *Tetranychus* sp., ventral aspect on the left side, dorsal aspect on the right, with separate view of seta *d*PFe as modified on adult male. (5, 6) Distal extremity of infracapitulum, ventral aspect: (5) *Tetranychus* sp, (6) *Bryobia* sp. (7, 8). Palpi, ventrolateral aspect: (7) *Tetranychus* sp., (8) *Bryobia* sp. Abbreviations: ch l, cheliceral lever; f ch, fixed cheliceral process; m s, median septum within stylophore. See text for setal notation.

at the stigmata to their apices (Figs. 1.1.1.49 and 1.1.1.50; see also figs. 15 and 17 in Blauvelt (1945) and SEM micrograph plate 18 in Jeppson et al. (1975)). In some genera, primarily in the Bryobiinae, the peritrematal apices are enlarged and emerge as horn-like processes from the prodorsum. As is found in other families of Tetranychoidea, the stylophore base is attached to a membranous structure which functions like a sheath in facilitating projection and retraction of the stylophore. However, this structure is not reinforced by a discernible longitudinal ribbing, which superficially resembles a 'turtleneck' sheath, as in the Linotetranidae and Tenuipalpidae. This may be because the stylophore is not so deeply retractable in spider mites as in these other families. A pair of inconspicuous slit-like structures, which André and Remacle (1984) regarded as a second pair of stigmata, opens ventrolaterally near the base of this sheath in spider mites. The inferior surface of the stylophore is keeled and fits into, and slides along, the longitudinally trough-like dorsomedial surface, the rostral gutter, or cervix, of the infracapitulum (Blauvelt, 1945; Summers et al., 1973). The functional morphology of the cervical region has been newly elucidated by André and Remacle (1984).

The ventral surface of the infracapitulum usually bears 1 pair of subcapitular setae, m, which first appear on the protonymph and are inserted subproximally at a level between and slightly behind the bases of the palpi (Fig. 1.1.1.4). Between these setae the sclerotized walls of the pharynx are readily evident internally. Distad of the bases of the palpi, the infracapitulum narrows abruptly into the buccal cone which bears the mouth, flanked by hyaline lateral lips and 3 pairs of usually inconspicuous adoral setae, or_{1-3}, apically. The membranous margins of the lateral lips may be fringed, so as to appear fimbriate or denticulate, amidst the adoral setae. The adoral setae are sometimes hyaline and modified in shape, rendering them even less recognizable as setae, though birefringency persists in their structure (Figs. 1.1.1.5 and 1.1.1.6). The ventral surface of the buccal cone bears a median hollow subapically, the rostral fossette; this structure is a manifestation of the inferior oral commissure (André and Remacle, 1984), and it is situated directly below the confluence of the pre-oral area and the lumen of the pharynx (Fig. 1.1.1.54; see also SEM micrograph figs. 13 and 14 in Summers et al. (1973) and fig. 1D in Hislop and Jeppson (1976)).

The pair of palpal supracoxal setae, e, evident dorsally at the bases of the palpi in other families of Tetranychoidea, is present but often difficult to discern on spider mites (Fig. 1.1.1.4). The palpi themselves are consistently well developed, with 4 or 5 segments including a conspicuous 'thumb-claw' process, in contrast to the tendency towards structural reduction evident among tenuipalpid mites. However, the trochanter is sometimes indistinctly separated from the femur in spider mites. The chaetotaxy of the palpal segments is constant in the Tetranychidae, except for the complement on the tarsal 'thumb'; the trochanter is glabrous, the femur has 1 dorsal seta, the genu has 1 posterolateral seta, and the tibia has a dorsal 'claw' and 3 setae (1 dorsal, 1 anterolateral, 1 posterolateral) (Figs. 1.1.1.7 and 1.1.1.8). The palptarsal 'thumb' bears 7 setiform structures or phaneres: 3 simple setae (a, b, c), 3 eupathidia (ul', ul'', su), and 1 solenidion (Figs. 1.1.1.7 and 1.1.1.8). In some genera such as *Bryobia*, the solenidion may be indistinguishable in form from the eupathidia when viewed under a transmitted light microscope, but it remains distinguishable by its lack of birefringence in polarized light. In bryobiine spider mites, which do not produce silk, the 3 eupathidia are similar in form (Fig. 1.1.1.8). However, in the web-producing tetranychines, eupathidium su is much thicker than the other 2 and a hollow tip is evident, through which a silken substance is extruded from the silk gland that occupies most of the space within the palpus (Fig. 1.1.1.7; see SEM micrograph fig. 16C in Robaux and Gutierrez (1973)). All of these palpal structures are present on the larva and remain unaltered during post-larval development except for femoral seta d, which sometimes changes into a short, stout, partly sunken spine-like process on the adult male (Fig.

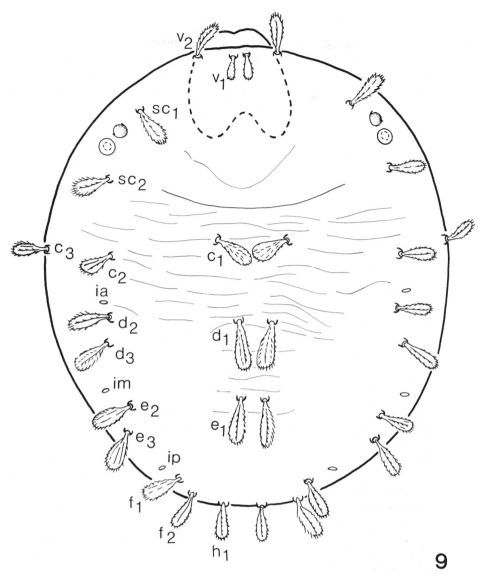

Fig. 1.1.1.9. *Bryobia* sp., larva, dorsal view of idiosoma, with notation for dorsal body setae after Oudemans (1928) for prodorsum and Grandjean (1939) for opisthosoma.

1.1.1.4). This dimorphism of palpfemoral seta *d* is generally present among the genera of Tetranychinae but is sporadic in the Bryobiinae, even within 1 genus (e.g., *Petrobia*).

PRODORSUM

The prodorsum of spider mites is rather simple in structure, bearing 3, or more rarely 4, pairs of setae and 2 pairs of eyes (Figs. 1.1.1.1 and 1.1.1.9). As in the other families of Tetranychoidea, none of these setae is modified as a trichobothrium, though they vary considerably in size and form. The anteriormost pair of setae, v_1 (commonly called the inner verticals in other prostigmatic mites), is present or absent among the bryobiine genera, or is rarely represented by an unpaired seta anteromedially, as in *Mezranobia*; this pair is consistently absent among the tetranychine genera. Neotrichy does not occur on the prodorsum of any spider mites, so far as is known.

The eyes are consistently present laterally between setae sc_1 and sc_2 (commonly called the inner and outer scapulars in other prostigmatic mites), which appear as the second and third pairs of prodorsal setae in the absence of v_1 in most tetranychids. The two eyes on either side, surrounded by striated integument, are nearly contiguous, the anterior one having a smooth, double-convex lens whereas the posterior one generally has a minutely striated surface which is nearly flat in the Tetranychinae (Fig. 1.1.1.51; see also SEM micrograph plate 20 in Jeppson et al. (1975)) but may retain a more convex surface in the Bryobiinae, as in the Tenuipalpidae and Tuckerellidae. Differences in ultrastructure as well as form of the eyes suggest that they also have different functions: the lens-like anterior eyes may sense differences in images and colours, while the posterior eyes may sense differences in light intensity (Nuzzaci, 1982).

The prodorsum may be simply striated or it may be more distinctively ornamented with puncta, reticula, and the like. A more or less distinct shield is present in some genera, and there may be from 1 to 4 anterior projections, which usually bear setae, among some of the genera of Bryobiinae; smaller seta-bearing tubercles may be present among the genera of both subfamilies. The setae and eyes are present on the larval and post-larval instars, but the shielding and projections may not appear until the adult stage. The prodorsum is delimited posteriorly from the opisthosoma by the sejugal furrow, which is at least evident by a change in pattern of the cuticular striae.

OPISTHOSOMAL DORSUM

Larvae and post-larval instars of all families of Tetranychoidea retain the setal and cupule elements of the 6 opisthosomal segments characteristic of larval acariform mites generally (Figs. 1.1.1.1 and 1.1.1.9). The standard notation of Grandjean (1939) is used here for these structures to clarify segmental, setal, and cupule homologies, and to facilitate comparison of these structures among the families of Tetranychoidea and in turn among other superfamilies of eleutherengone mites. This notation has been similarly applied recently by Quirós and Baker (1984) to the other families of Tetranychoidea. The symbols C, D, E, F, H, and PS indicate the 6 larval opisthosomal segments, with the pseudanal segment, PS, typically reduced in size and occupying a ventrocaudal position (Figs. 1.1.1.10 and 1.1.1.14). The cupules, which are stress receptors also called lyrifissures (Fig. 1.1.1.56), are often difficult to discern on ornamented integument but are consistent in presence and location relative to the setae of certain segments: cupules *ia* are in the region of segment D, usually anteriad of setae d_2; *im* are in the region of segment E, anteriad or anterolaterad of e_2; *ip*, in the region of segment F, are usually anterolaterad of f_2; *ih*, in the ventrolateral region of segment H, are laterad of ps_1; *ips* are absent, as is typical of taxa which do not add segments during post-larval development (Grandjean, 1939). Table 1.1.1.1 compares the notation of Grandjean with those which have traditionally been applied to tetranychid mites for the last 35 years, beginning with the revisionary works in English by Pritchard and Baker (1955) on the one hand and in Russian by Rekk (1947, 1959) on the other. These traditional notations, derived from an earlier scheme by Oudemans (1928) and used by Geijskes (1939), were applied primarily to tetranychine spider mites; they were inherently difficult to apply to bryobiine mites in which the 'sacral' and 'clunal' setae are not positioned so that they can readily be recognized as such.

The opisthosoma of spider mites generally bears a maximum of 14 pairs of 'dorsal' setae, i.e., setae belonging to segments C, D, E, F, H; usually only 12

TABLE 1.1.1.1

Comparison of systems of notation applied to equivalent setae of opisthosoma of tetranychid mites (all setae larval excepting aggenitals and genitals, as indicated)

Grandjean (1939, 1947)		Pritchard and Baker (1955)		Oudemans (1928) Geijskes (1939) Rekk (1947, 1959)	
C	c_1	1st Dorsocentrals	DC1	Inner humerals	Humerals
	c_2	1st (Dorso)laterals	L1	Middle humerals	
	c_3	Humerals or 1st Sublaterals	H	Outer humerals	
D	d_1	2nd Dorsocentrals	DC2	Inner dorsals or prelumbars	Dorsals or prelumbars
	d_2	2nd (Dorso)laterals	L2	Middle dorsals or prelumbars	
	d_3	2nd Sublaterals	—	Outer dorsals or prelumbars	
E	e_1	3rd Dorsocentrals	DC3	Inner lumbals or lumbars	Lumbars
	e_2	3rd (Dorso)laterals	L3	Middle lumbals or lumbars	
	e_3	3rd Sublaterals	—	Outer lumbals or lumbars	
F	f_1	Inner sacrals or 4th dorsocentrals	DC4	Inner sacrals	Sacrals
	f_2	Outer sacrals or 4th (dorso)laterals	L4	Outer sacrals	
H	$h_1(?f_3)$	Clunals or 5th dorsocentrals	DC5	Clunals or caudals	
	h_2	Postanals or Posterior para-anals	—	Posterior postanals	
	h_3	Anterior para-anals	—	Anterior postanals	
PS	ps_{1-3}	Anals	—	Anals	
	g_1(Dn)	1st Genitals (anteromedial)	—	Epigynials	
	g_2(Ad)	2nd Genitals (posterolateral)	—	Intermedials	
	ag(Pn)	Pregenitals	—	Pre-epigynials	

pairs of these setae are evident as dorsals, since setae h_{2-3} are smaller and inserted ventrocaudally near the anal opening (referred to as 'para-anals' or 'post-anals' in preceding works). This maximum is also the primitive, or plesiomorphic, condition and is found in various genera of Bryobiinae (Fig. 1.1.1.9). Excluding the *ps* setae, this basic complement of 14 pairs of dorsal opisthosomal setae considerably exceeds that found in raphignathoid and most other eleutherengonine and eupodine mites exclusive of the anystoid—parasitengone assemblage, which usually have 8—10 pairs including c_{1-2}, d_1 (and occasionally d_2), e_1 (and occasionally e_2), f_{1-2}, and h_{1-2}. Therefore, the basic condition in spider mites appears to exhibit neotrichy of the idionymous oligotrichous (denotable) type, with c_3, d_3, e_3, and possibly h_3 regarded as neotrichous setae. These additional setae are also basically present in the other families of Tetranychoidea, except that h_3 is absent in Tenuipalpidae. This trend of dorsal opisthosomal neotrichy is furthered in Linotetranidae, in which c_4, f_{3-4}, and h_4 are present, and more so in Tuckerellidae, in which c_{4-7}, d_{4-5}, e_4, and 5 pairs of additional *f* and *h* setae are present. These basic complements of setae are fully expressed on the larva in each family of Tetranychoidea.

Chapter 1.1.1. references, p. 26

Rarely, the basic number of 14 pairs of 'dorsal' setae in Tetranychidae is exceeded because of further neotrichy on segments C, D, E, as in *Dasyobia*, which has a doubling or trebling of the setae normally present in the *c, d, e* series; these extra setae are post-larval additions. Reductions from the 14 pairs occur in both subfamilies of spider mites. In Bryobiinae, reduction to as few as 8 pairs of setae (including ventrocaudal h_{2-3}) is found in *Marainobia*, in which c_3, d_3, e_1, e_3, f_1, h_1 apparently are absent. In Tetranychinae, the maximum number retained is 12 pairs, since d_3 and e_3 are consistently absent (Fig. 1.1.1.1); there is little further reduction among the genera of this subfamily, with only h_1 (the 'clunals' or 'caudals') being lost, as in *Tetranychus* spp. and, apparently, *Oligonychus* spp. Pritchard and Baker (1955) interpreted the clunal setae h_1 to be consistently present, and the caudal pair of para-anal (or post-anal) setae h_2 to be present or absent. Judging from the position and form of the pair of setae consistently present, these are apparently the post-anals h_2, leaving the clunal pair h_1 present or absent, as first interpreted by Oudemans (1930) and later by Pritchard and Baker (1952) and Attiah (1970). These interpretations of setal homologies were made by preceding authors primarily on the basis of their appearance on adult females, but their appearance on adult males is perhaps more convincing for the alternative interpretation that h_2 remain present.

The dorsal opisthosomal setae, except h_{2-3}, are usually similar to one another and to the prodorsal setae in size and shape; occasionally, the more posterior setae (especially f_{1-2}, h_1) are larger, or the central ones (especially d_1, e_1) are smaller, or the peripheral ones differ in form. These setae may be finely to coarsely barbed, tapered or not, and setiform, clavate, spatulate, or lanceolate; some or all of them may be inserted individually on small to large tubercles. By contrast, the ventrocaudal, or 'para-anal', setae h_{2-3} are relatively slender, setiform, finely barbed, and short, similar to the pseudanal setae *ps* and to the ventral body setae (Figs. 1.1.1.13 and 1.1.1.14). The integument of the opisthosomal dorsum is usually similar to that of the prodorsum, and it may simply be striated or otherwise more distinctively ornamented. It usually lacks shielding, but a single shield is occasionally present on the posterior half of the opisthosoma, as in *Notonychus* and *Neopetrobia*, or several plate-like areas may be demarcated over most of the opisthosoma, as in *Peltanobia* and *Neotrichobia*. In the tribe Tetranychini, the dorsal cuticle typically is striated, but the pattern or orientation of the striae, especially in the area between setae e_1 and f_1, may be of diagnostic importance for adult females. The ultrastructure of the individual striae is sometimes also of diagnostic value in distinguishing between closely related species or between diapausing and non-diapausing populations of the same species (Dosse and Boudreaux, 1963; see also TEM micrograph figs. 18 and 22 in Boudreaux and Dosse (1963) and SEM micrograph Fig. 1.1.1.56 herein and plate figs. 4 and 5 in Jeppson et al. (1975)).

PODOSOMAL VENTER

The podosomal venter is the most stable region, with the least number of character states, of the body of tetranychoid mites. Generally, it bears 3 pairs of prominent simple setae, known as ventral or intercoxal, setae, of which the first or anterior pair, *1a*, is inserted between the bases of legs I and II, the second or middle pair, *3a*, between the bases of legs III, and the posterior or third pair, *4a*, between the bases of legs IV. The first 2 pairs are present in the larva but the third pair, like many of the setae on leg IV itself (see below), does not appear until the deutonymphal instar (Figs. 1.1.1.10—

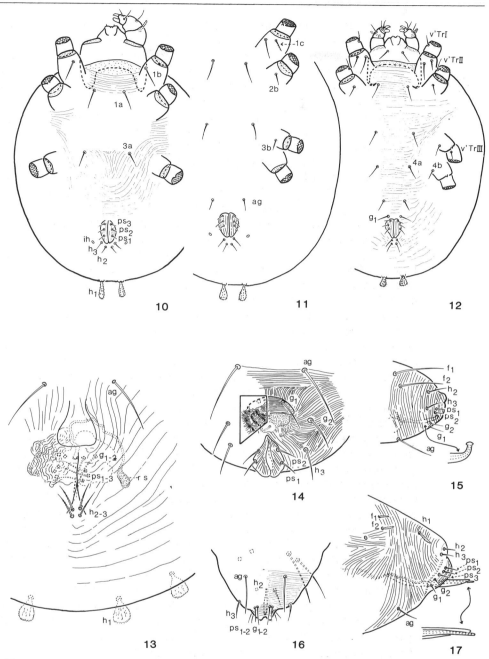

Figs. 1.1.1.10—1.1.1.17. (10—12) Ventral views of idiosoma, with setae denoted on instar of initial appearance: (10) Larva, (11) Protonymph, (12) Deutonymph. (13—14) Genito-anal region of adult female, ventral aspect: (13) *Bryobia* sp, (14) *Tetranychus* sp., with subsurface plicated cuticle of ovipositor shown in quadrangle. (15—17) Genito-anal region of adult male (15) *Tetranychus* sp., lateral aspect, with enlarged exposure of tip of aedeagus, (16) *Tetranychus* sp., ventral aspect, (17) *Lindquistiella* sp., lateral aspect, with enlarged exposure of distal half of aedeagus and paired accessory structures (1 of which is omitted to show aedeagus). Abbreviation: r s, seminal receptacle. See text for setal notation.

1.1.1.12). Rarely, additional setae may be present on the podosomal venter, reflecting derived neotrichous states either of the plethotrichous (non-denotable) type, as in the tetranychid genus *Neotrichobia*, or of the idionymous oligotrichous (denotable) type as in the tenuipalpid genus *Tenuipalpus* and the tetranychid genus *Tauriobia*. The cuticle of this region is membranous, simply striated, and is not of notable interest anatomically. Correlated with the loss of genital papillae in post-larval instars, Claparède organs (urstigmata) are absent between the bases of legs I and II on the larva of tetranychoid mites.

OPISTHOSOMAL VENTER

This region is simpler in the Tetranychidae than in the other families of Tetranychoidea. Adult females generally bear 1 pair of aggenital setae (*ag*, also known as pregenital setae), a membranous plicated genital opening which is an eversible ovipositor (discussed in detail by Van de Lustgraaf, 1977) that is devoid of valves or plates but is flanked by a small unpaired flap anteromedially and by 2 pairs of genital setae (g_{1-2}) laterally, and an anal opening flanked by a pair of weakly demarcated valves bearing 1—3 pairs of pseudanal setae (*ps*, commonly and inaccurately known as 'anal' setae) (Figs. 1.1.1.13, 1.1.1.14 and 1.1.1.55). Between the genital and anal openings, there is a tiny third opening which may easily be overlooked (Fig. 1.1.1.55; see also figures in Van de Lustgraaf, 1977). This is the orifice of the bursa copulatrix, which leads by way of a long, convoluted, fine canal to a small but often distinctively shaped seminal receptacle (receptaculum seminis, Fig. 1.1.1.13). This organ appears to be present in adult females, regardless of whether the species is bisexual or not. The number of pseudanal setae is fundamentally different between the 2 subfamilies of Tetranychidae: bryobiine mites retain the primitive number of 3 pairs (Figs. 1.1.1.10—1.1.1.13), as in the Tuckerellidae, Linotetranidae, and early derivative genera of Tenuipalpidae; tetranychine mites have 2 pairs (Fig. 1.1.1.14), or rarely 1 pair as in *Atrichoproctus*. Immediately in front of the genital opening is a differentiated, almost plate-like area or flap of cuticle on which the pattern of stria may be distinctive for some species. As in other tetranychoid families, there are no vestiges of genital papillae or discs in the nymphal and adult instars.

Adult males have the same complement of setae as adult females but in a different configuration (Figs. 1.1.1.15—1.1.1.17). Behind the aggenital (pregenital) pair of setae, the genital and pseudanal setae are grouped together, as a series of 5 pairs (in Bryobiinae) or 4 pairs (in Tetranychinae) on a set of genitoanal valves, caudally. Within these valves lies a sclerotized 'aedeagus', more precisely termed a 'penis' according to Alberti and Storch (1976), the shape of which is a critical diagnostic character for determining many species, especially of the tetranychine genera. In the Bryobiinae, the penis may be flanked by a pair of shorter, hyaline, pointed accessory structures (Fig. 1.1.1.17); these are *not* of setal origin (they are not optically birefringent when viewed under polarized light), and they are therefore not homologous with the pair of setigenous accessory genital stylets found in other families of Tetranychoidea. From the proximal portion of the penis of bryobiine males a long, conspicuous ejaculatory duct leads further into the body to a large spheroidal structure, the seminal vesicle (vesicula seminalis). In the Tetranychinae, a pair of accessory structures flanking the penis is generally not apparent (*Schizotetranychus cynodonis* McGregor appears to be an exception, according to Pritchard and Baker (1955)); the ejaculatory duct is relatively short and leads to a relatively small, inconspicuous seminal vesicle, in contrast with the bryobiine type. Whether these differences in structure of male genital organs are consistent between members of the 2 subfamilies has not been investigated.

The pair of aggenital setae first appears on the protonymph (Fig. 1.1.1.11), and 1 pair of genital setae first appears on the deutonymph (Fig. 1.1.1.12); the second pair of genital setae is expressed only on the adult. Flanking the anal valves in all active instars are 1 or 2 pairs of so-called 'para-anal' setae. Despite their ventrocaudal position, these are setae h_{2-3} and pertain to the dorsal opisthosomal setal series, as discussed above.

LEG SEGMENTATION AND COXISTERNAL PLATES

The segmentation of the 4 pairs of legs is consistent and unmodified throughout the Tetranychidae, as in the other families of Tetranychoidea. In all active instars, each leg has 5 articulating segments, namely the trochanter, femur, genu, tibia, and tarsus; none of these is subdivided, or secondarily fused or reduced. The femur represents a composite of the 2 primitive femoral segments that characterize post-larval instars of early derivative acariform mites. As discussed further below, setae representing verticils of each of these 2 segments are retained on tetranychoid mites as on many other acariform mites having undivided femora (Grandjean, 1940, 1952b).

The fundamental studies of the podosomal exoskeleton of oribatid mites by Grandjean (1952a) apply equally to the other major groups of acariform mites, including Tetranychoidea, and show that the most basal free leg segment is the trochanter. The trochanter attaches basally to a coxisternal plate, or epimeral region, which is well delimited laterally, but not medially, from the rest of the podosomal surface. It is, therefore, rather arbitrary that the ventral, or intercoxal, setae on the podosomal venter are not included among the 'coxal' setae in taxonomic treatments of tetranychoid mites, since all setae of this region are epimeral setae. A supracoxal seta, e, is evident above the base of leg I, though not of leg II (Fig. 1.1.1.18). The coxisternal regions of legs I and II are contiguous on either side, as are those of legs III and IV; but those of legs II and III are separated by a sejugal interval of striated cuticle (Fig. 1.1.1.12), more extensively so on elongated bodies as in *Monoceronychus*.

On adults of Tetranychidae and other families of Tetranychoidea, the primitive and usually maximum number (formula) of coxisternal setae on each of legs I—II—III—IV is 2—2—1—1, respectively. Rarely in spider mites, these counts are exceeded, owing to neotrichy of either the plethotrichous type, as in *Neotrichobia* and *Tauriobia*, or the oligotrichous type, as in *Brachynychus* in which the coxal formula is 4—3—2—2; based on out-group comparisons (see below for family names of the out-group taxa studied), these are secondarily derived conditions. More commonly, there is a derivative loss of coxisternal setae. This is usually restricted to coxisternal plate II, which has only 1 seta in various genera of Bryobiinae (Figs. 1.1.1.10—1.1.1.12). Rarely, as in the bryobiine genus *Tenuicrus*, a similar loss also occurs on plate I, such that the coxal formula is 1—1—1—1. The full expression of coxisternal setae is present on the deutonymph and adult (Fig. 1.1.1.12). In the protonymph, there is generally only 1 seta present on each of plates II, and none on plates IV, such that the coxal formula is 2—1—1—0 (Fig. 1.1.1.11). In the larva, the formula is generally 1—0—0—x (Fig. 1.1.1.10; the lack of legs IV is indicated by x). The absence or presence of setae on coxisternal plates IV is a reliable distinction between the protonymph and deutonymph of spider mites.

SETATION OF LEG SEGMENTS

Homologies and ontogenetic patterns of addition of setae have not been fully determined in a comparative manner, either within the Tetranychidae

Chapter 1.1.1. references, p. 26

or between the families of Tetranychoidea. The generalities and concepts discussed by Grandjean (1958) for oribatid mites apply equally well to spider mites and their relatives, with regard to the distinction between fundamental and accessory setae, to the sequential appearance of verticils of accessory setae on increasingly proximal areas of a given leg segment during ontogeny, and to the displacement of one member of a pseudosymmetrical pair of setae to a more distal position which is generally a prime ($'$) disjunction in the case of ventral (v'—v'') pairs and a second ($''$) disjunction in the case of lateral (l'—l'') pairs.

Few serious attempts have been made to determine setal homologies and their ontogenies, using Grandjean's notation, among spider mites. The comparative studies by Vainshtein (1958), on the homologies and ontogenies of the tarsal setae of tetranychids of a variety of tribes and genera, used this notation but his work did not appear to stimulate further such research by others. A preliminary study, comparing the leg chaetotaxy between each instar of 1 species of Bryobiinae and 1 species of Tetranychinae, was undertaken by Robaux and Gutierrez (1973). This study did not fully determine some of the setal homologies between the 2 species; some setal notations evidently were based on the actual positions of setae on instars of each species, rather than accounting for probable setal homologues occupying relatively *different positions* in each species. The positions of setal homologues often compensate for reductions or additions of other setae on the same segment or for modifications in the orientation and shape of the segment (particularly elongation). Recent observations by Quirós and Baker (1984), on the ontogeny of leg setae in the Tuckerellidae, have provided useful data for some of the comparative statements made below, though some of their setal designations reflect the same problem of positions as mentioned above for the work of Robaux and Gutierrez (1973). Until the important task of comparative homologization of leg setae is accomplished for the families of Tetranychoidea as a group, a great deal of character state data of potential importance to the systematics and classification of these groups will remain masked. For comparative morphological and out-group studies concerning leg setation here, material was examined representing the eleutherengonine families Raphignathidae (1 genus), Caligonellidae (4), Barbutiidae (1), Cryptognathidae (1), Camerobiidae (1), Stigmaeidae (4), Eupalopsellidae (2), Cheyletidae (2) and Tarsocheylidae (2).

Trochanters

The trochanter of each of the legs I—IV in adult spider mites generally has 1 seta, v', which may occupy an l' position in some cases. This basic pattern of 1—1—1—1 is a derived, or apomorphic, state for the Tetranychidae as a whole, since a basic pattern of 1—1—2—1 is common to the Tenuipalpidae and Tuckerellidae (as to the Raphignathidae), in which a second seta, l', is present on trochanter III (l' is also found on trochanter III in Linotetranidae, though v' is absent on trochanters III and IV). In Tetranychidae seta v' is absent on all legs of larvae and protonymphs (Figs. 1.1.1.10 and 1.1.1.11); it appears on trochanters I to III of deutonymphs, but not on trochanter IV until adulthood (Fig. 1.1.1.12). This ontogenetic pattern affords a ready distinction between protonymphs and deutonymphs, as well as adults, of spider mites. This pattern of expression of v' is similar to that found in Tenuipalpidae, except for the presence in the latter of the second seta of trochanter III, l', which first appears on the protonymph. The pattern shared by Tetranychidae and Tenuipalpidae is less advanced (derived) than that of

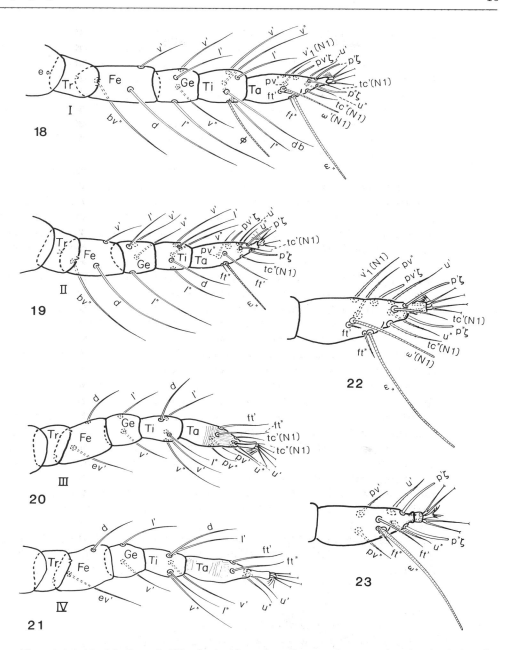

Figs. 1.1.1.18—23. Legs I—IV of a spider mite, *Tetranychus* sp., showing basic larval—protonymphal leg setation and notation in dorsal aspect. (18—21) Legs of protonymph, on same scale: (18) leg I, (19) leg II, (20) leg III, (21) leg IV. (22, 23) Enlarged views of tarsus I: (22) protonymph, (23) larva. Solenidia shown slightly less tapered than natural. Setae are larval on legs I—III unless denoted by N1 in parentheses, indicating protonymphal; setae are protonymphal on leg IV.

Tuckerellidae and Linotetranidae which, apart from the expression of the protonymphal seta l' on trochanter III, have seta v' suppressed on all legs of the protonymph *and* deutonymph.

Femora

In the larval and protonymphal instars a general pattern of femoral setation on legs I—IV apparently holds for both subfamilies of

Tetranychidae, the formula being 3—3—2—2 (Figs. 1.1.1.18—1.1.1.21). Femoral seta d subapically and a ventral seta subproximally (occupying a v'' position on legs I—II but a v' position on legs III—IV) are present on all legs; another seta, v' (often occupying a nearly l' position, and denoted as such by Robaux and Gutierrez (1973)), is present on legs I and II. This is similar to the basic formula of 3—3—2—1 for these instars in Tenuipalpidae and Linotetranidae, and 3—3—1—1 in Tuckerellidae. Apparently homologous setae, d, v' and bv'', are basically present on femora I and II in larvae and protonymphs of all families of Tetranychoidea; of these, v' is consistently absent, and the proximoventral seta occupies a v' position, on femora III and IV. A further loss, of femoral seta d, occurs on leg IV in the other tetranychoid families; this is considered an advanced condition over that in Tetranychidae, with the reductive trend furthered by suppression of d on leg III of the larvae and protonymphs of Tuckerellidae.

The subproximal femoral seta of legs I and II is denoted as bv'' (rather than v''), as introduced by Grandjean (1942, 1943b, 1947) for oribatid and certain prostigmatic mites, to indicate that it is a remnant of a basifemoral verticil of setae that originally belonged to the more basal of the 2 femoral segments on the legs of early derivative acariform mites. The leg femora of larval and post-larval tetranychoid mites, like those of their raphignathoid and cheyletoid relatives, are undivided, yet their setal complement is a composite of the 2 primitive, fundamental verticils representing the basal and distal femoral segments. Generally, bv'' is thought to be restricted to femora I and II, with another fundamental seta of the basifemoral verticil, ev', present instead on femora III and IV (Grandjean, 1942; Norton, 1977); this was noted also by Grandjean (1944) for some raphignathoid taxa. Although this interpretation is accepted here for the Tetranychoidea, it remains uncertain whether bv'' is truly replaced by ev' on femora III—IV or is simply displaced to a v' position on legs III and IV in order to maintain an antiaxial position, as on legs I and II.

On deutonymphs of the Tetranychidae, 3—6 femoral setae, including l', l'' and v'', are usually added on leg I (Figs. 1.1.1.24 and 1.1.1.28), but generally none is added on leg IV (Figs. 1.1.1.27 and 1.1.1.32); 1—3 of the femoral setae l', l'' and v'' are usually added on leg II and sometimes v' on leg III in the Bryobiinae (Figs. 1.1.1.30 and 1.1.1.31), but usually none is added on these legs in the Tetranychinae (Figs. 1.1.1.25 and 1.1.1.26). Of setae l'—l'' and v'', which are normally added on femur I in spider mite deutonymphs, l'—l'' are added, but v'' remains suppressed, in deutonymphs of Tuckerellidae and Linotetranidae, and none of these setae is added in Tenuipalpidae. Parallel trends towards suppression of additional femoral setae on legs II—IV of deutonymphs are found in the other families of Tetranychoidea, much as in the Tetranychinae: thus no setae are added in Linotetranidae and Tenuipalpidae; however, in the Tuckerellidae, much as in the bryobiine Tetranychidae, 2 setae, l' and l'', but not v'', are added on femur II as on femur I, and l' is also added on femur III.

In some bryobiine deutonymphs, several other setae may be added subproximally on femur I to augment the basifemoral verticil of setae represented otherwise only by bv''. In this chapter the notation of these setae is prefixed by b to indicate that they are part of the basifemoral verticil (Fig. 1.1.1.28). These unusual additions may be part of a neotrichous trend that is furthered by the addition of other, less proximal setal verticils on femur I in the adult (see below). Generally, setal verticils are added in sequence, increasingly more proximad from the apex of a given leg segment (Grandjean, 1943b); however, various phenomena, such as the regression or displacement of certain setae, may interfere with this pattern (Coineau,

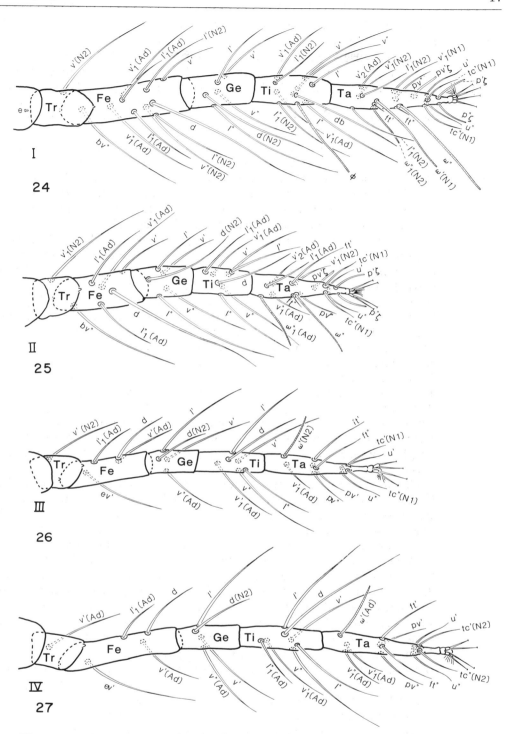

Figs. 1.1.1.24—27. Legs I—IV of an adult female tetranychine spider mite, *Tetranychus* sp., figured in dorsal aspect on same scale, showing leg setation and notation. (24) Leg I, (25) leg II, (26) leg III, (27) leg IV. Solenidia shown slightly less tapered than natural. Setae are larval on legs I—III and protonymphal on leg IV unless denoted otherwise in parentheses by N1, N2 or Ad, indicating protonymphal, deutonymphal or adult.

1974). The appearance of several setae, seemingly out of sequence, sub-proximally on femur I in some bryobiine deutonymphs may be explained by their origin on the basal region of the femur, which represents the more

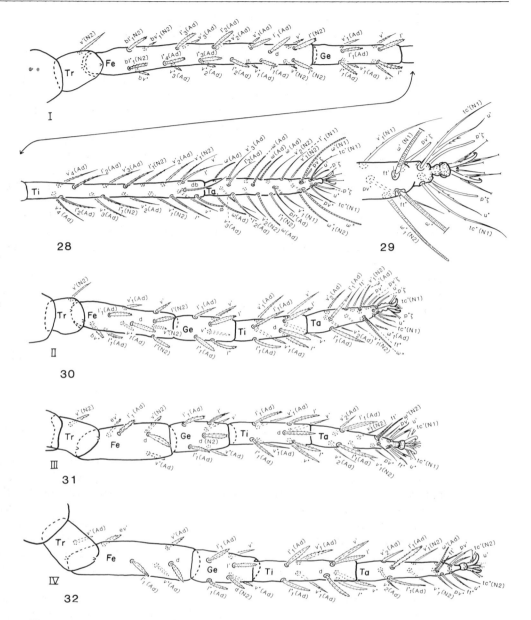

Figs. 1.1.1.28—32. Legs I—IV of an adult female bryobiine spider mite, *Bryobia* sp., figured in dorsal aspect on same scale (except (29)), showing leg setation and notation. (28) Leg I (artificially separated between genu and tibia), (29) enlarged view of distal extremity of tarsus I, (30) leg II, (31) leg III, (32) leg IV. Solenidia shown slightly less tapered than natural. Setae are larval on legs I—III and protonymphal on leg IV unless denoted otherwise in parentheses by N1, N2 or Ad, indicating protonymphal, deutonymphal or adult.

basal of the 2 ancestral femoral segments; the other femoral setae added on the adult originate on more distal regions of the femur which represent the more apical and elongated of these 2 segments.

Additional femoral setae are usually added on all legs of adult spider mites; their correct notation has not been determined up to the present. On femur I, it appears that $l'_1-l''_1$ and sometimes 1 or 2 ventral setae (v'_1, v''_1) may be added in some Tetranychinae (Fig. 1.1.1.24), and $l'_1-l''_1$, and up to 12 other ventral ($v'_1-v''_1$, $v'_2-v''_2$, $v'_3-v''_3$) and lateral ($l'_2-l''_2$,

$l'_3-l''_3$, $l'_4-l''_4$) setae in some Bryobiinae (Fig. 1.1.1.28). On femur II, $l'_1-l''_1$ and v''_1 are usually added to adult Tetranychinae (Fig. 1.1.1.25); these setae and v'_1 are sometimes added also on adult Bryobiinae (Fig. 1.1.1.30). On femora III and IV, it appears that l'_1 and v' are usually added in adults of both subfamilies (Figs. 1.1.1.26, 1.1.1.27, 1.1.1.31 and 1.1.1.32), though v' may already be present on femur III in Bryobiinae; in addition, v'' and apparently l''_1 are sometimes added on femora III—IV in Bryobiinae. Note that the metameric homonome of adult subproximal seta l'_1 of femora I—II may be expressed similarly on femora III—IV, even though the deutonymphal setae $l'-l''$ of femora I—II are suppressed on legs III—IV.

Although there tends basically to be an ontogenetically earlier expression, and richer setation on the leg femora in Bryobiinae than in Tetranychinae, there are reductive trends derived independently among some genera *within* the Bryobiinae. For example, adults of some *Marainobia* and *Neopetrobia* have only 2 setae, d and ev', on each of femora III and IV; this represents a suppression of setal additions beyond the protonymphal stage. Similarly, the tetranychine genus *Mixonychus* retains only d and ev' on femur IV, and these setae plus v' on femur III. In the other tetranychoid families, there is a general suppression of setal additions on femora III and IV after the protonymph; only d and ev' are present on femur III of adults of Linotetranidae and Tenuipalpidae. The presence of only 1 femoral seta, ev', on leg IV of adults of Linotetranidae, Tuckerellidae, and Tenuipalpidae represents suppression even of a protonymphal seta, d, which is a possible synapomorphy for these 3 families.

Genua

On larval and protonymphal spider mites, the standard genual setation on legs I—IV is 4—4—2—2 (Figs. 1.1.1.18—1.1.1.21); genua I and II each bear setae $l'-l''$ and $v'-v''$, whereas genua III and IV each bear only l' and v' (the latter often occupying a v'' position, especially on leg IV). This is a richer complement of setae than is found in larvae and protonymphs of all other families of Tetranychoidea, and is thought to be plesiomorphic, based on comparison with the states present in out-group (raphignathoid) taxa on the one hand and in the other tetranychoid families on the other. In all other tetranychoid families, seta v' at least is suppressed in these instars (except that v' is retained on genu I in Linotetranidae); this may be another apomorphy shared by these taxa.

On deutonymphs of both subfamilies of spider mites, seta d is added to the genual complement on legs III and IV (d may be delayed until adulthood on genu IV in some Bryobiinae); it is also added on genua I and II of deutonymphs of Tetranychinae but not of Bryobiinae, such that the standard deutonymphal genual setation on legs I—IV is 4—4—3—2 or 3 in Bryobiinae (Figs. 1.1.1.28—1.1.1.32) and 5—5—3—3 in Tetranychinae (Figs. 1.1.1.24—1.1.1.27). Since d is generally added on genua I and II in the deutonymphal stage of other tetranychoid families, as in Tetranychinae, the delay of its appearance until the adult (if it appears at all) in Bryobiinae seems to be apomorphic. Seta d is suppressed on genua III and IV in the deutonymphs of other families of Tetranychoidea, and this again may be a shared apomorphy.

Addition of setae on the genua of adult spider mites is more variable in the Bryobiinae than in the Tetranychinae. In the latter, generally no setae are added to genua I and II (Figs. 1.1.1.24 and 1.1.1.25), and only 1 seta, v'', is added to genua III and IV (Figs. 1.1.1.26 and 1.1.1.27), so that the standard adult genual setation is 5—5—4—4, with reduction down to 5—5—2—2 in

Chapter 1.1.1. references, p. 26

some taxa, which represents retention of the larval—protonymphal complement of setae on legs III and IV. In the Bryobiinae, this setal formula may
vary from 3—3—1—1 to 8—8—6—6. The minimal complement of genual setae
in bryobiine taxa may include only setae l'—l'' and v' on legs I—II in
Porcupinychus and *Afronobia*, and only seta v' on legs III —IV in *Marainobia*;
these examples involve losses of setae from the basic larval—protonymphal
complement given above. In other bryobiine taxa, at least l''_1 (which might
also be interpreted as d) is commonly added to the deutonymphal setal
complement of genua I and II, and at least v'' to that of genua III and IV; the
maximum complement of genual setae on bryobiine mites includes l', v'—v''
and l'_1—l''_1 on each of genua I—IV, plus l'' and v'_1—v''_1 (but apparently not
d) on genua I and II (Figs. 1.1.1.28 and 1.1.1.30), plus d (but apparently not
l'') on genua III and IV (Figs. 1.1.1.31 and 1.1.1.32). The lack of l'' on genua
III and IV appears to be constant in the Tetranychidae, as in other families
of Tetranychoidea. These patterns contrast in number with those of
Tenuipalpidae and Linotetranidae, in which no genual setae are added on the
adult; they contrast in content and sequence with those of Tuckerellidae, in
which setae l'_1 and either l''_1 or d are added to genua I and II on the adult
and v' and d are delayed additions to genua III and IV on the tritonymph.
The spider mites, therefore, show trends both in setal additions (oligotrichous neotrichy, according to the concepts of Grandjean (1965)) among
some of the Bryobiinae, and in setal reductions, independently among some
Bryobiinae and Tetranychinae. Based on out-group comparison, these
trends may be derivative in *both* directions.

Tibiae

On larval and protonymphal spider mites, the basic tibial setation on legs
I—IV is 5(+ 1 solenidion)—5—5—5; the tibia of each leg bears one complete
whorl or verticil of setae, d, l', l'', v', v'' (Figs. 1.1.1.18—1.1.1.21). This is
a slightly fuller, more plesiomorphic complement of setae (as found in
Raphignathoidea) than is found in larvae and protonymphs of the other
tetranychoid families: that of the Tuckerellidae is identical except that on
tibia IV, d is suppressed until the deutonymph and l' is completely suppressed;
that of the Linotetranidae lacks the solenidion on tibia I and seta l' on tibiae
II, III and IV; that of the Tenuipalpidae lacks the solenidion on tibia I and
setae l' and l'' on tibiae III and IV. In some bryobiines, tibia I of the
protonymph may add 2 setae, v'_1—v''_1, to the larval complement; in others,
these setae may be suppressed until the deutonymph or the adult; or one or
both may be suppressed entirely. On tibia I, as noted by Grandjean
(1943a), seta d sometimes has a trichobothridial aspect (denoted by db), particularly
among the bryobiine taxa; seta d of femora I and II sometimes has a similar
aspect.

On deutonymphs in Tetranychidae, tibial setae generally are added only
to leg I in the subfamily Tetranychinae, but sometimes they are added to all
the legs in the Bryobiinae, so that the standard deutonymphal tibial setation
is 7(+ 1 solenidion)—5—5—5 in tetranychines, but varies from the protonymphal complement of 5(+ 1ϕ)—5—5—5 to 9(+ 4ϕ)—7—7—7 in bryobiines.
The 2 setae sometimes added on tibiae II—IV in bryobiines are v'_1—v''_1,
although they may be suppressed until the adult. Tibia I generally adds setae
l'_1—l''_1 in both subfamilies (Figs. 1.1.1.24 and 1.1.1.28), as well as v'_1—v''_1
sometimes in Bryobiinae, if these are not already present. These additions of
tibial setae, particularly on leg I, in tetranychid deutonymphs contrast with
the complete lack of such additions in the other families of Tetranychoidea
(except for the delayed addition of d to tibia IV in Tuckerellidae).

As with the femora and genua, the complement of setae on the tibiae of adult spider mites is less variable in the Tetranychinae than in the Bryobiinae. In the former, the standard adult tibial setation is $9(+1\phi)$—7—6—7, with setae $v'_1-v''_1$ added to tibia I, v'_1 and l'_1 to tibia II, v'_1 alone to tibia III, and v'_1 and l''_1 to tibia IV (Figs. 1.1.1.24—1.1.1.27); however, as few as the regular deutonymphal complement of $7(+1\phi)$—5—5—5 may be retained on adults in some tetranychine genera, as in *Panonychus* and some *Oligonychus*. An additional 2—4 solenidia are usually added on tibia I of the adult male, in contrast to the 1 larval solenidion retained on the adult female. In the Bryobiinae, this setal formula may vary from $8(+1\phi)$—5—5—5 to $15(+15\phi)$—11—11—11; setae may or may not be added to tibia I (Figs. 1.1.1.28 and 1.1.1.33), generally in the sequence $v'_2-v''_2$ followed by $l'_2-l''_2$ followed by $v'_3-v''_3$; and some or all of setae $v'_1-v''_1$ (if not already present) followed by $l'_1-l''_1$ followed by $v'_2-v''_2$ may or may not be added to tibiae II—IV (Figs. 1.1.1.30—1.1.1.32). The bryobiine genera *Neopetrobia* and *Marainobia* are examples in which a nearly protonymphal complement of setae is retained on tibiae I—IV of the adult. In adult male bryobiines, as in tetranychines, solenidia are generally added on tibia I in addition to the complement present on the female; even when solenidial neotaxy is evident on the adult female, the number of solenidia on the adult male is greater. In the presence of many solenidia on adult males (and more rarely on females), some (up to 8 or 10 in *Lindquistiella* and *Petrobia*, for example) of the setae are smaller than in the preceding instar and are closely associated with solenidia as coupled, or duplex, sets, as on the tarsus (discussed below, Fig. 1.1.1.33). The basic pattern of setal additions to tibiae I—IV in both subfamilies of Tetranychidae again contrasts with the nearly complete lack of such additions in the other families of Tetranychoidea (setae $l'_1-l''_1$ are added to the basic 5 setae on tibia I in tritonymphal Tuckerellidae), and also in the raphignathoid families. The conditions in spider mites (especially the Bryobiinae), therefore, may be apomorphic trends of setal additions (oligotrichous neotrichy) rather than those in the other families being apomorphic trends of setal repression. Apparently, for the tibial setation as for the genual setation, there are derivative trends both in setal reduction, independently in each subfamily of Tetranychidae, and in setal addition, in the Bryobiinae.

Tarsi

Larval spider mites have a basic tarsal setation on legs I—III of $8(+1$ solenidion $\omega)$—$8(+1\omega)$—6. The tarsi of legs I and II generally each bear 1 solenidion, ω'', and 4 paired sets of setae, which, according to the special notation developed by Grandjean (1940, 1941, 1948) for the tarsal segment of the legs of acariform mites, are the prorals $p'-p''$ which are distalmost and eupathidial in form, fastigials $ft'-ft''$, unguinals $u'-u''$, and primiventrals $pv'-pv''$ (Fig. 1.1.1.23). Setae $ft'-ft''$, $u'-u''$ and $pv'-pv''$ are also generally present on tarsus III, but the solenidion and prorals are absent. Retention of setae $pv'-pv''$ on tarsi I—III is plesiomorphic, relative to the conditions found in these instars in other tetranychoid families: seta pv'' is absent in Tuckerellidae, and $pv'-pv''$ are absent in Tenuipalpidae and Linotetranidae. However, the tetranychid complement is apomorphic, along with that of Tenuipalpidae, in lacking a set of primilaterals $pl'-pl''$ on tarsi I and II; $pl'-pl''$ are retained on leg I, and pl'' on leg II, in Tuckerellidae and Linotetranidae. The larval tetranychid complement on tarsi I and II is uniquely apomorphic within Tetranychoidea in having solenidion ω'' (which is generally elongated and tapered like a seta) located closely beside seta ft''

Chapter 1.1.1. references, p. 26

(which is generally reduced in size relative to ft') to form a set of 'duplex setae', or 'chaetopair' (Fig. 1.1.1.23; see also SEM micrograph figs. 1A and 6F in Robaux and Gutierrez (1973)). Comparable sets are termed 'coupled setae' in oribatid mites (Grandjean, 1935; Norton, 1977), and in both cases the solenidion is usually distal to the companion seta (Fig. 1.1.1.53). This duplex arrangement is apparently disjointed secondarily *within* the Tetranychidae, among many members of the Eurytetranychini (Vainshtein (1958), however, regarded the disjunct and non-attenuated state of ω'' TaI—II and ω' TaI as a primitive condition and the primary basis for treating the eurytetranychines as a subfamily equivalent to, and separate from, the Bryobiinae and Tetranychinae). Rarely, in a few bryobiines (e.g., *Bryobiella*), the duplex arrangement is no longer discernible because the setal element ft'' has become so reduced secondarily as to be vestigial, thereby eliminating the duplex aspect. In all tetranychoid families, the complement of tarsal setae characteristic of larval leg III appears on leg IV when it is formed on the protonymph (except that pv' is suppressed until the deutonymph stage in Tuckerellidae).

On the protonymph of all families of Tetranychoidea, a set of tectal setae tc'—tc'' is generally added to the larval complement of setae on legs I to III (Figs. 1.1.1.18—1.1.1.20, except that they are suppressed on leg III in Lino- tetranidae); the tectal set is suppressed on leg IV (Fig. 1.1.1.21) until the deutonymph (except that they are absent, as on leg III, in Linotetranidae). Within the Tetranychidae, the tectals may be suppressed as a secondarily derived condition in some genera of both bryobiine and hystrichonychine Bryobiinae. In contrast to the other families, tetranychid protonymphs generally also add seta v'_1 (and sometimes v''_1 in bryobiines) and solenidion ω' (also added in Linotetranidae) on leg I (Fig. 1.1.1.22); seta ft' is smaller relative to its size on the larva, and is generally closely associated with ω' to form a second duplex set on leg I (except again, secondarily disjointed in many members of Eurytetranychini, and secondarily vestigial in a few members of Bryobiinae). Further, seta pv' generally transforms into a ventral eupathidium on legs I and II (Fig. 1.1.1.22); that this eupathidium is a transformation of larval seta pv' rather than seta v'_1 which is newly added on leg I in the protonymph, was discussed by Grandjean (1948). Rarely, this eupathidial transformation may be secondarily repressed, so that pv' remains setiform throughout is ontogeny, or pv' itself may be completely suppressed. These protonymphal additions result in setal formulas typically of 11 $(+2\omega)$—$10(+1\omega)$—8—6 for Tetranychinae, but varying from $10(+2\omega)$— $10(+1\omega)$—7—6 to $14(+2\omega)$—$10(+1\omega)$—9—6 for Bryobiinae, in which setae v'_1—v''_1 and l'_1—l''_1 may or may not be added on tarsus I. As companion setae of solenidia on tarsus I, the fastigials are generally more reduced in size on bryobiines (Fig. 1.1.1.29) than on tetranychines (Fig. 1.1.1.22).

Deutonymphs of spider mites are unique among the Tetranychoidea in generally adding setae on tarsi I and II, and a solenidion on tarsus III. In the Bryobiinae, setae v'_1—v''_1 and sometimes l'_1 are basically added to tarsi II— IV (Figs. 1.1.1.30—1.1.1.32), and l'_1—l''_1 (if not already added on the protonymph), v'_2—v''_2 and sometimes l'_2—l''_2 to tarsus I (Fig. 1.1.1.28); again some of these may be suppressed secondarily *within* the group, as in *Marainobia* and *Neopetrobia*, so that deutonymphal formulas may vary from $13(+3\omega)$—$8(+1\omega)$—7—8 to $16(+4\omega)$—$13(+1\omega)$—$11(+1\omega)$—11. Note that a solenidion is also added to tarsus I (ω''_1) and sometimes III (ω'), but generally not to tarsi II and IV. In the Tetranychinae, fewer deutonymphal structures are generally added (Figs. 1.1.1.24—1.1.1.27), so that the typical setal formula for this instar is $14(+3\omega)$—$11(+1\omega)$—$8(+1\omega)$—8: v'_1—v''_1 are usually suppressed on tarsi III and IV (though they may be added in

Eurytetranychus), and only v'_1 is added on tarsus II; v''_1 (already present on protonymphs in most bryobiines) and $l'_1-l''_1$ but not $v'_2-v''_2$ are added on tarsus I; and, as in bryobiines, a solenidion is added to legs I (ω''_1) and sometimes III (ω'), but not to legs II and IV. On deutonymphs of the other tetranychoid families, the only tarsal structure generally added on tarsi I—III is a solenidion on leg I in Tuckerellidae and some Tenuipalpidae, which is the homologue of ω' expressed on the protonymph in the Tetranychidae and Linotetranidae. The complete suppression of deutonymphal ω' on tarsus III is an apomorphy which is shared by members of the other tetranychoid families.

Spider mites are unique among the Tetranychoidea in generally having tarsal setae added to each of legs I—IV, and tarsal solenidia added to legs II and IV, in transforming from deutonymph to adult. These additions are few in the Tetranychinae (Figs. 1.1.1.24—1.1.1.27): setae v'_2 on leg I, v''_1, l'_1 and v'_2 on leg II, v'_1 and rarely v''_1 on leg III, $v'_1-v''_1$ on leg IV, and solenidia ω''_1 on leg II and ω' on leg IV, such that a full, typical formula is $15(+3\omega)$ $-14(+2\omega)-9$ or $10(+1\omega)-10(+1\omega)$, as found in some *Tetranychus* and *Eurytetranychus*. In addition, a few more solenidia are generally added to adult male tetranychines compared with adult females. Usually, the setae (not solenidia) added on tetranychine adults are ones already added on the deutonymph in Bryobiinae, excepting those added on tarsus II. The full tarsal formula for Tetranychinae may be reduced to $13(+3\omega)-11(+1\omega)$ $-9(+1\omega)-9(+1\omega)$ by suppression of setae l''_1 and v'_2 on tarsus I, l'_1, v''_1 and v'_2 on tarsus II, and v''_1 on tarsus IV, as in *Schizotetranychus* and *Mixonychus*. Adults of the tetranychine genus *Brevinychus* retain 10 setae on each of tarsi III—IV, but are unusual in having both v'_2 and v''_2 added, yet ft' and ft'' suppressed, on these segments. This is a good example of how simple setal counts may obscure significant character state data. The fastigial seta ft' is suppressed on tarsi III—IV in some *Oligonychus*, and both $ft'-ft''$ are delayed in expression until adulthood on tarsi III—IV in some *Eurytetranychus*. In the Bryobiinae, these additions are basically richer (Figs. 1.1.1.28—1.1.1.32), though subject again to secondary suppression *within* this group. Setae $v'_1-v''_1$ (if not already present on the deutonymph), $l'_1-l''_1$, and $v'_2-v''_2$ may be added on each of tarsi II—IV, as well as solenidia ω' (if not already present on the deutonymph) and ω''_1 on tarsus II and ω' on tarsus IV. Several sets of l or v setae, and up to 8 solenidia may be added to tarsus I of the adult female; the setae may include the retarded expression of apparently pl'' (in a d position between l'_1 and l''_1), $l'_2-l''_2$ (if not already present on the deutonymph), $v'_3-v''_3$, $l'_3-l''_3$, $v'_4-v''_4$, and $v'_5-v''_5$. As on the tibia, the number of solenidia on the tarsus of leg I of the adult male is greater than in the female; in the presence of many solenidia (up to 30), several of the setae are smaller than on the adult female and are closely associated with solenidia as duplex sets (up to 3 or 4 in *Lindquistiella* and *Petrobia*, for example, Fig. 1.1.1.33) in addition to $ft'-ft''$ with $\omega'-\omega''$. Such additional duplex sets are generally restricted to leg I. Because of the secondary trends towards suppression of tarsal setation within the Bryobiinae (e.g., *Neopetrobia*), formulae for adult females vary greatly, from $11(+2\omega)$ $-8(+1\omega)-7(+0\omega)-8$ or $9(+0\omega)$, as in some *Neopetrobia* and *Marainobia*, to 19 to $25(+3$ to $12\omega)-15$ to $20(+2$ or $3\omega)-14$ to $17(+1\omega)-14$ to $17(+1\omega)$, as in some species of *Bryobia*, *Bryocopsis*, *Petrobia* and *Lindquistiella*. As with the genual and tibial segments, the setal additions on the tarsi are usually oligotrichous and appear to be a derived condition peculiar to the ancestral stock of the Bryobiinae *within* the Tetranychidae, based on out-group comparison with other families of Tetranychoidea and Raphignathoidea. This form of neotrichy is not considered here to represent

Chapter 1.1.1. references, p. 26

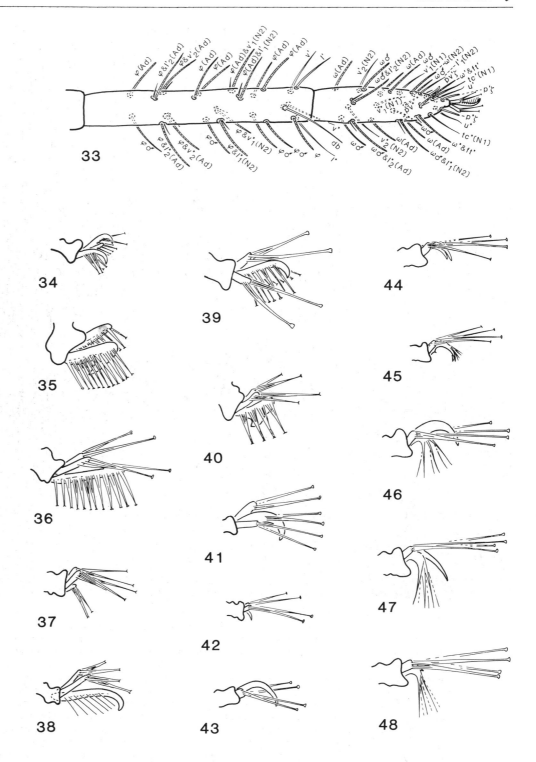

Figs. 1.1.1.33—48. (33) Tibia and tarsus of leg I, dorsal aspect, of an adult male bryobiine mite, *Lindquistiella* sp., showing accessory solenidia and duplex sets; solenidia shown slightly less tapered than natural; notation as in preceding figures, except that ♂ denotes accessory male structures. (34—38) Diversity of leg ambulacral structures among spider mites of various bryobiine (34—41) and tetranychine (42—48) genera. (34) *Bryobia* sp., (35) *Marainobia* sp. (1 of the paired uncinate claws omitted behind pad-like empodium), (36) *Dolichonobia* sp., (37) *Aplonobia* sp., (38) *Lindquistiella* sp., (39) *Petrobia* sp., (40) *Edella* sp., (41) *Schizonobia* sp., (42) *Eurytetranychus* sp., (43) *Anatetranychus* sp., (44, 45) *Schizotetranychus* spp., (46) *Panonychus* sp. (also *Oligonychus* spp.), (47, 48) *Tetranychus* spp. ((48) also *Eotetranychus* spp.).

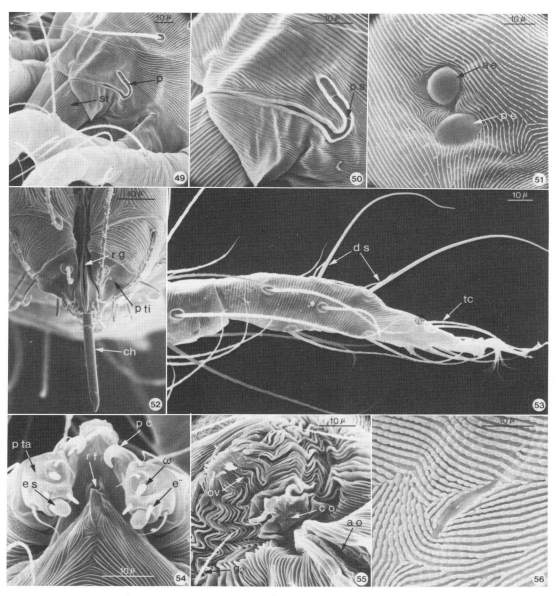

Figs. 1.1.1.49—56. SEM micrographs (courtesy of Dr A. Crooker) of external structures of *Tetranychus urticae* Koch, adult female: (49) Peritreme and base of stylophore, (50) peritreme enlarged to show inner structure, (51) eyes and surrounding prodorsal integument, (52) dorsal view of palpi, rostrum and cheliceral stylets, all in appressed position preparatory to feeding, (53) posterolateral view of tarsus I, (54) ventral view of rostral and palpal apices, (55) genito-anal region, (56) lyrifissure (cupule) *im* and surrounding opisthosomal integument, showing rounded cuticular lobes. Abbreviations: a e, anterior eye; a o, anal opening; ch, cheliceral stylets juxtaposed; c o, copulatory orifice of bursa copulatrix; d s, duplex sets, each a solenidion and companion seta; e″, posterolateral palptarsal eupathid; e s, palptarsal eupathidial spinneret; g_2, posterolateral genital seta; ov, plicated cuticle of ovipositor; p, peritreme; p c, palptibial claw; p e, posterior eye; p s, peritrematal septum; p ta, palpal tarsus; p ti, palpal tibia; r f, rostral fossette; r g, rostral gutter; st, stylophore; tc, tectal setae; ω, palptarsal solenidion.

a primitive condition of the Tetranychidae as a whole, as interpreted by Mitrofanov (1971). More rarely, because of even more extensive setal additions to tarsi I—IV, the setal formula may be as high as 56 (+ 5ω)—43 (+ 3ω)—38 (+ 3ω)—37 (+ 2ω), as in *Tauriobia*. This appears to be correlated with some neotrichy among the setae of the podosomal venter on adults of this genus, a condition which also is derived.

Chapter 1.1.1. references, p. 26

Ambulacra

Generally, the leg tarsal appendages of all active instars of spider mites each consist of a short flexible pretarsus bearing a pair of structures derived from true claws laterally and an unpaired empodium medially. The ancestral conditions of these structures for the tetranychoid stock as a whole are uncinate claws, equipped with a series or row of pseudosymmetrically paired tenent hairs, or chaetoids, and a pad-like empodium, similarly equipped with a series of paired tenent hairs (Fig. 1.1.1.34). These ancestral conditions are retained in the Tuckerellidae, most Tenuipalpidae, and in a few early derivative bryobiine Tetranychidae (e.g., *Bryobia*), but a diversity of derived states is found otherwise in the Tetranychidae. Within the bryobiine sub-family stock of Tetranychidae, the paired claws usually are reduced distally so as to form slender pad-like structures, and the tenent hairs are often united at their apices so as to form 1 compound pair on each 'claw' (Figs. 1.1.1.36 and 1.1.1.37); the empodium commonly remains pad-like and with a series of tenent hairs, but it is typically modified into a claw-like form, with a series or a compound pair of tenent hairs, in the petrobiine Bryobiinae (Figs. 1.1.1.38 and 1.1.1.39). Within the tetranychine subfamily stock, modifications are more varied: the paired claws are generally more strongly reduced to short rod-like structures from which a compound pair of tenent hairs arises distally (Figs. 1.1.1.42—1.1.1.48; see also SEM micrograph plate fig. 16B in Jeppson et al. (1975)); and the empodium is not pad-like and does not bear tenent hairs, but instead has a basic claw-like form from which several other states may be derived. The claw-like form of the empodium in Tetranychinae may be simple, as in *Anatetranychus* (Fig. 1.1.1.43), or split bilaterally into 2 claw-like processes as in *Schizotetranychus* (Figs. 1.1.1.44 and 1.1.1.45), or it may be further modified by the elaboration of (usually) 3 pairs of minute, attenuate hair-like processes (apparently not the homologues of tenent hairs) proximoventrally (Fig. 1.1.1.46; see also SEM micrograph plate fig. 16B in Jeppson et al. (1975)). From the latter state a reduction of the claw-like part may be derived, leaving the empodium simply as the proximoventrally directed cluster of hair-like processes, as in *Eotetranychus* and *Tetranychus* (Figs. 1.1.1.47—1.1.1.48). A separate reductive trend may derive directly from the basic, unmodified claw-like form of the empodium, and lead to a simple minute claw, as in *Eurytetranychus* (Fig. 1.1.1.42), and to a complete disappearance of this structure, as in *Eutetranychus* and *Aponychus*.

CONCLUSIONS

The above discussions and their accompanying figures suffice to demonstrate what a wealth of potential data may be available from the chaetotactic and ambulacral structures of the legs of spider mites, once the homologies and ontogenies of these structures have been determined from a wide variety of taxa. We have hardly begun to scratch the surface of this mine of anatomical information, and the impact that it will have on our understanding of the systematics, phylogeny and classification of species and supraspecific taxa within the family Tetranychidae is indeed an exciting prospect.

REFERENCES

Alberti, G. and Storch, V., 1976. Ultrastruktur-Untersuchungen am männlichen Genital-trakt und an Spermien von *Tetranychus urticae* (Tetranychidae, Acari). Zoomorphologie, 83: 283—296.

André, H.M. and Remacle, C., 1984. Comparative and functional morphology of the gnathosoma of *Tetranychus urticae* (Acari: Tetranychidae). Acarologia, 25: 179—190.

Attiah, H.H., 1970. The Tetranychini of the U.A.R. I. The genus *Tetranychus* Dufour (Acarina, Tetranychidae). Acarologia, 11: 733—741.

Blauvelt, W.E., 1945. The internal morphology of the common red spider mite (*Tetranychus telarius* Linn.). Mem. Cornell Univ. Agric. Exp. Stn., Ithaca, NY, 270: 1—35.

Boudreaux, H.B. and Dosse, G., 1963. The usefulness of new taxonomic characters in females of the genus *Tetranychus* Dufour (Acari: Tetranychidae). Acarologia, 5:13—33.

Coineau, Y., 1974. Éléments pour une monographie morphologique, écologique et biologique des Caeculidae (Acariens). Mém. Mus. Natl. Hist. Nat., Sér. A, Zool., 81: 1—299.

Dosse, G. and Boudreaux, H.B., 1963. Some problems of spider mite taxonomy involving genetics and morphology. In: J.A. Naegele (Editor), Advances in Acarology, Cornell University Vol. 1. pp. 343—349.

Geijskes, D.C., 1939. Beiträge zur Kenntnis der europäischen Spinnmilben (Acari, Tetranychidae), mit besonderer Berücksichtigung der niederländischen Arten. Meded. Landbouwhogesch. Wageningen, 42: 1—68.

Grandjean, F., 1935. Les poils et les organes sensitifs portés par les pattes et le palpe chez les Oribates. Première partie. Bull. Soc. Zool. Fr., 60: 6—39.

Grandjean, F., 1939. Les segments post-larvaires de l'hystérosoma chez les Oribates (Acariens). Bull. Soc. Zool. Fr., 64: 273—284.

Grandjean, F., 1940. Les poils et les organes sensitifs portés par les pattes et le palpe chez les Oribates, 2e partie. Bull. Soc. Zool. Fr., 65: 32—44.

Grandjean, F., 1941. La chaetotaxie comparée des pattes chez les Oribates (1re série). Bull. Soc. Zool. Fr., 66: 33—50.

Grandjean, F., 1942. Observations sur les Labidostommidae (3e série). Bull. Mus. Natl. Hist. Nat., Paris, 2e série, 14: 319—326.

Grandjean, F., 1943a. Les trichobothries pédieuses des Acariens et leur priorité chez les Bdelles. C. R. Séances Soc. Phys. Hist. Nat. Genève, 60: 241—246.

Grandjean, F., 1943b. Quelques genres d'Acariens appartenant au groupe des Endeostigmata (2e série). 1ere partie. Ann. Sci. Nat., Zool. Biol. Anim., 11e sér., 4: 85—135.

Grandjean, F., 1944. Observations sur les Acariens de la famille des Stigmaeidae. Arch. Sci. Phys. Nat., 5me période, 26: 103—131.

Grandjean, F., 1947. Les Enarthronota (Acariens). 1e série. Ann. Sci. Nat., Zool. Biol. Anim., 11e sér., 8: 213—248.

Grandjean, F., 1948. Quelques caractères des Tétranyques. Bull. Mus. Natl. Hist. Nat., Paris, 2e sér., 20: 517—524.

Grandjean, F., 1952a. Au sujet de l'ectosquelette du podosoma chez les Oribates supérieurs et de sa terminologie. Bull. Soc. Zool. Fr., 77: 13—36.

Grandjean, F., 1952b. Sur les articles des appendices chez les Acariens actinochitineux. C. R. Séances Acad. Sci., 235: 560—564.

Grandjean, F., 1958. Sur le comportement et la notation des poils accessoires postérieurs aux tarses des Nothroïdes et d'autres Acariens. Arch. Zool. Exp. Gén., 96: 277—308.

Grandjean, F., 1965. Complément a mon travail de 1953 sur la classification des Oribates. Acarologia, 7: 713—734.

Hislop, R.G. and Jeppson, L.R., 1976. Morphology of the mouthparts of several species of phytophagous mites. Ann. Entomol. Soc. Am., 69: 1125—1135.

Jeppson, L.R., Keifer, H.H. and Baker, E.W., 1975. Mites Injurious to Economic Plants. University of California Press, Berkeley and Los Angeles, CA, xxiv + 614 pp., 64 plates.

Mitrofanov, V.I., 1971. O khetome konechnostei u tetranikhovykh kleshchei (Acariformes, Tetranychoidea). Nauch. Dokl. Vyssh. Shk., Biol. Nauki, 1971(4): 11—17 (in Russian).

Norton, R.A., 1977. A review of F. Grandjean's system of leg chaetotaxy in the Oribatei and its application to the Damaeidae. In: D.L. Dindal (Editor), Biology of Oribatid Mites. Publ., State University of New York College of Environmental Science and Forestry, Syracuse, NY, pp. 32—62.

Nuzzaci, G., 1982. Osservazioni ultrastrutturali sugli organi fotorecettori del *Tetranychus urticae* Koch (Acarina Tetranychidae). Mem. Soc. Entomol. Ital., 60: 269—272.

Oudemans, A.C., 1928. Acarologische Aanteekeningen LXXXIX. Entomol. Ber., 7: 285—293.

Oudemans, A.C., 1930. Acarologische Aanteekeningen CV. Entomol. Ber., 8: 157—172.

Pritchard, A.E. and Baker, E.W., 1952. A guide to the spider mites of deciduous fruit trees. Hilgardia, 21: 253—287.

Pritchard, A.E. and Baker, E.W., 1955. A revision of the spider mite family Tetranychidae. Mem. Pac. Coast Entomol. Soc., 2: 1—472.

Quirós-Gonzalez, M.J. and Baker, E.W., 1984. Idiosomal and leg chaetotaxy in the Tuckerellidae Baker and Pritchard: ontogeny and nomenclature. In: D.A. Griffiths and C.E. Bowman (Editors), Acarology VI (Proc. VI Int. Congr. Acarology, Edinburgh, Scotland, Sept. 1982), Ellis Horwood Ltd., Chichester, England, Vol. 1, pp. 166—173.

Rekk, G.F., 1947. O znachenin tulovishchnykh shchetinok v sistematike pautinnykh kleshchei. Tr. Zool. Inst. Akad. Nauk Gruz. SSR, 7: 199—203 (in Russian).

Rekk, G.F., 1959. Opredelitel' tetranikhovykh kleshchei. Izv. Akad. Nauk Gruz. SSR, Tbilisi, 1—151 (in Russian).

Robaux, P. and Gutierrez, J., 1973. Les phanères des pattes et des palpes chez deux espèces de Tetranychidae: nomenclature et évolution au cours de l'ontogénèse. Acarologia, 15: 616—643.

Summers, F.M., Gonzalez, R.H. and Witt, R.L., 1973. The mouthparts of *Bryobia rubrioculus* (Sch.) (Acarina: Tetranychidae). Proc. Entomol. Soc. Wash., 75: 96—111.

Tanigoshi, L.K. and Davis, R.W., 1978. An ultrastructural study of *Tetranychus mcdanieli* feeding injury to the leaves of 'Red Delicious' apple (Acari: Tetranychidae). Int. J. Acarol., 4: 47—56.

Vainshtein, B.A., 1958. Khetom konechnostei pautinnykh kleshchei (Acariformes, Tetranychidae) i sistema semeistva. Zool. Zh., 37: 1476—1487 (in Russian).

Van de Lustgraaf, B., 1977. L'ovipositeur des tétraniques. Acarologia, 18: 642—650.

1.1.2 Internal Anatomy

G. ALBERTI and A.R. CROOKER

INTRODUCTION

Spider mites are one of the most intensively investigated groups of Acari; this also applies to internal anatomy, including histology and cytology. However, as these fields of investigation are only slowly expanding within acarology it is the study by Blauvelt (1945), now about 40 years old, which is still the basic work for anyone dealing with tetranychid anatomy. The present review therefore draws a considerable amount of information from this source and the reader is likewise referred to it when looking for earlier literature.

As comparison is one of the fundamental methods used in the biological sciences the outstanding works by Vitzthum (1943) on Acari and by Kaestner (1968) on Arachnida should be mentioned here. Much recent information which may also be of interest to acarologists is included in the book by Foelix (1982) dealing with Araneae.

Since Blauvelt's (1945) study, *Tetranychus urticae* Koch is the species which has been investigated most often. Thus our study is largely based on results obtained from this species. If other species have been used this is indicated or may be seen from the references. Only a few authors have tried a comparative approach (see, e.g., Anwarullah, 1963; Alberti and Storch, 1974).

Tetranychidae show some reductions which are to some extent shared with other Actinedida: there is no circulatory system and apparently no connective tissue. The haemocoel is merely represented by narrow spaces, thus allowing the organs to move against each other.

In 1959, Gibbs and Morrison published a paper on the cuticle of the two-spotted spider mite. This study may be the first using transmission electron microscopy to investigate tetranychids. Since then, this method together with scanning electron microscopy has brought considerable progress to the understanding of spider mites. However, a new field of problems and questions has likewise been opened, leaving much work for the future.

INTEGUMENT

As in all arthropods, the integument of tetranychids comprises an epidermis (hypodermis) covered by a cuticle. The integument is thin, flexible and transparent (Blauvelt, 1945). The cuticle is finely ridged, the ridges bearing lobes at their apices. The epithelium is slightly flattened in a plane parallel to the cuticle.

Chapter 1.1.2. references, p. 58

The cuticle affords support and protection through its rigidity and hardness and is of primary importance in restricting water loss through the body surface. The cuticle comprises an outer epicuticle and an inner procuticle (Fig. 1.1.2.1). The procuticle consists of several similar lamellae resulting from a complicated arrangement of chitin fibrils embedded in a proteinaceous matrix. The fibrils are probably oriented according to the model of Bouligand (1965), thus in cross-section a parabolic pattern is seen (for further details, see Neville, 1975). The outer part of the procuticle may show no layers and has been referred to as exocuticle by Mothes and Seitz (1982), but the clear separation into exo- and endocuticle which is common in more sclerotized mites (Alberti et al., 1981) cannot readily be observed. Because the procuticle is not well differentiated into exo- and endocuticle, the prominent electron-dense epicuticle has occasionally been misinterpreted as exocuticle.

The epicuticle lacks chitin microfibrils. It consists of at least 4 layers (Mayr, 1971) probably comparable to the dense inner epicuticle, cuticulin, wax and cement layers of insects (Locke, 1974). As in insects, the outermost layer of the epicuticle is the cement layer. The epicuticle may be responsible for transpiration resistance and hydrofuge properties of the cuticle (Gibbs and Morrison, 1959; McEnroe, 1961a).

Pore canals are found only rarely in the Tetranychidae (Alberti et al., 1981). Canals cross the procuticle vertically to end blindly under a small indentation of the epicuticle which is not penetrated (Fig. 1.1.2.1a, c). From studies on insects (Locke, 1974), it can be inferred that pore canals are responsible for secreting the wax layer.

Mothes and Seitz (1982) found the epicuticle of *Tetranychus urticae* to be 0.05—0.15 µm thick and the procuticle to be 0.25—2.0 µm thick; the average cuticle thickness (including lobes) was 1.5 µm. Observations by other authors (Gibbs and Morrison, 1959; Henneberry et al., 1965; Alberti and Storch, 1976) are in general agreement concerning total cuticle thickness but there are differences in the measurement of epicuticle and procuticle thickness depending on the interpretation of these layers. The same authors have also noted regional cuticle thickness differences, perhaps due to varying thickness

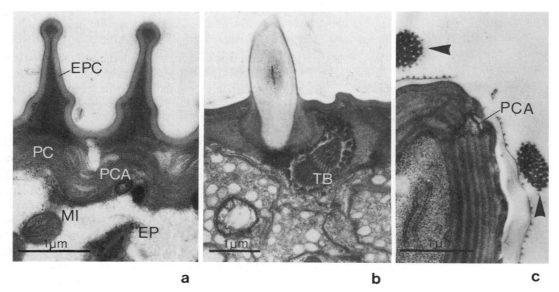

a   **b**   **c**

Fig. 1.1.2.1. (a) Integument of *T. urticae* showing lobes in cross-section. (b) Mechano-sensory seta from palp of *Bryobia praetiosa*. (c) Integument of *Bryobia praetiosa*. Note pore canal and extensive cement layer with prominent globules (arrowed). Abbreviations: EP = epidermis; EPC = epicuticle; MI = mitochondrion; PC = procuticle; PCA = pore canal; TB = tubular body.

of the procuticle and/or differences in the number of procuticular lamellae (4—16 in *T. urticae*). Epicuticle thickness remains fairly constant over the entire body.

Temporal variations in cuticular morphology have been demonstrated for summer and diapause forms of Tetranychidae (Pritchard and Baker, 1952; Boudreaux, 1958; Dosse, 1964). Actively feeding mites have cuticular ridges thickly studded with fine apical lobes whose shape varies approximately according to species; diapausing females lack these lobes over much or all of their body. One hypothesis is that these lobes may serve as evaporative structures in feeding mites; the absence of these lobes in diapausing (non-feeding) females would help conserve water. Crowe (1975) suggested that the ridges function in increasing the rigidity of the cuticle.

There are remarkable differences in cuticle structure among the tetranychids. The ridges and lobes which are characteristic and of taxonomic importance in most Tetranychidae (Anwarullah, 1963; Boudreaux and Dosse, 1963) are absent in *Bryobia*. In contrast to *Tetranychus* or *Panonychus*, *Bryobia* species have a more prominent cement layer with globules (Fig. 1.1.2.1c) which is similar to the so-called cerotegument of oribatids (Alberti et al., 1981). The lack of ridges and lobes in *Bryobia* may be compensated for by their thicker cement layer. This observation may have functional implications since surface sculpture plays an important role in maintaining the hydrophobic nature of the cuticle (Pal, 1950).

The epidermis consists of a single layer of flat epithelial cells (1.5—3.0 μm thick) containing the usual organelles and electron-dense granules, presumably pigment. A structureless thin basal lamina, 0.01 μm thick, subtends the epidermis.

From studies on other arthropods, it can be inferred that the epidermis is most active during moulting and cuticle formation. In the early phase of moulting, deposition of epicuticle occurs on top of small epidermal microvilli. Later, procuticle deposition occurs in the spaces between epicuticle and epidermal cell surfaces, during which time the old cuticle persists (Mothes and Seitz, 1981a).

SENSE ORGANS

Sense organs assist the central nervous system in obtaining information about the environment and are therefore of great importance in animal behaviour. Few workers, however, have examined the morphology of the sense organs of the Tetranychidae.

Setae

Using pore distribution as a criterion for classification, Bostanian and Morrison (1973) divided body and appendage setae of *T. urticae* into 3 groups (Fig. 1.1.2.2a, b). The first group, located on the palps, exhibited no pores and were considered to be mechanoreceptors. The second group, located on the palps and legs, included smooth, thick- and thin-walled setae with longitudinal striations and a single apical pore, and were considered to be contact chemoreceptors. Most body and leg setae belonged to the third group and were characterized by a spiculated appearance. Each spicule is said to have an apical pore, thus the whole seta is provided with numerous tiny pores. From their innervation and pore distribution it was determined that these setae function as mechano- and olfactory receptors.

Mills (1973b) studied the fine structure of several dorsal propodosomal setae, completing an earlier investigation by McEnroe (1969a). Mills noted

Chapter 1.1.2. references, p. 58

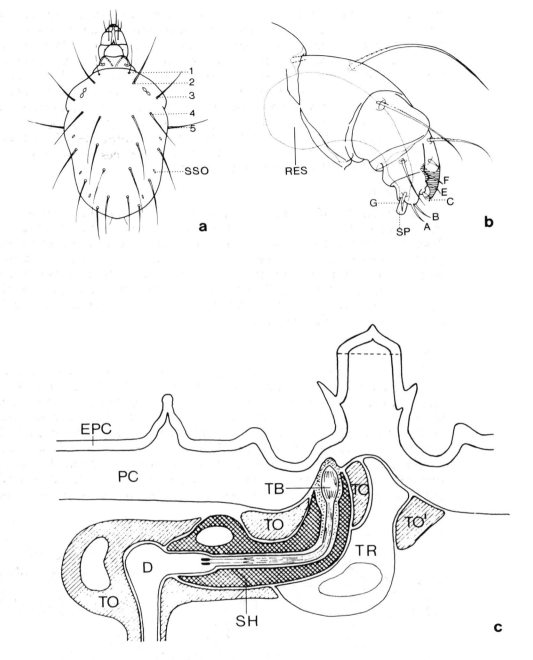

Fig. 1.1.2.2. Sense organs. (a) Arrangement of dorsal setae, eyes and slit sense organs in
T. urticae. Terminology of setae according to Mills (1973b). (b) Palpal setae of
Tetranychus lintearius (redrawn from Grandjean (1948) with the permission of the
Bulletin du Museum Nationale Histoire Naturelle, Paris). Terminology of setae according
to Bostanian and Morrison (1973): setae A, B = thick-walled contact chemoreceptors;
C = thin-walled contact chemoreceptor; E,F,G = mechanoreceptors. (c) Diagrammatic
drawing of seta 2 and its sensory apparatus of *T. urticae* (modified from Mills (1973b)
with the permission of the Museum Nationale Histoire Naturelle, Paris). Abbreviations:
D = dendrite of 1 of the 2 sensory cells; EPC = epicuticle; PC = procuticle; RES =
reservoir of silk gland; SH = sheath cell; SP = spinneret; SSO = slit sense organ;
TB = tubular body; TO, TO′ = tormogen cells; TR = trichogen cell.

that the spiculated setae of Bostanian and Morrison (1973) comprise 2
sensory types, one purely mechanosensory, the other having a dual chemo-
and mechanosensory function. The mechanoreceptors are innervated by 2

sensory cells whose dendrites terminate in the articulating membrane of the seta as a tubular body. A sheath cell surrounds the distal parts of the 2 dendrites in the ciliary region; the setal cilia have an 8 + 0 pattern of doublet microtubules. The seta itself is free of cellular processes. Two tormogen cells and 1 trichogen cell complete the sensory apparatus (Fig. 1.1.2.2c). These mechanoreceptors were found in rows 1, 2 and 5. Mills (1973b) believes that the setae of row 4 may be both chemo- and mechanosensory. These setae are provided with more than 2 sensory cells (probably 4), 2 of which form tubular bodies (characteristic for mechanoreceptors in arthropods) (Fig. 1.1.2.1b). The setal lumen contains a cellular process which may represent the dendrite(s) of the other sensory cell(s).

Slit sense organs

Anwarullah (1963) described 3 pairs of dermal glands located dorsally on the hysterosoma (Fig. 1.1.2.2a). In contrast, Penman and Cone (1974) have described these structures as lyrifissures functioning as mechanoreceptors or possibly release sites for sex attractants. Recently, A.R. Crooker (this work) observed an innervation. A fourth pair of these structures, which are often termed 'cupules', is located in a ventrolateral position (see chapter 1.1.1). The present authors prefer the term slit sense organ, when referring to the most probable function.

Eyes

Spider mites possess a pair of non-faceted eyes on each side of the dorsum of the propodosoma in both males and females (Fig. 1.1.2.2a). Mills (1974) studied the fine structure of these eyes in *T. urticae* (Fig. 1.1.2.3). According to him, the eyes are similar in structure in both sexes. The anterior and posterior eye of each side lie close together in a single eye-manifold. The anterior, circular cornea or lens is biconvex; the posterior, elliptical cornea is simple convex. Both corneae are provided with an epicuticle having fine ridges positioned about 180 nm apart (Gibbs and Morrison, 1959; McEnroe, 1969b) and both show pronounced procuticular lamellae. The eye-manifold

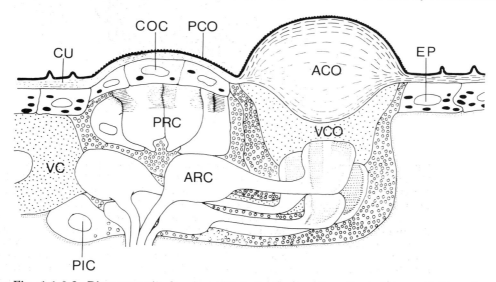

Fig. 1.1.2.3. Diagrammatic drawing of longitudinal section through eyes of *T. urticae* (modified from Mills (1974) with the permission of Pergamon Press, Oxford). Abbreviations: ACO = anterior cornea; ARC = anterior retinular cells; COC = cornea cells of posterior eye; CU = cuticle; EP = epidermis; PCO = posterior cornea; PIC = pigment cells; PRC = posterior retinular cells; VC = vitreous cell; VCO = vitreous cone.

contains 15 retinular cells, 6 pigment cells, 6 corneal cells and 1 'vitreous' cell. The latter presumably is a specialized epidermal cell which has a projection, the 'vitreous cone', immediately beneath the anterior cornea; the posterior cornea is located above the 6 corneal cells. The anterior eye comprises 5 retinular cells, the posterior eye 10. Their receptor poles possess regularly arranged microvilli which constitute rhabdomeres. The 5 rhabdomeres of the anterior eye are arranged in cup-shaped receptor poles; thus they are separated from each other. In contrast, 9 retinular cells of the posterior eye contribute to 1 large fused rhabdom. Further, 2 retinular cells of this rhabdom possess small 'accessory' rhabdomeres which are not included in the large fused rhabdom. The remaining 10th retinular cell also bears a separate small rhabdomere. There are thus 12 rhabdomeres belonging to the posterior eye.

The cell bodies of the 6 pigment cells lie together with those of other cells of the eye-manifold, more or less beneath the posterior eye. Processes of the pigment cells form a thin sheath under the posterior eye and surround the photoreceptive structures (receptor poles) of the retinular cells of the anterior eye, except for the dorsal side where the vitreous cone is located. Processes of the pigment cells contain 2 kinds of pigment. One is composed of small droplets which appear light grey to black in electron micrographs. These droplets presumably represent the red lipid responsible for the red colour of the eyes seen in live specimens. The second kind is similar to the black staining granules of epidermal cells.

A basal lamina subtends the whole eye-manifold. Axons of the 15 retinular cells of each pair of eyes are connected by an optic nerve to their respective optic lobes in the central nervous mass.

Apodemal organ

The apodemal organ of Mills (1974) is located in front of and about $2\,\mu$m below the dorsal apodeme attached to a tendon of muscle DV 9. The cells, presumably neurons, of this organ have 2 very large mitochondria, in addition to mitochondria of normal size. The 2 axons of these cells join the optic nerve but do not terminate in the optic neuropile. Presumably, the apodemal organ functions as a proprioceptor (Mills, 1974).

NERVOUS SYSTEM

Blauvelt (1945) stated:
'There is practically no literature on the central nervous system of the Tetranychidae.'

This statement still holds true, with the exception of his own work, a dissertation by Crooker (1981) and some observations by Mills (1974). Our knowledge of the spider mite nervous system is rather superficial compared with that of the nervous system of ticks and some other Arachnida (Babu, 1965; Pound and Oliver, 1982; Juberthie, 1983).

As in other Acari, the central nervous mass (Kaestner, 1968), or synganglion, of *Tetranychus* consists of the fused supraoesophageal and suboesophageal ganglia. Other ganglia such as the optic, cheliceral, pedipalpal, rostral, and those of the leg also make up the synganglion. The relatively large synganglion lies close to the ventral body surface, surrounded by the silk gland, dorsal podocephalic gland, midgut, and in females, the ovary (Fig. 1.1.2.18). The oesophagus passes through the thickened central part of the synganglion and divides it into a dorsal supraoesophageal and a ventral suboesophageal mass (Fig. 1.1.2.46). Blauvelt (1945) described a great

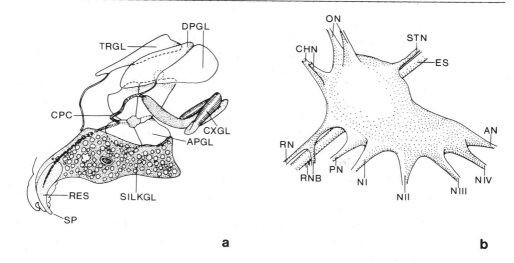

Fig. 1.1.2.4. (a) Schematic representation of the prosomal glands of *T. urticae* viewed laterally (redrawn from Alberti and Storch (1974) with the permission of Springer Verlag, Heidelberg). (b) Central nervous mass and bases of nerves of *T. urticae* (redrawn from Blauvelt, 1945). Abbreviations: AN = abdominal nerve; APGL = anterior podocephalic gland; CHN = cheliceral nerves; CPC = podocephalic canal; CXGL = coxal gland; DPGL = dorsal podocephalic gland; ES = oesophagus; NI—NIV = nerves of leg I—IV, respectively; ON = optic nerves; PN = pedipalpal nerve; RES = reservoir of silk gland; RN = rostral (infracapitular) nerve; RNB = lateral branches of RN; SILKGL = silk gland; SP = spinneret; STN = stomodeal nerve; TRGL = tracheal gland.

commissure joining the paired ganglia of the suboesophageal mass and the cheliceral ganglia; a cheliceral commissure could not be distinguished. A central body and stomodeal bridge characteristic of other arachnids (Hanström, 1919, 1928; Babu, 1965) was not found.

In common with other arthropods the synganglion is divided into an outer cortical region and an inner neuropile. The cortex is composed of pericarya and glial cells which comprise a peri- and subperineurium. The entire synganglion is surrounded by an extracellular sheath comparable to a basal lamina (Crooker, 1981).

The origin of the nerves arising from the synganglion provides an external indication of the relative positions of ganglia fused into the central nervous mass. Blauvelt (1945) described the following nerves of the peripheral nervous system (Fig. 1.1.2.4b):

Nerves of the supraoesophageal ganglion

There are a pair of optic nerves, a pair of cheliceral nerves, an unpaired rostral nerve (i.e. infracapitular nerve) and an unpaired stomodeal (stomatogastric) nerve arising from the supraoesophageal ganglion. The exit of the optic nerves from the periphery of the central nervous mass is marked by a distinct enlargement, the optic lobes. Mills (1974) found that each of the optic lobes contained an optic neuropile having a volume of about $440\,\mu\mathrm{m}^3$. He estimated that each optic neuropile was composed of several dozen cells which contained no more than a few hundred synaptic junctions. A bundle of nerve fibres leaves the optic lobe and passes into the main neuropile of the brain (Mills, 1974). In each of the optic lobes there is 1 enlarged cell ($6\,\mu$m across, compared with $3-4\,\mu$m in typical neurons) containing a high concentration of electron-dense particles up to $0.11\,\mu$m in diameter (Mills, 1974). These cells were described as argentiphilic cells by McEnroe (1969b). Other large specialized cells with conspicuous cytoplasm

Chapter 1.1.2. references, p. 58

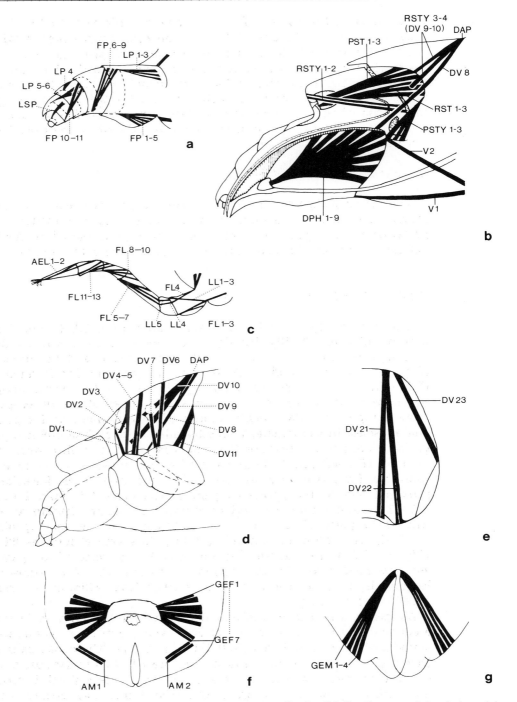

Fig. 1.1.2.5. Muscular system of *T. urticae*. (a) Pedipalp. (b) Gnathosoma, lateral view. (c) Leg I. (d) Propodosoma, lateral view. (e) Hind end of opisthosoma, lateral view. (f) Genital and anal muscles of female, ventral view. (g) Genital muscles of male, ventral view (redrawn with modifications, from Akimov and Jastrebtsov (1981) (a–c), whose permission is appreciated, and from Blauvelt (1945) (d–g). Abbreviations: AEL = levator of apotele; AM 1–2 = anal muscles 1–2; DAP = dorsal apodeme; DPH 1–9 = dilatator of pharynx 1–9; DV 1–11 and DV 21–23 = dorsoventral muscles 1–11 and 21–23; FL 1–13 = flexors of leg 1–13; FP 1–11 = flexors of palp 1–11; GEF 1–7 = genital muscles of female 1–7; GEM 1–4 = genital muscles of male 1–4; LL 1–5 = levators of leg 1–5; LP 1–6 = levators of palp 1–6; LSP = levator of spinneret; PST 1–3 = protractors of stylet 1–3; PSTY 1–3 = protractors of stylophore 1–3; RST 1–3 = retractors of stylet 1–3; RSTY 1–4 = retractors of stylophore 1–4; V 1–2 = ventral muscles 1–2.

were found in the cortex of the synganglion (Blauvelt, 1945). Below and slightly in front of the optic lobes the cheliceral ganglia branch off into the cheliceral nerves which run up and forward toward the stylophore. The rostral (infracapitular) nerve arises from the brain directly above the entrance of the oesophagus and gives off 2 lateral branches at some distance from the brain. The resulting anterior branch runs into the tip of the infracapitulum; the course of the 2 lateral branches was not traced. The stomodeal nerve leaves the posterior surface of the brain directly above the oesophagus and runs toward the ventriculus.

Nerves of the suboesophageal mass

The paired pedipalpal nerves and 4 pairs of leg nerves arise from their own ganglia. An unpaired abdominal nerve leaves the synganglion from the posterior median ganglion (which presumably was paired originally) and extends back along the median ventral surface beneath the gonads.

MUSCULAR SYSTEM

The main study of spider mite musculature (Fig. 1.1.2.5 and 1.1.2.6) is that by Blauvelt (1945). His figures are redrawn here. Blauvelt described 3 groups of body muscles: ventral, dorsoventral, and dorsolongitudinal (Fig. 1.1.2.6). The origin and insertion of the dorsolongitudinal muscles (DL1—DL5) are a pair of dorsal apodemes and the dorsal integument, respectively; the origins and insertions of muscles of the other groups are more complicated. Two smaller groups of muscles make up the genital and anal muscles. In the female, 7 muscles arise from each lateral margin of the genital plate and fan towards the ventrolateral body surface where they are inserted (Fig. 1.1.2.5f). Four pairs of genital muscles in the male are responsible for movements of the penis (aedeagus). Their insertion is the proximal end of the apodemal part of the penis (cf. Figs. 1.1.2.5g and 1.1.2.14) from where they extend posteroventrad and laterad to the body wall. Two pairs of anal muscles are found near the anal opening (Fig. 1.1.2.5f). According to Blauvelt (1945), dorsoventral muscles 22 and 23 (Fig. 1.1.2.5e) also function as anal muscles. Muscles of the gnathosoma are described below in the section on mouthparts. The arrangement of leg muscles may be seen in a figure redrawn from Akimov and Jastrebtsov (1981) (Fig. 1.1.2.5c).

Muscle attachment to the exoskeleton of tetranychids, as in insects, is mediated by specialized epidermal cells whose cytoplasm is marked by an abundance of microtubules (Fig. 1.1.2.7a, inset). These flat cells are also traversed by tonofilaments. Muscle fibres are attached to the specialized cells by desmosomes arranged in a cone shape; the epidermal cell is attached to the cuticle by hemidesmosomes. Cuticular fibres may deeply indent the apical region of the cell (Crooker, 1981). According to Blauvelt (1945) 2 pairs of apodemes located on the sejugal furrow (terminology of Van der Hammen, 1980) — 1 pair ventrally immediately behind coxa II and 1 pair dorsolaterally — also serve as muscle attachment sites. Some leg muscles are attached to the integument by way of specialized epidermal cells, whereas others are attached by long tendons which are cuticular apodemes covered with an epithelium.

The role of muscles in mediating movement is poorly known. Hydrostatic pressure, important in soft-bodied animals, may be exerted by the body contents when body muscles contract (Manton, 1958). Support of muscles is

Chapter 1.1.2. references, p. 58

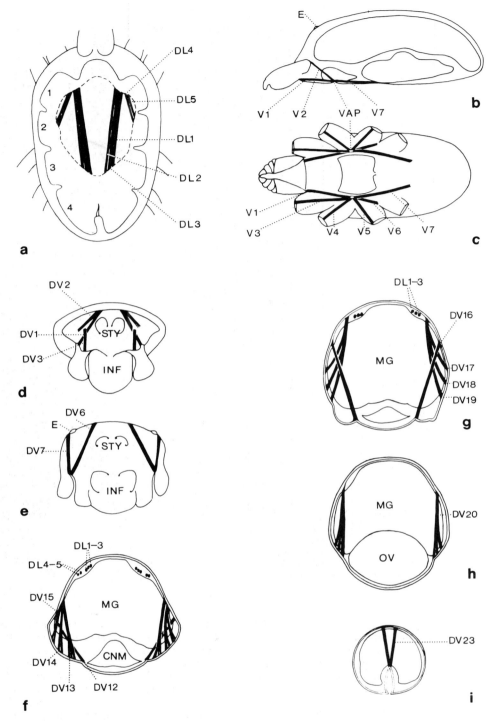

Fig. 1.1.2.6. Muscular system of *T. urticae*. (a) Dorsal longitudinal muscles, dorsal view with integument partly removed. (b) Sagittal section through ventral apodeme. (c) Ventral view on ventral muscles. (d)—(i) Cross-sections through: (d) basis of coxa I, (e) basis of coxa II, (f) ventral apodeme, (g) between 2nd and 3rd lateral pouch of midgut (cf. (a)), (h) between the 3rd and 4th lateral pouch of midgut (cf. (a)), (i) through posterior extremity of body (redrawn with slight modifications from Blauvelt, 1945). Abbreviations: CNM = central nervous mass; DL 1—5 = dorsolongitudinal muscles 1—5; DV 1—3, 6, 12—20, 23 = dorsoventral muscles 1—3, 6, 12—20 and 23; E = eye; INF = infracapitulum; MG = midgut; OV = ovary; STY = stylophore; V 1—7 = ventral muscles 1—7; VAP = ventral apodeme.

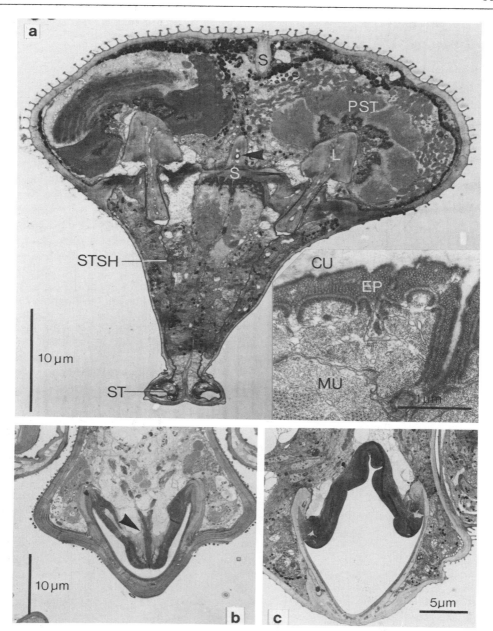

Fig. 1.1.2.7. (a) Cross-section through stylophore of *T. urticae* showing articulation of the levers with apodemes. Arrow: duct of tracheal gland inside cuticular septum. Inset: myocuticular junction between stylet muscle cell and posterior part of stylophore. (b) Cross-section through pharyngeal pump of *Bryobia praetiosa* with the dilatator muscles relaxed, resulting in a U-shaped, narrowed lumen. Note muscle attachment via cuticular tendon (arrowed) and different appearance of cuticle of plunger compared with that of floor and lateral walls of pump. (c) Pharyngeal pump of *T. urticae* plunger being withdrawn by infolding its flexible cuticle, thus enlarging the lumen of the pharyngeal pump. Note in (b) and (c) that the floor of the pharyngeal pump is almost entirely composed of rigid cuticle, epidermal components having largely been withdrawn. Abbreviations: CU = cuticle; EP = epidermal cell modified with high amounts of microtubuli embedded in dark substance; L = lever; MU = muscle cell attached to epidermal cell via desmosomes; PST = protractors of stylet; S = septum; ST = stylet; STSH = stylet sheath.

not dependent upon haemocoelic pressure because of the apparent lack of a haemocoel in spider mites. Some muscles may pull on the flexible parts of the integument; tendon and apodemal systems probably also play a role.

Chapter 1.1.2. references, p. 58

As in other arthropods, muscles of the Tetranychidae are of the transversely striated type.

PROSOMAL GLANDS

All Actinotrichida (Acariformes), to which the tetranychids belong possess a pair of lateral podocephalic canals on the prosoma (Van der Hammen, 1980). These canals meet at the basis of the gnathosoma to give rise to an unpaired canal (open groove), the median salivary duct, which runs the length of the infracapitulum to the tip of the gnathosoma. Each podocephalic canal in Tetranychidae receives the secretion of 3 glands (Fig. 1.1.2.4a). Posteriorly there is a tubular gland homologous with the coxal glands of other Actinotrichida (Alberti and Storch, 1977). Immediately in front of this an acinous gland, the dorsal gland, debouches via a short duct and finally, near the beginning of the unpaired median salivary duct, the duct of another acinous gland, the anterior gland, joins the podocephalic canal.

In addition to these 'podocephalic glands', there is an unpaired tracheal gland located dorsally in the anterior part of the propodosoma. Its duct runs medioventrad through the stylophore, to discharge its secretion between the cheliceral stylets or onto the infracapitulum (Figs. 1.1.2.4a, 1.1.2.7a, 1.1.2.8a and 1.1.2.18). In those spider mites capable of silk secretion, a large

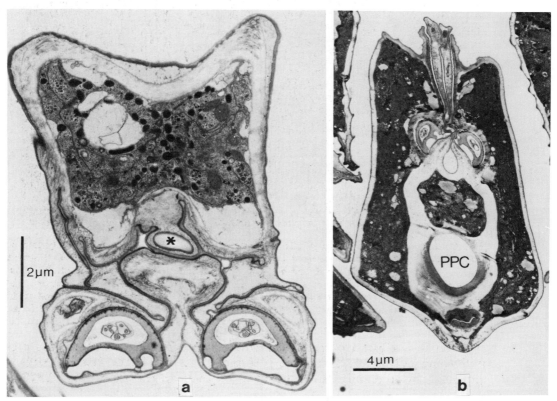

Fig. 1.1.2.8. (a) Cross-section through anterior part of stylophore of *T. urticae*. Note complicated arrangement of stylet sheath; nerve fibres inside the stylets, stylets provided with a ridge and groove, respectively. Asterisk: duct of tracheal gland. (b) Transverse section through infracapitulum of *Panonychus ulmi*, slightly in front of the pharyngeal pump. The following structures are to be seen from dorsal to ventral: fixed digits of stylophore bordered by elevated lips, stylet channels with stylets paralleling the median salivary duct. Between this duct and the prepharyngeal canal is the position from which the labrum emerges more anteriorly, between the prepharyngeal canal and the ventral surface the small channel leading to the rostral fossette is cross-sectioned. Abbreviation: PPC = prepharyngeal canal.

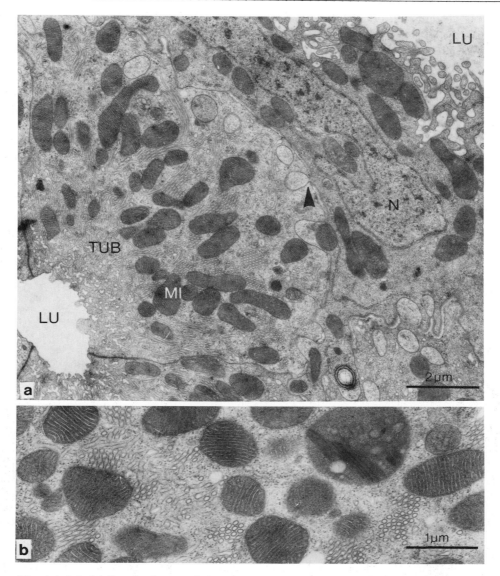

Fig. 1.1.2.9. (a) Proximal part of the coxal gland of *Bryobia praetiosa* closely adjacent to the midgut epithelium. The arrow points to interdigitation between the cell bases of both epithelia where the basal lamina is interrupted. Note the large amounts of mito-chondria in the cells of the coxal gland close to masses of tubular invaginations. (b) Tubules and mitochondria at higher magnification. Abbreviations: LU = lumen of coxal gland (lower left) and midgut (upper right); MI = mitochondria; N = nucleus of midgut cell; TUB = tubules.

unicellular silk gland extends from the tip of each palp back into the pro-podosoma (Figs. 1.1.2.4a, 1.1.2.10b and 1.1.2.18). Paired tracheal organs surround the ascending part of the loops of the tracheal trunk (Mills, 1973a).

Coxal glands

In contrast to other Actinotrichida, where the primitive state is retained (Alberti and Storch, 1977), the tubular or coxal gland of spider mites lacks a sacculus; only the tubulus is present. The latter is divided into proximal, middle (intercalated), and distal parts (Fig. 1.1.2.4a). The distal part leads via a short duct to the podocephalic canal; the proximal and middle parts double back on each other. The proximal part is adjacent to the midgut

Chapter 1.1.2. references, p. 58

epithelium which is modified in this region, but contrary to the statements of Mothes and Seitz (1980), there is no direct connection between these 2 regions (Fig. 1.1.2.9). The epithelium of the proximal part comprises columnar cells possessing a peculiar vertically arranged system of grouped parallel tubules which open at the cell apex. They extend nearly to the cell base, where the tubules are highly convoluted, those of the same group being connected with each other. Numerous large crista-type mitochondria are present between the parallel tubules (Fig. 1.1.2.9).

The middle part of the coxal gland has no unusual features, but the basal plasmalemma of the distal part is highly infolded and, together with numerous mitochondria, constitutes a basal labyrinth. The apex shows numerous electron-dense microtubules. From observations of coxal glands in other mites, it can be expected that the cuticular duct leading from the distal segment is closed passively and opened actively (Alberti and Storch, 1977).

Morphological observations and comparisons with other Actinotrichida possessing a sacculus indicate that the coxal glands are involved in regulating water and ion balance. No solid excretory products have been observed. Mills (1973a) proposed that in the female, these glands also produce a pheromone additive which attracts nearby males to the female webbing.

Dorsal and anterior podocephalic gland

These acinous podocephalic glands contain numerous granules, probably of a proteinaceous nature (Figs. 1.1.2.10a, c and 18). The cells are equipped with organelles characteristic of protein-secreting cells (large nuclei with prominent nucleus and numerous ribosomes, some attached to ER cisternae). The dorsal gland is composed of 2 cells, the anterior gland of 4 (*T. urticae*).

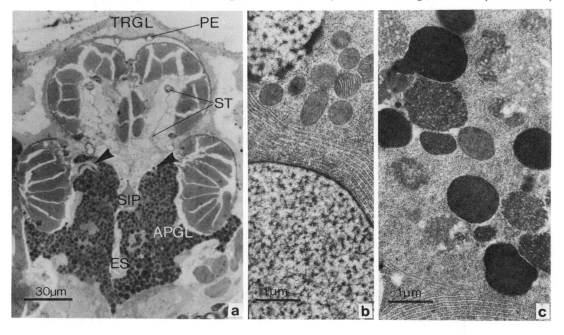

Fig. 1.1.2.10. (a) Cross-section through posterior part of gnathosome of *B. praetiosa*. The arrows point to ducts of the anterior podocephalic glands (from Alberti and Storch (1974) with the permission of Springer-Verlag, Heidelberg). (b) Detail of silk gland of *T. urticae* showing large secretory droplets, rough ER and mitochondria. (c) Anterior podocephalic gland of *T. urticae* with polymorphous granules. Abbreviations: APGL = anterior podocephalic gland; ES = oesophagus; PE = peritrema; SIP = sigmoid piece; ST = stylet; TRGL = tracheal gland.

Mothes and Seitz (1981e) believe these glands to be the main salivary glands. Summers et al. (1973), who were the first to describe the anterior podocephalic glands (termed oesophageal salivary glands by these authors) in a *Bryobia* species, stated that these glands correspond in position to the fat body described by Blauvelt (1945) for *Tetranychus*. However, as in *Bryobia*, there is no fat body present in *Tetranychus* (Alberti and Storch, 1974).

Tracheal gland

This unpaired gland consists of numerous cells connected by inter-digitations (Fig. 1.1.2.18). The cells contain large amounts of electron-lucent droplets, probably lipids; rough ER and smooth ER cisternae have also been observed. Whereas Blauvelt (1945) and Mills (1973a) refer to this gland as the dorsal salivary gland, Mothes and Seitz (1981e) assume that it may produce a secretion which facilitates the movements of the stylets in the infracapitular gutter.

Silk glands

In *T. urticae* a unicellular gland extends from each palp back to the central nervous mass (Alberti and Storch, 1974) (Figs. 1.1.2.4a, 1.1.2.10b and 1.1.2.18). The same observation was repeated recently by Mothes and Seitz (1981e). These very large cells are almost completely filled with vacuoles containing a proteinaceous secretory product derived from rough ER. In the distal joints of the pedipalps the droplets are discharged into a reservoir-like region of the cell apex which narrows to a final outlet at the tip of a hollow seta or spinneret. This reservoir was depicted by Blauvelt (1945) and Grandjean (1948) (Fig. 1.1.2.2b), who did not realize its true nature. Because of structural similarity to a fat body, these large glands have repeatedly been mistaken for a fat body in *T. urticae* and other species.

Morphological observations indicate that the unicellular glands are responsible for the silk characteristic of higher Tetranychidae (Alberti and Storch, 1974; Mothes and Seitz, 1981e). It is possible, however, that secretory products of other glands may also be involved in silk production. Mills (1973a) considered the entire podocephalic complex of *T. urticae* to be involved in silk production. *Bryobia* and *Tetranychina*, which do not spin silk, have no unicellular glands in the palps (Alberti and Storch 1974; G. Alberti, this work).

In *Panonychus ulmi* (Koch), a species which spins considerably less than *T. urticae*, (Gerson, 1979), the silk glands are restricted to the distal segments of the palps (G. Alberti, this work).

Tracheal organs

A tracheal organ completely surrounds the middle segment of the ascending loop of the tracheal trunks on each side of the mite (Mills, 1973a; Mothes and Seitz, 1981e). These paired organs consist of 5 or 6 cells containing many rough ER cisternae and vesicles. A cuticle-lined labyrinth of channels is in connection with the lumen of the appropriate tracheal trunk. Mills (1973a) suggested that the tracheal organs may function in humidity detection, pheromone production or detection, or ecdysis, but the exact function of these structures is unknown.

Chapter 1.1.2. references, p. 58

MOUTHPARTS

The gnathosoma of spider mites is highly advanced compared with the basic structure in Actinedida (Prostigmata and Endeostigmata) (Van der Hammen, 1970). The 3 main elements, chelicera, infracapitulum (rostrum) and pedipalps, are adapted to plant feeding. Becker (1935) and Blauvelt (1945) described the mouthparts of the two-spotted mite. Snodgrass (1948) included these studies in his comparative work on Chelicerata mouthparts. Baker and Connell (1963) presented information on *Tetranychus turkestani* (Ugarou & Nikolski); Summers et al. (1973) examined the mouthparts of *Bryobia rubrioculus* (Scheuten).

There is good agreement among these papers concerning the general features of the mouthparts. The external morphology of the mouthparts is discussed in Chapter 1.1.1.*

Chelicerae

The chelicerae are derived from typical prehensile euchelae which have their proximal joints fused to form the stylophore (mandibular plate of earlier authors). The stylophore bears a dorsomedian cleft in its basal part which extends inside to a median cuticular septum, the latter being an indication of the originally paired nature of the stylophore. This septum contains the tracheal gland duct (cf. previous section) (Figs. 1.1.2.7a and 1.1.2.8a). At the distal part of the stylophore close to the median line of the infracapitulum there are 2 projections which are thought to be homologous with the fixed digits of typical euchelae (Fig. 1.1.2.8b). The movable digits (distal joints of the chelicerae) are modified as long stylets adapted for piercing the host plant cells (Figs. 1.1.2.5b, 1.1.2.7a, 1.1.2.8 and 1.1.2.18). The articulation of these stylets with the stylophore is similar to that found in simple euchelae. The stylets are composed of 2 segments, a basal spatulate sclerite and a long, narrow terminal segment. The basal sclerite ('lever' of Summers et al., 1973) is suspended between 2 knob-like apodemes on which it pivots (Fig. 1.1.2.7a). The apodemes are developed from the exterior cuticle and from the median vertical septum which divides the stylophore. The terminal segments go backward from the basal segments, curve sharply downward, then go forward to emerge from the stylophore onto the dorsal surface of the infracapitulum, where they become enclosed in 2 channels. The terminal segments of the stylets and a portion of the basal segments are embedded in a deep, narrow invagination of the external cuticle, the stylet sheath (Figs. 1.1.2.7a and 1.1.2.8a). In cross-section, the stylets appear sickle-shaped; one is provided with a ridge, the other with a groove (Fig. 1.1.2.8a). Summers et al. (1973) thought that the protruded stylets were juxtaposed to form a hollow probe. Nerve fibres have been found inside the stylets (Mothes and Seitz, 1981c).

Infracapitulum

The infracapitulum comprises a broad basal portion which is prolonged into an apical rigid beak with the mouth at its tip. The basal portion contains the muscles of the pharyngeal pump and pedipalps and part of the silk glands (if present). The stylet channels and an unpaired furrow, the median salivary duct, are located on the dorsal surface of the infracapitulum. The stylet channels are bordered by elevated rims or lips; the median salivary duct is embedded in a strong sclerotized longitudinal ridge (Fig. 1.1.2.8b). The ridge is a prolongation of the aliform apodeme (Summers et al., 1973), a part of

---

*See *Note added in proof* on page 62.

the sigmoid piece (cf. section on the respiratory system). These sclerites, the posterior apodemes of the stylophore, and the basal part of the infracapitulum are the main structures to which muscles are attached (Fig. 1.1.2.5b). The median salivary duct and the stylet channels extend toward the tip of the infracapitulum to the labrum located just above the mouth. The ventral and lateral margins of the mouth are provided with lobes or flaps. Inside the ventral flaps several oral sensillae have been observed (Summers et al., 1973; Hislop and Jeppson, 1976; Mothes and Seitz, 1981c). On the anteroventral portion of the infracapitulum, the 'rostral fossette' (McGregor, 1950; Summers et al., 1973) (Fig. 1.1.2.18) communicates with the prepharyngeal canal by way of a short chitinous tube, as has been confirmed by the present authors (Fig. 1.1.2.8b).

Pedipalps

The palps of tetranychids are five-segmented. They contain the silk gland reservoir distally and a portion of the unicellular silk glands basally, in those species capable of silk production.

Spider mites presumably utilize information from receptors on their legs, palps, oral lobes and chelicerae in locating a suitable feeding site. Feeding may take place in 3 stages (Akimov and Jastrebtsov, 1981). First, protraction of the stylophore is initiated by contraction of the oblique protractor muscles (Fig. 1.1.2.5b). Hydrostatic pressure may also aid the protraction. The plant is then penetrated by the stylet. Stylet movement is achieved by protractors and retractors (levator and flexor of euchela) originating from the proximal apodeme of the stylophore and inserted on the basal segment of the stylets. Presumably, the mouth is pressed against the plant surface when the mite elevates its body by lifting the posterior legs. The puncture of the leaf cuticle is then sealed off by the mouth lobes and the pharyngeal pump is able to act. Hislop and Jeppson (1976) noted that the stylophore, which slides back and forth in the V-shaped trough of the infracapitulum, is protracted and retracted several times in order to puncture the same area. Retraction of the stylets may allow plant fluids to flow to the surface for suction into the pharynx; the connection between the rostral fossette and pharynx may help the animal avoid injury by excess plant cell turgor pressure. Both stylets slide back and forth on each other in a very rapid movement when probing.

During feeding, the stylets are interlocked to form a hollow tube. Summers et al. (1973) believed plant material to be sucked through this hollow tube. This implies a connection between the stylet tube and mouth; no such connection has been found. Another explanation for the interlocked stylets, which seems more likely to the present authors, was also proposed by Summers et al. (1973) and further commented upon by Hislop and Jeppson (1976). The interlocked stylets, which protrude in front of the median salivary duct, may convey expelled saliva to the feeding puncture.

DIGESTIVE TRACT

The digestive system consists of the foregut (mouth, pharynx, oesophagus), midgut (ventriculus and coecae) and hindgut (excretory organ, rectum and anus) (Fig. 1.1.2.18). Organs of the digestive system occupy the largest volume of any body system.

Foregut

The mouth is located at the tip of the infracapitulum. It is bordered ventrally and laterally by 4 outer and 4 inner oral lobes or flaps capable of

Chapter 1.1.2. references, p. 58

closing the mouth when pressed against the dorsal labrum and elevated lips of the stylet channels. The stylet channels terminate immediately above the oral orifice (Summers et al., 1973; Mothes and Seitz, 1981c). The mouth is connected with the large pharyngeal pump by a short prepharyngeal canal which curves slightly dorsad. The pharynx is in turn directed posterioventrad. In cross-section, the pharyngeal pump appears U-shaped, the dorsal wall being equipped with a strong cuticular plunger. Nine to 12 pairs of dilator muscles, depending on the species, are attached to the plunger (Blauvelt, 1945; Anwarullah, 1963; Akimov and Jastrebtsov, 1981). These muscles originate from the sigmoid piece and the wall of the infracapitulum. When contracted, the muscles force the plunger to infold along its longitudinal axis and by this means an enlargement of the lumen of the pharyngeal pump is achieved (Fig. 1.1.2.7b, c). Since no transverse constrictor muscles have been observed, it may be assumed that the elasticity of the cuticular structures of the pharynx may work antagonistically to that of the dilator muscles (Anwarullah, 1963). As a result of the sucking and pumping activities of the pharyngeal pump, a rhythmic inflow of plant cell contents can be observed during feeding (Liesering, 1960). The rostral fossette (McGregor, 1950; Summers et al., 1973), located on the anteroventral surface of the infracapitulum, is unique to spider mites. It is a small opening connected to the prepharyngeal canal by a short duct (Figs. 1.1.2.8b and 1.1.2.18). The pharynx is followed posteriorly by a narrow oesophagus which passes through the anterior salivary glands (*Bryobia*) or silk glands (*Tetranychus*) and the central nervous mass (Figs. 1.1.2.4b, 1.1.2.10a and 1.1.2.18). The oesophageal lumen is lined by a thin cuticle; the epithelium which secretes this cuticle closely resembles the epithelium of the integument and is subtended by a thin basal lamina. A valve is located at the entrance of the oesophagus into the unpaired portion of the midgut, the ventriculus. Cells of the valve are provided with numerous microtubules.

Midgut

The midgut is composed of an unpaired ventriculus (the term ventriculus is sometimes used for the entire midgut) located immediately behind and above the synganglion and 2 lateral coecae. The ventriculus is connected anteroventrally with the oesophagus and posteriodorsally with the excretory organ. Two large coecae extend from the ventriculus posteriolaterally to occupy most of the lateral body cavity; the coecae are separated along the midline by the excretory organ. The outer walls of the coecae and the ventriculus are indented by the groups of dorsoventral muscles (Fig. 1.1.2.6a). The resulting pouches are also often termed coecae. A similar arrangement of gut pouches is often found in Actinotrichida (Bader, 1954; Alberti, 1973). Midgut contents are often responsible for the colour of the live animal.

The midgut wall is provided with a muscular layer which is responsible for peristalsis. Contractions of the dorsoventral muscles may contribute to mixing of midgut contents. A pathway for food transport has been described by Wiesmann (1968) and Mothes and Seitz (1981c).

Considerable changes are to be expected in the appearance of the midgut epithelium, according to the physiological condition or activity of the animal, e.g., feeding, diapausing, egg production, life stage, age (Fig. 1.1.2.11). This results in difficulties of interpretation of histological or cytological observations. According to Blauvelt (1945), Ehara (1960), Wiesmann (1968) and Mothes and Seitz (1981c) most of the midgut epithelium consists of 1 cell type found in the ventriculus and the lateroventral walls of the coecae. These

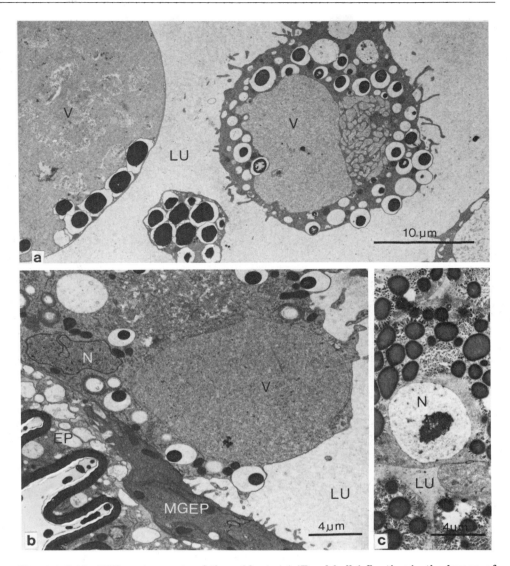

Fig. 1.1.2.11. Different aspects of the midgut. (a) 'Food balls' floating in the lumen of the midgut of *B. praetiosa*. Note the large central vacuole surrounded only by a very thin cytoplasmic margin (especially at the left). (b) *B. praetiosa*: one midgut cell bulging into the lumen, other cells showing a flat appearance. (c) *T. urticae*: inflated midgut cells with nucleus near to the cell apex, cells containing lipid droplets and glycogen, lumen of midgut reduced. Abbreviations: EP = epidermis; LU = lumen of midgut; MGEP = midgut epithelium; N = nucleus of midgut cell; V = central vacuole.

cells may range from flat ones, with large amounts of rough ER, large nuclei, crista-type mitochondria and some microvilli, to bulbous ones protruding into the gut lumen and containing large vacuoles (heterolysosomes) (Fig. 1.1.2.11b). Large amounts of glycogen may be present. The transition from flat to bulbous cells may be due to resorption of gut contents by phagocytosis and pinocytosis. Portions of the cell or the entire cell may then be pinched off from the remaining epithelium. These free or 'floating cells' (often called 'food balls') are characterized by a large central vacuole and a narrow cytoplasmic margin containing small electron-dense inclusions (Fig. 1.1.2.11a). As these cells near the excretory organ, the cytoplasmic components appear reduced, the vacuole becoming filled with excretory products which may be released into the lumen of the ventriculus (Mothes and Seitz, 1981c). Since these cells function in digestion (Akimov and Barabanova, 1977) resorption

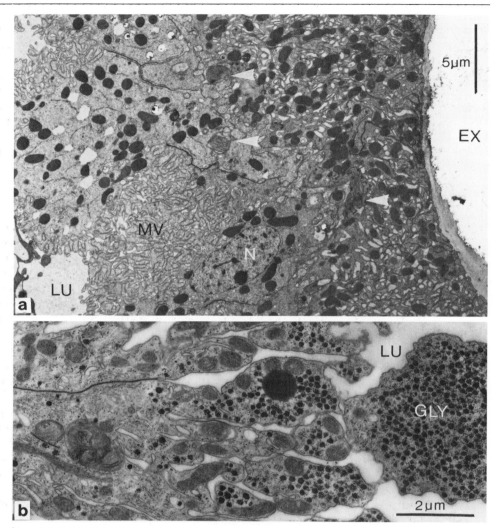

Fig. 1.1.2.12. (a) Excretory organ of *B. praetiosa* containing excretory ball (at EX) (largely dissolved during preparation). Epithelia of the excretory organ and the midgut are intimately interconnected, showing apical indentations (excretory organ) and irregular microvilli (midgut) and amounts of mitochondria. The arrows point to muscle cells lying between both epithelia. (b) *T. urticae*: Apical part of cells of the excretory organ with indentations and irregular projections containing masses of glycogen and mitochondria. Abbreviations: EX = excretory organ; GLY = glycogen; LU = lumen of midgut (a) and excretory organ (b); MV = microvilli of midgut cells; N = nucleus of midgut cell.

(phagocytosis and pinocytosis), and excretion, their appearance can be expected to vary. An additional function may be lipid storage during diapause (McEnroe, 1970). According to Mothes and Seitz (1981c), there is differentiation of this cell type during the phagocytic phase into cells ingesting mainly thylakoid granules and those specialized in starch resorption. Regeneration of these midgut cells has not been observed.

The epithelium neighbouring the excretory organ (Fig. 1.1.2.12a) differs from that described above. The cells are flat or cuboidal and often contain large amounts of rough ER, and sometimes glycogen. They are provided with irregularly arranged microvilli. According to Mothes and Seitz (1981c) these cells may contain refractile bodies of concentric appearance which are thought to be excretory products. Investigations of comparable inclusions occurring in virtually every phylum of animal have demonstrated that these are inorganic deposits (Simkiss, 1981).

Another specialized part of the midgut epithelium is found adjacent to the coxal glands (Fig. 1.1.2.9a). The few cells are similar to those neighbouring the excretory organ — flat or cuboidal with apical microvilli of irregular shape. The cytoplasm contains a considerable amount of glycogen and large crista-type mitochondria. The cells apparently do not participate in the cell cycle described above. Presumably, there is an exchange of substance, possibly excess water, between the specialized cells of the midgut and the excretory organ or the coxal glands (Alberti and Storch, 1977). Secretory cells similar to those described for mites or other arachnids (Alberti and Storch, 1983) do not seem to be present in the midgut of spider mites.

Hindgut

Blauvelt (1945) was the first to describe an open connection between the ventriculus and the excretory organ in *T. urticae*. Prior to this observation it was believed that the midgut of trombidiform mites, to which the Tetranychidae belong, ended blindly (Reuter, 1909). More recent studies (Ehara, 1960; Anwarullah, 1963) have confirmed the observations of Blauvelt (1945). The connection through which food passes is located close to the dorsal integument at the posterior margin of the ventriculus (Fig. 1.1.2.18). As this opening is usually closed, presumably by constrictor muscles, it is difficult to find.

The excretory organ is V-shaped in cross-section, its lateral walls consisting of cuboidal cells with apical indentations, and irregularly shaped processes projecting into the lumen (Fig. 1.1.2.12). The apical regions of these cells contain numerous crista-type mitochondria and large amounts of glycogen; the basal region is deeply infolded and is closely adjacent to the midgut epithelium. Muscle cells are found between the 2 epithelia. The dorsal roof of the excretory organ is characterized by cells possessing a few small mitochondria. Large amounts of rough ER are present and the cells are vacuolated; cell processes are lacking. According to the present authors' observations, this part of the epithelium is not syncytial as stated by Mothes and Seitz (1980).

Blauvelt (1945) believed the excretory organ to be a combined hindgut and excretory organ responsible both for the transportation of food residue and for excretion. The excretory organ does excrete guanine (McEnroe, 1961b) in the form of birefringent spherules, thus performing a function analogous to the Malpighian tubules of other arthropods. The lateral cells of the excretory organ may excrete guanine, whereas the modified dorsal cells may produce a matrix in which the excretory crystals are embedded, forming large droplets or balls. Moreover, it seems likely that the lateral cells are also involved in osmoregulatory processes since these cells show characteristics typical of cells capable of transcellular transport (microvilli-like processes, infoldings of the plasmalemma, numerous mitochondria, glycogen). The close connection with the modified midgut epithelium may be of importance with respect to ion and water balance.

Two types of excrement pass through the hindgut. Excrement having the appearance of black balls comes from the midgut; white guanine-containing residues are produced by the excretory organ. These 2 kinds of excrement are of different size (Gasser, 1951; Wiesmann, 1968; Hazan et al., 1974). These same types have also been observed in other Actinedida (Alberti, 1973).

The excretory organ leads posteriorly into the cuticle-lined rectum. The lateral walls of the rectum are plicated, the folds of the opposite sides interdigitating distally, thus closing the anus (or uroporus, respectively). Strong muscles which underlie the epithelium of the rectum may push the excrement through the posterior (male) or posterioventral (female) anus.

RESPIRATORY SYSTEM

The respiratory system of Actinedida (Prostigmata and Endeostigmata) is composed of cuticle-lined tracheae which open via a primary stigma at the cheliceral bases. From there, 2 furrows or channels may run between the chelicerae to a secondary neostigma (Grandjean 1938; Van der Hammen, 1980). In spider mites (Fig. 1.1.2.13) and some other Actinedida this basic structure is further modified by fusion of the proximal parts of the chelicerae to form the stylophore. This results in complete separation of the vertical channels (main tracheal trunks) from the surrounding air, leaving the dorsal neostigma as the definitive stigma. The stigmata are associated with discrete sclerotized grooves called peritremes (Figs. 1.1.2.2a and 1.1.2.10a), which extend from the cheliceral base onto a flexible fold of the integument encircling the stylophore.

Presumably, the peritremes function as extended stigmatal openings to ensure that obstruction of the stigmata will not necessarily interfere with respiration (Radovsky, 1969; Hinton, 1971). Peritremes may also function in plastron respiration (Hinton, 1971; Krantz, 1974; Alberti et al., 1981). The peritremes of tetranychids can be withdrawn into the cuticular collar ensheathing the base of the stylophore when the latter is retracted. The mite is thus able to block communication of the tracheal system with the surrounding air. Blauvelt (1945) believed that this ability may explain the resistance of spider mites to certain toxic gases.

The main or dorsoventral tracheal trunks descend ventrally from their opening at the cheliceral bases to a point approximately in line with the dorsal surface of the base of the infracapitulum. The trunks lie close together, separated only by a thin septum and enclosed by a common thick chitinous wall which has a ridge projecting to the rear. The ridge is probably homologous with the sigmoid piece stated by Thor (1905) to be common in Prostigmata. The ridge or sigmoid piece curves and runs forward to attach to the base of the infracapitulum at the posterior extremity of the cheliceral stylet channels. It then runs anteroventrally as the aliform apodeme of Summers et al. (1973). The tracheal trunks in *T. urticae* leave the sigmoid

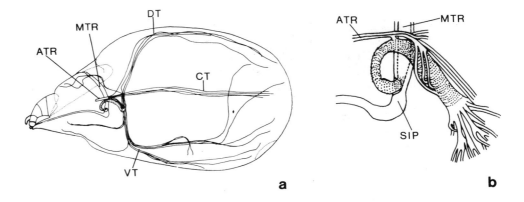

Fig. 1.1.2.13. Respiratory system of *T. urticae*. (a) General course of tracheae. (b) Main tracheal trunk, accessory tracheal trunk and sigmoid piece (redrawn from Blauvelt, 1945). Abbreviations: ATR = accessory tracheal trunk; CT = central tracheae; DT = dorsal tracheae; MTR = main tracheal trunk; SIP = sigmoid piece; VT = ventral tracheae.

piece slightly above the aliform apodeme, then curve forward, upward and then backward to form a small loop (Fig. 1.1.2.13a, b). Here, the tracheae are provided with creased and wrinkled walls which are probably flexible along their axis (Blauvelt, 1945). Movements of the stylophore may aid in ventilation by alternately compressing and relaxing the flexible trunks which pivot on the basal part of the sigmoid piece.

Each of the main tracheal trunks gives off bundles of tracheae which run caudally in the ventral, dorsal and central planes (Fig. 1.1.2.13a). These bundles give off smaller groups of tracheae and finally individual tracheae. Accessory tracheal trunks run from the lower portion of each of the main tracheal trunks and open into the cavity between the stylophore and infracapitulum. Grandjean (1938) has described tracheae similar to the accessory tracheae in other Actinedida.

Tracheal fine structure is comparable to that of insects. The tracheae have taenidia and are lined with a thin cuticle surrounded by a simple squamous epithelium.

EXCRETORY SYSTEM

The excretory system comprises the coxal glands (Figs. 1.1.2.4a and 1.1.2.9) and the excretory organ (Figs. 1.1.2.12 and 1.1.2.18). The coxal glands may be important in ion and water regulation (cf. section on the prosomal glands). The excretory organ appears to have several functions: excretion of guanine, evacuation of food residues, resorption of ions and resorption or excretion of water (cf. section on the digestive tract). As sapfeeders these animals are confronted with the problem of excess water. It has been assumed that the entrance of the oesophagus into the ventriculus may be brought into contact with the open connection between midgut and excretory organ by contraction of the dorsoventral muscles (McEnroe, 1963). This action should cause fluid to pass directly from the oesophagus into the

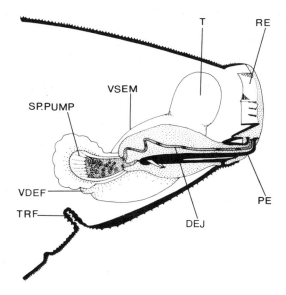

Fig. 1.1.2.14. Schematic representation of male genital system of *T. urticae*. The arrow indicates insertion of genital muscles (redrawn from Alberti and Storch (1976) with the permission of Springer-Verlag, Heidelberg). Abbreviations: DEJ = ductus ejaculatorius; PE = penis; RE = rectum; SP.PUMP = sperm pump; T = testis; TRF = transverse furrow; VDEF = vas deferens; VSEM = vesicula seminalis.

Chapter 1.1.2. references, p. 58

excretory organ. However, there would then be no opportunity for the midgut to digest the ingested food. It seems more likely to the present authors that there is direct transport of water via the modified epithelia of the midgut to the excretory organ and the coxal glands. Comparable combinations of excretory—osmoregulatory organs involving the intestine occur in plant-sucking insects, e.g., the filter chambers of Homoptera (Marshall and Cheung, 1974).

Mothes and Seitz (1980) described a Malpighian complex located between the rectum and genital tract in *T. urticae*. This observation has not been made for any other Tetranychidae. The possible derivation of the hindgut from Malpighian tubules is discussed by Blauvelt (1945) for *T. urticae* and for other Actinedida by Alberti (1973), Vistorin-Theiss (1977), Vistorin (1979) and Ehrnsberger (1981).

Nephrocytes, present in some other Actinedida, have not been found in spider mites.

MALE REPRODUCTIVE ORGANS

The male genital tract is composed of paired testes, 2 vesiculae seminales connected by short vasa deferentia to an unpaired sperm pump, and an unpaired ductus ejaculatorius and penis (Fig. 1.1.2.14). These organs are located posteriorly in the body and are surrounded dorsally by the hindgut, and laterally and ventrally by the body wall and part of the midgut. The genital opening is posterior, directly below the anus.

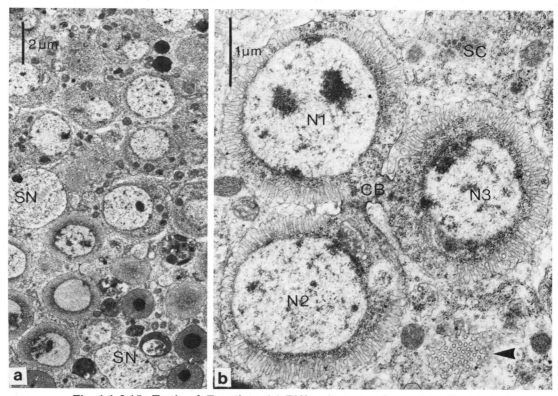

Fig. 1.1.2.15. Testis of *T. urticae*. (a) Different stages of sperm development, the germ cells being embedded in a large multinuclear somatic cell. (b) Spermatids still combined by cell bridges. The plasmalemmae are provided with radial, tubular indentations which are seen in tangential section (arrowed) (from Alberti and Storch (1976) with the permission of Springer-Verlag, Heidelberg). Abbreviations: CB = cell bridge; N1—N3 = nuclei of combined spermatids; SC = somatic cell; SN = somatic cell nuclei.

The paired testes are derived ontogenetically from an unpaired cell mass (Pijnacker and Drenth-Diephuis, 1973). Each testis comprises germ cells and 1 multinucleate somatic cell which envelops the developing germ cells (Fig. 1.1.2.15a). The somatic cell appears to be a supportive and nutritive structure in which spermiogenesis takes place. The nuclei of the supportive cell are easily distinguished from those of germ cells, since they are provided with tubular structures (Alberti and Storch, 1976). Spermatogonia, irregularly shaped and densely packed, are located in the posterior-most portion of the testis. They develop into round spermatids (Pijnacker and Drenth-Diephuis, 1973) provided with peripheral radial invaginations. The spermatozoa which develop from the spermatids are among the simplest acarine germ cells known (Alberti, 1983). They possess a rounded chromatin body with no nuclear envelope and have no mitochondria, cilia or acrosomal structures (Fig. 1.1.2.15b). The peripheral invaginations detach from the periphery, forming radial vesicles. Spermiogenesis is discussed in more detail in Chapter 1.2.1.

Spermatozoa are discharged into the lumen of the vesiculae seminales. The vesiculae seminales are thought to be glandular parts of the testis by Blauvelt (1945), Alberti and Storch (1976) and Mothes and Seitz (1981d) or vasa deferentia by Ehara (1960). The glandular cells of the vesiculae seminales are cuboidal or flat. Their nuclei contain tubular structures similar to those found in the nuclei of the somatic cells in the testes. The basal region of these cells possesses infoldings which are invaded by extensions of muscle cells; rough ER and Golgi fields are well developed. The lumen contains spermatozoa embedded in a secretory product.

The genital tract narrows to become the short vasa deferentia which open into the unpaired sperm pump. The sperm pump is a large, globular structure equipped with a strong muscular sheath surrounding a columnar, distally flattened epithelium (Fig. 1.1.2.14). The pump is filled with spermatozoa and the same secretory material which is present in the vesiculae seminales. A narrow ejaculatory duct extends from the sperm pump into the penis. Cells of the duct are equipped with an unusual tubular system and contain extraordinary amounts of glycogen, of unknown significance.

During copulation, the male positions himself front-first under the venter of the female's opisthosoma and flexes his posterior upward at nearly a right angle to the rest of his body by contracting the dorsal longitudinal muscles. A ventral transverse furrow of the male's integument (Fig. 1.1.2.14) unfolds to allow the extra flexion. Then, while holding the female with his front 2 pairs of legs, the tip of the male's opisthosoma is brought into contact with the region of the female's copulatory pore (cf. next section). The penis is made to protrude by muscles inserted at the proximal end of the apodemal part of the penis; the convoluted ejaculatory duct unfolds as the penis is inserted into the copulatory pore. During ejaculation, contractions of the muscular sheath of the sperm pump force spermatozoa into the lumen of the ejaculatory duct where they take on an elongated shape. The sperm are deposited directly into the seminal receptacle of the female.

FEMALE REPRODUCTIVE ORGANS

The female reproductive system consists of an ovary, a single oviduct comprising an anterior and posterior portion, vagina and a receptaculum seminis (seminal receptacle) (Fig. 1.1.2.18). The reproductive system opens on the posteroventral body surface at the hinged, anterior end of the genital flap. Unlike many other Actinotrichida, tetranychids have no genital papillae. The reproductive organs fill the ventral part of the body cavity, extending

Chapter 1.1.2. references, p. 58

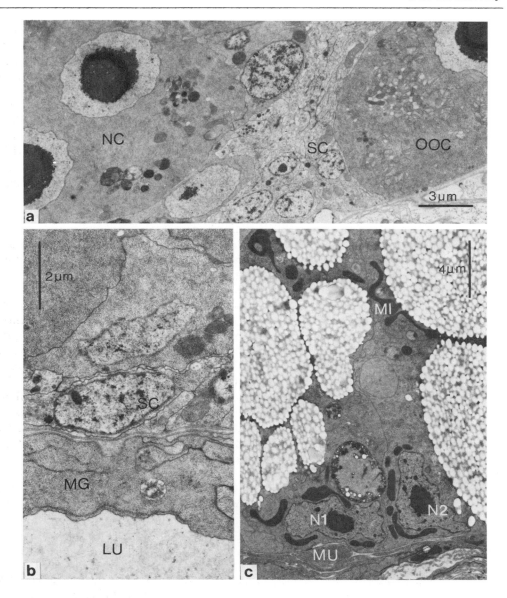

Fig. 1.1.2.16. Ovary of *Bryobia praetiosa*. (a) Nurse cell (with 2 nuclei in this section) and oocyte partly 'outside' the ovary. Further there are several somatic cells to be seen. (b) Ovary wall adjacent to midgut epithelium. (c) Part of the glandular anterior oviduct of *Bryobia praetiosa* showing its prominent secretory droplets, which are probably of lipid nature. Abbreviations: LU = lumen of midgut; MG = midgut cell; MI = mitochondria; N1, N2 = nuclei of anterior oviduct cells; NC = nurse cell; OOC = oocyte; SC = somatic cell.

from the anal region forward to the central nervous mass. The volume occupied by the organs is second only to that of the digestive system and varies depending on the developmental stage of eggs in the ovary of the adult female. For oogenesis, see Chapter 1.2.2.

The ovary is located medioventrially in the body cavity immediately posterior to the central nervous mass (Figs. 1.1.2.16a, b and 1.1.2.18). It is surrounded laterally by the midgut, dorsally by the hindgut and ventrally by the body wall. Blauvelt (1945) believed the single ovary to be derived from originally paired ovaries. However, Langenscheidt (1973) has described the development of the reproductive system from a single mass of cells.

The ovary comprises both somatic and germ cells. The somatic cells include the ovary wall and supportive tissue located among the germ cells. The ovary wall is a one-cell-thick epithelium of slightly flattened cells which encircles the posterior parts of the ovary; in the anterior parts of the ovary this epithelium travels into the ovary to ramify among the developing eggs, thus constituting the supportive tissue.

Besides these supportive cells there are so-called nutritive or nurse cells (Blauvelt, 1945; Seifert, 1961). These cells have 3 nuclei and are developed as sister cells of oogonia (Langenscheidt, 1973; Weyda, 1980). The maturing oocytes penetrate the ovary wall protruding from the basal lamina of the epithelium and are finally located 'outside' the ovary, at least partly covered only by the extended basal lamina of the ovary wall. Oocytes are connected to the nutritive cells by a nutritive stalk (Feiertag-Koppen and Pijnacker, 1982). Similar observations have been reported by Pijnacker et al. (1981) for the related *Brevipalpus* (Tenuipalpidae). Dosse and Langenscheidt (1964) and Mothes and Seitz (1981b) report that the nutritive cells gradually degenerate.

Maturation and differentiation of germ cells results in a division of the ovary into an anterior germarium, a median previtellogenic region, and a posterior vitellogenic portion (Fig. 1.1.2.18). These divisions are not well defined and well-developed eggs are occasionally found in anterior parts of the ovary. The plasmalemma of vitellogenic oocytes shows pinocytotic activity and is provided with numerous microvilli. Nutrients are supplied by the nurse cells and the adjacent midgut (Weyda, 1980). The storage of proteinaceous yolk spheres, lipid droplets and glycogen occurs concurrently with egg shell formation. Small pores in the developing shell allow substances to pass into the developing egg. When the egg has reached its final size, the pores are closed and the egg is ready for oviposition. The mature egg will most likely leave the pouch made up from the basal lamina of the ovary wall and will reach the anterior oviduct via an intercalated part (Weyda, 1980). However, this migration has never been observed.

Vitellogenic oocytes have been observed only in mature females (Langenscheidt, 1973). Only 1 or 2 oocytes develop to final size at a time (Blauvelt, 1945; Beament, 1951; Seifert, 1961; Anwarullah, 1963; Dosse and Langenscheidt, 1964).

Anterior oviduct

The anterior oviduct connects the ovarian wall with the posterior oviduct (Fig. 1.1.2.18). It is surrounded by the hindgut dorsally, the midgut laterally and the body wall ventrally. The anterior oviduct is a large, tubular convoluted sac comprising a single layer of columnar epithelial cells (Blauvelt, 1945; Anwarullah, 1963). In mature females, the cells appear glandular (Fig. 1.1.2.16c). Apical portions of the cells may appear vacuolated due to small globules, presumably lipid (Beament, 1951). Anwarullah (1963) believed a secretion of the anterior oviduct to be responsible for lubrication of the lumen during egg passage. The lumen of this portion of the oviduct is not visible unless distended by an egg. Small muscle cells are found below the basal lamina of the epithelium.

Posterior oviduct

From its connection with the anterior oviduct near the ventral body wall, the posterior oviduct runs posteriorly and dorsally to meet the dorsal portion of the vagina. It is surrounded by the midgut dorsally and laterally, the body wall ventrally and the seminal receptacle posteriorly (Fig. 1.1.2.18). The lumen of the posterior oviduct is lined with an epithelium which is

Chapter 1.1.2. references, p. 58

vacuolated and possesses invaginations and extensions, forming cell processes. Electron-dense droplets can be seen in apical portions of the epithelium and an electron-dense substance is often present along the apical plasma membrane. The function of this material is unknown. A well-developed muscular layer underlies the epithelium (probably the 'ring muscle' of Snieder-Berkenbosch, 1955). At its posterior junction with the vagina, the posterior oviduct has a distinct rearward projection which is in contact with the seminal receptacle to the rear and the midgut dorsally (Crooker and Cone, 1979).

Vagina

The vagina is a short, robust tube leading from the genital flap dorsally to open into the posterior oviduct (Fig. 1.1.2.18). Its highly folded walls comprise a single layer of hypodermal cells overlaid by a thin chitinous intima. A thin basal lamina subtends the hypodermis.

A portion of the female reproductive system, including the vagina, protrudes during oviposition to form a simple ovipositor (Beament, 1951; Van de Lustgraaf, 1972; Weyda, 1980; Mothes and Seitz, 1981b).

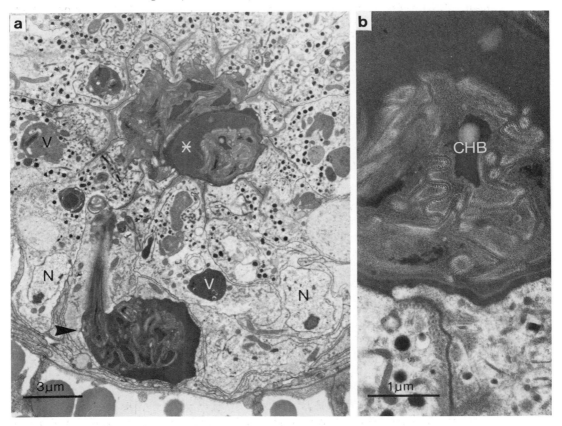

Fig. 1.1.2.17. Receptaculum seminis of *T. urticae*. (a) Columnar cells surrounding the lumen containing transformed spermatozoa partly grouped together by a secretory product (asterisk). Note irregular cell apices, large vacuoles and small electron-dense droplets. A group of spermatozoa has deeply invaded the epithelium (arrowed). (b) Detail of (a), showing characteristics of altered spermatozoa: irregular shape, chromatin body and, especially, finger-like processes containing filaments. Part of the cell apex of the epithelium of the receptaculum is seen with its microtubuli near to cell junctions. Abbreviations: CHB = chromatin body; N = nucleus; V = vacuole.

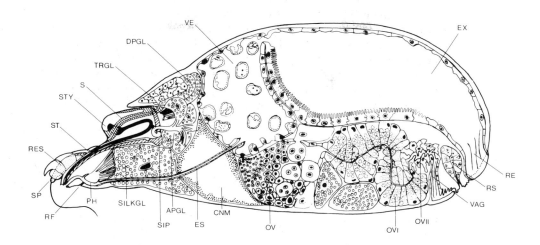

Fig. 1.1.2.18. Schematic representation of a longitudinal section through a female *T. urticae*. Abbreviations: APGL = anterior podocephalic gland; CNM = central nervous mass; DPGL = dorsal podocephalic gland; ES = oesophagus; EX = excretory organ; OV = ovary with vitellogenic oocytes 'outside' the ovary; OVI, OVII = anterior and posterior oviduct; PH = pharyngeal pump; RE = rectum; RES = reservoir of silk gland; RF = rostral fossette; RS = receptaculum seminis; S = septum inside stylophore; SIP = sigmoid piece; SILKGL = silk gland; SP = spinneret; ST = stylet; STY = stylophore; TRGL = tracheal gland; VAG = vagina; VE = ventriculus.

Seminal receptacle (receptaculum seminis)

The posteriormost portion of the female reproductive system is the seminal receptacle. It is a small, ovoid sac-like structure with a basal cone-shaped pedicel and is located ventrally between the posterior oviduct and the more caudal anus (Figs. 1.1.2.17 and 1.1.2.18). Several studies have indicated that this organ is attached by a short curved duct to the vagina (Blauvelt, 1945; Ehara, 1960; Anwarullah, 1963; Dosse and Langenscheidt, 1964). More recent studies (Helle, 1967; Van Eyndhoven, 1972; Smith and Boudreaux, 1972; Pijnacker and Drenth-Diephuis, 1973; Van de Lustgraaf, 1977; Crooker and Cone, 1979; Mothes and Seitz, 1981d) have confirmed that the seminal receptacle is not connected to the oviduct or vagina. Instead, a copulatory pore positioned on a slight elevation of the integument between the genital opening and anus leads into the receptacle.

The pedicel portion of the seminal receptacle possesses an inner chitinous tube which leads from the copulatory pore to the lumen of the sac-like receptacle. A simple epithelium encircles the tube and is subtended by a basal lamina. The sac of the seminal receptacle comprises a simple columnar epithelium. The basal portions of the cells are pentagonal or hexagonal and contain the nucleus; the cells are deeply invaginated apically. Many small vacuoles are present throughout the epithelium, especially basally. A peculiarity of these cells is an apical ring of microtubules (Alberti and Storch, 1976; Crooker and Cone, 1979). Small, round, electron-dense droplets are also found in apical portions of the cells.

The lumen of the receptaculum of an inseminated female contains numerous sperm. For *T. urticae*, this number is 242 ± 28 (Pijnacker and Drenth-Diephuis, 1973). In contrast with the spherical sperm found in the male, sperm in the seminal receptacle of the female are ovoid to elongate-ovoid with long finger-like processes and numerous filaments or tubules

(Fig. 1.1.2.17a, b) (Alberti and Storch, 1976; Crooker and Cone, 1979; Mothes and Seitz 1981d). It has been assumed that the radial vesicles of the spherical sperm are incorporated in the plasmalemma, thus providing the cell with membraneous material needed for rapid outgrowth of the finger-like processes (Alberti and Storch, 1976). The observations of Pijnacker and Drenth-Diephuis (1973) suggest that these changes do not occur before sperm reach the receptaculum, since these authors also found unaltered sperm in this organ. Perhaps the putative secretory product of the receptacle epithelium (electron-dense droplets) is inducing these changes (capacitation) of spermatozoa. The spermatozoa are also found in the deep invaginations of the epithelial cells (Fig. 1.1.2.17a). A receptaculum is lacking in those species reproducing exclusively parthenogenetically (Anwarullah, 1963). However, this statement has been questioned by Van Eyndhoven (1972).

CONCLUSIONS

Although our knowledge of tetranychid anatomy has increased considerably since Blauvelt's (1945) work, there is almost no system which does not need further investigations. This is especially true of the nervous system, including neurohemal and endocrine functions, for which the recent review of Juberthie (1983) provides basic information on arachnid relatives. Keeping in mind the importance of nervous system and sense organs for animal behaviour, further research should also focus on these systems. Up to now, only Mills (1973b, 1974) has taken a closer view of these structures; however, this is merely considered to be a starting point for the investigation of, e.g. setal sensillae. When compared with the bulk of information available on other invertebrates, especially insects (Altner and Prillinger, 1980) it is obvious that our knowledge of these receptors in spider mites is only provisional. Almost nothing is known concerning those organs we have considered as slit sense organs, whereas they have been studied extensively in Araneae (Barth, 1971). Some recent researchers, cited in the appropriate sections, have concentrated on reproductive organs and the digestive tract. Whereas the male reproductive system seems to be understood quite well, there are some problems with that of the female. What is the interrelationship between the cells constituting the ovary (somatic cells, germ cells, nurse cells) and the surrounding tissues concerning egg development? What happens shortly before and during oviposition? More information is needed about the sperm track in the female. The digestive tract appears more and more to be a very complex and peculiar system which underlines the adaptation of spider mites to a very specialized way of nutrition. Changes occurring during different physiological states should be of interest in future studies. Even the mouthparts of spider mites, being perhaps the most frequently investigated parts of these animals, provide surprising results (e.g. palpal silk gland, stylets with nerve fibres, rostral fossette).

Another point is that anatomical investigations of tetranychids has largely been concentrated on only a few species. A comparative study, including not only other species of spider mites but also representatives of related families (Krantz and Lindquist, 1979) would almost certainly result in a more comprehensive view of the evolution of these successful tiny animals.

REFERENCES

Akimov, I.A. and Barabanova, V.V., 1977. Morphological and functional peculiarities of the digestive system of tetranychid mites (Trombidiformes, Tetranychidae). Rev. Entomol. URSS, 56: 912—922 (in Russian, with English summary).
Akimov, I.A. and Jastrebtsov, A.V., 1981. Structure and function of muscles of mouth-

parts and appendages in spider mites. Vestn. Zool. (Kiev), 3: 54—59 (in Russian, with English summary).

Alberti, G., 1973. Ernährungsbiologie und Spinnvermögen der Schnabelmilben (Bdellidae, Trombidiformes). Z. Morphol. Tiere, 76: 285—338.

Alberti, G., 1984. Contributions of comparative spermatology to acarine systematics. In: D.A. Griffiths and C.A. Bowman (Editors), Acarology VI, Vols. 1 and 2. Ellis Horwood, Chichester.

Alberti, G. and Storch, V., 1974. Über Bau und Funktion der Prosoma-Drüsen von Spinnmilben (Tetranychidae, Trombidiformes). Z. Morphol. Tiere, 79: 133—153.

Alberti, G. and Storch, V., 1976. Ultrastruktur-Untersuchungen am männlichen Genitaltrakt und an Spermien von *Tetranychus urticae* (Tetranychidae, Acari). Zoomorphologie, 83: 283—296.

Alberti, G. and Storch, V., 1977. Zur Ultrastruktur der Coxaldrüsen actinotricher Milben (Acari, Actinotrichida). Zool. Jahrb. Abt. Anat. Ontog. Tiere, 98: 394—425.

Alberti, G. and Storch, V., 1983. Zur Ultrastruktur der Mitteldarmdrüsen von Spinnentieren (Scorpiones, Araneae, Acari) unter verschiedenen Ernährungsbedingungen. Zool. Anz., (Jena), 211: 145—160.

Alberti, G., Storch, V. and Renner, H., 1981. Über den feinstrukturellen Aufbau der Milbencuticula (Acari, Arachnida). Zool. Jahrb. Abt. Anat. Ontog. Tiere, 105: 183—236.

Altner, H. and Prillinger, L., 1980. Ultrastructure of invertebrate chemo-, thermo- and hygroreceptors and its functional significance. Int. Rev. Cytol., 67: 69—139.

Anwarullah, M., 1963. Beiträge zur Morphologie und Anatomie einiger Tetranychiden (Acari, Tetranychidae). Z. Angew. Zool., 50: 385—426.

Babu, K.S., 1965. Anatomy of the central nervous system of arachnids. Zool. Jahrb. Abt. Anat. Ontog. Tiere, 82: 1—154.

Bader, C., 1954. Das Darmsystem der Hydracarinen. Rev. Suisse Zool., 61: 505—548.

Baker, J.E. and Connell, W.A., 1963. The morphology of the mouthparts of *Tetranychus atlanticus* and observations on feeding by this mite on soybeans. Ann. Entomol. Soc. Am., 56: 733—736.

Barth, F.G., 1971. Der sensorische Apparat der Spaltsinnesorgane (*Cupiennius salei* Keys., Araneae). Z. Zellforsch. Mikrosk. Anat., 112: 212—246.

Beament, J.W.L., 1951. The structure and formation of the egg of the fruit tree red spider mite, *Metatetranychus ulmi* Koch. Ann. Appl. Biol., 38: 1—24.

Becker, E., 1935. Die Mundwerkzeuge des *Tetranychus telarius* (L.) und deren Funktion in Beziehung zur chemischen Bekämpfung des letzteren. Rev. Zool. Russe, 14: 637—654 (in Russian, with German summary).

Blauvelt, W.E., 1945. The internal anatomy of the common red spider mite (*Tetranychus telarius* Linn.). Mem. Cornell Univ. Agric. Exp. Stn., Ithaca, NY, 270: 1—35.

Bostanian, N.J. and Morrison, F.O., 1973. Morphology and ultrastructure of sense organs in the two-spotted spider mite (Acarina: Tetranychidae). Ann. Entomol. Soc. Am., 66: 379—383.

Boudreaux, H.B., 1958. The effect of relative humidity on egg-laying, hatching, and survival in various spider mites. J. Insect Physiol., 2(1): 65—72.

Boudreaux, H.B. and Dosse, G., 1963. The usefulness of new taxonomic characters in females of the genus *Tetranychus* Dufour (Acari: Tetranychidae). Acarologia, 5: 13—33.

Bouligand, Y., 1965. Sur une architecture torsadée répandue dans de nombreuses cuticles d'arthropodes. C. R. Acad. Sci., 261: 3665—3668.

Crooker, A.R., 1981. Internal Morphology and Morphological Observation of the Sperm Path in the Adult Female Two-spotted Spider Mite *Tetranychus urticae* Koch (Acarina: Tetranychidae). Ph.D. Dissertation. Washington State University, Pullman, WA, xiv, 122 pp.

Crooker, A.R., Jr. and Cone, W.W., 1979. Structure of the reproductive system of the adult female two-spotted spider mite *Tetranychus urticae*. Rec. Adv. Acarol., 2: 405—409.

Crowe, J.H., 1975. Studies on acarine cuticles. III. Cuticular ridges in the citrus red mite. Trans. Amer. Microsc. Soc., 94: 98—108.

Dosse, G., 1964. Beobachtungen über Entwicklungstendenzen im *Tetranychus urticae—cinnabarinus*-Komplex (Acari, Tetranychidae). Pflanzenschutzberichte, 31: 113—128.

Dosse, G. and Langenscheidt, M., 1964. Morphologische, biologische und histologische Untersuchungen an Hybriden aus dem *Tetranychus urticae*-Komplex (Acari, Tetranychidae). Z. Angew. Entomol., 54: 349—359.

Ehara, S., 1960. Comparative studies on the internal anatomy of three Japanese trombidiform acarinids. J. Fac. Sci., Hokkaido Univ., Ser. 6, 14: 410—434.

Ehrnsberger, R. 1981. Ernährungsbiologie der bodenbewohnenden Milbe *Rhagidia* (Trombidiformes). Osnabrücker Naturwiss. Mitt., 8: 127—132.

Feiertag-Koppen, C.C.M. and Pijnacker, L.P., 1982. Development of the female germ cells and process of internal fertilization in the two-spotted spider mite *Tetranychus urticae* Koch (Acariformes: Tetranychidae). Int. J. Insect Morphol. Embryol., 11: 271—284.

Foelix, R.F., 1982. Biology of Spiders, Harvard University Press, Cambridge, London, 306 pp.

Gasser, R., 1951. Zur Kenntnis der gemeinen Spinnmilbe *Tetranychus urticae* Koch. 1. Mitteilung: Morphologie, Anatomie, Biologie und Ökologie. Mitt. Schweiz. Entomol. Ges., 24: 217—262.

Gerson, U., 1979. Silk production in *Tetranychus* (Acari: Tetranychidae). Rec. Adv. Acarol., 1: 177—188.

Gibbs, K.E. and Morrison, F.O., 1959. The cuticle of the two-spotted spider mite, *Tetranychus telarius* (L.) (Acarina: Tetranychidae). Can. J. Zool., 37: 633—637.

Grandjean, F., 1938. *Retetydeus* et les stigmates mandibulaires des acariens prostigmatiques. Bull. Mus. Natl. Hist. Nat. Zool., 10: 279—286.

Grandjean, F., 1948. Quelques caractères des Tétranyques. Bull. Mus. Natl. Hist. Nat. Zool., 20: 517—524.

Hanström, B., 1919. Zur Kenntnis des Centralen Nervensystems der Arachnoiden und Pantopoden Nebst Schlussfolgerungen Betreffs der Phylogenie der Genannten Gruppen. Inauguraldissertation. Zootomisches Institut der Hochschule zu Stockholm, Stockholm, 191 pp.

Hanström, B., 1928. Vergleichende Anatomie des Nervensystems der wirbellosen Tiere unter Berücksichtigung seiner Funktion. Springer, Berlin, 628 pp.

Hazan, A., Gerson, U. and Tahori, A.S., 1974. Spider mite webbing. I. The production of webbing under various environmental conditions. Acarologia, 16: 68—84.

Helle, W., 1967. Fertilization in the two-spotted spider mite (*Tetranychus urticae*: Acari). Entomol. Exp. Appl., 10: 103—110.

Henneberry, T.J., Adams, J.R. and Cantwell, G.E., 1965. Fine structure of the integument of the two-spotted spider mite, *Tetranychus telarius* (Acarina: Tetranychidae). Ann. Entomol. Soc. Am., 58: 532—535.

Hinton, H.E., 1971. Plastron respiration in the mite *Platyseius italicus*. J. Insect Physiol., 17: 1185—1199.

Hislop, R.G. and Jeppson, L.R., 1976. Morphology of the mouthparts of several species of phytophagous mites. Ann. Entomol. Soc. Am., 69: 1125—1135.

Juberthie, C., 1983. Neurosecretory systems and neurohemal organs of terrestrial Chelicerata (Arachnida). In: A.P. Gupta (Editor), Neurohemal Organs of Arthropods. Charles C. Thomas, Springfield, IL, pp. 149—203.

Kaestner, A., 1968. Lehrbuch der Speziellen Zoologie. Bd. I. Wirbellose. 1. Teil. Fischer, Stuttgart, 898 pp.

Krantz, G.W., 1974. *Phaulodinychus mitis* Leonardi (1899) (Acari Uropodidae), an intertidal mite exhibiting plastron respiration. Acarologia, 16: 11—20.

Krantz, G.W. and Lindquist, E.E., 1979. Evolution of phytophagous mites (Acari). Ann. Rev. Entomol., 24: 121—158.

Langenscheidt, M., 1973. Zur Wirkungsweise von Sterilität erzeugenden Stoffen bei *Tetranychus urticae* Koch (Acari, Tetranychidae). I. Entwicklung der weiblichen Genitalorgane und Oogenese bei unbehandelten Spinnmilben. Z. Angew. Entomol., 73: 103—106.

Liesering, R., 1960. Beitrag zum phytopathologischen Wirkungsmechanismus von *Tetranychus urticae* Koch (*Tetranychus*, Acari). Z. Pflanzenkr. (Pflanzenpathol.) Pflanzenschutz, 67: 524—542.

Locke, M., 1974. The structure and formation of the integument of insects. In: M. Rockstein (Editor), The Physiology of Insecta, Vol. VI. pp. 124—213.

Manton, S.M., 1958. Hydrostatic pressure and leg extension in arthropods, with special reference to arachnids. Ann. Mag. Nat. Hist. (Zool., Bot. Geol.), 1: 161—182.

Marshall, A.T. and Cheung, W.K., 1974. Studies on water and ion transport in Homopteran insects: Ultrastructure and cytochemistry of the cicadoid and cercopoid Malpighian tubules and filter chamber. Tissue Cell, 6: 153—171.

Mayr, L., 1971. Untersuchungen zur Feinstruktur der Cuticula der Bohnenspinnmilbe *Tetranychus urticae* (Koch). Z. Angew. Zool., 58: 271—278.

McEnroe, W.D., 1961a. The control of water loss by the two-spotted spider mite (*Tetranychus telarius*). Ann. Entomol. Soc. Am., 54: 883—887.

McEnroe, W.D., 1961b. Guanine excretion by the two-spotted spider mite (*Tetranychus telarius* (L.)). Ann. Entomol. Soc. Am., 54: 925—926.

McEnroe, W.D., 1963. The role of the digestive system in the water balance of the two-spotted spider mite. Adv. Acarol., 1: 225—231.

McEnroe, W.D., 1969a. A tactile seta of the two-spotted spider mite, *Tetranychus urticae* K. (Acarina: Tetranychidae). Acarologia, 11(1): 29—31.

McEnroe, W.D., 1969b. Eyes of the two-spotted spider mite, *Tetranychus urticae*. I. Morphology. Ann. Entomol. Soc. Am., 62: 461—466.

McEnroe, W.D., 1970. Fat in the midgut wall of the overwintering two-spotted spider mite, *Tetranychus urticae*. Ann. Entomol. Soc. Am., 63: 912—913.

McGregor, E.A., 1950. Mites of the family Tetranychidae. Am. Midl. Nat., 44: 257—420.

Mills, L.R., 1973a. Morphology of glands and ducts in the two-spotted spider mite, *Tetranychus urticae* Koch, 1836. Acarologia, 15: 218—236.

Mills, L.R., 1973b. Structure of dorsal setae in the two-spotted spider mite *Tetranychus urticae* Koch, 1836. Acarologia, 15: 649—658.

Mills, L.R., 1974. Structure of the visual system of the two-spotted spider mite, *Tetranychus urticae*. J. Insect Physiol., 20: 795—808.

Mothes, U. and Seitz, K.-A., 1980. Licht- und elektronenmikroskopische Untersuchungen zur Funktionsmorphologie von *Tetranychus urticae* (Acari, Tetranychidae). I. Exkretionssysteme. Zool. Jahrb. Abt. Anat. Ontog. Tiere, 104: 500—529.

Mothes, U. and Seitz, K.-A., 1981a. A possible pathway of chitin synthesis as revealed by electron microscopy in *Tetranychus urticae* (Acari, Tetranychidae). Cell Tissue Res., 214: 443—448.

Mothes, U. and Seitz, K.-A., 1981b. Licht- und elektronenmikroskopische Untersuchungen zur Funktionsmorphologie von *Tetranychus urticae* (Acari, Tetranychidae). II. Weibliches Geschlechtssystem und Oogenese. Zool. Jahrb. Abt. Anat. Ontog. Tiere, 105: 106—134.

Mothes, U. and Seitz, K.-A., 1981c. Functional microscopic anatomy of the digestive system of *Tetranychus urticae* (Acari, Tetranychidae). Acarologia, 22: 257—270.

Mothes, U. and Seitz, K.-A., 1981d. The transformation of male sex cells of *Tetranychus urticae* K. (Acari, Tetranychidae) during passage from the testis to the oocytes: an electron microscopic study. Int. J. Invertebr. Reprod., 4: 81—94.

Mothes, U. and Seitz, K.-A., 1981e. Fine structure and function of the prosomal glands of the two-spotted spider mite, *Tetranychus urticae* (Acari, Tetranychidae). Cell Tissue Res., 221: 339—349.

Mothes, U. and Seitz, K.-A., 1982. Action of the microbial metabolite and chitin synthesis inhibitor Nikkomycin on the mite *Tetranychus urticae*; an electron microscope study. Pesticide Sci., 13: 426—441.

Neville, A.C., 1975. Biology of the Arthropod Cuticle. Springer-Verlag, Heidelberg, 448 pp.

Pal, R., 1950. The wetting of insect cuticle. Bull. Entomol. Res., 41: 121—139.

Penman, D.R. and Cone, W.W., 1974. Structure of cuticular lyrifissures in *Tetranychus urticae*. Ann. Entomol. Soc. Am., 67: 1—4.

Pijnacker, L.P. and Drenth-Diephuis, L.J., 1973. Cytological investigations on the male reproductive system and the sperm track in the spider mite *Tetranychus urticae* Koch (Tetranychidae, Acarina). Neth. J. Zool., 23: 446—464.

Pijnacker, L.P., Ferwerda, M.A. and Helle, W., 1981. Cytological investigations on the female and male reproductive system of the parthenogenetic privet mite, *Brevipalpus obovatus* (Donnadieu) (Phytoptipalpidae, Acari). Acarologia, 22: 157—163.

Pound, J.M. and Oliver, J.H., Jr., 1982. Synganglial and neurosecretory morphology of female *Ornithodoros parkeri* (Cooley) (Acari: Argasidae). J. Morphol., 173: 159—177.

Pritchard, A.E. and Baker, E.W., 1952. A guide to the spider mites of deciduous fruit trees. Hilgardia, 21(9): 253—387.

Radovsky, F.J., 1969. Adaptive radiation in the parasitic Mesostigmata. Acarologia, 11(3): 450—483.

Reuter, E., 1909. Zur Morphologie und Ontogenie der Acariden, mit besonderer Berücksichtigung von *Pediculopsis gramineum* (E. Reuter). Acta Soc. Sci. Fenn., 36: 287.

Seifert, G., 1961. Der Einfluss von DDT auf die Eiproduktion von *Metatetranychus ulmi* Koch (Acari, Tetranychidae). Z. Angew. Zool., 48: 441—452.

Simkiss, K., 1981. Calcium, pyrophosphate and cellular pollution. Trends Biochem. Sci., 6: 3—5.

Smith, J.W. and Boudreaux, H.B., 1972. An autoradiographic search for the site of fertilization in spider mites. Ann. Entomol. Soc. Am., 65: 69—74.

Snieder-Berkenbosch, L., 1955. Some remarks about the digestive tract and the female reproductive system of the fruit tree red spider mite, *Metatetranychus ulmi* (Koch). Proc. K. Ned. Akad. Wet., Ser. C, 58: 489—494.

Snodgrass, R.E., 1948. The feeding organs of Arachnida, including mites and ticks. Smithson. Misc. Collect., 110(10): 1—93.

Summers, F.M., Gonzales-R., R.H. and Witt, R.L., 1973. The mouthparts of *Bryobia rubrioculus* (Sch.) (Acarina: Tetranychidae). Proc. Entomol. Soc. Wash., 75: 96—111.

Thor, S., 1905. Recherches sur l'anatomie comparée des Acariens prostigmatiques. Ann. Sci. Nat. Zool., Paris, Ser. 8, 19: 1—190.

Van Eyndhoven, G.L., 1972. Some details about the genitalia of Tetranychidae (Acari). Notulae ad Tetranychidae 12. Zesz. Probl. Postepow Nauk Roln., 129: 27—42.

Van de Lustgraaf, B., 1977. L'ovipositeur des tétraniques. Acarologia, 4: 642—650.

Van der Hammen, L., 1970. Remarques générales sur la structure fondamentale du gnathosoma. Acarologia, 12: 16—22.

Van der Hammen, L., 1980. Glossary of acarological terminology. Vol. 1. General Terminology. Dr. W. Junk, The Hague, 244 pp.

Vistorin, H.E., 1979. Ernährungsbiologie und Anatomie des Verdauungstraktes der Nicoletiellidae (Acari, Trombidiformes). Acarologia, 21: 204—215.

Vistorin-Theis, G., 1977. Anatomische Untersuchungen an Calyptostomiden (Acari, Trombidiformes). Acarologia, 19: 242—257.

Vitzthum, H. Graf von, 1943. Acarina. In: Bronns Klassen und Ordnungen des Tierreiches, 5, Abt. 4, Buch 5. Akademische Verlagsgesellschaft, Leipzig, 1011 pp.

Weyda, F., 1980. Reproductive system and oogenesis in active females of *Tetranychus urticae* (Acari, Tetranychidae). Acta Entomol. Bohemoslov., 77: 375—377.

Wiesmann, R., 1968. Untersuchungen über die Verdauungsvorgänge bei der gemeinen Spinnmilbe, *Tetranychus urticae* Koch. Z. Angew. Entomol., 61: 457—465.

Note added in proof

In a recent paper André and Remacle (Acarologia, XXV (1984) 179—190) have further contributed to comparative and functional morphology of the gnathosoma of *Tetranychus urticae*.

1.1.3 Diagnosis and Phylogenetic Relationships

E.E. LINDQUIST

DIAGNOSIS

Spider mites, along with other members of the superfamily Tetranychoidea, have the cheliceral bases fused to form a retractable stylophore (see Chapter 1.1.1, Figs. 1—3); the movable cheliceral digits are stylet-like, greatly elongated and retractable so as to be strongly recurved basally within the stylophore. The palpi are five-segmented and terminate in a 'thumb-claw' process, i.e. a thumb-like palptarsus in apposition to a palptibial 'claw' (Figs. 1.1.1.7, 8). A pair of well-developed peritremes arises from the base of the stylophore and is embedded on the membranous anterior surface of the prodorsum. The prodorsum bears 3 or occasionally 4 pairs of setae, none of which is bothridial, and 2 pairs of eyes. The opisthosoma bears 6—14 pairs of setae dorsally (rarely more, owing to neotrichy), of which the more caudal pairs are not strongly modified or differently arranged from the others (Figs. 1.1.1.1 and 1.1.1.9). The adult female genital opening is transverse, lacks genital discs (papillae) and any sort of genital plate or valves, and is surrounded by characteristically wrinkled, transversely striated, membranous cuticle (Figs. 1.1.1.13, 14 and 55). The adult male has a sclerotized aedeagus which usually lacks a pair of accessory stylets flanking the aedeagal shaft (Figs. 1.1.1.15—17). The tarsi of all legs terminate with a pretarsus bearing paired claws and an unpaired empodium which are much more variable in form than in related families (Figs. 1.1.1.34—48). The paired claws, or their remnants, bear a series of minute nail-like tenent hairs or, more often, these form 1 compound pair of such hairs. The empodium may be similarly equipped with tenent hairs; if not, it may be claw-like, or split distally into a series to fine projections, or it may be vestigial or absent. A unique characteristic of the Tetranychidae as a whole within the Tetranychoidea is the usual presence of 2 sets of 'duplex setae', or 'chaetopairs', on tarsus I and 1 such set on tarsus II, each consisting of a usually attenuated solenidion and a clearly associated and usually much shorter seta (Figs. 1.1.1.22, 29 and 53); rarely, the elements of a duplex set are not closely associated, but this is evidently a secondary condition that has arisen in a small group (tribe Eurytetranychini) within the family; more rarely, a duplex association is not evident because the setal element has become so secondarily reduced as to be indiscernible, such as in a few species of Bryobiinae. There are only 2 nymphal instars in development from the larva to the adult, and the nymphs and adults consistently have 4 pairs of legs. In contrast to members of other tetranychoid families, many tetranychids produce silk, and hence the common name 'spider mite', despite their obligate phytophagous habits; however, the capability to produce silk is not a general characteristic of the family as a whole, since it is restricted to the subfamily Tetranychinae.

Chapter 1.1.3. references, p. 73

PHYLOGENETIC RELATIONSHIPS

Family-level limits

As treated here, the family Tetranychidae includes *both* the bryobiine and the tetranychine groups, and is definable by the characteristic uniquely derived among the families of Tetranychoidea of having 'duplex setae' on the tarsi of legs I and II. This has been the most generally accepted concept of Tetranychidae for the last 30 years (see, e.g., Baker and Pritchard, 1953; Pritchard and Baker, 1955; Smith Meyer, 1974; and Vainshtein, 1958, 1960, except that the latter regarded the family as comprising 3 subfamilies, as discussed below). Several other characteristics may be used to support this natural grouping, as discussed further below. The concept of the Russian school of acarology (Rekk, 1952, 1959 and, among others, Mitrofanov, 1972, 1977) of giving family status to each of the bryobiine and tetranychine groups, is not accepted here, for several reasons.

First, the bryobiine grouping has not been defined by previous authors on the basis of uniquely derived character states (synapomorphies). All of the characteristics used at present to define this group are clearly primitive (plesiomorphic) and are found among members of other families of Tetranychoidea, e.g., retention of 3 pairs of anal setae; retention of tenent hairs (or chaetoids) on the empodium; frequent retention of a greater number of dorsal opisthosomal setae and of leg setae; and lacking the capability to produce silk. As such, this grouping hitherto has not been demonstrated to be natural, or monophyletic.

Secondly, although reference has been made to some bryobiines having structures more similar to those of the Tenuipalpidae than those of the Tetranychidae (Mitrofanov, 1972), no evidence has been given to support the concept that the bryobiines as a group are more closely related (i.e., have a sister-group relationship) to another family of Tetranychoidea than to the tetranychine Tetranychidae.

Thirdly, from the viewpoint of classification, the elevation of Bryobiinae and Tetranychinae to families disrupts the balance of family-level groupings elsewhere in the Tetranychoidea, and also confuses the superfamily concept itself. For example, one might also elevate the subfamily groupings of Tenuipalpidae (the Brevipalpinae and Tenuipalpinae *sensu* Mitrofanov, 1973) to a comparable family level. At the same time, 'Tetranychoidea' would acquire a dual meaning — the traditional one encompassing several families as discussed below, and one restricted to the apparent sister grouping of 'Bryobiidae' and 'Tetranychidae'.

The reader is referred to Chapter 1.1.1, on external anatomy, for review and illustrations of the structures and their notations which are used in the discussions on phylogenetic relationships below. The reader is also referred to the textbook by Wiley (1981) for an excellent treatment of the phylogenetic systematic, or cladistic, methodology which is espoused here. For cladistic out-group analysis, representatives of 3 superfamilies of Eleutherengona other than Tetranychoidea were examined (see Chapter 1.1.1). Unless there was compelling evidence to the contrary within each superfamily, *maximum* setal complements were hypothesized to be plesiomorphic.

Evidence for monophyly of the Tetranychidae

Several shared, derived character states lend support to the hypothesis that Tetranychidae is a monophyletic group, with each of its 2 subfamilies being more closely related to each other (having a more recent common ancestor) than either is to any other group in the Tetranychoidea:

(1) On tarsi I and II, seta ft'' and solenidion ω'' form a duplex set on the larva and subsequent instars; a similar duplex set is formed by ft' and ω' (which is not present on the larva) on tarsus I of the protonymph; separation of the elements of these sets occurs secondarily, in a few genera of the Tetranychinae; further reduction of the setal element of these sets to a vestige occurs secondarily in a few species of Bryobiinae, so as to eliminate the duplex aspect.

(2) On tarsi I and II, pv', which is setiform like pv'' on the larva, transforms into a ventral eupathidium on the protonymph and subsequent instars; suppression of the eupathidial form of pv' or complete suppression of this seta occur secondarily in species of a few genera.

(3) The trochanter of leg III lacks seta l', which is generally added on the protonymph in the other families of Tetranychoidea.

Character states (1) and (2) are autapomorphic (uniquely derived) for this family; they are progressive, rather than regressive, conditions, which lend them further weight as indicators of a natural group. Character state (3) is a simple regressive loss that is not unique to this family (e.g., it occurs within the raphignathoid family Caligonellidae). In addition, there are several shared character states for which assignment of polarity is problematic, as follows:

(4) Seta v'_1 is generally added on tarsus I of the protonymph, on tarsus II of the deutonymph, and on tarsi III—IV either of the deutonymph (Bryobiinae) or of the adult (Tetranychinae). This seta is absent in the other families of Tetranychoidea. The ontogenetic *pattern of appearance* of this seta, if not its appearance in general, may be synapomorphic. The appearance of v'_1 on tarsi III and IV, as on II, of the deutonymph probably is the basic pattern for the ancestral tetranychid stock, which has been retained in the ancestral bryobiine substock. Delay of its appearance on tarsi III—IV until adulthood in Tetranychinae is 1 transformation step closer to complete suppression, as in the other tetranychoid families.

(5) Correlated with (4) is the expression of seta v''_1 on tarsus I of the protonymph in Bryobiinae or of the deutonymph in Tetranychinae, and on tarsi II—IV either of the deutonymph in Bryobiinae or of the adult in Tetranychinae. As in (4), the earlier pattern of appearance is assumed to characterize the ancestral family stock and to be retained in the bryobiine substock.

(6) Setae $v'_1-v''_1$ are added basically on tibia I in the deutonymph of Bryobiinae but in the adult of Tetranychinae; they are added on tibiae II to IV in either the deutonymph or the adult of Bryobiinae, and in the adult of Tetranychinae. These tibial setae are absent in other families of Tetranychoidea and in the out-group Raphignathoidea. Possibly, therefore, their presence is part of an oligotrichous trend toward tibial neotrichy and a synapomorphy, rather than prototrichy and a symplesiomorphy. If this is tibial neotrichy, the states in Tetranychinae probably represent the basic condition in the ancestral family stock, with the states in Bryobiinae being a more derivative intensification of the neotrichous trend. If this is tibial prototrichy, the states in Bryobiinae probably represent the basic condition in the ancestral family stock, with the states in Tetranychinae being a more derived step closer to complete suppression of $v'_1-v''_1$.

(7) Setae $l'_1-l''_1$ are added basically on tibia I in the deutonymph of both bryobiine and tetranychine stocks of Tetranychidae. These tibial setae are suppressed until the tritonymph of Tuckerellidae, and are totally suppressed in the Linotetranidae and Tenuipalpidae. Again, whether these setae

represent part of a synapomorphic trend toward neotrichy, correlated with the appearance of setae $v'_1-v''_1$ in (6), or are a symplesiomorphic retention of prototrichy, is uncertain.

Subfamilial relationships within the Tetranychidae

Traditionally (Rekk, 1952; Pritchard and Baker, 1955), 2 major groupings have been recognized as comprising the spider mites, the Bryobiinae Berlese and the Tetranychinae Donnadieu, though Vainshtein (1958) treated the eurytetranychines as a third subfamily, which is not accepted here for reasons discussed below. Various Russian specialists (see, e.g., Rekk, 1952; Mitrofanov, 1972) have given these 2 groupings full family status which, for reasons discussed above, is not accepted here. The biological characteristics presented by Mitrofanov (1972) to distinguish between 'Bryobiidae' and 'Tetranychidae' are too fraught with discrepancies or exceptions to be significant as generalizations, except for the capability to produce silk, which is accounted for below. The states of 3 characters have been used to distinguish between members of these 2 groups (the alternative in brackets for each characterizes the Tetranychinae):

(1) Empodia of legs with [without] tenent hairs.

(2) 3 [2 (or less)] pairs of pseudanal setae.

(3) The unmodified [enlarged] form of 1 of the 3 palptarsal eupathidia, which is correlated with the incapacity [potential capacity] to produce silk.

The states of 3 other characters as expressed in the more ancestral or 'generalized' taxa of each group, have been used secondarily to support these groupings, though they are not useful in a practical sense (i.e., in keys) because they are subject to further modification among the more derived taxa within each group, as follows:

(4) Paired claws of legs well formed as pads or claws [strongly reduced as vestigial pads].

(5) Prodorsum and opisthosomal dorsum with a maximum of 4 and 12 pairs [3 and 10 pairs] of setae.

(6) Tarsi I and II with the primary (larval—protonymphal) duplex sets placed apically on the abruptly declivate area, with the setal member of each duplex minute [placed somewhat less apically, with the setal member of each duplex less reduced].

From a phylogenetic standpoint, analysis of the first 3 characters given above, using out-group comparisons, indicates that all of the states are derived or apomorphic for the Tetranychinae, but primitive or plesiomorphic for the Bryobiinae. This leaves the Bryobiinae as not being definable as a natural (monophyletic) group. The 4th and 5th characters similarly have plesiomorphic states for the Bryobiinae and apomorphic states for the Tetranychinae. However, the state of part of the 6th character in the Bryobiinae — the extreme reduction in size of the setal element of each primary duplex set on tarsi I and II — appears to be a more specialized condition than that in the Tetranychinae. The Bryobiinae as a monophyletic group may also be supported by the following states:

(a) Seta d on genua I and II is generally suppressed until adulthood or absent in Bryobiinae, whereas it first appears on the deutonymph in Tetranychinae, as in the other families of Tetranychoidea.

(b) Adults and, to a lesser extent, deutonymphs of relatively early derivative taxa of Bryobiinae (e.g., *Bryobia, Bryocopsis*) have more setae added to tibia I ($v'_2-v''_2$, $l'_2-l''_2$, and sometimes $v'_3-v''_3$), tibiae II—IV ($v'_1-v''_1$, $l'_1-l''_1$, and sometimes $v'_2-v''_2$), tarsus I ($l'_2-l''_2$, $v'_3-v''_3$, and sometimes $l'_3-l''_3$, $v'_4-v''_4$), tarsus II (l''_1, v''_2) and tarsi III and IV ($l'_1-l''_1$, $v'_2-v''_2$) than in the Tetranychinae or in any other group of Tetranychoidea. This has been regarded by various authors as retention of a primitive condition (or prototrichy according to the concept of Grandjean (1965)), which may be the case for the tarsal setation since it compares with that found in early derivative taxa (e.g., *Raphignathus*) of the out-group superfamily Raphignathoidea. However, some of these additions may be derived secondarily, as neotrichy, *within* the Tetranychidae, especially the tibial setal additions, for which there are no comparable conditions among early derivative taxa, either elsewhere in Tetranychoidea or in Raphignathoidea. If this is the case, then the ancestral stock of Bryobiinae may be further defined apomorphically by tibial, and possibly tarsal, neotrichy (of the oligotrichous type, according to the concepts of Grandjean (1965)), as discussed in Chapter 1.1.1.

As a monophyletic group, the subfamily Tetranychinae is already well supported by 5 apomorphic character states mentioned above. The 5th apomorphy, concerning the number of setae on the opisthosomal dorsum, should refer more precisely to the *loss* of setae d_3 and e_3. Parallel losses of these setae occur within the Bryobiinae, but these are not apomorphic for that subfamily as a whole. A 6th apomorphy for the Tetranychinae is evident from the ontogeny of seta v' on femur III. This seta is generally suppressed until adulthood in Tetranychinae, whereas it generally appears first on the deutonymph in Bryobiinae, as in Tuckerellidae. This ontogenetic delay is an intermediate state in a transformation series leading to complete suppression of v'FeIII, which occurs in the Linotetranidae and Tenuipalpidae.

There may be other significant differences in ontogenetic patterns of setal development on the legs, which may be synapomorphic for the subfamilial, tribal, or generic level of groupings within the Tetranychidae. These cannot be discovered so long as the great majority of genera remain known only from adult representatives.

The group Eurytetranychini is retained here as a tribe within the Tetranychinae, rather than as a separate subfamily as proposed by Vainshtein (1958), because its members are characterized by the above-mentioned synapomorphies that define Tetranychinae as a monophyletic taxon (noting that the tarsal duplex sets are considered to have undergone secondary change in the eurytetranychine stock). The Eurytetranychini, therefore, arose as a group *within* the Tetranychinae; to treat this group separately would leave the Tetranychinae as a paraphyletic taxon. Various authors (see, e.g., Rekk, 1950, 1952; Pritchard and Baker, 1955; Vainshtein, 1958, 1960; Tuttle and Baker, 1968; Gutierrez et al., 1971; Mitrofanov, 1977) have proposed or recognized tribes and other groups of genera, and have attempted to propose phylogenetic relationships among these taxa, within the subfamilies of Tetranychidae. These have been unsatisfactory from a cladistic point of view, because some of the tribes or other taxa are based partly or entirely on primitive character states. Some of these studies refer to a given tribe as being 'intermediate' between 2 others or as having given rise to another, more evolved tribe. These concepts are cladistically even more unsatisfactory, because they imply that such tribes are paraphyletic. A cladistic approach, similar to the one used in only a preliminary way here to analyse family-level groupings, should also be applied to the tribes and lower taxa within the subfamilies of Tetranychidae.

Relationships of Tetranychidae with other families of Tetranychoidea

Phylogenetic relationships among the families of Tetranychoidea have not been resolved convincingly. Traditionally, the Linotetranidae and Tuckerellidae have been regarded as the more early derivative or ancient families,

Chapter 1.1.3. references, p. 73

apparently because of their richer dorsal body chaetotaxy; and the Tenuipalpidae and Tetranychidae have been regarded as the more recent derivative families, partly because of their reduced dorsal body setation and partly because of their more highly specialized associations with plant hosts. However, there is no reason to assume that more recent derivative taxa must be more highly modified morphologically and be more highly specialized ecologically than earlier derivative taxa. Moreover, based on a comparison of the leg chaetotaxy of various groups of tetranychoid mites, the Tetranychidae have the richest setation.

A cladistic analysis and search for sister-group relationships between the families of Tetranychoidea readily indicates a *lack* of evidence for such a relationship between Tuckerellidae and Tenuipalpidae, between Tuckerellidae and Tetranychidae, and between Linotetranidae and Tetranychidae. The solution to formulating hypotheses of relationships among tetranychoid families appears to lie in the determination of (a) the sister group of the Tenuipalpidae, and (b) the relative recency of derivation of the ancestral stock of the Tetranychidae.

There is synapomorphic evidence for 2 alternative hypotheses for the sister family of the Tenuipalpidae. One supports the traditional concept of the Tetranychidae and Tenuipalpidae as sister groups. This may be based on the following synapomorphies:

(a) Opisthosomal dorsum lacking larval setae c_4.
(b) Opisthosomal dorsum lacking larval setae f_3 (or f_1 — see below).
(c) Opisthosomal dorsum lacking larval setae h_4.
(d) Tarsi I and II lacking larval seta pl''.
(e) Tarsus I lacking larval seta pl'.
(f) Trochanter IV with deutonymphal seta v' suppressed until adult.

These states provide weak support, at best, for a common ancestry of Tenuipalpidae and Tetranychidae. They are few; all are simple regressions that may be subject to homoplasy; and none is uniquely derived. Moreover, characters (a), (b) and (c) may be correlated with each other, as may (d) and (e). Whether the loss of an *f*-series seta involves a homologous structure, is uncertain: f_3 is apparently lost in Tetranychidae (though it superficially appears as if f_1 is absent in Bryobiinae, and f_3 is absent in Tetranychinae); but whether the loss is f_3 or f_1 is not clear in Tenuipalpidae. It is also uncertain whether the *loss* of c_4 and h_4, or the *addition* of these setae, is apomorphic; these setae are not present in the out-group superfamily Raphignathoidea. Character (f) represents an intermediate state towards complete suppression of v' on trochanter IV, as occurs in Linotetranidae; therefore, this state does not support a grouping of Tenuipalpidae and Tetranychidae that excludes Linotetranidae.

The alternative hypothesis relates the Tenuipalpidae to the Linotetranidae, based on the following considerable array of synapomorphies:

(a) Propodosoma with a longitudinally ribbed invagination of the membranous cuticle which surrounds the cheliceral stylophore when retracted.

(b) Palptarsus lacking at least 1 (apparently *b*) of the 3 setiform setae (eupathidia and solenidion excluded).

(c) Tarsi I—IV lacking larval seta pv' (protonymphal on IV).
(d) Tarsus I with deutonymphal seta l''_1 suppressed at least until adult.
(e) Tibia I lacking deutonymphal setae l'_1—l''_1.
(f) Tibia III lacking larval seta l'.
(g) Genua I and II with larval seta v'' suppressed at least until deutonymph.
(h) Genua II—IV lacking larval seta v' (protonymphal on IV).
(i) Genua III and IV lacking deutonymphal seta d.

(j) Genua III and IV lacking adult seta v''.

(k) Femora I and II lacking adult seta v''_1.

These states provide moderately strong support for a common ancestry between the tenuipalpid and linotetranid stocks[1]. Character (a) is a progressive state that appears to be uniquely derived. The polarity of character (e) is uncertain: rather than the *lack* of setae $l'_1 - l''_1$ being apomorphic, their *addition* may be apomorphic, as discussed previously. Characters (h) and (i) may be correlated insofar as they apply to legs III and IV; setae v' and d have undergone a similar suppressive sequence, with an intermediate step of suppression until the tritonymph on legs III and IV in the Tuckerellidae, followed by complete suppression. Despite these possible correlations (which are by no means certain), there remains a diversity of shared regressive characters, on a variety of palpal and leg segments, which lessens the possibility of this all being a result of homoplasy. These regressions are at least partly independent of each other, and have resulted in a notably reduced leg setation, especially on femur II and on genua II—IV, relative to the Tuckerellidae and Tetranychidae.

A case can also be made for the Linotetranidae and Tuckerellidae as sister groups, based on the following possible synapomorphies:

(a) Opisthosomal dorsum with additional larval setae c_4, f_4 and h_4.

(b) Tarsus I with deutonymphal seta l''_1 suppressed until the tritonymph (Tuckerellidae) or adult (Linotetranidae).

(c) Tibia IV with larval seta d suppressed until the deutonymph.

(d) Genu IV lacking protonymphal seta l'.

(e) Trochanters I and II with deutonymphal seta v' suppressed until the tritonymph (Tuckerellidae) or adult (Linotetranidae).

These states provide weak support for common ancestry between Linotetranidae and Tuckerellidae. For character (a), as mentioned earlier, it remains uncertain whether the presence of setae c_4, f_4 and h_4 is apomorphic or plesiomorphic. Character (b) represents an intermediate state towards complete suppression of l''_1 on tarsus I, as occurs in Tenuipalpidae; therefore, this state does not support a grouping that excludes Tenuipalpidae. Character (d) is subject to homoplasy; seta l' is completely suppressed in various taxa within the family Tenuipalpidae and, rarely, within the Tetranychidae (*Marainobia*). Whether character (e) is a result of homoplasy is uncertain, since the complement of trochanter setae prior to adulthood is unknown for the majority of taxa in Tenuipalpidae and Tetranychidae.

As indicated earlier, there is no evidence apparent, in the way of shared derived character states, to support a sister-group relationship between the Tetranychidae and Linotetranidae. Similarly, a sister-group relationship between Tenuipalpidae and Tuckerellidae is not supported (the shared suppression of the protonymphal solenidion ω' of tarsus I until the deutonymph is insufficient). Having hypothesized that Tenuipalpidae and Linotetranidae are probably sister families, and in the absence of synapomorphic evidence for a sister relationship between the Tetranychidae and Tuckerellidae, there remains the question as to which of the latter 2 families is the sister group of Tenuipalpidae + Linotetranidae, and which is the most early derivative (ancient) stock or out-group of the other 3. The

---

[1] This grouping also contains the monobasic Allochaetophoridae Krantz (1970), described from a nymph which possibly is an early derivative tenuipalpid taxon that retains a palptibial claw. If more detailed analysis of *Allochaetophora californica* McGregor (1950) supports its separate familial status, then Allochaetophoridae would probably be the sister family of Tenuipalpidae, and Linotetranidae would become the sister group of Tenuipalpidae + Allochaetophoridae (as shown in Fig. 1.1.3.1).

Chapter 1.1.3. references, p. 73

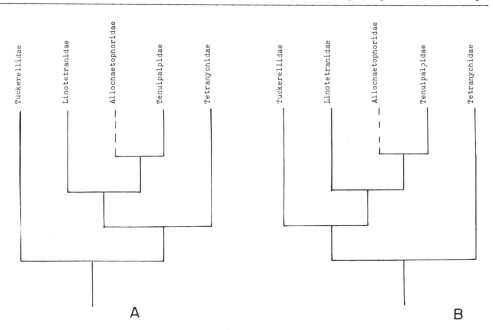

Fig. 1.1.3.1. Alternative cladograms of phylogenetic relationships among families of Tetranychoidea (see text for discussion).

alternatives are shown in the dendrograms in Fig. 1.1.3.1, and evidence in support of each alternative is given below.

The case for Tetranychidae being the sister group of Tenuipalpidae + Linotetranidae, and for Tuckerellidae being the out-group of these 3 families (Fig. 1.1.3.1A) rests on few characters:

(a) Tritonymphal stage suppressed in these 3 families.

(b) Genu IV with formerly tritonymphal seta v'' suppressed until adult.

(c) Trochanter IV with formerly tritonymphal seta v' suppressed until adult.

This case rests essentially on a single character state, (a), since states (b) and (c) are correlated with the loss of the tritonymph of character (a)[2]. Suppression of a developmental stage, though a regression, is far more significant than simple character suppressions, e.g., of setae. Nevertheless, the loss of the last nymphal stase *is* susceptible to homoplasy: the same suppression apparently has occurred in the ancestral stock of the Raphignathoidea and either in the ancestral stock or within early derivative stocks of the Cheyletoidea; it has also occurred independently within substocks of the Tydeidae (Kuznetsov, 1980). It is not improbable, therefore, that tritonymphal suppression could evolve twice, rather than once, within the Tetranychoidea. Furthermore, it is uncertain whether the tritonymphal stage is suppressed in the Linotetranidae.

The case for Tuckerellidae being the sister group of Tenuipalpidae + Linotetranidae, and for Tetranychidae being the most early derivative of the extant families of Tetranychoidea (Fig. 1.1.3.1B), rests on a surprising array of characters:

(a) Palptarsus lacking 1 (apparently *su*) of the 3 eupathidial setae (setiform setae and solenidion excluded).

---

[2] The retention of an active tritonymphal instar in Tuckerellidae, which was suggested by previous observations of Baker and Tuttle (1975), has recently been confirmed independently by Quirós and Baker (1984) and by Rasmy and Abou-Awad (1984).

(b) Tarsus I lacking deutonymphal solenidion ω''_1.

(c) Tarsi II—IV lacking deutonymphal (leg III) or adult (legs II, IV) solenidion ω'.

(d) Tarsi I—III lacking larval seta pv''.

(e) Tarsus I lacking protonymphal setae $v'_1-v''_1$.

(f) Tarsus I lacking deutonymphal seta l'_1.

(g) Tarsus I with deutonymphal seta l''_1 suppressed at least until tritonymph or adult.

(h) Tarsi II—IV lacking deutonymphal setae $v'_1-v''_1$.

(i) Tibia I with deutonymphal setae $l'_1-l''_1$ suppressed at least until tritonymph.

(j) Tibia IV lacking protonymphal seta l'.

(k) Genua II—IV with larval seta v' (protonymphal on IV) suppressed at least until deutonymph.

(l) Genua III and IV with deutonymphal seta d suppressed at least until tritonymph or adult.

(m) Genu IV with protonymphal seta l' suppressed at least until deutonymph.

(n) Femur I with deutonymphal seta v'' suppressed at least until tritonymph.

(o) Femora I—IV lacking adult seta l'_1.

(p) Femur IV lacking protonymphal seta d.

(q) Femur IV lacking adult seta v'.

All of these character states are simple regressions, of which probably some are correlated with each other and some are results of homoplasy. For example, states (e) and (h) are probably highly correlated, as are (f) and (g). Nevertheless, there remains a substantial diversity of shared regressive characters, on a variety of leg segments and with different ontogenetic patterns, which lessens the possibility of them all being a result of homoplasy. In addition, the transformation series for each of 6 characters (g, i, k, l, m, n) indicates that the state found in Tuckerellidae is intermediate between a more primitive condition found in Tetranychidae and a more specialized condition found in Linotetranidae and Tenuipalpidae. Such transformations further support the hypothesis of Tuckerellidae being the immediate out-group of Linotetranidae + Tenuipalpidae, and Tetranychidae being the out-group of these 3 families. Note again that this hypothesis requires that suppression of the tritonymph has occurred twice within Tetranychoidea — once in the stock ancestral to Tetranychidae and once in that ancestral to Linotetranidae + Tenuipalpidae.

Several other characters may possibly support this hypothesis:

(r) Tarsus I lacking deutonymphal setae $v'_2-v''_2$.

(s) Tibiae I—IV lacking protonymphal (leg I) or deutonymphal (legs II—IV) setae $v'_1-v''_1$.

(t) Tibia I lacking adult setae $v'_2-v''_2$, $l'_2-l''_2$, $v'_3-v''_3$, and tibiae II—IV lacking adult setae $v'_2-v''_2$.

These states are probably correlated with one another to some extent. Moreover, as discussed earlier, the polarities of these transformation series remain uncertain: if the *absence* of these setae represents setal *repressions*, the above states are apomorphic and may support the above hypothesis; if the *presence* of these setae (as found in Tetranychidae) represents setal *additions* (oligotrichous neotrichy), the above states are plesiomorphic and would neither support nor repudiate the above hypothesis.

On the basis of the present data, a rather unexpected hypothesis is newly proposed: not only are the spider mites apparently *not* the sister group of the Tenuipalpidae, but they are the sister group of the ancestral stock of

Chapter 1.1.3. references, p. 73

Tuckerellidae + Linotetranidae + Tenuipalpidae. As such, the Tetranychidae may represent the most early derivative, or ancient, family lineage within the Tetranychoidea (Fig. 1.1.3.1B). Elucidation of the ontogenies of setal systems, based on studies of the immature instars of a wide variety of tetranychoid mites, is needed to test the validity of the phylogenetic relationships suggested here.

The superfamily Tetranychoidea: definition and family membership

The superfamily Tetranychoidea remains as originally conceived by Baker and Pritchard (1953) some 30 years ago, and consists of the extant families Tetranychidae, Tuckerellidae, Linotetranidae, and Tenuipalpidae (including 'Allochaetophoridae'). Together, these families share several derived character states which strongly support their forming a monophyletic group:

(1) The stylophore of the fused cheliceral bases is movable independently of the infracapitulum, and can be retracted deeply into the propodosoma.

(2) The cheliceral stylets are greatly elongated, and strongly recurved basally within the stylophore, thus allowing for a greater degree of retraction and protraction than in mites of other superfamilies having a stylophore; when protruded for feeding, the stylets are juxtaposed to form a single hollow probe.

(3) A pair of well-developed peritremes are embedded on the membranous anterior surface of the prodorsum, where they end in somewhat protuberant enlargements.

(4) The basic larval setation of the opisthosomal dorsum includes c_3, d_3, e_3, and h_3, which are evidently idionymous neotrichous setae.

(5) The true, paired claws of all legs are equipped with tenent hairs, or chaetoids.

(6) Proral setae $p'-p''$ occur only on tarsi I and II, where they are formed as smooth eupathidia beginning with the larval instar (these are secondarily absent on tarsus II in Linotetranidae).

(7) Tectal setae $tc'-tc''$ are not eupathidial, and are suppressed until the protonymphal instar on tarsi I—III and until the deutonymphal instar on tarsus IV (these are secondarily absent on tarsi III and IV in Linotetranidae).

(8) All active instars are obligately phytophagous on vascular plants.

The states of the first 5 characters are progressive, and those of (1) and (2) appear to be uniquely derived; states (4) and (5) may not be unique, but they are certainly unusual among the superfamilies of eleutherengone mites. Those of (6) and (7) are regressive, and their uniqueness is uncertain. There may be other regressive synapomorphies characteristic of Tetranychoidea, but these cannot be elucidated without comparable information on the ontogeny of homologous setae on the appendages of raphignathoid and cheyletoid mites, which are evidently the closest out-group superfamilies of Tetranychoidea (Lindquist, 1976).

There has been some speculation about a relationship between the monobasic family Iolinidae and the Tetranychoidea and Raphignathoidea (Pritchard, 1956; Vainshtein, 1978). Iolinid mites are not characterized by any of the 8 apomorphies mentioned above. Their stylophore is not deeply retractable, and their stylets are neither so greatly elongated nor designed apparently to function as a single probe as in tetranychoid mites. Iolinid gnathosomal structures can be derived more readily from further specialization of tydeoid structures than from a 'despecialization' or regression of tetranychoid structures. In iolinids, the peritremes are not strongly developed and embedded on the prodorsum; the true claws lack tenent hairs; prodorsal setae sc_1 are bothridial; and there is no evidence that their palpi are derived from an ancestral 'thumb-claw' type. That Iolinidae probably belongs in the Tydeoidea, as suggested by Krantz (1978) and André (1980), is also supported by other characteristics, including: the inverted T-shaped form of a sclerotized part of the pharyngeal complex in the infracapitulum; the position of prodorsal setae v_1 posteromediad of bothridial setae sc_1, correlated with a procurved shape of the dehiscence line, on the prodorsum; the lack of humeral setae c_3 on the opisthosoma; the presence of tibial famulus k'' on leg I; and their feeding habits, as parasites of insects. Further, André (1984) has recently shown that iolinid mites share some of the ontogenetic patterns of development among the leg setae which are characteristic for the Tydeidae as described by André (1981).

REFERENCES

André, H.M., 1980. A generic revision of the family Tydeidae (Acari: Actinedida). IV. Generic descriptions, keys and conclusions. Bull. Ann. Soc. R. Belge Entomol., 116: 103—130, 139—168.

André, H.M., 1981. A generic revision of the family Tydeidae. III. Organotaxy of the legs. Acarologia, 22: 165—178.

André, H.M., 1984. Redefinition of the Iolinidae (Acari: Actinedida) with a discussion of their familial and superfamilial status. In: D.A. Griffiths and C.E. Bowman (Editors), Acarology VI (Proc. VI Int. Congr. Acarology, Edinburgh, Scotland, Sept. 1982), Ellis Horwood, Chichester, England, Vol. 1, pp. 180—185.

Baker, E.W. and Pritchard, A.E., 1953. The family categories of tetranychoid mites, with a review of the new families Linotetranidae and Tuckerellidae. Ann. Entomol. Soc. Am., 46: 243—258.

Grandjean, F., 1965. Complément a mon travail de 1953 sur la classification des Oribates. Acarologia, 7: 713—734.

Gutierrez, J., Helle, W. and Bolland, H.R., 1971. Étude cytogénétique et réflexions phylogénétiques sur la famille des Tetranychidae Donnadieu. Acarologia, 12: 732—751.

Krantz, G.W., 1970. A Manual of Acarology. Oregon State University Book Stores, Corvallis, OR, xi + 335 pp.

Krantz, G.W., 1978. A Manual of Acarology, second edition. Oregon State University Book Stores, Corvallis, OR, viii + 509 pp.

Kuznetzov, N.N., 1980. Adaptivnye osobennosti ontogeneza kleshchei Tydeidae (Acariformes). Zool. Zh., 59: 1018—1024 (in Russian).

Lindquist, E.E., 1976. Transfer of the Tarsocheylidae to the Heterostigmata, and reassignment of the Tarsonemina and Heterostigmata to lower hierarchic status in the Prostigmata (Acari). Can. Entomol., 108: 23—48.

McGregor, E.A., 1950. Mites of the family Tetranychidae. Am. Midl. Nat., 44: 257—420.

Mitrofanov, V.I., 1972. Sistema kleshchei semeistva Bryobiidae (Acariformes, Trombidiformes). Zool. Zh., 51: 1171—1179 (in Russian).

Mitrofanov, V.I., 1973. Reviziya sistemy rastitel'noyadnykh kleshchei podsemeistva Brevipalpinae (Trombidiformes, Tenuipalpidae). Zool. Zh., 52: 507—512 (in Russian).

Mitrofanov, V.I., 1977. Sistema semeistva Tetranychidae (Acariformes, Trombidiformes). Zool. Zh., 56: 1797—1804 (in Russian).

Pritchard, A.E., 1956. A new superfamily of trombidiform mites, with the description of a new family, genus, and species (Acarina: Iolinoidea: Iolinidae: Iolina nana). Ann. Entomol. Soc. Am., 49: 204—206.

Pritchard, A.E. and Baker, E.W., 1955. A revision of the spider mite family Tetranychidae. Mem. Pac. Coast Entomol. Soc., 2: 1—472.

Quirós-Gonzalez, M.J. and Baker, E.W., 1984. Idiosomal and leg chaetotaxy in the Tuckerellidae Baker and Pritchard: ontogeny and nomenclature. In: D.A. Griffiths and C.E. Bowman (Editors), Acarology VI (Proc. VI Int. Congr. Acarology, Edinburgh, Scotland, Sept. 1982), Ellis Horwood Ltd., Chichester, England, Vol. 1, pp. 166—173.

Rasmy, A.H. and Abou-Awad, B.A. (1984). A redescription of *Tuckerella nilotica* Zaher and Rasmy (Acarina: Tuckerellidae), with descriptions of the immature stages. Acarologia, 25: 337—340.

Rekk, G.F., 1950. Materiali k faune pautinnikh kleshchei Gruzii (Tetranychidae: Acarina). Tr. Inst. Zool. Akad. Nauk Gruz. SSR, 9: 117—134 (in Russian).

Rekk, G.F., 1952. O nekotorykh osnovakh klassifikatsii tetranikhovykh kleshchei. Soobshch. Akad. Nauk Gruz. SSR, 13: 419—425 (in Russian).

Rekk, G.F., 1959. Opredelitel' tetranikhovykh kleshchei. Izv. Akad. Nauk Gruz. SSR, Tbilisi, 1—151 (in Russian).

Smith Meyer, M.K.P., 1974. A revision of the Tetranychidae of Africa (Acari), with a key to the genera of the world. Repub. S. Afr. Dep. Agric. Tech. Serv., Entomol. Mem., 36: 1—291.

Tuttle, D.M. and Baker, E.W., 1968. Spider mites of southwestern United States and a revision of the family Tetranychidae. University of Arizona Press, Tucson, AZ, vii + 143 pp.

Tuttle, D.M. and Baker, E.W., 1975. A new species of *Tuckerella* (Acarina: Tuckerellidae) from Thailand. U. S. Dep. Agric., Coop. Econ. Insect Rep., 25: 337—340.

Vainshtein, B.A., 1958. Khetom konechnostei pautinnykh kleshchei (Acariformes, Tetranychidae) i sistema semeistva. Zool. Zh., 37: 1476—1487 (in Russian).

Vainshtein, B.A., 1960. Tetranikhovye kleshchei Kazakhstana (s reviziei semestva). Tr. Naukhno-Issled. Inst. Zashch. Rast., Kaz. Fil. Vsesoyuz. Akad. Sel'sk. Nauk Imeni V.I. Lenina, Alma-Ata, 5: 1—275 (in Russian).

Vainshtein, B.A., 1978. Sistema, evolyutsiya i filogeniya trombidiformnykh kleshchei. In: M.S. Gilyarov (Editor), Opredelitel' Obitayushchikh v Pochve Kleshchei Trombidiformes. Nauka, Moscow, pp. 228—245 (in Russian).

Wiley, E.O., 1981. Phylogenetics. The Theory and Practice of Phylogenetic Systematics. John Wiley and Sons, New York, NY, xv + 439 pp.

1.1.4 Systematics

J. GUTIERREZ

INTRODUCTION

Considerable progress has been made in the systematics of Tetranychidae since 1955, the date of publication of the revision of the family carried out by Pritchard and Baker. This work also provides the first modern definition of this group, as well as its division into two large sub-families. The number of known genera which, before 1955, totalled 20, has now risen to 63, while the number of valid species has increased from 185 to almost 900. Numerous species, however, are still to be discovered. In 1979 Baker estimated that approximately 70% of the world fauna was still unnamed.

After a period during which most descriptions were unclear, valuable progress was made in 1913 by Ewing, who pointed out the taxonomic value of the male genital armature. This character was used by McGregor, particularly in one of his last publications (McGregor, 1950), but previously many authors had neglected this piece of information, just as Grandjean's basic work (Grandjean, 1948) had also gone unnoticed.

The economic significance of red spider mites was the reason why research on them was approached from a practical angle. For a considerable length of time, taxonomists limited investigations and bibliography to their immediate geographical region. This happened to such an extent that a cosmopolitan species like *Tetranychus urticae* Koch managed to end up with more than 50 different names.

Nowadays, there is greater communication between scientists, and the number of features recorded by taxonomists has increased. Gradually, a phase of synthesis, integrating factors from several disciplines, is being reached. Also, attempts are being made to formulate phylogenetic hypotheses for a number of tribes. There is now a strong possibility of a transition from a phenetic classification, which attributes equal value to available characters, to a cladistic classification based on phylogeny.

PECULIARITIES AND TRENDS IN THE SYSTEMATICS OF SPIDER MITES

Ideally, the systematics of tetranychid mites should be based on excellent descriptions. The descriptions should be potentially extensive enough to replace the examination of the type material. This demand is even more crucial and delicate than in insects because of the transient nature of acarine preparations. Microscopic slides will degenerate and are likely to break down after a relatively short period of time. A large number of old slides have disappeared and are no longer usable. An additional difficulty interfering

---

Chapter 1.1.4. references, p. 89

with the study of the type material is that collections are scattered through-out the world. The Berlese collection, for example, may only be consulted on the spot in Florence.

There are other drawbacks, related to the bad practices of workers in tetranychid systematics. Most descriptions are based on drawings made from specimens flattened out between slide and cover-slide, which give little idea of the actual physical appearance of the mite and of the relative position of the setae. This is particularly important for those groups which are differentiated by the latter character. It would therefore be preferable to make such descriptions from specimens prepared in cavity slides.

Another point to mention here concerns the stases studied. Whereas academic research includes all stases in the mite's life-cycle, systematicists actually working on red spider mites generally only use the adult stase.

As is the case for many other animal groups, the papers of first authors allow identification of tetranychid mites at genus level only. Descriptions should now be extended to include the largest possible number of features and scientists should not be satisfied with a mere drawing of an aedeagus or dorsum, even if these elements seem to be sufficient to identify a species at the time of its discovery. This attitude will make future research a more complicated matter. Systematic studies should include clearly drawn plates, presented uniformly to increase legibility. For example, right-sided tibiae and tarsi should always be drawn from the outer side, as Pritchard and Baker did in 1955, as well as in later publications. Likewise, the aedeagus should always be drawn in perfect profile, with the distal end towards the right.

Since morphological criteria are often small in number and the differences between them often subtle, there is an increasing tendency to call on several other disciplines to complete this data:

— Biological information is the most simple to note and from the beginning, authors have given the name of the host plant. The latter should always be indicated by its latin name, not by its common name. Although most economic species are polyphagous, a certain number are restricted to a single botanical species or family. One may be even more specific in noting also the preferential position occupied by the mites on the plant (upper or lower surface of the leaf) and by describing the damage done. It is also of interest to indicate the colour of adults and to provide information on the structure and appearance of the webs spun, as well as on the shape and position of eggs, and position of exuviae and faeces. Without going into a detailed study, it is sometimes possible to note the existence of quiescent or diapausing stages (eggs or adults), under the effect of climatic factors (photoperiod, cold or drought).

— A considerable amount of data have been collected on the cytogenetics of Tetranychidae, in particular by Helle and Bolland, in a series of papers published since 1967. The diversity of the results obtained is promising and by cross-checking these data against other morphological and biological information, it is possible to make a certain number of comparisons between species.

— The assessment of reproductive barriers between species usually represents important information. However, the interpretation of data on the genetic affinities may sometimes pose problems for the student. It has been shown for the *T. urticae* complex that morphologically indistinguishable populations having the same colour may exhibit complete reproductive barriers (see Chapter 1.3.3). On the other hand, gene flow could be demonstrated between certain green and carmine-coloured populations of this complex (Dupont, 1979).

— Further developments may be anticipated by studies on enzyme polymorphism in tetranychid mites (see Chapter 1.3.2 and Ward et al., 1982). Zymograms may reveal patterns of relationships between members of the large genera, such as *Eotetranychus* Oudemans and *Tetranychus* Dufour, which possibly will be of significance for systematic studies.

RELATIVE IMPORTANCE OF CHARACTERS

The division of tetranychid mites into sub-families, tribes and genera is based mainly on the examination of females, in which the morphology of empodium, the chaetotaxy of the dorsum and the position of duplex setae have been studied. The first 2 characters are clearly of phylogenetical importance. The shape of the empodium is linked to the mite's life type and the nature of the surface on which it moves (stalk, upper or lower surface of leaves) or marks an adaptation to locomotion along silken strands or on a web that is of varying density (see evolution of the ambulacrum in Chapter 1.1.5).

The hypothesis may be put forward that the position, length and shape of dorsal setae is connected with the protection of the mites from predators, especially in non-spinning species (most Bryobiinae Berlese) and in species secreting only little silk (Tenuipalpoidini Pritchard and Baker, Eurytetranychini Reck). In certain genera, as in *Eutetranychus* Banks, for instance, the length of dorsal setae may vary from one specimen to another. Saito and Takahashi (1980) found that in the case of *Schizotetranychus celarius* (Banks) a correlation exists between the length of some dorsal setae and the height of the web above the leaf surface.

In the tribe Tetranychini Reck, it would be interesting to know whether or not the lengthening of the solenidia of the duplex setae can be linked to web spinning. In these species, which often live in dense colonies, the lengthening of the solenidia may also be correlated with the frequency of social contacts.

Other features used in distinguishing genera and sub-genera appear to constitute nothing more than a convenient method of separating groups and species. Such is the case for the pattern of dorsum and dorsal striation of the female opisthosoma. The division of the genus *Oligonychus* Berlese into separate sub-genera based solely on this last criterion seems to conform less to phylogeny, than to a classification based on the form of the empodial claw or of the aedeagus (Gutierrez et al., 1979).

As for the division into species, the study of females must be complemented by that of males, if and when they exist. This study is often indispensable, e.g. for the identification of members of the tribe Tetranychini, where the females are morphologically very similar within each genus.

Besides those characters which have led to the determination of the genus, the following are also studied: (1) the shape of the male aedeagus; (2) the chaetotaxy of legs; (3) the shape of the peritreme (which may end in a simple bulb, a distal hook or an anastomosing system); (4) the shape of the spinning eupathidium of the palpal tarsus; (5) the aspect of the dorsal integument and the dorsal striation pattern of the female opisthosoma; (6) the ventral chaetotaxy of the body. The characteristics of the integumental striae on and just anterior to the genital flap are also considered, especially in the genus *Eotetranychus* (Pritchard and Baker, 1955).

Hopes were held of distinguishing the green and red forms of the *T. urticae* complex by examining the form of the lobes on the dorsal cuticular ridges of the female mites. However, the significance of this character for the distinction of 'the' red species in this species complex is questionable (Dupont, 1979).

Chapter 1.1.4. references, p. 89

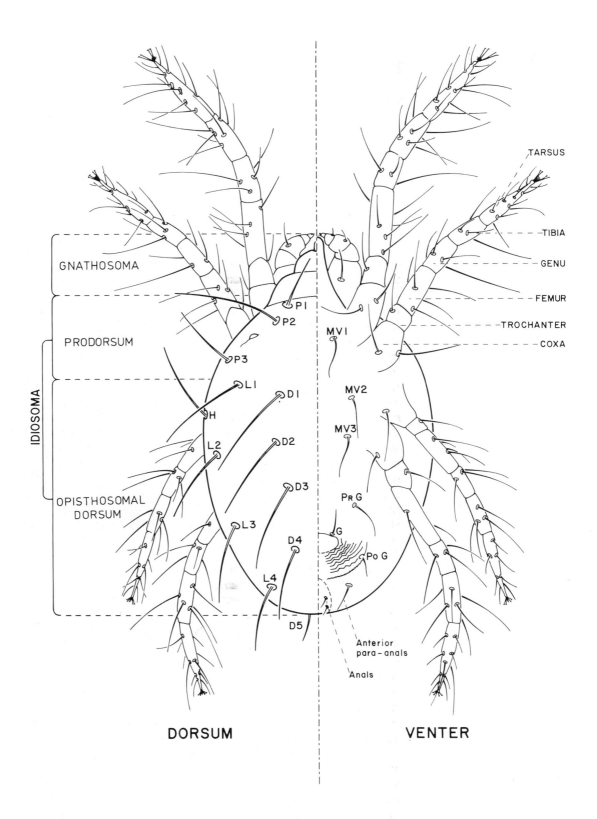

Fig. 1.1.4.1. *Tetranychus urticae* Koch: dorsoventral aspect of the female showing the nomenclature of body setae.

NOMENCLATURE OF THE PHANERES STUDIED IN TAXONOMY

The nomenclature of phaneres often varies from one author to another; consequently it has become necessary to update the terms used in this field in order to facilitate the reading of determination keys. In the present paragraph, a brief survey is given of the terms used.

Body setae

Figure 1.1.4.1, which represents a dorsoventral aspect of the female of *Tetranychus urticae*, shows the system most commonly used to designate setae.

Dorsum

In the Tetranychinae Berlese, the prodorsum generally bears 3 pairs of setae (P_1, P_2, P_3). Their designation according to Grandjean's notation is still uncertain because of the strong reduction in their original number (L. van der Hammen, personal communication, 1984).

The opisthosomal dorsum has 5 pairs of dorsal setae (D_1, D_2, D_3, D_4, D_5), 4 pairs of lateral setae (L_1, L_2, L_3, L_4) and 1 pair of humeral setae (H). In Grandjean's notation, the dorsals should be named c_1, d_1, e_1, f_1, h_1, the laterals c_2, d_2, e_2, f_2, and the humerals c_3.

D_1, D_2 and D_3 are also termed dorsocentral hysterosomals; L_1, L_2 and L_3, dorsolateral hysterosomals; D_4 and L_4, inner and outer sacrals; D_5, clunals. Reck (1959) and Mitrofanov (1971a) named the dorsal setae as follows: parietals (P_1), oculars (P_2, P_3), scapulars (D_1, L_1, H), prelumbals (D_2, L_2), lumbals (D_3, L_3), sacrals (D_4, L_4) and caudals (D_5).

In the Bryobiinae Berlese, another pair of inner setae may occur together with P_1, the former pair occasionally being reduced to a single seta (genus *Septobia* Zaher et al.). L_2 and L_3 are also often doubled, resulting in 12 pairs of dorsal hysterosomal setae instead of 10. D_4 may be set further back to occupy a marginal position. In rare cases, the number of opisthosomal setae may be as few as 6 pairs (genus *Marainobia* Meyer), or as many as 34 pairs (genus *Dasyobia* Strunkova) owing to replacement of several dorsal setae by tufts of 4 setae (neotrichy).

Venter

In the females of Tetranychinae, there are generally 3 pairs of medioventral setae (Mv_1, Mv_2, Mv_3), 1 pair of pregenitals (Pr G), 1 pair of genitals (G), 1 pair of post-genitals (Po G), 2 pairs of anals and 2 pairs of para-anals (anterior and posterior para-anals). The males have 4 pairs of genito-anal setae.

According to Grandjean's notation, the medioventrals should be named 1a, 3a, 4a, the pregenitals ag_1, the genitals g_1, the post-genitals ag_2, the anals (or pseudanals) ps_1, ps_2 and the para-anals h_1, h_2.

Reck (1959) and Wainstein (1960) named the para-anal setae as post-anals.

In 5 genera (*Eurytetranychoides* Reck, *Oligonychus*, *Hellenychus* Gutierrez, *Atrichoproctus* Flechtmann and *Tetranychus*), the posterior para-anals are absent. There is only 1 pair of anal setae in the genera *Aponychus* Rimando, *Paraponychus* Gonzalez and Flechtmann, *Acanthonychus* Wang, *Palmanychus* Baker and Pritchard, and *Atrichoproctus*.

In the Bryobiinae, the females have 3 pairs of anal setae and the males 5 pairs of genito-anal setae.

Chapter 1.1.4. references, p. 89

The para-anals may be in a dorsal position in the genera *Bryobiella* Tuttle and Baker, and *Edella* Meyer. There are 2 pairs of pregenital setae instead of 1 in *Strunkobia pamirica* Livshitz and Mitrofanov. Instead of the usual medioventral and pregenital setae, about 19 pairs of plumose setae may be present in *Neotrichobia* Tuttle and Baker.

Phaneres of legs and palps

The chaetotaxy of legs and palps of several species of Tetranychidae has been studied in detail by Grandjean (1948), Wainstein (1958), Mitrofanov (1971b), then Robaux and Gutierrez (1973). These works, based on the study of all stases under the polarizing microscope, enables a distinction to be made between anisotropic phaneres (ordinary setae, bothridial setae and eupathidia) and isotropic phaneres (solenidia). However, most authors have used the nomenclature popularized by Pritchard and Baker (1955), which is based on the observation of the adult stase under an ordinary microscope. This system uses only the terms sensory setae, tactile setae and duplex setae. The tactile setae are in fact anisotropic phaneres, while the sensory setae are solenidia. The duplex setae are composed of 1 proximal ordinary seta and 1 distal solenidion.

Legs

Figure 1.1.4.2 represents the tibiae and tarsi I and II of *Tetranychus neocaledonicus* André, each phanere being named according to the notation of Grandjean.

Anisotropic phaneres. Ordinary setae are named according to the file and the verticil to which they belong. The file may be: dorsal (d), laterodorsal (l' and l"), or lateroventral (v' and v"). The verticil is indicated by a number. The whole is followed by the name of the particular stase after the larval stase (L), at which the seta appeared: N_1, N_2 or adult (Ad). When no stase is mentioned, the base level is larval.

For the setae on the distal end of the tarsus, specific terms are used: prorals (p), unguinals (u), tectals (tc), fastigials (ft) and primiventrals (pv).

There is only 1 bothridial seta on tibia I: db.

The legs I and II each have 3 eupathidia: p'ζ, p"ζ and pv'ζ.

Isotropic phaneres. There are solenidia represented by the greek letters ω for the tarsi and φ for the tibiae. Males often have additional solenidia represented by ωδ and φδ.

With the exception of a few species which have already been studied in depth, it would be impossible in practice to undertake such a detailed description of each taxon. Nonetheless, Grandjean's research should be taken into consideration. The differences between various 'tactile setae' should be indicated and the term 'sensory setae' replaced by solenidia, especially since the exact role of these setae is unknown. The term 'duplex setae', used by systematists and retained in the present study should be replaced by the term 'coupled phaneres' (L. van der Hammen, personal communication, 1984).

Palps

Taxonomists basically use the chaetotaxy of palpal tarsus. According to Pritchard and Baker (1955), 7 setae occur on this segment: 3 tactile setae and 4 sensory setae. In the latter 'the proximal is fusiform, 2 are tapering, the

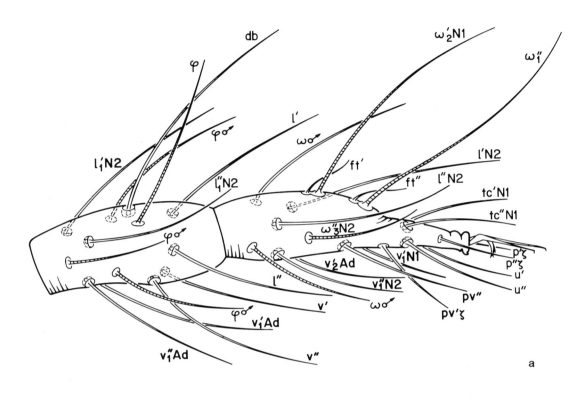

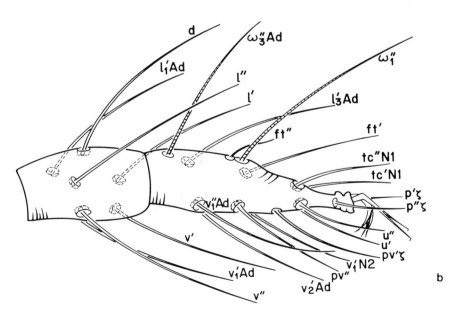

Fig. 1.1.4.2. *Tetranychus neocaledonicus* André, male, designation of setae according to Grandjean's nomenclature. (a) Tibia and tarsus I; (b) tibia and tarsus II.

Chapter 1.1.4. references, p. 89

terminal sensillum is usually well developed and rounded at the tip'. In fact, as shown by Grandjean's research, the palpal tarsus does bear 3 ordinary setae, but the 4 sensory setae must not be interpreted as such. The 'fusiform sensory seta' is, in fact, a solenidion, whereas the 3 others are eupathidia. The 'terminal sensillum' is, in fact, in the Tetranychinae a spinning eupathidium. At the tip of this seta are several tiny orifices through which flow the secretion from the silk gland. The size of this eupathidium varies considerably from one species to another and also according to sex.

DIVISION OF THE TETRANYCHIDAE INTO SUB-FAMILIES, TRIBES AND GENERA

After having distinguished between the 2 sub-families of Tetranychidae, several keys lead to 6 tribes and 63 present genera. The following system has been set up using the work of Smith Meyer (1974) and that of Jeppson et al. (1975). The system has been revised and completed with the addition of 12 new genera recently described: *Acanthonychus* Wang, 1982; *Atetranychus* Tuttle et al., 1974; *Crotonella* Tuttle et al., 1974; *Eremobryobia* Strunkova and Mitrofanov, 1982; *Lindquistiella* Mitrofanov, 1976; *Meyernychus* Mitrofanov, 1977; *Paraponychus* Gonzalez and Flechtmann, 1977; *Septobia* Zaher et al., 1982; *Sonotetranychus* Tuttle et al., 1976; *Strunkobia* Livshitz and Mitrofanov, 1972; *Tenuipalponychus* Channabasavanna and Lakkundi, 1977; *Yezonychus* Ehara, 1978. Moreover, *Eurytetranychoides* Reck, 1950, has been restored.

McGregorella Baker and Tuttle, 1972 has been considered to be a synonym of *Beerella* Wainstein, 1961, since these 2 taxa have the same

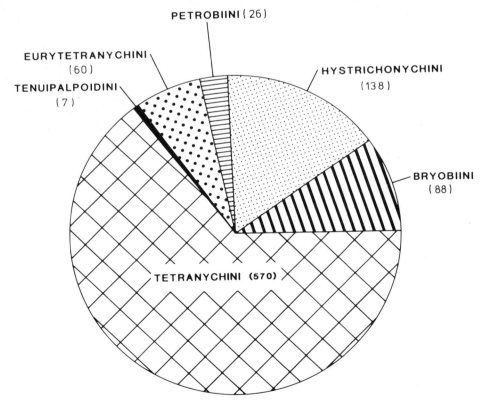

Fig. 1.1.4.3. Distribution of the number of species presently known, between the 6 tribes of Tetranychidae.

characters (prodorsum with 3 pairs of setae; opisthosoma with 9 pairs of dorsal setae; dorsocentral setae set on strong tubercles, the 5th being absent).

Georgiobia Wainstein, 1960 has been considered to be a synonym of *Aplonobia* Womersley, 1940 since the distinguishing characters between these 2 groups (D_2, D_3 and D_4 contiguous or well-separated) are not clear enough.

Sinotetranychus Ma and Yuan, 1980 and *Chinotetranychus* Ma and Yuan, 1982 have been considered to be synonyms of *Aponychus* Rimando, 1966, since slight differences in the length of dorsal setae, in the present author's opinion, is not a sufficient criterion for the separation of genera of Eurytetranychini. In this tribe, the length of dorsal setae is unreliable even for species separation.

For practical considerations, the genus *Bryocopsis* Meyer with lack of hooked claws, has been transferred to the tribe Hystrichonychini Pritchard and Baker, although the dorsal aspect of the female is similar to that of the members of *Bryobia* Koch in the Bryobiini Reck.

Several genera have been reduced in sub-generic rank: *Reckia* Wainstein, *Langella* Wainstein, *Anaplonobia* Wainstein, *Brachynychus* Mitrofanov and Strunkova, *Tylonychus* Miller and *Bakerina* Chaudhri. The division into sub-genera of the genus *Petrobia* Murray, made by Wainstein (1960), has been restored.

Figure 1.1.4.3 shows the relative size of the 6 tribes: the Tetranychini alone, with 570 species, comprise more than half of the members of the family, whereas the Tenuipalpoidini have only 7 known species.

The different genera are also extremely varied in size: 38 genera, of which 34 date from after 1955, include less than 5 species. Figure 1.1.4.4 indicates the number of species for the genera with more than 20 known members; only 3 genera have more than 100 species: *Eotetranychus*, *Oligonychus* and *Tetranychus*.

Within each tribe, the use of dichotomic keys has not permitted the presentation of genera according to sequences in line with current phylogenetic hypotheses. Thus, for example, in the genera of Bryobiini with 8 setae on the prodorsum, instead of the series *Marainobia*, *Bryobia*, *Pseudobryobia*, *Strunkobia*, one should have the sequence *Bryobia*, *Pseudobryobia*, *Strunkobia*, *Marainobia*. For the last 4 genera of Tetranychini, instead of the series *Tetranychus*, *Atrichoproctus*, *Oligonychus*, *Hellenychus*, one should have the sequence *Oligonychus*, *Hellenychus*, *Tetranychus*, *Atrichoproctus*.

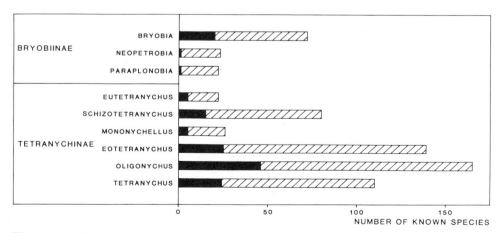

Fig. 1.1.4.4. Number of species of the genera represented by more than 20 known taxa. The fraction indicated in black corresponds to the number of species described before 1955.

Chapter 1.1.4. references, p. 89

KEYS

TETRANYCHIDAE Donnadieu

Tetranycidés Donnadieu, 1875: 9.
Tetranychidae Murray, 1877: 93; Pritchard and Baker, 1955: 4; Tuttle and Baker, 1968: 1.

Key to the sub-families

— Empodium with tenent hairs; females with 3 pairs of anal setae and males with 5 pairs of genito-anal setae . . . . BRYOBIINAE Berlese
— Empodium absent, or if present without tenent hairs; females with 1 or 2 pairs of anal setae and males with 4 pairs of genito-anal setae . . . .
. . . . TETRANYCHINAE Berlese

1. BRYOBIINAE Berlese
Bryobiini Berlese, 1913: 17
Bryobiinae Reck, 1950: 122; Pritchard and Baker, 1955: 12.

Key to the tribes

1 — True claws uncinate, empodium pad-like . . . . BRYOBIINI Reck
— True claws pad-like, empodium pad-like or uncinate 2
2 — Claws and empodium pad-like . . . .
. . . . HYSTRICHONYCHINI Pritchard and Baker
— Claws pad-like and empodium uncinate . . . . .PETROBIINI Reck

1.1 BRYOBIINI Reck
Bryobiinae Reck, 1952: 423.
Bryobiini Reck, Pritchard and Baker, 1955: 14.

Key to the genera based on the females

1 — Prodorsum with 8 setae (4 pairs) 2
— Prodorsum with 6 or 7 setae. 5
2 — Opisthosoma with 12 pairs of dorsal setae .3
— Opisthosoma with 6 pairs of dorsal setae . . . . MARAINOBIA Meyer
3 — Prodorsum with prominent lobes over the gnathosoma. .BRYOBIA Koch
— Prodorsum without prominent lobes over the gnathosoma 4
4 — Venter with 1 pair of setae on the mentum and 1 pair of pregenitals . . . .
. . . . PSEUDOBRYOBIA McGregor
— Venter with 2 pairs of setae on the mentum and 2 pairs of pregenitals. . .
. . . . STRUNKOBIA Livshitz and Mitrofanov
5 — Prodorsum with 6 setae (3 pairs) 6
— Prodorsum with 7 setae . . . . SEPTOBIA Zaher et al.
6 — Opisthosoma with 12 pairs of dorsal setae .7
— Opisthosoma with 10 pairs of dorsal setae . . . .
. . . . EREMOBRYOBIA Strunkova and Mitrofanov.
7 — Tarsus I without duplex setae, para-anal setae dorsal . . . .
. . . . BRYOBIELLA Tuttle and Baker
— Tarsus I with 2 duplex setae, para-anal setae in ventral position . . . . . .
. . . . .HEMIBRYOBIA Tuttle and Baker

1.2 HYSTRICHONYCHINI Pritchard and Baker
Hystrichonychini Pritchard and Baker, 1955: 35.

Key to the genera based on the females

1 — Prodorsum with 4 pairs of setae 2
  — Prodorsum with 3 pairs of setae 4
2 — With 4 prominent lobes over the gnathosoma. . . .BRYOCOPSIS Meyer
  — Without prominent lobes over the gnathosoma. 3
3 — Opisthosoma with 12 pairs of dorsal setae . . . .
   . . . . TETRANYCOPSIS Canestrini
  — Opisthosoma with 9 pairs of dorsal setae . . . . . . NOTONYCHUS Davis
4 — Opisthosoma with 10 or more pairs of dorsal setae. 5
  — Opisthosoma with 8 to 9 pairs of dorsal setae. 14
5 — Body at least twice as long as broad; prodorsum with lobes over the
   gnathosoma more or less developed 6
  — Body not elongate; prodorsum without lobes over the gnathosoma . . .8
6 — Prodorsal lobes well developed 7
  — Prodorsal lobes poorly developed . . . . DOLICHONOBIA Meyer
7 — Prodorsum with 3 lobes over the gnathosoma. . . . .
   . . . . MONOCERONYCHUS McGregor
  — Prodorsum with 2 lobes over the gnathosoma. . . . .
   . . . . MESOBRYOBIA Wainstein
8 — D4 in normal position. 9
  — D4 in marginal position or nearly so . . . . . NEOPETROBIA Wainstein[1]
9 — Dorsum with 12 pairs of opisthosomal setae. . . . .
   . . . . HYSTRICHONYCHUS McGregor
  — Dorsum with 10 pairs of opisthosomal setae. 10
10 — Two sets of duplex setae on tarsus I 11
  — Four sets of duplex setae on tarsus I. . . . .
   . . . . PARAPETROBIA Meyer and Ryke
11 — D1, D2, D3 and D4 located on cushion-like plates. .PELTANOBIA Meyer
  — Opisthosoma without plates 12
12 — Coxal formula not exceeding 4—3—2—2. 13
  — Many ventral and coxal setae. . TAURIOBIA Livshitz and Mitrofanov
13 — Some or all dorsal setae set on strong tubercles . . . .
   . . . . APLONOBIA Womersley
  — Dorsal setae not set on strong tubercles. . .PARAPLONOBIA Wainstein[2]
14 — Opisthosoma with 9 pairs of dorsal setae; dorsocentral setae set on
   strong tubercles . . . . BEERELLA Wainstein
  — Opisthosoma with 8 pairs of dorsal setae .15
15 — Humeral setae contiguous with L_1; dorsal body setae set on strong
   tubercles and mostly contiguous. . . . . .PORCUPINYCHUS Anwarullah
  — Humeral setae and other dorsal body setae well separated and set on
   small tubercles . . . . AFRONOBIA Meyer

1.3 PETROBIINI Reck
  Petrobiinae Reck, 1952: 423.
  Petrobiini Reck, Pritchard and Baker, 1955: 42; Wainstein, 1960: 131;
  Tuttle and Baker, 1968: 71.

Key to the genera based on the females

1 — 3 pairs of medioventral body setae 2
  — Many medioventral body setae . . . . NEOTRICHOBIA Tuttle and Baker[3]
2 — 1 set of duplex setae on tarsus I . . . . SCHIZONOBIELLA Beer and Lang
  — 2 or more sets of duplex setae on tarsus I. .3
3 — Prodorsum with 3 prominent lobes over the gnathosoma . . . .
   . . . . MEZRANOBIA Athias-Henriot
  — Prodorsum without prominent lobes over the gnathosoma. 4

4 — Empodium with 2 rows of ventrally directed tenent hairs 5
— Empodium with a single pair of tenent hairs. . . . .
. . . . SCHIZONOBIA Womersley
5 — Dorsum with not more than 15 pairs of dorsal body setae 6
— Dorsum with many dorsal body setae. . . . . .DASYOBIA Strunkova
6 — True claws without hair-like processes 7
— True claws with 2 short hair-like processes . . . .
. . . . LINDQUISTIELLA Mitrofanov
7 — Para-anal setae in dorsal position, empodium with a strong medioventral
claw . . . . EDELLA Meyer
— Para-anal setae in ventral position, empodium curved distally. . . . .
. . . . PETROBIA Murray[4]

2. TETRANYCHINAE Berlese
Tetranychini Berlese, 1913: 17
Tetranychinae Reck, 1950: 123; Pritchard and Baker, 1955: 96.

Key to the tribes

1 — Empodium claw-like when present, tarsus I with loosely 'associated
setae' or with 1 pair of duplex setae; when 2 pairs of duplex setae on
tarsus I then no pairs on tarsus II . . . . . . EURYTETRANYCHINI Reck
— Empodium claw-like or split distally, tarsus I with 2 pairs of duplex
setae and tarsus II with 1 pair. 2
2 — Opisthosoma with D_4 in marginal position or absent . . . .
. . . . TENUIPALPOIDINI Pritchard and Baker
— Opisthosoma with D_4 in normal position . . . . . . TETRANYCHINI Reck

2.1 TENUIPALPOIDINI Pritchard and Baker
Tenuipalpoidini Pritchard and Baker, 1955: 97; Wainstein, 1960: 145—
146; Tuttle and Baker, 1968: 83.

Key to the genera based on the females

1 — D_4 in marginal position. 2
— D_4 absent . . . . .EONYCHUS Gutierrez
2 — Empodium a simple claw 3
— Empodium split distally . . . . CROTONELLA Tuttle et al.
3 — Tarsus II with distal member of duplex setae a short solenidion. . . . .
. . . . TENUIPALPOIDES Reck and Bagdasarian
— Tarsus II with distal member of duplex setae a long solenidion . . . .
. . . . TENUIPALPONYCHUS Channabasavanna and Lakkundi

2.2 EURYTETRANYCHINI Reck
Eurytetranychinae Reck, 1950: 123; Wainstein, 1960: 223.
Eurytetranychini Pritchard and Baker, 1955: 100; Tuttle and Baker,
1968: 81.

Key to the genera based on the females

1 — Tarsus I with or without 'associated setae', or with 2 pairs of duplex
setae. 2
— Tarsus I with 1 pair of duplex setae. . . ATETRANYCHUS Tuttle et al.
2 — Tarsus I with or without 1 pair of 'associated setae' 3
— Tarsus I with 2 pairs of 'associated setae' or 2 pairs of duplex setae 9

3 — Empodial claw large . . . . SYNONYCHUS Miller
  — Empodial claw small or absent 4
4 — Empodial claw small. 5
  — Empodial claw apparently absent 6
5 — 2 pairs of para-anal setae. . . . . EURYTETRANYCHUS Oudemans
  — 1 pair of para-anal setae . . . . EURYTETRANYCHOIDES Reck
6 — 2 pairs of anal setae 7
  — 1 pair of anal setae 8
7 — Opisthosoma with 10 pairs of dorsal setae . . EUTETRANYCHUS Banks
  — Opisthosoma with 9 pairs of dorsal setae (L_1 absent) . . . .
  . . . . MEYERNYCHUS Mitrofanov
8 — Opisthosoma with 10 pairs of dorsal setae . . . . . APONYCHUS Rimando
  — Opisthosoma with 9 pairs of dorsal setae (H absent). . . . .
  . . . . PARAPONYCHUS Gonzalez and Flechtmann
9 — Empodium claw long and slender . . . . ANATETRANYCHUS Womersley
  — Empodium apparently absent. . . . . .DUPLANYCHUS Meyer

2.3 TETRANYCHINI Reck
  Tetranychinae Reck, 1950: 123.
  Tetranychini Pritchard and Baker, 1955: 124; Wainstein, 1960: 223;
  Tuttle and Baker, 1968: 83.

Key to the genera based on the females

1 — 2 pairs of para-anal setae. 2
  — 1 pair of para-anal setae 14
2 — Empodium claw-like. 3
  — Empodium split distally or ending in tuft of hairs 9
3 — Empodium a single claw-like structure 4
  — Empodium split into 2 claw-like structures, usually with appendant
  hairs 8
4 — Empodium without proximoventral hairs. 5
  — Empodium with proximoventral hairs 7
5 — Empodial claw much longer than the pads of the true claws. 6
  — Empodial claw very short, about as long as the pads of the true claws . . . .
  . . . . BREVINYCHUS Meyer
6 — Empodial claw strong; dorsal setae stout; integument forming reticulate
  pattern . . . . MIXONYCHUS Meyer and Ryke[5]
  — Empodial claw thin; dorsal setae fine; integument with simple striations
  . . . . SONOTETRANYCHUS Tuttle et al.
7 — Empodial claw as long or longer than proximoventral hairs, which are at
  right angles to the claw. . . . . PANONYCHUS Yokoyama
  — Empodial claw shorter than proximoventral hairs, which are at less
  than right angles to the claw. . . . . ALLONYCHUS Pritchard and Baker
8 — Opisthosoma with 10 dorsal setae. . SCHIZOTETRANYCHUS Trägårdh
  — Opisthosoma with 9 dorsal setae (L_4 absent) . . . YEZONYCHUS Ehara
9 — Empodium split distally; dorsal body setae set on tubercles 10
  — Empodium split near the middle into 3 pairs of hairs. 11
10 — 2 pairs of anal setae . . . . .NEOTETRANYCHUS Trägårdh
  — 1 pair of anal setae . . . . ACANTHONYCHUS Wang
11 — Opisthosoma with longitudinal striae between the D_3 setae; dorsal
  body setae serrate . . . . MONONYCHELLUS Wainstein
  — Opisthosoma with transverse striae. 12
12 — Dorsal body setae much shorter than the intervals between their bases
  . . . . PLATYTETRANYCHUS Oudemans
  — Dorsal body setae as long or longer than the intervals between their
  bases. 13

Chapter 1.1.4. references, p. 89

13 — 2 pairs of anal setae . . . . .EOTETRANYCHUS Oudemans
 — 1 pair of anal setae . . . . .PALMANYCHUS Baker and Pritchard
14 — Empodium claw-like with proximoventral hairs; duplex setae of tarsus
  I distal and adjacent 15
 — Empodium split distally, usually into 3 pairs of hairs; duplex setae of
  tarsus I well separated . . . . TETRANYCHUS Dufour[6]
15 — 2 pairs of anal setae 16
 — 1 pair of anal setae . . . . ATRICHOPROCTUS Flechtmann
16 — All the legs or most of them with empodial claws as long or longer than
  the proximoventral hairs. . . . . OLIGONYCHUS Berlese[7]
 — All the legs of most of them with empodial claws nearly as long as the
  proximoventral hairs. . . . . HELLENYCHUS Gutierrez

----

[1] It is proposed to divide the genus *Neopetrobia* Wainstein into 3 sub-genera, separated according to the presence or absence of tubercles on the dorsum of females and the aspect of the dorsal setae:
Neopetrobia Wainstein: type species *Neopetrobia dubinini* Wainstein, 1956.
Reckia Wainstein: type species *Mesotetranychus samgoriensis* Reck, 1960.
Langella Wainstein: type species *Aplonobia dyschima* Beer & Lang, 1958.
  1 — Integument with tuberculate pattern. . . . . *Reckia*
  — Integument without tuberculate pattern. 2
  2 — Dorsal setae generally flattened or foliate. . . . . *Langella*
  — Dorsal setae rounded or spindle-shaped . . . .*Neopetrobia s.str.*

[2] It is proposed to divide the genus *Paraplonobia* Wainstein into 3 sub-genera recognizable by the number of coxal setae and the aspect of peritremes:
Paraplonobia Wainstein: type species *Aplonobia (Paraplonobia) echinopsili* Wainstein, 1960.
Anaplonobia Wainstein: type species *Aplonobia calame* Pritchard and Baker, 1955.
Brachynychus Mitrofanov & Strunkova: type species *Brachynychus cousiniae* Mitrofanov and Strunkova, 1971.
  1 — Coxal formula not exceeding 3—3—1—1. 2
  — Coxal formula 4—3—2—2 . . . . *Brachynychus*
  2 — Anastomosing peritremes . . . . *Anaplonobia*
  — Simple peritremes . . . . *Paraplonobia s.str.*

[3] Tuttle and Baker (1968) proposed the name Neotrichobiini as a new tribe for the monospecific genus *Neotrichobia*, but the present author does not agree and considers *Neotrichobia arizonensis* Tuttle and Baker, 1968 as a case of plethotrichy, a phenomenon which, in other Acari, is generally regarded as of secondary importance inside a genus.

[4] The genus *Petrobia* Murray was divided by Wainstein (1960) into 3 sub-genera namely *Petrobia* Murray, *Mesotetranychus* Reck and *Tetranychina* Banks.
  1 — Some or all the body setae set on tubercles. . . . .*Tetranychina*
  — Dorsal body setae not set on tubercles. 2
  2 — Anastomosing peritremes . . . . *Petrobia s.str.*
  — Simple peritremes . . . . *Mesotetranychus*

[5] It is proposed to divide the genus *Mixonychus* Ryke and Meyer into 3 sub-genera separated according to the aspect of the dorsal integument:
Mixonychus Ryke and Meyer: type species *Mixonychus acaciae* Ryke and Meyer, 1960.
Tylonychus Miller: type species *Tylonychus tasmaniensis* Miller, 1966.
Bakerina Chaudhri: type species *Bakerina lepidus* Chaudhri, 1971.
  1 — Dorsum with lumps, forming a reticulate pattern. . . . .*Mixonychus s.str.*
  — Dorsum without lumps, with striae only. 2
  2 — Dorsal striae with spinules. . . . .*Tylonychus*
  — Dorsal striae without spinules . . . .*Bakerina*

[6] The genus *Tetranychus* Dufour was divided by Tuttle and Baker (1968) into 7 sub-genera: *Tetranychus* Dufour, *Polynychus* Wainstein, *Armenychus* Wainstein, *Pentanychus* Wainstein, *Septanychus* McGregor, *Pseudonychus* Wainstein, *Amphitetranychus* Oudemans.

[7] The genus *Oligonychus* Berlese was divided by Tuttle and Baker (1968) into 6 subgenera: *Oligonychus* Berlese, *Wainsteiniella* Tuttle and Baker, *Homonychus* Wainstein, *Metatetranychus* Wainstein, *Reckiella* Tuttle and Baker, *Pritchardinychus* Wainstein.

REFERENCES

Baker, E.W., 1979. Spider mites revisited — A review. In: J.G. Rodriguez (Editor), Recent Advances in Acarology, Vol. 2. Academic Press, New York, NY, pp. 387—394.

Beer, R.E. and Lang, D.S., 1958. The Tetranychidae of Mexico. Univ. Kansas Sci. Bull., 38: 1231—1259.

Berlese, A., 1913. Acarotheca Italica. Firenze, 221 pp.

Channabasavanna, G.P. and Lakkundi, N.H., 1977. A new genus of Tenuipalpoidini (Acarina: Tetranychidae) from Karnataka, India. Indian J. Acarol., 1: 23—27.

Chaudhri, W.M., 1971. A new spider mite genus, *Bakerina* (Acarina, Tenuipalpoidini) with the descriptions of four new species. Pak. J. Zool., 3: 195—202.

Donnadieu, A.L., 1875. Recherches pour servir à l'histoire des Tétranyques. Soc. Linn. Lyon: 1—134.

Dupont, L.M., 1979. On gene flow between *Tetranychus urticae* Koch, 1836 and *Tetranychus cinnabarinus* (Boisduval) Boudreaux, 1956 (Acari: Tetranychidae): synonymy between the two species. Entomol. Exp. Appl., 25: 297—303.

Ehara, S., 1978. A new genus and a new subgenus of spider mites from Northern Japan (Acarina: Tetranychidae). J. Fac. Educ. Tottori Univ., Nat. Sci., 28: 87—93.

Ewing. H.E., 1913. The taxonomic value of characters of the male genital armature in the genus *Tetranychus* Dufour. Ann. Entomol. Soc. Am., 6: 453—460.

Grandjean, F., 1948. Quelques caractères des Tétranyques. Bull. Mus. Natl. Hist. Nat., Paris, 20: 517—524.

Gonzalez, R.H. and Flechtmann, C.H.W., 1977. Revision de los acaros fitofagos en El Peru y description de un nuevo genero de Tetranychidae (Acari). Rev. Peru. Entomol., 20: 67—71.

Gutierrez, J., Bolland, H.R. and Helle, W., 1979. Karyotypes of the Tetranychidae and the significance for taxonomy. In: J.G. Rodriguez (Editor), Recent Advances in Acarology, Vol. 2. Academic Press, New York, NY, pp. 399—404.

Helle, W. and Bolland, H.R., 1967. Karyotypes and sex determination in spider mites (Tetranychidae). Genetica, 38: 43—53.

Jeppson, L.R., Keifer, H.H. and Baker, E.W., 1975. Mites Injurious to Economic Plants. University of California Press, Berkeley, CA, 614 pp.

Livshitz, I.Z. and Mitrofanov, V.I., 1972. Contribution to the knowledge of the mites of the family Bryobiidae. Tr. Gos. Nikitsk. Bot. Sad., 61: 5—12 (in Russian).

Ma, E.P. and Yuan, Y.L., 1980. A new genus and a new species of spider mites from south China (Acarina: Tetranychidae). Acta Entomol. Sinica, 23: 441—442 (in Chinese).

Ma, E.P. and Yuan, Y.L., 1982. A new genus and five new species of Tetranychidae from China (Acari: Tetranychidae). Entomotaxonomia, 4: 109—114 (in Chinese).

McGregor, E.A., 1950. Mites of the family Tetranychidae. Am. Midl. Nat., 2: 257—420.

Miller, L.W., 1966. The Tetranychid mites of Tasmania. Pap. Proc. R. Soc. Tasmania, 100: 53—67.

Mitrofanov, V.I., 1971a. On the taxonomy of the Bryobiidae (Acariformes, Tetranychoidea). In: Milan Daniel and Bohumir Rosicky (Editors), Proc. 3rd Int. Congr. Acarol. W. Junk, Amsterdam, pp. 297—298.

Mitrofanov, V.I., 1971b. O khetome konechnostei u tetranikhovykh kleshchei (Acariformes, Tetranychoidea) Nauchn. Dokl. Vyssh. Shk. Biol. Nauki, 4: 11—17 (in Russian).

Mitrofanov, V.I., 1976. *Lindquistiella* gen. (Trombidiformes, Bryobiidae, Petrobiini) from the eastern coast of Canada. Zool. J., 55: 1256—1257 (in Russian).

Mitrofanov, I.I. and Strunkova, Z.I., 1971. New genera and species of phytophagous acarina from Pamir. Nauchn. Dokl. Vyssh. Shk. Biol. Nauki, 12: 17—18 (in Russian).

Murray, A., 1877. Economic Entomology, Aptera. Chapman and Hall, London, 433 pp.

Pritchard, A.E. and Baker, E.W., 1955. A revision of the spider mite family Tetranychidae. Pac. Coast Entomol. Soc., Mem. Ser., 2: 1—472.

Reck, G.F., 1950. Contributions to the fauna of spider mites in Georgia (Tetranychidae: Acari). Tr. Inst. Zool. Akad. Nauk Gruz. SSR, 9: 117—134 (in Russian).

Reck, G.F., 1952. Some principles on the classification of tetranychid mites. Soobshch. Akad. Nauk Gruz. SSR, 13: 420—425 (in Russian).

Reck, G.F., 1959. Identification of tetranychid mites. Isdatel. Akad. Nauk Gruz. SSR Tbilisi: 1—150 (in Russian).

Robaux, P. and Gutierrez, J., 1973. Les phanères des pattes et des palpes chex deux espèces de Tetranychidae: nomenclature et évolution au cours de l'ontogenèse. Acarologia, 15: 616—643.

Ryke, P.A.J. and Meyer, M.K.P., 1960. The parasitic and predacious mite fauna (Acarina) associated with *Acacia karoo* Hayne in the Western Transvaal. Inst. Politech. Nac. Escuela Nac. Cienc. Biol., Mexico: 559—569.

Saito, Y. and Takahashi, K., 1980. Study on variation of *Schizotetranychus celarius* (Banks). I. Preliminary descriptions of morphological and life type variation. Japan. J. Appl. Entomol. Zool., 24: 62—70 (in Japanese).

Smith Meyer, M.K.P., 1974. A revision of the Tetranychidae (Acari) of Africa with a key to the genera of the world. Repub. S. Afr., Dep. Agric. Tech. Serv., Entomol. Mem., 36: 1—291.

Strunkova, Z.I. and Mitrofanov, V.I., 1982. A new genus and two new species of plant-eating mites (Acariformes, Bryobiidae) from Turkmenia and Uzbekistan. Zool. J., 61: 1909—1911 (in Russian).

Tuttle, D.M. and Baker, E.W., 1968. Spider Mites of Southwestern United States and a Revision of the Family Tetranychidae. The University of Arizona Press, Tucson, AZ, pp. 1—143.

Tuttle, D.M., Baker, E.W. and Abbatiello, M., 1974. Spider mites from Northwestern and North Central Mexico (Acarina: Tetranychidae). Smithson. Contrib. Zool., 171: 1—18.

Tuttle, D.M., Baker, E.W. and Abbatiello, M., 1976. Spider mites of Mexico (Acari: Tetranychidae). Int. J. Acarol., 2: 1—102.

Wainstein, B.A., 1956. Contribution to the fauna of tetranychid mites of Kazakhstan. Tr. Resp. Sta. Kaz. Filial Vaskh., 3: 70—83 (in Russian).

Wainstein, B.A., 1958. Chaetom of the Tetranychidae (Acariformes) and the systematics of this family. Zool. J., 37: 1476—1487 (in Russian).

Wainstein, B.A., 1960. Tetranychoid mites of Kazakhstan (with revision of the families). Kaz. Akad. Selsk. Nauk Tr. Nauchn. Issled. Inst. Zasch. Rast., 5: 1—276 (in Russian).

Wainstein, B.A., 1961. On the systematic position of two species of Tetranychidae mites (Acariformes) with a description of two new genera and a tribe. Zool. Zh. Akad. Nauk SSR, 40: 606—608 (in Russian).

Wang, H.F., 1982. A new genus and species of Tetranychidae (Acarina). Acta Zootaxonom. Sinica, 7: 57—60 (in Chinese).

Ward, P.S., Boussy, I.A. and Swincer, D.E., 1982. Electrophoretic detection of enzyme polymorphism and differentiation in three species of spider mites (*Tetranychus*) (Acari: Tetranychidae). Ann. Entomol. Soc. Am., 75: 595—598.

Zaher, M.A., Gomaa, E.A. and El-Enany, M.A., 1982. Spider mites of Egypt. Int. J. Acarol., 8: 91—114.

1.1.5 Evolutionary Changes in the Tetranychidae

J. GUTIERREZ and W. HELLE

INTRODUCTION

Spider mite taxonomists are usually reticent about speculating on evolutionary pathways within the family of the Tetranychidae. In the literature, views and comments on phylogenetic structures and lineages occur only piecemeal, and must very often be deduced from the way classification is presented. This indifference on the part of the taxonomist is understandable. There is little doubt that only a minor part of the tetranychid taxa on earth have been described and in the near future new taxa will appear. This expectation makes the prospect of speculations on phylogenetic resemblances somewhat unattractive and a waiting attitude is usually adopted.

This does not mean that no patterns in the evolution of the Tetranychidae can be recognized by comparisons of adult morphology. Indeed, data which would make it possible to study the ontogeny of these acarines are lacking, since the examination of developmental stages has, until now, been neglected by taxonomists (however, see Chapters 1.1.1, 1.1.3 and 1.1.4). A study on the relationship between different types of ambulacra was carried out as early as 1915 by Trägårdh, but Pritchard and Baker (1955), in their revision of the family, were the first to consider phylogeny. A tentative synthesis including biological and cytogenetic data has also been proposed by Gutierrez et al. (1970).

Especially with regard to the higher categories within Tetranychidae, it is useful to review the evolutionary changes that may be of phylogenetic significance. These involve not only morphological characters (ambulacrum, chaetotaxy of body and legs, dorsal integument pattern, shape of peritremes and aedeagus, duplex setae displacement), but also biological features, mainly those dealing with adaptations to the host plant, to defense against predators and to dispersal (habitat, spinning behaviour, population dynamics, cytogenetics, etc.). The separation of the family into Bryobiinae Berlese and Tetranychinae Berlese on the basis of the presence or absence of tenent hairs on the empodium, implies many other changes, both morphological and biological.

EVOLUTION OF THE AMBULACRUM

The ambulacrum may have evolved from a type that is functional in clinging to the upperside of leaves and twigs, to a type that is better suited to locomotion along, or on, silken strands or web structures on the underside of leaves.

Chapter 1.1.5. references, p. 106

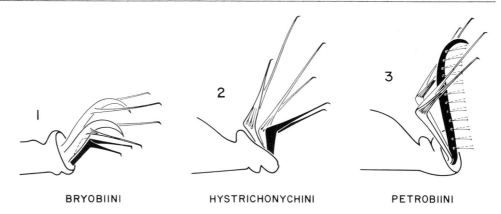

BRYOBIINI   HYSTRICHONYCHINI   PETROBIINI

Fig. 1.1.5.1. Evolution of the ambulacrum in the Bryobiinae. Comparison of the ambulacra I of females in the 3 tribes (the empodia are shaded in black): 1, *Bryobia praetiosa* Koch; 2, *Hystrichonychus gracilipes* (Banks); 3, *Petrobia (Tetranychina) harti* (Ewing).

Bryobiinae (Fig. 1.1.5.1)

There is a gradual reduction of ambulacral claws, which are claw-like in the Bryobiini Reck, but pad-like in the Hystrichonychini Pritchard and Baker, and in the Petrobiini Reck. Concomitantly the empodium, which is pad-like in both the Bryobiini and the Hystrichonychini, apparently takes over the task of the ambulacral claws and becomes claw-like in the Petrobiini. This last tribe seems therefore to have the most evolved type of ambulacrum and to occupy a transitional position with the Tetranychinae sub-family. This point will be discussed in the last paragraph of this chapter.

The functional significance of the conspicuous changes of the ambulacrum in the Bryobiinae is not very well understood. The regression of the ambulacral claws, together with the amalgamation of the tenent hairs and the synchronous evolution of the empodium cannot be associated with locomotion on silken strands, since members of this sub-family are nearly all unable to produce silk. It seems reasonable, however, to interpret the changes in terms of an improvement in locomotion on the leaf. Most bryobiine species reside on the upper surface of the leaf, visiting the undersurface only rarely. Nevertheless, the leaf undersurface offers a challenging niche to the bryobiines. Which mechanical handicaps prevent occupation and full exploration of the leaf undersurface in so many species? Are rows or bundles of tenent hairs of any use? It is apparent that this question can be answered by experimental studies on the locomotion of the various species, using artificial substrates in different spatial positions. Such experiments may lead to the necessary understanding of the function of the various tarsal appendages, and of the evolution of the ambulacrum in the Bryobiinae.

Tetranychinae

All members of this sub-family show the same type of ambulacral claws, which are reduced to pads with 1 pair of solid, compound tenent hairs. These tenent hairs are very movable instruments and render lateral support to the empodial claw (see Fig. 1.1.5.6). The following comparisons deal only with the morphology of the empodium.

In the Tenuipalpoidini Pritchard and Baker, the empodium consists of a simple stout claw, the most elaborate form being that observed in the genus *Crotonella* Tuttle et al., where the empodial claw is split distally.

In the Eurytetranychini Reck, the empodium, when present, has the

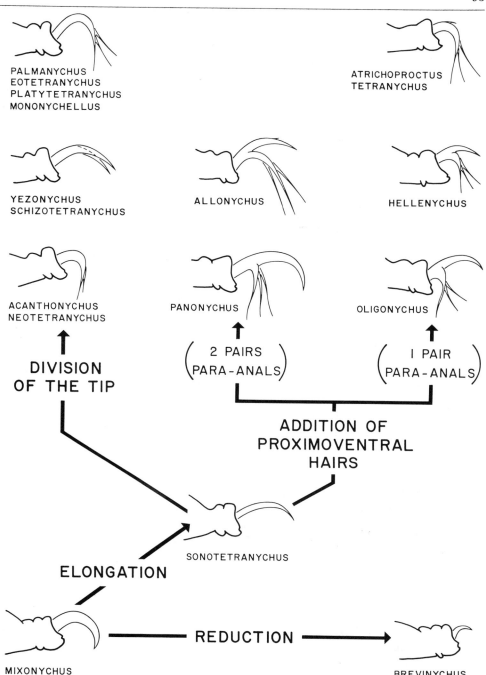

Fig. 1.1.5.2. Some lineages in the evolution of the empodium in the Tetranychini. Distinction between the different trends by comparison of the ambulacra I of females.

appearance of a simple claw. A derived form is represented in *Anatetranychus* Womersley, whose long slender empodium appears to be of the same type as that of *Oligonychus* Berlese, since it bears 2 extremely fine proximoventral hairs (H.B. Boudreaux, personal communication, 1971). Another line of evolution in this tribe leads to complete loss of the empodium, as is found in 5 genera (*Eutetranychus* Banks, *Meyernychus* Mitrofanov, *Aponychus* Rimando, *Paraponychus* Gonzalez and Flechtmann, *Duplanychus* Meyer). The largest number of different types occurs in the Tetranychini Reck, where the empodium, evolved from the simple claw type, has given rise to all forms used by systematicists for the separation into genera (Fig. 1.1.5.2).

Chapter 1.1.5. references, p. 106

The most primitive type of empodium would be that of the genus *Mixonychus* Meyer and Ryke, which is simple and stout. From this last type, one can observe either reduction (*Brevinychus* Meyer), or elongation (*Sonotetranychus* Tuttle et al.). The elongation may be followed either by the division of the tip of the empodial claw itself, which results in an empodium like that of *Eotetranychus* Oudemans, or by the addition of proximoventral hairs, which leads to an empodial claw of the type *Tetranychus* Dufour. In both cases, the development of these setae is apparently associated with locomotion on silken strands or webbing structures. Here too, experiments to study the extent to which various Tetranychinae are adapted to different web structures are eagerly anticipated.

REDUCTION IN SETAE NUMBER

These reductions apparently are evolutionary trends and concern prodorsal, dorsal opisthosomal and genito-anal setae.

The most primitive dorsal chaetotaxy would be that of 3 genera of Bryobiini (*Bryobia* Koch, *Strunkobia* Livshitz & Mitrofanov, and *Pseudobryobia* McGregor) with 4 pairs of propodosomal setae and 12 pairs of opisthosomal setae in unmodified positions.

The most evolved dorsal chaetotaxy would seem to consist of 3 pairs of prodorsal setae and 10 pairs of opisthosomal setae, as may be observed in the Hystrichonychini (9 genera out of 16), and again, with the exception of 2 genera (*Dasyobia* Strunkova and *Edella* Meyer) out of 8 in the Petrobiini, and with the exception of 4 genera (*Eonychus* Gutierrez, *Meyernychus*, *Paraponychus*, *Yezonychus* Ehara), out of 31 in the Tetranychinae.

The number of dorsal setae appears not to undergo modifications in the course of development, since it remains unchanged from larva to adult in the Bryobiinae (Mathys, 1957; Manson, 1967; Zein-Eldin, 1956), as well as in the Tetranychinae (English and Snetsinger, 1957; Singer, 1966).

Prodorsal setae

Genera with 4 pairs of prodorsal setae are found only amongst the Bryobiini and the Hystrichonychini. The genus *Septobia* Zaher et al., with 7 prodorsal setae, would constitute an intermediary step between the former genera and those with 3 pairs of prodorsal setae (Fig. 1.1.5.3). All the other tribes of the family have only 3 pairs of prodorsal setae.

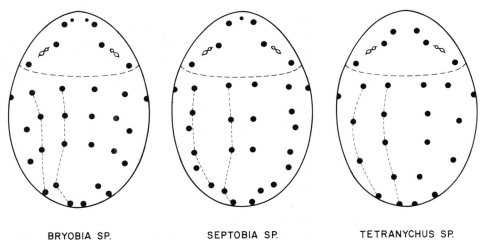

BRYOBIA SP.   SEPTOBIA SP.   TETRANYCHUS SP.

Fig. 1.1.5.3. Evolutionary trends in dorsal chaetotaxy.

Dorsal opisthosomal setae

Their number varies more, since considerable reductions are noted amongst the Bryobiini themselves with, for instance, 6 pairs of opisthosomal setae in *Marainobia* Meyer. On the other hand, a total of more than 12 pairs of setae may be observed in 2 genera: *Bryobiella* Tuttle and Baker (Bryobiini), and *Dasyobia* (Petrobiini). In *Bryobiella*, the total number of 14 pairs is in fact artificial, since it is only obtained by including the 2 pairs of para-anal setae, visible dorsally on specimens flattened out between slide and cover slide. The genus *Dasyobia*, with 25—29 pairs of opisthosomal setae, would represent a case of neotrichy.

Anal and para-anal setae

The reduction of the number of anal setae from 3 pairs in the Bryobiinae to 2 pairs in the Tetranychinae is followed by a further reduction from 2 to 1 pair in 2 genera of Eurytetranychini (*Aponychus* and *Paraponychus*) and independently in 3 genera of Tetranychini (*Acanthonychus* Wang, *Palmanychus* Baker and Pritchard, and *Atrichoproctus* Flechtmann).

The loss of 1 pair of para-anal setae appears only in the Eurytetranychini (*Eurytetranychoides* Reck). In the Tetranychini the species with 1 pair of para-anal setae would constitute a grouping parallel to that of taxa with 2 pairs and form the sequence: *Oligonychus—Hellenychus—Tetranychus—Atrichoproctus*.

VARIATIONS IN THE OUTLINE OF THE DORSAL SETAE

In the Bryobiinae, of which only 1 species is reported to produce silk, as well as in the Tenuipalpoidini and the Eurytetranychini, which usually spin webs only to protect their eggs, the dorsal setae take on a wide variety of forms. These setae are rather thick; they may be subspatulate or blunt distally, sometimes coarsely serrate. When they are long, they are often set on strong tubercles.

In the Tetranychini, on the other hand, apart from 2 genera which in any case have a rudimentary empodium (*Brevinychus* and *Mixonychus*) and a few species belonging to the genera *Atrichoproctus*, *Neotetranychus* Trägårdh, *Acanthonychus* and *Mononychellus* Wainstein, the dorsal setae become more uniform and are slender. This last type of seta is found in species living in web structures, which may be more or less complicated.

In the Bryobiinae, the shape and arrangement of setae may act as a warning or protective screen against predators. This kind of defence is probably less stringent for the web-inhabiting species. In these more evolved tetranychids, the dorsal setae could play a role in controlling spinning and in regulating the height of webs above the substrate. The problem has been approached by analysis of the anatomic structure of the base of dorsal setae in *Tetranychus urticae* Koch (Mills, 1973) and through the study of variations in the morphology and behaviour of different strains of *Schizotetranychus celarius* (Banks) (Saito and Takahashi, 1980).

LEG LENGTH

In the more advanced Tetranychinae, legs are never longer than the body. Long legs are found in certain Bryobiinae and in less derived Tetranychinae such as Eurytetranychini. In the other tetranychoid families, a moderate leg

Chapter 1.1.5. references, p. 106

length is the rule. This also applies to most raphignatid taxa, although the stilt-legged mites (Camerobiidae Southcott) are the obvious exception.

In the Bryobiinae it is above all those species living on low-growing plants and moving about frequently on the soil which have the longest legs and a correspondingly higher number of verticils of setae on the different segments: *Bryobia cristata* (Duges), *B. repensi* Manson, *B. watersi* Manson, and all members of the genera *Schizonobia* Womersley and *Petrobia* Murray.

In the Eurytetranychini, the phenomenon is noticeable in the females which, when resting, lie flat on the leaf surface and stretch their legs out in front and behind, but it is even clearer in the males, which have long frail legs. In this particular case, the lengthening of the legs would seem to constitute a means of protection from predators, similar to that ensured by the development of dorsal setae.

Without concomitant lengthening of the legs, this defensive attitude persists in certain Tetranychini (*Oligonychus* spp.) living on the upperside of leaves, in the open or protected by a few strands of silk. This behaviour is absent in species living in webs.

DORSAL BODY INTEGUMENT PATTERN

The overall trend with regard to the dorsal body integument pattern is from the wide variety of types, such as present in the Bryobiinae, towards a more or less uniform striation pattern of the web-inhabiting Tetranychini. Different kinds of striation patterns are equally found in the Tenuipalpidae Berlese; a reticulate pattern is predominant in several genera of this family. In the Bryobiinae and in the Tenuipalpoidini, the integument is generally irregularly striated or punctuated, or covered with lumps or with a reticulum, at least on the central part of the prodorsum. Certain Bryobiinae may even show dorsal shields, as in the following genera: *Peltanobia* Meyer, *Notonychus* Davis and *Monoceronychus* McGregor.

The dorsum of 5 genera of Eurytetranychini is of similar appearance to that of the Bryobiinae, but in the genus *Eutetranychus*, the majority of species are entirely covered with regular striations, as are all the members of the genera *Synonychus* Miller, *Eurytetranychus* Oudemans, *Eurytetranychoides* and *Meyernychus*.

In the Tetranychini, the integument is almost always covered with fine parallel striations. The exceptions are the genera *Brevinychus*, *Mixonychus*, *Neotetranychus* and *Acanthonychus*, and also a few species belonging to the genera *Mononychellus* and *Schizotetranychus* Trägårdh: *Mononychellus planki* (McGregor), *M. hyptis* Tuttle et al., *M. waltheria* Tuttle et al., and *Schizotetranychus reticulatus* Baker and Pritchard. In the tribe as a whole, the non-striated pattern remains only in 25 taxa out of 570, that is 4.4% of known species.

SHAPE OF THE PERITREMES

Peritremes of the less derived tetranychid mites end in anastomosing sacs and contain many chambers. They may even protrude, as in *Schizonobia sycophanta* Womersley. As indicated by Pritchard and Baker (1955), the more advanced species of the family have peritremes ending in a simple bulb or in a single hook formed by 4—6 chambers. Anastomosing peritremes are reported to occur in several tenuipalpid species (Pritchard & Baker, 1951),

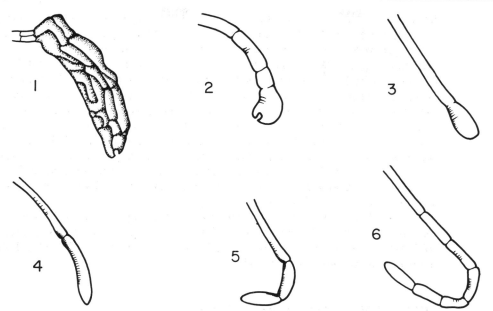

Fig. 1.1.5.4. Evolutionary trends in the shape of the peritremes: 1, *Bryobia praetiosa* Koch; 2, *Petrobia (Tetranychina) harti* (Ewing); 3, *Eutetranychus africanus* (Tucker); 4, *Oligonychus coffeae* (Nietner); 5, *Eotetranychus smithi* Pritchard and Baker; 6, *Tetranychus neocaledonicus* André.

and are equally present in several raphignatoid families. Anastomosing peritremes are considered to represent the ancestral state.

Bryobiinae (Fig. 1.1.5.4; parts 1 and 2)

Peritremes are anastomosed distally except in certain genera in which they end in a simple bulb. The Bryobiini all have anastomosed peritremes except the genus *Bryobiella*, which represents no more than 2 species out of 88 (2.3%). In the Hystrichonychini, the ancient form may be observed in 2 genera only (*Dolichonobia* Meyer and *Paraplonobia* Wainstein *s. str.*), that is in 11 species out of 138 (8%).

For the Petrobiini, peritremes end in a simple bulb in the genera *Edella* and *Petrobia* (Subg. *Mesotetranychus* Reck), that is in 4 species out of 26 (15%).

Tetranychinae (Fig. 1.1.5.4; parts 3,4,5 and 6)

In the Tenuipalpoidini there is as yet only 1 species with simple peritremes (*Crotonella mazatlana* Tuttle et al.), that is 1 species out of 7 (14%), while in the 2 other tribes the proportions are reversed.

In the Eurytetranychini, 59 species out of 60 have simple peritremes (98.3%).

For the Tetranychini, the corresponding figure is 564 out of 570 (98.9%), with the exception of 2 species (*Acanthonychus jianfengensis* Wang and *Tetranychus viennensis* Zacher) and a few strains within 4 species (*Schizotetranychus garmani* Pritchard and Baker, *Mononychellus georgicus* (Reck), *Eotetranychus populi* (Koch) and *Tetranychus savenkoae* Reck). In Tetranychini, amongst the genera *Schizotetranychus*, *Eotetranychus* and *Oligonychus*, in a certain number of species, one may note the appearance of hook-shaped peritremes, made up of several adjacent chambers. This type of peritreme is found in the majority of species in the genus *Tetranychus*.

Chapter 1.1.5. references, p. 106

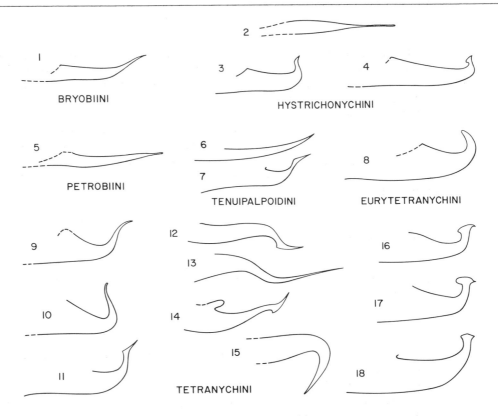

Fig. 1.1.5.5. Evolutionary trends in the shape of the aedeagus: 1, *Bryobia imbricata* Meyer; 2, *Monoceronychus californicus* McGregor; 3, *Porcupinychus insularis* (Gutierrez); 4, *Afronobia januae* Meyer; 5, *Petrobia (Tetranychina) apicalis* (Banks); 6, *Eonychus grewiae* Gutierrez; 7, *Tenuipalpoides dorychaeta* Pritchard and Baker; 8, *Eutetranychus africanus* (Tucker); 9, *Panonychus ulmi* (Koch); 10, *Allonychus braziliensis* (McGregor); 11, *Schizotetranychus schizopus* (Zacher); 12, *Platytetranychus multidigituli* (Ewing); 13, *Eotetranychus pruni* (Oudemans); 14, *Eotetranychus ancora* Baker and Pritchard; 15, *Oligonychus milleri* (McGregor); 16, *Oligonychus pratensis* (Banks); 17, *Tetranychus kanzawai* Kishida; 18, *Tetranychus urticae* Koch.

SHAPE OF THE AEDEAGUS

The aedeagus, as found in the Bryobiinae, is generally of a simple shape. In the Tetranychinae, the aedeagus becomes more complicated and is formed of a shaft and a knob. The shape of the knob, together with the copulation pore, probably becomes a critical key—lock system and an effective mating barrier between species living on the same host plant (Fig. 1.1.5.5). In the other tetranychoid families, the aedeagus is usually a straight, rather simple rod, and this shape represents the ancestral condition. The key—lock system is a derived condition.

Bryobiinae

In the Bryobiini and in the Petrobiini, when males are known, the aedeagus is simply composed of a shaft narrowing to form a slender stylet. This also occurs in the majority of the Hystrichonychini; however, the shaft may be bent dorsad in a sigmoid curve in the genera *Hystrichonychus* McGregor and *Porcupinychus* Anwarullah. One taxon only, *Afronobia januae* Meyer, has a shaft prolonged by a small terminal knob.

Tetranychinae

The same aedeagus, shaped like a straight or curved stylet, is found in the Tenuipalpoidini and in 3 genera of Eurytetranychini: *Atetranychus* Tuttle et al., *Synonychus* and *Anatetranychus*. In the majority of the Eurytetranychini, however, the aedeagus is made of a simple hook, bent upwards and rounded at the tip.

In the Tetranychini, the stylet-shaped aedeagus disappears almost completely, to be replaced by a wide variety of forms. It may be sigmoid with a slender tip pointing up or downwards, but often ends in a knob, as in a few *Eotetranychus* and *Schizotetranychus*, in a number of *Oligonychus* and in almost all *Tetranychus*.

DUPLEX SETAE DISPLACEMENT

These setae usually number 2 sets on tarsus I and 1 set on tarsus II. They are composed of 1 proximal ordinary seta and 1 distal solenidion, but their role is largely unknown.

Bryobiinae

Duplex setae may be 'absent' in the Bryobiini (*Bryobiella*); (actually the setal element on this species is simply so reduced as to be vestigial, while the solenidial element is attenuated), while on tarsus I, one may find 3 sets in the Hystrichonychini (*Parapetrobia* Meyer and Ryke). In the Petrobiini, there may be only 1 set (*Schizonobiella* Beer & Lang), or up to 10 sets (male of *Schizonobia sycophanta*).

Tetranychinae

In the Eurytetranychini, duplex setae are transformed into loosely 'associated setae', except in genus *Atetranychus*, which has only 1 set on tarsi I and II.

In the Tetranychini, duplex setae are generally distal and adjacent on tarsus I. The 2 sets begin to separate in the *Pratensis* group Pritchard and Baker (1955) of *Oligonychus* and are distinctly apart in *Tetranychus*. At the same time, the solenidion of each set tends to become longer. It is possible that the increase in the distance between the sets improves localization of information received.

HABITAT AND SPINNING BEHAVIOUR

The production of silk and webbing has had a revolutionary significance for the evolution of spider mites and many of the great differences between Bryobiinae and Tetranychinae are connected with this particularity. Silk secretion is more abundant in the most advanced species and, at the same time, the mites tend to move from the leaf upperside to live on the leaf underside.

Bryobiinae

Most species live on twigs or on the upperside of leaves, but a few cases of very elaborate adaptation to the structure of the host plant have been

---

Chapter 1.1.5. references, p. 106

observed, where mites make themselves almost imperceptible and reduce the chances of capture by predators to a minimum, e.g. *Bryobia sarothamni* Geijskes, which flattens itself out along the striations of broom twigs (*Cytisus scoparius* (L.) Link).

The only references to the secretion of silk in this group concern the larva of *Petrobia (Tetranychina) apicalis* (Banks) (Smith and Weber, 1954; Zein-Eldin, 1956).

In the Hystrichonychini tribe, several species live on the leaf undersurface, e.g. *Hystrichonychus sidae* Pritchard and Baker, and *H. gracilipes* (Banks) on *Sida hederacea* (Dougl.) Torr. The *Sida* leaves are provided with star hairs, forming a continuous canopy on the undersurface which is comparable with the web structure of the advanced Tetranychinae.

From the Bryobiini to the Petrobiini, the system of protection of eggs from climatic factors as well as from predators, becomes more elaborate. Whereas for the *Bryobia*, winter eggs are spherical, relatively large and protected merely by the choice of their position on the plant (underside of twigs, forks of branches, crevices in trunks), diapausing eggs of several Petrobiini have a wax structure with air chambers, as described for *Petrobia (Petrobia) latens* (Muller) (Lees, 1961), or for *Petrobia apicalis* (Zein-Eldin, 1956). Similar means of protection also occur in the genus *Schizonobia.*

Tetranychinae

The least evolved species live on the upperside of leaves and secrete very little silk, the more advanced live preferably on the leaf underside in dense webs. However, this trend does not apply to spider mites living on parts of plants with grass-like leaves or on needles. During the course of this evolution, the Tetranychinae pass from a solitary to a community life pattern. The silk secretion and the spinning of webs play an essential part in the protection of all stages from predators, but they are also used for dispersal, in mating behaviour, in interspecific relationships and in protection from climatic factors (several references are reviewed by Gerson, 1979; see also Chapter 1.4.1). The shifting of the habitat from leaf upperside to underside offers several additional advantages: it reduces temperature variations and provides good protection against heavy rains, which may sweep off the mites (several references are reviewed by Van de Vrie et al., 1972).

Tenuipalpoidini and Eurytetranychini

These species live preferably on the upper surface of the leaves. A few of them still use the structure of the host plant, e.g. *Tenuipalpoides dorychaeta* Pritchard & Baker, which lives wedged in the bark and along stems of *Gleditschia triacanthos* (Singer, 1966), or *Aponychus grandidieri* (Gutierrez), which stretches itself out along the veins of *Phragmites mauritianus.* The majority of the *Eurytetranychus* and the *Eutetranychus* live on the upper surface of smooth and waxy leaves (*Buxus, Citrus, Ficus, Artocarpus,* etc.), on which they lie flat, but run away when directly threatened.

The females of these 2 tribes spin only a protective network over each egg that tends to flatten against the leaf surface. They spend a considerable amount of time in this operation, which is carried out with great precision.

Tetranychini

In the genus *Panonychus* Yokoyama, immatures feed on the leaf undersurface whereas adults are found on both surfaces. In the genera *Oligonychus* and *Schizotetranychus*, some members live on the upper surface, while others live on the lower surface. Members of the most advanced genera, such as *Eotetranychus* and *Tetranychus*, live exclusively on the underside, where moreover, veins offer fixations for silk.

Webs spun by the species of the genera *Panonychus* and *Oligonychus* are sparse. Those of the genus *Schizotetranychus* are dense and are usually made up of 2 layers, the first designed to cover the eggs and young larvae, the second to protect the adults. It is not uncommon to observe the formation of 'nest' structures as found in *Schizotetranychus schizopus* (Zacher) or *S. celarius*. For the genera *Eotetranychus* and *Tetranychus*, webs are abundant but loose, eggs and resting stages often being suspended above the substrate.

Panonychus ulmi (Koch) lays summer eggs on the underside of the leaves. At the end of the laying time, eggs are protected with an outer wax layer, which is drawn out into a spike at the top, and the females spin guy ropes from the spike to the substrate (Beament, 1951). The eggs of *Panonychus citri* (McGregor) are of similar appearance and are protected in the same way. Numerous species in the genera *Schizotetranychus*, *Eotetranychus* and *Oligonychus*, also have eggs with a thick chorion, flattened against the substrate and ending in an apical spike and which are covered with individual or collective webs. Finally in some *Eotetranychus* and all *Tetranychus*, eggs have a thinner chorion, are spherical, relatively small and suspended in webs, as soon as the community reaches a certain density.

The webs of *Tetranychus* represent a handicap for the movement of predators and those other species which secrete little or no silk (e.g. *Bryobia* spp.). Males of *Tetranychus urticae* are likely to attack, web down and kill females of other species, e.g. those of *Panonychus ulmi* (Lee, 1969).

POPULATION DYNAMICS

The potential of an increase in a population can be estimated in terms of r_m, the intrinsic rate of natural increase as defined by Birch (1948). This parameter, which is treated in detail in Chapters 1.2.6 and 1.4.5, is large when the duration of the generation is short and fertility high. For the Tetranychidae, its value is dependent on the breeding conditions; basically these are: the nature of the host plant, the surface available to each individual, temperature and humidity. It is difficult to make comparisons between the various results shown in the literature, as breeding techniques vary from author to author. Moreover, depending on their geographical origin, each species has a maximal value of r_m for a particular combination of temperature and humidity.

Laboratory studies on the *Bryobia* are few, but it is known that for *Bryobia rubrioculus* (Scheuten), the incubation period for eggs at 25 °C is 10 days. At 20 °C, fecundity averages 20 eggs (Mathys, 1957). As 10 days represent the total developmental period for the majority of *Tetranychus* species at 25 °C, this information suggests that the r_m of *B. rubrioculus* is low in comparison with that of the other species. Table 1 in Chapter 1.4.5 shows r_m values of several species from different genera.

The highest r_m is found in the Tetranychini; the species of the genus *Tetranychus* are the ones which appear to be the most prolific. At 5 days old, females lay up to 10 eggs a day, which phenomenon is probably linked to the reduction in the size of the egg in relation to that of the female, and to the simplification of the laying process. An increase in r_m leads to an increase in the number of annual generations.

Tetranychus, which infests annuals above all, succeeds in destroying its

Chapter 1.1.5. references, p. 106

host plant. This strategy implies the existence of reliable and effective means of dispersal, which are lacking in most of the Bryobiinae. As quoted by Saito (1979), the higher r_m of the *Tetranychus* species is the result of a successful adaptation to an originally unstable habitat. On the other hand, species living on perennial plants have a more stable habitat and a lower r_m.

Saito and Ueno (1979), who have compared the r_m values of 2 species with a stable habitat (*Aponychus corpuzae* Rimando and *Schizotetranychus celarius*), consider that the former species has a high rate of natural increase in order to compensate for its vulnerability to predators, while the second has a more sophisticated system of protection.

CYTOGENETICS

Table 1 of Chapter 1.2.3 gives a list of all karyotype examinations and chromosome numbers established for Tetranychidae. This list involves 17 Bryobiinae (6.7% of known species) and 109 Tetranychinae (17.1% of known species). It appears that there is considerable variation in chromosome number, varying from $n = 2$ to $n = 7$. The most common number for the Bryobiinae is $n = 4$; the Tetranychinae has a common number of $n = 3$. There is only one number, ($n = 2$), which is present in all 6 tribes, i.e. in the Bryobiini (1 species), Hystrichonychini (1), Petrobiini (1), Eurytetranychini (3), Tenuipalpoidini (2) and Tetranychini (18). For a comparison with related groups, the Tenuipalpidae are of some interest, since several species of this tetranychoid family have been karyotyped. In this family, little variation in chromosome numbers was found to occur; the haploid numbers are $n = 2$ and $n = 3$. An apparent phylogenetic trend in the Tenuipalpidae is the reduction in the number of palpal segments: a number of 5 represents the ancestral state, while the lower numbers of segments are derived (cf. Pritchard and Baker, 1951). It appears that the species with 5 and 4 palpal segments have $n = 2$ chromosomes (Bolland & Helle, 1981). This gives additional support to the supposition that $n = 2$ is the ancestral number of the Tetranychoidea.

The number $n = 2$ is rather frequently found in many other Actinedida (Helle et al., 1984), and is also regularly found in gall mites (Helle and Wysoki, 1983). The significance of a particular chromosome number within the range $n = 2$ to $n = 7$ is not understood and remains obscure. As is apparent from the list in Chapter 1.2.3, the number $n = 3$ becomes widespread in the large tetranychine genera. It should be noted that 10 *Oligonychus* out of 32 (31%) have $n = 3$, while this ratio is 9 out of 25 (36%) for *Eotetranychus* and 18 out of 24 (75%) for *Tetranychus*.

The *Oligonychus* species living on monocotyledons (*Pratensis* group) have the same chromosome number as the *Tetranychus* living on the same plants (*Tetranychus panici* Gutierrez, *T. roseus* Gutierrez, *T. tchadi* Gutierrez & Bolland) and these 2 groups present very clear morphological similarities (Gutierrez et al., 1979). It has been postulated (Gutierrez et al., 1970) that polyploidy may have contributed to speciation in spider mites, particularly in *Schizotetranychus* and *Tetranychus*, in which some species have $n = 6$, while the common number is $n = 3$, but Helle et al. (1983) demonstrated that the DNA content of sperm of *Tetranychus tumidus* Banks ($n = 6$) is the same as that of *T. urticae* ($n = 3$). The number of 6 chromosomes in this case would result from fragmentation of chromosomes rather than from polyploidization.

OTHER BIOLOGICAL FEATURES

Thelytokous parthenogenesis

In the Bryobiinae different genera contain both arrhenotokous and thelytokous species. Thelytokous parthenogenesis has been demonstrated by breeding and by karyotype determination in 5 taxa (Helle et al., 1981). The number of species with this mode of reproduction is certainly higher, since males are unknown in nearly 40% of taxa described from specimens collected from the natural environment.

In the Tetranychinae, however, arrhenotokous parthenogenesis is the rule, except for *Oligonychus thelytokus* Gutierrez (1977). Four bisexual species may also have thelytokous strains: *Eurytetranychus buxi* (Garman) (Ries, 1935), *Oligonychus ilicis* (McGregor) (Flechtmann and Flechtmann, 1982), *Tetranychus pacificus* McGregor (Helle and Bolland, 1967) and *T. urticae* (Boudreaux, 1963) (see also Chapter 1.2.3).

Host plants

Most species live on only a few taxonomically close plant species; a few are monophagous, others are extremely polyphagous. The Tetranychidae as a whole infest a wide range of plants belonging to many different botanical groups.

Nine species of Bryobiinae out of 252 (3.6%) were collected from plants considered primitive: Pteridophytes (*Selaginella* sp.), Gymnosperms and Chlamydosperms (*Ephedra* spp.). They belong to 3 genera of Bryobiini (*Bryobia, Pseudobryobia, Hemibryobia* Tuttle and Baker), 1 genus of Hystrichonychini (*Monoceronychus*) and 1 genus of Petrobiini (*Petrobia* subg. *Mesotetranychus*).

In the Tetranychinae, 24 species out of 636 (3.8%) were collected from Gymnosperms. They belong to 2 genera of Eurytetranychini (*Eurytetranychus* and *Eurytetranychoides*), and to 3 genera of Tetranychini. Out of the 20 Tetranychini living on Gymnosperms there were: 15 *Oligonychus* out of 165 (9%), 4 *Platytetranychus* Oudemans out of 8 (50%) and a single *Tetranychus* (*T. ezoensis* Ehara) out of 110 (1%). On the other hand, the *Schizotetranychus* almost all live on monocotyledons, considered as a derived group.

The host plant factor is difficult to interpret. In *Monoceronychus*, for instance, most species live on grasses, but 2 species are found living on *Pinus* as well as on grass. For the *Oligonychus*, all species living on Gymnosperms belong to the group *Ununguis* Pritchard and Baker, while all species living on grasses belong to the group *Pratensis*.

Polyphagy is a tendency which may also be observed in species living on herbaceous plants, being an unstable habitat. It occurs in several Bryobiinae: *Bryobia praetiosa* Koch, *Petrobia (Petrobia) latens*, *Petrobia (Tetranychina) apicalis*, but appears to reach its highest level in the Tetranychinae, in the genus *Tetranychus*.

Species living on a single herbaceous plant often show a remarkable adaptation to the annual rhythm of development of the host plant, as in *Petrobia (Tetranychina) harti* (Ewing) for instance, which spends the dry season in the egg form on the aestivating bulbs of *Oxalis* spp.

Diapause

An elaborate account of references and observations dealing with the occurrence of diapause in different tetranychid species is presented in

Chapter 1.1.5. references, p. 106

Table 1 of Chapter 1.4.6. Diapause appears to be regular with at least 40 species of Tetranychidae living in climates with an adverse season, and both aestivation and hibernation diapause have been reported for Bryobiinae, as well as for Tetranychinae. The ability to diapause is already present in Bryobiini, and the evolutionary roots of this ability may reach the earliest terrestrial arthropods.

An aspect of diapause considered here is the stage which is alloted for the diapause. In Tetranychidae, this can be either the egg or the adult female. Diapausing eggs are reported for 8 species of the Bryobiinae (genera *Bryobia*, *Aplonobia* Womersley, *Schizonobia* and *Petrobia*); the special structures of the diapausing egg of the Bryobiinae have been mentioned in Chapter 1.4.6. Diapause affecting the adult stage has not been reported thus far to occur in any bryobiine species: all reports on diapausing females concern exclusively the Tetranychinae. In this group, adult diapause is found in species of the genera *Platytetranychus* (1 species), *Neotetranychus* (1 species), *Eotetranychus* (11 species) and *Tetranychus* (10 species). For the genera *Eurytetranychus*, *Panonychus*, *Schizotetranychus*, all reports deal with egg diapause, which is also common in *Oligonychus* (4 species). It is noteworthy that the grass-inhabiting species *O. pratensis* (Banks) diapauses as an adult, which is an additional argument for the special position of the *Pratensis* group in the genus *Oligonychus*. Diapausing females usually hibernate in groups in sheltered places, such as crevices in the trunk or in the bark, obturated by a web. Possibly, they are less vulnerable to predation, compared with diapausing eggs. Another advantage of the female diapause as compared to the egg diapause may be that after reactivation the adult females have better chances of survival and a broader dispersal area, compared with the larvae hatching from the diapausing eggs.

Considering the facts known for the 2 sub-families, it would seem plausible to advocate that the egg diapause is the more ancestral, and adult diapause the derived state. However, out-group information does not support this view. The Tenuipalpidae, although mainly tropical in distribution, have some representatives in the temperate zones, belonging to the genera *Brevipalpus* Donnadieu and *Cenopalpus* Pritchard and Baker. The winter is passed in shelters in the barks of trees by the adult females, and it is conceivable that a normal diapause occurs.

CONCLUSIONS

The characteristics used for the classification of the Tetranychidae, basically the morphology of the ambulacrum and dorsal chaetotaxy, appear to be of good phylogenetic value. Leg chaetotaxy and its ontogeny offers additional potential. From the preceding paragraphs, several evolutionary changes have become apparent. The data as a whole tend to show that the Bryobiinae have in general retained more primitive characteristics than the Tetranychinae. For the Bryobiinae, one may conclude that the least evolved tribe would be that of the Bryobiini, the most evolved that of the Petrobiini. Two events are essential with respect to the evolutionary changes within the Tetranychidae. The first event is the production of silk. The second, and even more important, one is that the silk becomes 'recognized' as an element for the construction of webbing and that, using this webbing a micro-habitat can be created in which the mite colony can reside and flourish in a comfortable way.

To evaluate the changes adequately, a short comparison of the main characteristics is given between a 'modal' bryobiine and tetranychine. The bryobiine representative, then, is a creature rather poorly adapted to the life on the host plant. It is obliged to reside on the upper surface of the leaf, and it is fully exposed to the climatic uncertainties. It is a rather easy prey for various kinds of predators, and it must arm itself, or must try to avoid or escape from predators. Also, the eggs need special protection, and have to be hidden carefully in shelters present in petioles or branches. The bryobiine colony is made up of individuals with loose social relations. With regard to the host plant, the bryobiine mite cannot allow itself to exhaust its host plant, since the possibilities for dispersal are limited, and extinction is a continuous threat.

The advanced tetranychine (Fig. 1.1.5.6) has solved most of those problems satisfactorily. As a consequence of life-lines and webbings, it resides on the underside of the leaf and this side is fully explored. The webbing forms a micro-habitat which enables reproductive and feeding activities to occur even under adverse climatic conditions. The colonies under the webbing have developed a kind of social behaviour, and this trend results in certain taxa which constitute communities of varying degrees of organization. This, and the presence of the webbing, gives protection to many

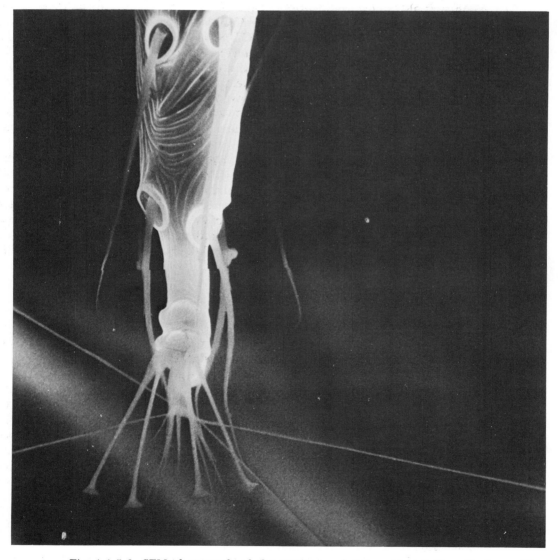

Fig. 1.1.5.6. SEM photograph of the pretarsus of *Tetranychus urticae* Koch on silken strands (Courtesy of M. Sabelis and F. Thiel, TFDL, Wageningen, Holland.)

Chapter 1.1.5. references, p. 106

predator species. With the production of silken ropes, the possibilities for dispersal have greatly improved. The R_0 can reach high values, since host plant destruction is no longer followed by extinction.

With regard to the evolution of the capability for silk production, there seems to exist a transitional step via the Petrobiini between the Bryobiinae and Tetranychinae. In other words, it seems that the Tetranychinae has been evolved out of the Petrobiini. There are 2 arguments for this. The first is that the evolution of the tarsal appendages in the Petrobiini has arrived at a point which is very close to that of the Tetranychinae: the ambulacral claws have been reduced to a compound pair, and the empodium is already transformed into a claw which is still provided with rows or tufts of amalgamated tenent hairs. The second argument for the transitional position of the Petrobiini is the nature of the spinning abilities of larvae of the legume spider mite, *Petrobia apicalis*. In Chapter 1.1.3, the monophyletic states of both the Bryobiinae and the Tetranychinae have been discussed and synapomorphies have been presented. The conclusion that Bryobiinae and Tetranychinae are sister groups leaves us with the inevitable assumption that a striking parallel evolution of the ambulacral appendages occurred in both Bryobiinae and Tetranychinae, and that spinning capabilities also developed independently in both Bryobiinae and Tetranychinae. This seems rather improbable. A further discussion of this matter is prevented by the fact that more detailed knowledge about the source of the silk in *Petrobia apicalis* is lacking: it is not known whether or not eupathidial spinneret is involved. Undoubtedly, this essential information will be obtained in the near future.

Finally, short comments are given on the lineages in the Tetranychini. Within the Tetranychini, 2 lineages may be distinguished, one leading to *Eotetranychus*, the other to *Tetranychus*. In the latter lineage, it is conceivable that *Oligonychus* is less derived than *Tetranychus*. In that of *Eotetranychus*, on the other hand, the situation is more complex and it is difficult to put forward a hypothesis about the evolution of the genera.

In the future, the study of the evolution of the various features may lead to a split in the heterogenous genera represented by a large number of species (*Eotetranychus*, *Oligonychus* and *Tetranychus*) into several more natural units. Proposals for division based on the striation of the dorsal or caudoventral integument appear to be of little or no phylogenetic value, and this is true for the division of *Eotetranychus* into groups proposed by Pritchard and Baker (1955), as well as for Tuttle and Baker's proposal (1968) for the division of *Oligonychus* and *Tetranychus* into sub-genera. On the other hand, Pritchard and Baker's proposal (1955) that the genera *Oligonychus* and *Tetranychus* be divided into groups based on the shape of their empodium and aedeagus, while bearing in mind the nature of host plants, appears a good deal more realistic (Gutierrez et al., 1979). This last division would need to be revised but at present certain information indispensable to the completion of such a task is still lacking.

REFERENCES

Beament, J.W.L., 1951. The structure and formation of the egg of the fruit tree red spider mite, *Metatetranychus ulmi* Koch. Ann. Appl. Biol., 38: 1—24.

Birch, L.C., 1948. The intrinsic rate of natural increase of an insect population. J. Anim. Ecol., 17: 15—26.

Bolland, H.R. and Helle, W., 1981. A survey of chromosome complements in the Tenui-palpidae. Int. J. Acarol., 7: 157—160.

Boudreaux, H.B., 1963. Biological aspects of some phytophagous mites. Ann. Rev. Entomol., 8: 137—154.

English, L.L. and Snetsinger, R., 1957. The biology and control of *Eotetranychus multi-digituli* (Ewing), a spider mite of honey locust. J. Econ. Entomol., 50: 784—788.

Flechtmann, N.N.B. and Flechtmann, W.H.C., 1982. Observações sobre a reprodução do ácaro vermelho do cafeeiro. II Congresso Brasileiro de iniciação cientifica em ciências agrárias, 8—10 setembro. ESALQ-USP, Piracicaba S.P., pp. 109—111.

Gerson, U., 1979. Silk production in *Tetranychus* (Acari: Tetranychidae). In: J.G. Rodriguez (Editor), Recent Advances in Acarology, Vol. 1. Academic Press, New York, NY, pp. 177—188.

Gutierrez, J., 1977. Un acarien polyphage de la zone intertropicale: *Oligonychus thelytokus* n.sp. (Acariens, Tetranychidae). Description et premières données biologiques. Cah. ORSTOM, Sér. Biol., 12: 65—72.

Gutierrez, J., Bolland, H.R. and Helle, W., 1979. Karyotypes of the Tetranychidae and the significance for taxonomy. In: J.G. Rodriguez (Editor), Recent Advances in Acarology. Vol. 2. Academic Press, New York, NY, pp. 399—404.

Gutierrez, J., Helle, W. and Bolland, H.R., 1970. Etude cytogénétique et réflexions phylogénétiques sur la famille des Tetranychidae Donnadieu. Acarologia, 12: 732—751.

Helle, W. and Bolland, H.R., 1967. Karyotypes and sex determination in spider mites (Tetranychidae). Genetica, 38: 43—53.

Helle, W., Bolland, H.R. and Heitmans, W.R.B., 1981. A survey of chromosome complements in the Tetranychidae. Int. J. Acarol., 7: 147—155.

Helle, W., Bolland, M.R., Jeurissen, S.H.M. and van Seventer, G.A., 1984. Chromosome Data of Aetinedida, Tarsonemida and Oribatida. Proc. VIth Int. Congr. Acarol., Edinburgh, 1982. pp. 449—454.

Helle, W., Tempelaar, M.J. and Drenth-Diephuis, L.J., 1983. DNA contents in spider mites (Tetranychidae) in relation to karyotype evolution. Int. J. Acarol., 9: 127—129.

Helle, W. and Wysoki, M.F., 1983. The chromosomes and sex determination of some actinotrichid taxa (Acari), with special reference to Eriophyidae. Int. J. Acarol., 9: 67—71.

Lee, B., 1969. Cannibalism and predation by adult males of the two-spotted mite *Tetranychus urticae* (Koch) (Acarina: Tetranychidae). J. Aust. Entomol. Soc., 8: 1—210.

Lees, A.D., 1961. On the structure of the egg shell in the mite *Petrobia latens* Muller (Acarina: Tetranychidae). J. Insect Physiol., 6: 146—151.

Manson, D.C.M., 1967. The spider mite family Tetranychidae in New Zealand I. The genus *Bryobia*. Acarologia, 9: 76—123.

Mathys, G., 1957. Contribution à la connaissance de la systématique et de la biologie du genre *Bryobia* en Suisse romande. Bull. Soc. Entomol. Suisse, 30: 189—284.

Mills, L.R., 1973. Structure of dorsal setae in the two-spotted spider mite *Tetranychus urticae* Koch, 1836. Acarologia, 15: 649—658.

Pritchard, A.E. and Baker, E.W., 1951. The false spider mites of California (Acarina: Phytoptipalpidae). Univ. Calif. Berkeley, Publ. Entomol. 9: 1—94.

Pritchard, A.E. and Baker, E.W., 1955. A revision of the spider mite family Tetranychidae. Pac. Coast Entomol. Soc., Mem. Ser., 2: 1—472.

Ries, D.T., 1935. A new mite (*Neotetranychus buxi* n.s. Garman) on boxwood. J. Econ. Entomol., 28: 55—62.

Saito, Y., 1979. Comparative studies on life histories of three species of spider mites (Acarina: Tetranychidae). Appl. Entomol. Zool., 14: 83—94.

Saito, Y. and Takahashi, K., 1980. Study on variation of *Schizotetranychus celarius* (Banks). I. Preliminary descriptions of morphological and life type variation Jap. J. Appl. Entomol. Zool., 24: 62—70 (in Japanese).

Saito, Y. and Ueno, J., 1979. Life history studies on *Schizotetranychus celarius* (Banks) and *Aponychus corpuzae* Rimando as compared with other Tetranychid mite species (Acarina: Tetranychidae). Appl. Entomol. Zool., 14: 445—452.

Singer, G., 1966. The bionomics of *Tenuipalpoides dorychaeta* Pritchard and Baker (1955) (Acarina, Trombidiformes, Tetranychidae). Univ. Kans. Sci. Bull., 46: 625—645.

Smith, C.E. and Weber, J.C., 1954. The legume mite *Petrobia (Tetranychina) apicalis* (Banks), a pest on several winter growing legumes. La. State Univ., Agric. Exp. Stn. Tech. Bull., 493: 1—25.

Trägårdh, I., 1915. Morphologische und Systematische Untersuchungen über die Spinmilben, *Tetranychus* Dufour. Z. Angew. Entomol., 2: 158—163.

Tuttle, D.M. and Baker, E.W., 1968. Spider Mites of Southwestern United States and a Revision of the Family Tetranychidae. The University of Arizona Press, Tucson, AZ, pp. 1—143.

Van de Vrie, M., McMurtry, J.A. and Huffaker, C.B., 1972. Ecology of Tetranychid mites and their natural enemies: a review. III. Biology, ecology and pest status, and host plant relations of Tetranychids. Hilgardia, 41: 343—432.

Zein-Eldin, E.A., 1956. Studies on the legume mite, *Petrobia apicalis*. J. Econ. Entomol., 49: 291—296.

Chapter 1.2 Reproduction and Development

1.2.1 Spermatogenesis

L.P. PIJNACKER

INTERNAL GENITALIA

The first researches on the male internal genitalia of *Tetranychus urticae* Koch date back over 100 years. Claparède (1868) could distinguish a pair of testes, a vesicula seminalis and a penis in the caudal part of whole males. Detailed histo-cytological information was given in the classic paper of Blauvelt (1945) and later by Pijnacker and Drenth-Diephuis (1973), Alberti and Storch (1976) and Mothes and Seitz (1981). As for the organs, the terminology used by these authors varies (Table 1.2.1.1). The following description includes a proposal for a uniform nomenclature: the internal genitalia of an adult male consist of a pair of testes, a pair of vesiculae seminalis, a pair of vasa deferentia, a single sperm pump, a single ductus ejaculatorius, and a single penis (Fig. 1.2.1.1). These organs are found in the ventral part of the body between the central nerve mass and the anus (Chapter 1.1.2). It may be stated that they are rather similar to the organs of the only other tetranychid species investigated thus far (Ehara, 1960), namely *Bryobia eharai* Pritchard and Keifer.

TABLE 1.2.1.1

Nomenclature of male reproductive organs of *Tetranychus urticae*

Claparède (1868)	Blauvelt (1945)	Pijnacker and Drenth-Diephuis (1973)	Alberti and Storch (1976)	Mothes and Seitz (1981)	Proposal	
Hoden	Testis (2 lobes)	Testis	Hoden (Keimteil)	Testis (germ region)	Testis	Paired
		Vesicula seminalis	Hoden (Drüsenteil)	Testis (glandular region)	Vesicula seminalis (seminal vesicle)	
		Short duct	Vas deferens	Vas deferens	Vas deferens	
Samen-blase	Seminal vesicle	Sperm pump	Vesicula seminalis	Seminal vesicle (sperm pump)	Sperm pump	Single
	Ejaculatory duct	Ductus ejaculatorius	Ductus ejaculatorius	Ductus ejaculatorius	Ductus ejaculatorius (ejaculatory duct)	
Ruthe	Penis	Aedeagus	Penis	Penis	Penis	

Chapter 1.2.1. references, p. 115

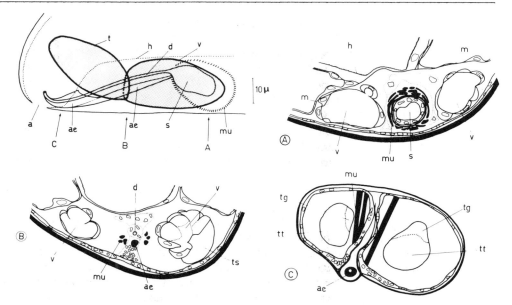

Fig. 1.2.1.1. Genital region of an adult of *T. urticae*. Diagram of a median-sagittal section with projection of right seminal vesicle and right half of the testis. A—C, transverse sections at points indicated by arrows in the diagram. Abbreviations: a = anus; aa = anlage penis; ad = anlage ejaculatory duct; ae = penis; am = anlage muscles; d = ejaculatory duct; h = hindgut; i = invagination; m = midgut; mu = muscles; s = sperm pump; t = testis; tg = part of the testis with spermatogonia; ts = part of the testis with spermatozoa; tt = part of the testis with spermatids; v = seminal vesicle. (Reprinted with the permission of E.J. Brill, Leiden.)

Each testis is more or less sack-shaped and filled with germ cells only. The wall is a thin layer of cells, the nuclei of which contain a group of characteristic tubular structures as in the epithelium of the ovary (Chapter 1.2.2). The testis opens at its anterior end in an ovoid vesicula seminalis. Its wall is a single layer of cells with microvilli and is provided with small muscle fibres, also in the transition region with the testis. The nuclei of these cells also exhibit tubules. The cells seem to be able to envelop sperm and, as a consequence, divide the vesicula into compartments. Sperm is embedded in some matrix, indicating a glandular function of the cells. When low numbers of sperm are present, the vesicula is lobate owing to the peculiar shape of the cells. The vesicula seminalis has also been considered as a part of the testis. However, it develops separately from the testis (cf. next section) and becomes filled with mature sperm from the testis. This organ thus may be a dilation of a vas deferens and as such, the term vesicula seminalis is more correct. The short connection running through muscular tissue between each vesicula and the anterio-lateral parts of the sperm pump has remained as the duct of each vas deferens anlage.

The sperm pump is pear-shaped. The lumen is paved with an epithelium on a thick membrane, which is connected to a complex outer layer of muscle cells. Both layers of cells are thinner in the posterior direction. The sperm pump has been named the vesicula seminalis. Since it develops separately from the vesiculae (cf. next section), stores only few sperm (cf. final section), and has one-sided well-developed muscles, the term sperm pump is appropriate. The dorso-posterior end of the lumen of the sperm pump opens on the narrow ductus ejaculatorius. The latter is a somewhat curved, elastic tube which is surrounded, together with the penis, by high cells. It continues as the penial canal in the posterior one-third of the penis, where it ends in a hooked tip. This part (shaft) is heavily sclerotized and is continuous with the integument, so it can be extruded during copulation. The anterior two-thirds (stem), also hardened and round to oval in cross-section, has been provided with muscles, apparently for copulation, and is connected to the sperm pump underneath the ejaculatory duct. In mite literature the term aedeagus instead of penis is met with rather frequently. The ontogeny of an insect aedeagus is quite different from that of the intromittent organ of mites (cf. next section) and therefore the term penis is less ambiguous.

The internal genitalia are equipped to transfer the non-flagellate sperm (cf. final section) to the receptaculum seminis of the female during copulation. They may act as follows. Sperm very likely enter the vesiculae seminalis by pressure of the proliferating germ cells, since the testes do not have muscles. Turning back is not possible because the opening can be closed by muscles. The muscles in the wall of the vesiculae seminalis are needed to contract the vesiculae and, consequently, to press sperm through the vasa deferentia into the sperm pump. The latter also occurs during insemination (cf. final section). Contraction of the muscles of the sperm pump presses sperm caudad into the ductus ejaculatorius and penial canal. The latter straightens when the anterior part of the penis is extruded with the aid of the attached muscles during copulation. Pressure must be exercised by the sperm pump because the diameter of a spermatozoon is wider than that of the ductus and of the penial canal. The latter diameters allow the sperm to leave the male only one after another. Interruption of a copulation consequently results in a transfer of a lower number of sperm (Boudreaux, 1963; Overmeer, 1972). How far the matrix plasm in which the sperm are embedded in vesiculae seminalis and sperm pump has a function, for instance in sperm transport, is unknown.

DEVELOPMENT OF INTERNAL GENITALIA

The development of the internal genitalia of the post-embryonic life stages of *Tetranychus urticae* goes on rather constantly, also during the chrysalid stages, though variation exists between individuals (Figs. 1.2.1.1—1.2.1.3;

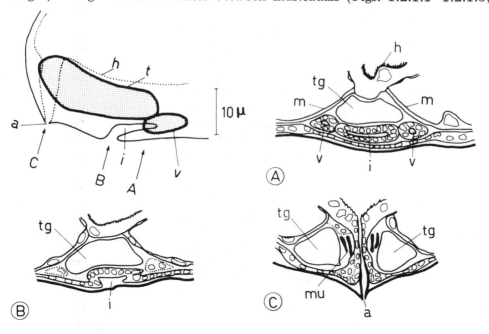

Fig. 1.2.1.2. Genital region of a larva of *T. urticae*. For legend, see Fig. 1.2.1.1. (Reprinted with the permission of E.J. Brill, Leiden.)

Chapter 1.2.1. references, p. 115

Pijnacker and Drenth-Diephuis, 1973). Immediately after hatching, the larval testis appears to be single (volume about $3300\,\mu m^3$). The posterior half, however, may be split along the median line through the hindgut. With proceeding growth, division of the testis into 2 halves becomes more and more prominent, and this is generally established after the deutonymphal stage. Both halves remain enclosed by a thin epithelium. Each half attains its maximum volume (at least $10\,000\,\mu m^3$) in deutonymphs and teleiochrysalids. In adults its size varies considerably.

In newly emerged larvae a flat epidermal invagination exists underneath the centre of the testis, just anterior to the anus. This invagination grows cephalad. Against each lateral side of the anterior part of the invagination a group of cells develop into a vesicle. The vesicles open to the cavity of the invagination at the end of the larval instar and make contact caudad with the split anterior margin of the testis during the protonymphal stage. Each vesicle expands through cell multiplication until the teleiochrysalis stage and through swelling afterwards. These vesicles, called the vesiculae seminalis, already become filled with sperm in deutonymphs. The cells take on their peculiar shape as described above in early adulthood, when the vesicles also attain their maximum size (about $10\,000\,\mu m^3$ each). Each vesicle remains larger than a testis half.

During the protonymphal and deutochrysalis stages the cells of the distal

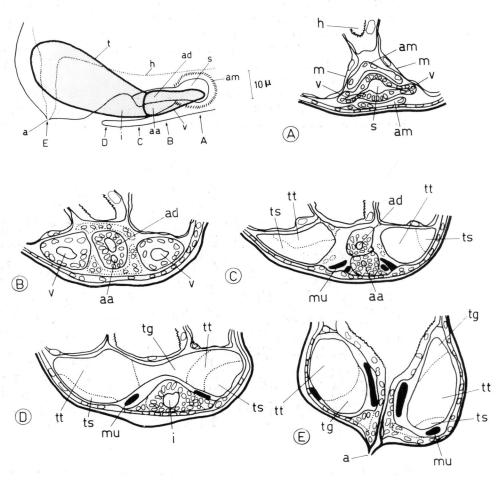

Fig. 1.2.1.3. Genital region of a deutonymph of *T. urticae*. For legend, see Fig. 1.2.1.1. (Reprinted with the permission of E.J. Brill, Leiden.)

third part of the invagination take the shape of a sphere. This becomes sur-rounded by an additional layer of cells, originating from cells lying on top of the invagination since hatching of the larva. This sphere develops into the sperm pump in deutonymphs and teleiochrysalids. The cells of the invagination form the epithelial inner layer, the additional cells differentiate into the muscle layer. Sperm enter the lumen as early as the teleiochrysalis stage through the paired short ducts, called vasa deferentia, which differen-tiated from the lateral connections between vesicles and invagination. Together with development of the sperm pump the remainder of the invagination differentiates into a dorsal and a ventral duct, which lengthen by cell multiplication. The dorsal duct remains continuous with the lumen of the sperm pump, the ventral duct connects with the wall of the sperm pump beneath the dorsal duct. The entrance of the invagination expands till it is about twice as high and then its anterior wall connects with the dorsal duct. At the end of deutonymphal stage and in teleiochrysalids this dorsal duct forms the elastic ductus ejaculatorius and the ventral duct the sclerotized penis. Sperm may be found in the ductus and the penial canal as early as the teleiochrysalis stage.

SPERMATOGENESIS

Male haploidy of tetranychid mites (Chapter 1.3.2) did imply that spermatogenesis might evolve at the haploid level. The spermatogenesis of *T. urticae* indeed shows only haploid equational divisions (Fig. 1.2.1.4; Schrader, 1923; Pijnacker and Drenth-Diephuis, 1973). A possible diploid meiosis, due to an additional chromosome reduplication before the onset of meiosis, does not occur.

The testis of a newly hatched larva contains approximately 150 closely packed and irregularly shaped spermatogonia. The gonia multiply by asyn-chronous mitoses and then show the haploid number of (3) chromosomes

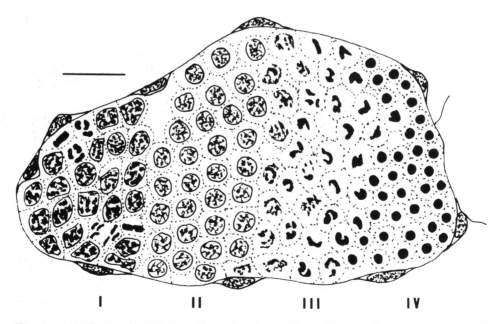

I   II   III   IV

Fig. 1.2.1.4. Testis of adult *T. urticae*. Frontal section with zone I: spermatogonia with metaphases and anaphase; zone II: spermatids; zone III: spermiogenesis; zone IV: spermatozoa. Bar = 5 µm.

Chapter 1.2.1. references, p. 115

(Chapter 1.2.3). Though clusters of mitoses may occur, cysts are not formed. During the nymphochrysalis stage, spermatogonia in the anterior and ventrolateral parts of the testis differentiate into more or less spherical cells through which the packing becomes regular. The chromatin threads become thinner and more clear-cut than in spermatogonia, nucleoli apparently disappear. These cells do not multiply and change into spermatozoa and consequently may be considered spermatids. Any sign of meiosis is absent, which may also be inferred from electron microscopic investigations (Alberti and Storch, 1976; Mothes and Seitz, 1981). The spermatids are interconnected by plasm bridges in groups of 4 or 8 cells. This syncytional development apparently arises when the spermatogonia divide in clusters without cytokinesis taking place and is rather similar to that occurring in the females (Chapter 1.2.2).

Spermatids demonstrating spermiogenesis appear in deutochrysalids, again in the anterior and ventrolateral parts of the testis. Spermiogenesis does not take place synchronously, though clusters may be in the same phase. The chromatin spiralizes and chromosomal structures become visible. The structures form a line and then fuse into a compact sphere. Simultaneously the cytoplasm changes into 2 equally thick concentric hyaline layers, of which the peripheral one is somewhat more transparent than the inner layer. The ripe spermatozoon is thus spherical or, due to overcrowding, somewhat ovoid. In the ductus ejaculatorius and penial canal, however, it is elongated.

Spermatogonia, and their mitoses, are found in all the post-embryonic stages. Their number is maximal (500 cells) in protonymphs, deutochrysalids and deutonymphs; in deutonymphs the gonia occupy the caudal quarter of each testis. Spermatids are present in the median region of the testis from the beginning of deutochrysalis stage. Their number is maximal in deutonymphs and teleiochrysalids; afterwards they generally fill at least half of the testis. The number of sperm is maximal in teleiochrysalids and they then occupy the anterior quarter of the testis. These conditions of testis and germ cells must be taken into consideration when the reproductive material is treated by mutagens.

The morphology of a spermatozoon needs further attention since the spherical shape and absence of organelles are not characteristic for most

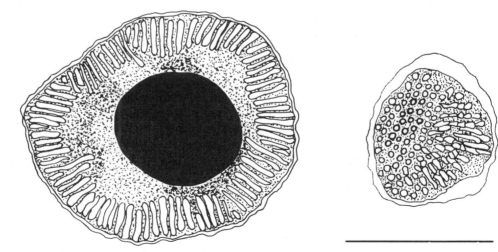

Fig. 1.2.1.5. Mature spermatozoon from testis of *T. urticae*. Drawing of ultra-thin section showing the spherical nucleus surrounded by granular cytoplasm, the layer with infoldings, and the 2 membranes. Right-hand side: section through vesicles. Bar = 1 μm.

Acari (Fig. 1.2.1.5; Pijnacker and Drenth-Diephuis, 1973; Alberti and Storch, 1976; Mothes and Seitz, 1981). The onset of spermiogenesis is characterized by the development of peripheral infoldings of the cell membrane, resulting in a layer of radial tubular infoldings. During this development, Golgi apparatus, multivesicular bodies, mitochondria and ribosomes disappear from the cytoplasm. Moreover, the chromatin condenses, the nuclear membrane dissolves, and the interconnecting plasm bridges are closed. The mature sperm ultimately consists of a homogeneously electron-dense sphere of chromatin, surrounded by a dark granular layer of cytoplasm and a layer with tubular membrane infoldings, together enclosed by an envelope consisting of 2 membranes; an acrosome complex, cyto-plasmic organelles and flagellum are thus absent. According to Alberti and Storch (1976) and Mothes and Seitz (1981) the outer membrane is of somatic origin, which remains behind when the sperm enters the vesicula seminalis (see also Chapter 1.1.2). The structure of the sperm hardly changes in the vesiculae seminalis and sperm pump; the infoldings are not always exactly radial and may become vesicle like. Mothes and Seitz (1981) found, however, that sperm become amoeboid, demonstrating pseudopodia, at the end of their stay in the vesicles.

The number of sperm found in each vesicula seminalis and the sperm pump of adults vary considerably (Pijnacker and Drenth-Diephuis, 1973). In one-day old virgin males the numbers (mean ± standard deviation, $n = 10$) are: vesicula left 643 ± 591, vesicula right 653 ± 481, sperm pump 70 ± 50; total 1366 ± 1049. In eight-day old males from the stock population, i.e. males which very likely had copulated, the numbers are respectively: 1063 ± 321, 1027 ± 349, 151 ± 49, 2241 ± 478. Immediately after copulation the sperm pump contains 33 ± 20 sperm ($n = 10$); in the receptaculum seminis of the females are then found 242 ± 28 sperm ($n = 10$). This means that during copulation sperm are released from the vesiculae seminalis and that females get a sufficient number of sperm per insemination to fertilize the (< 100) eggs. A population of mites has an excess of sperm, notwithstanding the fact that females outnumber males by 2—3 times.

REFERENCES

Alberti, G. and Storch, V., 1976. Ultrastrukturuntersuchungen am männlichen Genital-trakt und an Spermien von *Tetranychus urticae* (Tetranychidae, Acari). Zoomorpho-logie, 83: 283—296.

Blauvelt, W.E., 1945. The internal morphology of the common red spider mite (*Tetranychus telarius* Linn.). Mem. Cornell Univ. Agric. Exp. Station, Ithaca, NY, 270: 1—35.

Boudreaux, H.B., 1963. Biological aspects of some phytophagous mites. Ann. Rev. Entomol., 8: 137—154.

Claparède, E., 1868. Studien an Acariden. Z. Wiss. Zool., 18: 445—546.

Ehara, S., 1960. Comparative Studies on the Internal Anatomy of Three Japanese Trombidiform Acarinids. J. Fac. Sci., Hokkaido Univ., Ser. 6, 14: 410—434.

Mothes, U. and Seitz, K.-A., 1981. The transformation of male sex cells of *Tetranychus urticae* K. (Acari, Tetranychidae) during passage from the testis to the oocytes: an electron microscopic study. Int. J. Invertebr. Reprod., 4: 81—94.

Overmeer, W.P.J., 1972. Notes on mating behaviour and sex ratio control of *Tetranychus urticae* Koch (Acarina: Tetranychidae). Entomol. Ber. (Amsterdam), 32: 240—244.

Pijnacker, L.P. and Drenth-Diephuis, L.J., 1973. Cytological investigations on the male reproductive system and the sperm track in the spider mite *Tetranychus urticae* Koch (Tetranychidae, Acarina). Neth. J. Zool., 23: 446—464.

Schrader, F., 1923. Haploidie bei einer Spinnmilbe. Arch. Mikrosk. Anat., 97: 610—622.

1.2.2 Oogenesis

C.C.M. FEIERTAG-KOPPEN and L.P. PIJNACKER

INTRODUCTION

The reproductive system of female tetranychid mites is composed of a single meroistic ovary, an oviduct, a vagina and a receptaculum seminis (spermatheca). These organs occupy the ventral part of the body between the central nerve mass and the anus (see also Chapter 1.1.2).

The internal anatomy of the common red spider mite *Tetranychus urticae* Koch was described by Blauvelt (1945), Gasser (1951), Jalil and Morrison (1969) and Jalil (1969). During the last decade elaborate studies have been carried out on *T. urticae* at the light microscopic as well as at the electron microscopic level. Aspects of the oogenesis, internal fertilization, development of the ovary and the functional morphology were investigated by Langenscheidt (1973a), Pijnacker and Drenth-Diephuis (1973), Feiertag-Koppen (1976, 1980), Alberti and Storch (1976), Crooker and Cone (1979), Weyda (1980), Mothes and Seitz (1981a, b), Feiertag-Koppen and Pijnacker (1982), and Koppen (1982). Dosse and Langenscheidt (1964) studied the internal genitalia of hybrids of the *Tetranychus urticae—cinnabarinus* complex. The general morphology and special topics of some other species were considered by Beament (1951), Snieder-Berkenbosch (1955), Ehara (1960), Seifert (1961), Anwarullah (1963), Smith and Boudreaux (1972), Jeppson et al. (1975) and Sirsikar and Nagbhushanam (1983).

Since there are apparently no fundamental differences between the reproductive systems of the tetranychid species, the results on *T. urticae* will mainly be considered in this chapter.

DEVELOPMENT OF THE OVARY

Larva (Figs. 1.2.2.1 and 1.2.2.2)

The larval ovary contains a monolayer of oogonia (a) at the periphery of the anterior two-thirds. The diploid number of 6 chromosomes appears during mitosis. The 2 final oogonial divisions give rise to groups of 4 cells which remain connected mutually by cytoplasmic bridges. These cells become primary oocytes which pass after the premeiotic interphase (b), the meiotic stages leptotene, zygotene and pachytene, the last 2 in bouquet stage (c). They are found in the centre of the ovary. The oocytes become larger in the posterior region (d) and again obtain an interphase-like nucleus. The ovary is covered by a very thin epithelial sheath.

Chapter 1.2.2. references, p. 125

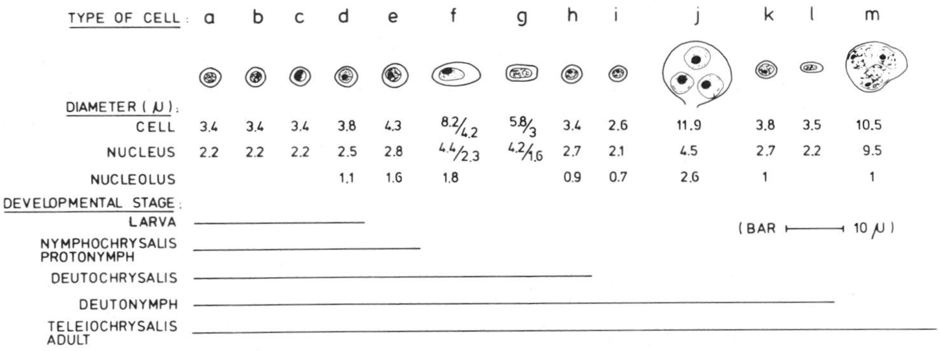

TYPE OF CELL:	a	b	c	d	e	f	g	h	i	j	k	l	m
DIAMETER (µ):													
CELL	3.4	3.4	3.4	3.8	4.3	8.2/4.2	5.8/3	3.4	2.6	11.9	3.8	3.5	10.5
NUCLEUS	2.2	2.2	2.2	2.5	2.8	4.4/2.3	4.2/1.6	2.7	2.1	4.5	2.7	2.2	9.5
NUCLEOLUS				1.1	1.6	1.8		0.9	0.7	2.6	1		1

DEVELOPMENTAL STAGE:
LARVA
NYMPHOCHRYSALIS PROTONYMPH
DEUTOCHRYSALIS
DEUTONYMPH
TELEIOCHRYSALIS ADULT

(BAR ⊢——→ 10 µ)

Fig. 1.2.2.1. Semi-schematic drawing of the ovarian cells of *T. urticae*. Dimensions have been determined at the stage in which the cells/organelles could first be identified. Abbreviations: a = oogonial cell; b = cell in interphase; c = meiotic cell in bouquet stage; d—e = meiotic cells with interphase-like chromatin; f = growing oocyte; g = angular cell; h = terminal cell; i = cell in interphase; j = three-nucleate nurse cell; k = chromatin-rich cell; l = border cell; m = chromatin-rich cell. (After Feiertag-Koppen and Pijnacker (1982) with the permission of Pergamon Press, Oxford.)

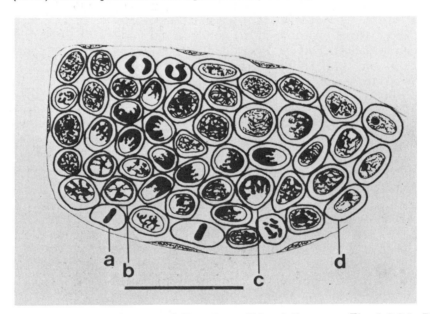

Fig. 1.2.2.2. Larval ovary of *T. urticae*. Abbreviations: see Fig. 1.2.2.1. Bar = 10 µm. (After Feiertag-Koppen and Pijnacker (1982) with the permission of Pergamon Press, Oxford.)

Nymphochrysalis and protonymph (Fig. 1.2.2.1)

The numbers of oogonia (a) and of oocytes (b—d) increase. Oocytes with larger dimensions (e) border the posterior part.

Deutochrysalis (Fig. 1.2.2.1)

Maximum numbers of oogonia (a) and of primary oocytes (b—e) have developed. Growing oocytes (f), in the diffuse stage of meiosis following pachytene, can be recognized for the first time at the periphery of the ovary. Rows of angular, mesodermal cells (g) are present between and posterior to the growing oocytes. In the posterior direction, they are continuous with the columnar cells of the oviduct.

Deutonymph (Figs. 1.2.2.1 and 1.2.2.3)

The ovary becomes separated into 2 histologically distinct parts, the anterior cap, containing the oogonia (a) and the primary oocytes (b—e), and the anterior part, characterized by the presence of larger cells, i.e. growing

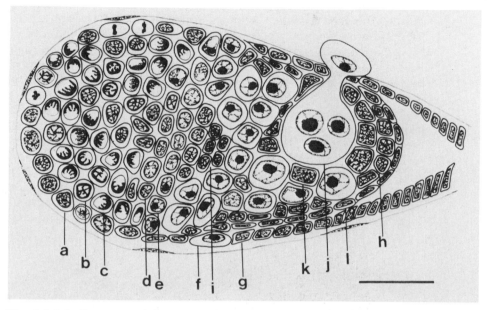

Fig. 1.2.2.3. Deutonymphal ovary of *T. urticae*. Abbreviations: see Fig. 1.2.2.1. Bar = 10 μm. (After Feiertag-Koppen and Pijnacker (1982) with the permission of Pergamon Press, Oxford.)

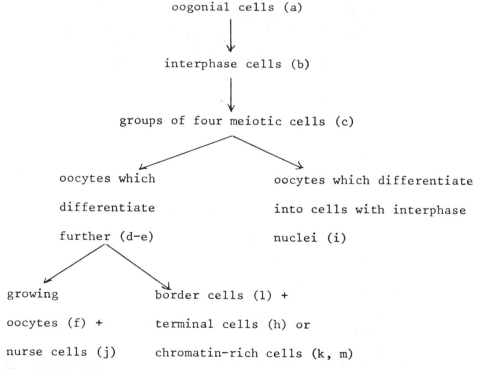

Fig. 1.2.2.4. Differentiation of ovarian cells from oogonia of *T. urticae*. Abbreviations: see Fig. 1.2.2.1. (After Feiertag-Koppen and Pijnacker (1982) with the permission of Pergamon Press, Oxford.)

Chapter 1.2.2. references, p. 125

oocytes, nurse cells and chromatin-rich cells. All the various types of cells found in the posterior part originate from the primary oocytes (Fig. 1.2.2.4). One cell of each group of 4 primary oocytes (d, e), which becomes situated on the surface of the ovary, differentiates into a growing oocyte (f); the 3 complementary cells become a three-nucleate nurse cell (j), which remains inside the ovary. Other groups of cells of the population (d, e), differentiate into border cells (l) which form the inner layer of the periphery of the ovary; their 3 complementary cells differentiate into the terminal cells (h) or into the large chromatin-rich cells (k) which are situated between the nurse cells. Primary oocytes (c), which do not reach the peripheral area of the ovary remain in the centre of the ovary between the cap and the posterior part (i). The function of the apparently abortive germ cells (h, i, k and l), apart from the fact that they sustain the morphology of the ovary, is unknown. The growing oocytes, developing in close contact with the midgut and the hemolymph, form together with the mesodermal cells (g) the outer layer of the ovary. A cytoplasmic bridge, connecting the growing oocyte and the nurse cell, becomes very distinct when the oocyte has attained a diameter c f ± 9 µm. The nucleus of the oocyte occupies an eccentric position in the immediate vicinity of this cytoplasmic bridge. The posterior angular cells (g) form the enclosure of a cavity between the ovary and the oviduct.

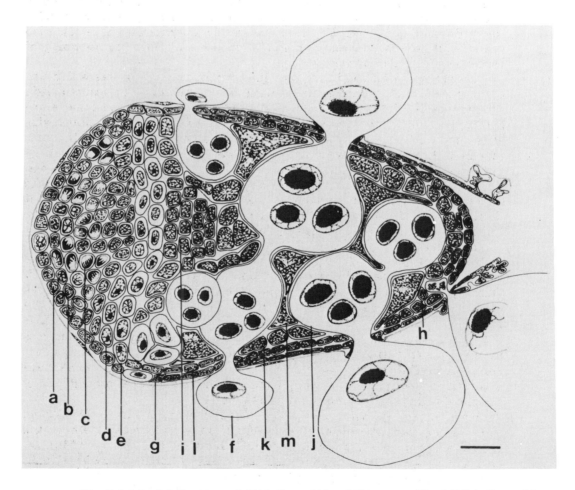

Fig. 1.2.2.5. Adult ovary of *T. urticae*. Abbreviations: see Fig. 1.2.2.1. Bar = 10 µm. (After Feiertag-Koppen and Pijnacker (1982) with the permission of Pergamon Press, Oxford.)

Teleiochrysalis (Fig. 1.2.2.1)

The numbers of oogonia (a) and of primary oocytes (b—e) decrease. The posterior part of the ovary contains increased numbers of the other types of cells. Large, irregularly shaped, chromatin-rich cells (m) are situated adjacent to the nurse cells. They have arisen from fusion of the chromatin-rich cells (k).

Adult (Figs. 1.2.2.1 and 1.2.2.5)

The ovarian cap shows a more compact arrangement of cells. It is still covered by the epithelial sheath, the nuclei of which contain tubular structures as in the epithelium covering the testis (Chapter 1.2.1). An increased number of chromatin-rich cells (m) and of nurse cells (j) are present in the posterior part of the ovary. The three-nucleate nurse cells are interconnected, thus forming a syncytium; the chromatin-rich cells (m) also are interconnected and contain bacteria in their cytoplasm. The space occupied by the ovary, as well as by the exit duct, is much influenced by the formation of the eggs from growing oocytes. The oocytes are arranged in successive stages of development in the posterior direction. They are in close contact with the midgut, often separated from the latter only by a basal membrane.

The previtellogenic oocytes show a steady growth in the amount of cytoplasm, forming bulbs on the ovarian surface. The dense cytoplasm contains rough ER, numerous free ribosomes and mitochondria. The nurse cells also show rapid growth and reach their maximum size before the beginning of vitellogenesis in the attached oocyte.

Yolk sphere formation is first seen in the cytoplasm of oocytes with a size of about 30 μm. In general, 2 growing oocytes show yolk accumulation simultaneously. It goes together with morphological changes of the mitochondria, but endocytotic processes are not visible. Shell formation begins after the onset of vitellogenesis. It takes place by the oocyte itself. Pores remain in the shell until the end of yolk formation.

During the accumulation of yolk, the nurse cells diminish in size and shrivel more and more until the final stage of vitellogenesis. When the oocyte has reached its maximum size (diameter at least 110 μm), it detaches from the shrivelled nurse cell and enters, in an amoeboid way, into the cavity. The shrivelled nurse cell detaches from the nurse cell syncytium and degenerates.

The growth period of the previtellogenic oocytes and their attached nurse cells is also accompanied by an increase in size of their diffuse stage nuclei. During vitellogenesis, the oocyte nucleus increases still more. RNA-positive nucleolar-like particles appear at random in the nucleoplasm, while the nucleolus becomes vacuolated. When yolk accumulation is nearly complete, 3 diplotene bivalents become visible. By this time the nucleolar-like particles have disappeared and the nucleolus has become smaller. When the oocyte enters into the cavity, the nucleolus and the nuclear membrane have disappeared and the chromosomes are in the first meiotic prometaphase. Ultimately one oocyte is mature (egg) at a time.

During oocyte growth the adjacent mesodermal cells bulge to the outside, thus forming pedicels. The pedicel is added to the cavity envelope when the oocyte enters the cavity. Whether in the cavity and the oviduct additional cement and wax layers are added to the shell, as in *Panonychus ulmi* (Koch) (Beament, 1951) still needs to be investigated.

During the development of the ovary the numbers of germ cells vary. The numbers of oogonia and of primary oocytes increase during the larval, nymphochrysalid and protonymphal stases, are highest in deutochrysalis and deutonymph, and are lower later on. Growing oocytes are seen first in deuto-chrysalis. Their number increases in deutonymph, is highest in teleiochrysalis and in young adults, and decreases in older adults. The increase and subse-quent decrease of oogonia and of growing oocytes in the female occur in the same stages as the increase and subsequent decrease of spermatogonia and of sperm in the male life-cycle (Chapter 1.2.1). The female deutonymphs show a maximum development of germ cells and, consequently, are an important stage for experiments on germ cells with regard to biological control. The results of investigations on the mode of action of sterilants in post-embryonic stages and in adult females of *T. urticae* (Jalil and Morrison, 1969; Langen-scheidt, 1973b) illustrate that each stage reacts according to the composition of the germ cells.

NUTRITION OF THE GROWING OOCYTE

The ovary of spider mites is of the meroistic type, meaning the presence of an oocyte—nurse cell complex. The oocytes develop on the surface of the ovary in close contact with the midgut and the haemolymph. These features are also characteristic for other arachnids (Oliver, 1971; Seitz, 1971).

The previtellogenic oocytes of *T. urticae* derive proteins and ribosomal RNA from the nurse cells. The supply of this nutritive material decreases from the beginning of the vitellogenic phase onwards. During vitellogenesis, the oocytes absorb material from the midgut/haemolymph. This material is of low molecular nature since the absorption has not been observed structurally. Proteins, and most likely, glycogen and lipid precursors, are taken up through perforations in the shell. Proteins associated with carbohydrates are trans-formed to yolk spheres; lipids appear as droplets adjacent to the spheres.

An arresting aspect of the oocyte growth is that the oocyte nucleus also plays a role in the storage of proteins. The nucleus shows the features of ribo-somal DNA amplification and magnified ribosomal RNA synthesis because the nuclei contain a well-developed nucleolus and nucleolar particles. In meroistic ovaries these syntheses are carried out by nurse cells only; they are characteristic for panoistic ovaries (Bier et al., 1967; Telfer, 1975). This means that the ovary of *T. urticae* shows a feature of the panoistic ovary. In this type of ovary, which occurs in several insect orders (De Wilde and De Loof, 1973) the oocytes are accompanied by follicle cells which form the only trophic tissue of the vitellarium. The ovary of *T. urticae* thus represents a transition between these 2 types of ovaries. It is comparable to that found in some species of Coleoptera (Urbani, 1970). The question of why DNA ampli-fication and magnified RNA synthesis occur in the growing oocytes of *T. urticae* cannot be fully answered. Bier et al. (1967) assume that the capacity of an increased RNA synthesis will not only shorten the period needed for oocyte maturation in the ovary, but will presumably result in a much larger number of oocytes. Since the reproductive potential of *T. urticae* is extremely high, this explanation is probably applicable to this species. DNA amplification in the oocytes of *Dytiscus* has been interpreted as a strong development of the phylogenetically primitive pattern of multiplication of the nucleolar region (Bier et al., 1967). On this assumption, DNA amplification would be significant phylogenetically and for classification of the Tetranychidae.

INVERTED MEIOSIS

Chromosomes with diffuse kinetochores usually follow a so-called inverted or post-reductional meiosis. In this type of meiosis the first division is equational, i.e. sister chromatids separate at the first anaphase, and the second division is reductional, i.e. homologous (non-sister) chromatids separate at the second anaphase (Battaglia and Boyes, 1955; Sybenga, 1981). The terms 'sister' and 'homologous' are applicable only to the non-crossover regions of the chromosomes. The holokinetic chromosomes of *T. urticae* (Chapter 1.2.4) also follow an inverted sequence of the meiotic divisions.

As described above, 3 diplotene bivalents become visible at the end of yolk accumulation. Each of them shows at least 2—3 chiasmata. In all likelihood, each of the 4 chromatids of a bivalent is involved in chiasma formation, since there is no reason why multiple-strand double crossovers would not occur (Helle and Van Zon, 1970). During early prometaphase the bivalents become arranged in a spindle with chromosomal and continuous fibres. The chromosomes are then double lengthways and lie parallel to each other. Complex movements at prometaphase takes place when the oocyte moves into the ovarian cavity (Fig. 1.2.2.6). The homologous chromosomes of the axially orientated bivalents make turning movements in opposite directions until they have attained a tandem arrangement with the long axis perpendicular to the spindle fibres. Subsequently, the homologous chromosomes make a lateral movement, apparently forced by attracting forces between the homologous parts. The ultimate arrangement is an equatorial pairing of the chromosomes through which the metaphase shape of the bivalents is attained. This arrangement leads to the equational first meiotic division. The metaphase bivalents have this ultimate shape when still in the ovarian cavity.

The further course of meiosis, from first metaphase onwards, takes place in eggs within a couple of hours after oviposition. It has also been described by Schrader (1923) (Fig. 1.2.2.7; for photographs see Chapter 1.2.4). Just

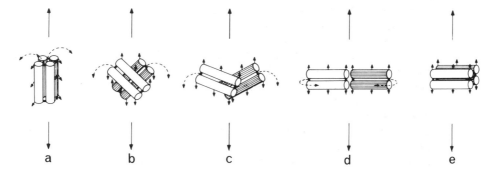

Fig. 1.2.2.6. Schematic outlines of the bivalents during prometaphase I of *T. urticae*: (a) axial position; (b, c) crossed position; (d) equatorial tandem arrangement; (e) equatorial paired arrangement. (After Feiertag-Koppen (1980) with the permission of Dr. W. Junk BV, The Hague.)

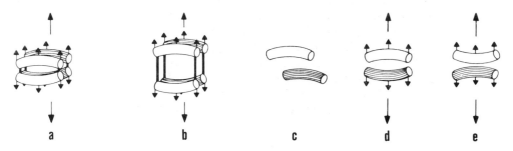

Fig. 1.2.2.7. Inverted meiosis in *T. urticae*: (a) metaphase I; (b) anaphase I; (c) telophase I; (d) metaphase II; (e) anaphase II. (After Feiertag-Koppen (1980) with the permission of Dr. W. Junk BV, The Hague.)

Chapter 1.2.2. references, p. 125

after oviposition, the 3 metaphase bivalents, being of identical size and shape, are situated in a cytoplasmic island in the egg cortex, in the upper half of the egg with respect to the surface they are deposited on. The cylindrical spindle runs parallel to the egg surface. The bivalents are seen in frontal and lateral view as rings and in polar view as 2 separated rods. The sister chromatids, which separate parallel to each other at the first anaphase, show connective threads between their terminal ends, which have been interpreted as remaining chiasmata. The first maturation division results in 2 groups of 3 double-structured telophase chromosomes. One of the groups becomes the first polar body, which remains in the periphery of the egg. The other group enters directly into the second meiotic metaphase with the spindle obliquely disposed with respect to the egg surface. The second maturation division results in 2 groups of 3 chromosomes. The group of chromosomes situated most centrally in the egg becomes the female pronucleus, each chromosome of which has changed into a karyomere (Chapter 1.2.4). The 3 karyomeres, situated close to each other, migrate towards the centre of the egg. The 3 chromosomes, which are situated at the periphery, become the second polar body.

In fertilized eggs, a torpedo-shaped spermatozoon surrounded by a thin layer of bright plasm is situated in the cortex immediately after oviposition. The distance between the spermatozoon and the bivalents is about a quarter of the circumference of the egg. Polyspermy is not involved. Just before syngamy, there are 3 karyomeres of female and 3 of male origin in the egg centre. The karyomeres then change into 6 chromosomes and enter the metaphase of the first cleavage division. This situation exists about 2—2.5 h after oviposition.

OVIDUCT AND VAGINA

The oviduct and the vagina have been investigated in a collapsed situation only, since a passing egg is laid during fixation for cytological work (Fig. 1.2.2.8). In cross-section the oviduct is twice as wide as it is thick; the vagina is round (see also Chapter 1.1.2).

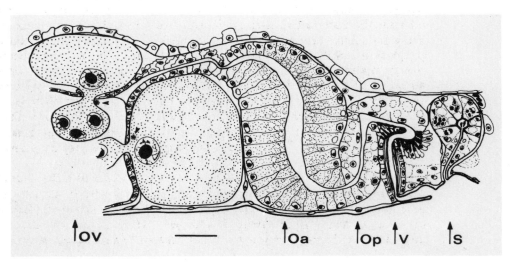

Fig. 1.2.2.8. Median-sagittal section through ovary (Ov), anterior and posterior part of oviduct (Oa, Op), vagina (V), and seminal receptacle (S) of adult *T. urticae* female. Only 2 growing oocyte/nurse cell syncytia of the ovary are drawn; from between the ovary and the growing oocytes towards the oviduct, cells of the ovarian cavity can be seen; arrowheads point to one or more spermatozoa. Bar = 10 μm.

The oviduct is a large tube, extending from the ovarian cavity to the anterior part of the vagina. It is divided into 2 histologically distinct regions. The epithelium of the anterior part is formed of large, vacuolated columnar cells. Emission of secretion into the central duct takes place from the vacuoles in the apical portion of the cells. The posterior part of the oviduct is a tube which is formed of secretory cells, each of them containing 1 large vacuole. The glandular cells of both parts of the oviduct play a role in the formation of a hygroscopic egg shell.

The vagina is a short tube, leading to the genital opening. The walls are composed of a single layer of epidermal cells with a pleated chitinous intima and are provided with muscles.

RECEPTACULUM SEMINIS AND SPERM TRACK

The receptaculum seminis (seminal receptacle) is a blind erect ovoid sac situated posterior to the vagina (Fig. 1.2.2.8). It opens by a narrow seminal duct in the epidermis caudad to the vaginal opening (Van Eyndhoven, 1972; Van de Lustgraaf, 1977) and not, as often described, into the vagina. The duct is sclerotized with a wall of compact cells. The sac-like part is composed of a monolayer of columnar cells around a lumen. These cells have a complex structure; the membranes are characteristic, having zonulae adherentes, microtubules, and vacuoles. A connection between receptacle and vagina or oviduct has not yet been observed; however, they can touch each other (see also Chapter 1.1.2).

During copulation, about 242 aflagellate sperm (Chapter 1.2.1) are deposited with the shaft of the penis inserted directly into the lumen, the latter being filled with granular material. These sperm vary in shape from spherical to oblong and up to 6 may stick together in packet form. The peripheral infoldings have disappeared and instead a few large projections are formed which may get entangled (possibly forming packets). Most of the sperm enter the columnar cells during the 24 h following insemination. Then the sperm enlarges a little; the chromatin begins to diffuse and rods or microtubules appear underneath the membrane. They are generally organized in packets of between 3 and 10 sperm. One day after insemination, sperm are also found outside the receptacle, mainly dorsally to the oviduct and near the ovary. They lie singly, are more or less spherical again and apparently have lost the projections. Still later, sperm enter the ovarian cavity and scatter in the ovary, mainly in the dorsal region. In the ovary some sperm come to lie adjacent to the nutritive cords. When the accumulation of yolk has nearly ended, 1 spermatozoon enters the growing oocyte, apparently via the nutritive cord. The spermatozoon becomes torpedo-shaped and remains close to the oocyte nucleus until oviposition (cf. above). Sperm do not reach the first developing oocytes in time for fertilization; consequently these oocytes remain male-determined.

How the sperm reaches the oocyte without the aid of a flagellum has been a matter for discussion. How they move through the columnar cells of the receptacle, then from the receptacle towards the ovary, and ultimately into the oocyte itself is far from elucidated. Whether or not the changes in morphology of the sperm have a function also remains unanswered.

REFERENCES

Alberti, G. and Storch, V., 1976. Ultrastrukturuntersuchungen an männlichen Genitaltrakt und an Spermien von *Tetranychus urticae*. Zoomorphologie, 83: 283—296.

Anwarullah, M., 1963. Beiträge zur Morphologie und Anatomie einiger Tetranychiden (Acari, Tetranychidae). Z. Angew. Zool., 50: 385—426.

Battaglia, E. and Boyes, J.W., 1955. Post-reductional meiosis: its mechanism and causes. Caryologia, 8: 87—134.

Beament, J.W.L., 1951. The structure and formation of the egg of the fruit tree red spider mite *Metatetranychus ulmi* (Koch). Ann. Appl. Biol., 38: 1—24.

Bier, K., Kunz, W. and Ribbert, D., 1967. Struktur und Funktion der Oocytenchromosomen und Nukleolen sowie der Extra-DNS während der Oogenese panoistischer und meroistischer Insekten. Chromosoma, 23: 214—254.

Blauvelt, W.E., 1945. The internal morphology of the common red spider mite *Tetranychus telarius* Linn. Mem. Cornell Univ. Agric. Exp. Station, Ithaca, NY, 270: 1—35.

Crooker, A.R. and Cone, W.W., 1979. Structure of the reproductive system of the adult female two-spotted spider mite *Tetranychus urticae*. Rec. Adv. Acarol., 2: 405—409.

De Wilde, J. and De Loof, A., 1973. Reproduction. In: M. Rockstein (Editor), The Physiology of Insects, Vol. 1. Academic Press, New York, NY, London, pp. 11—95.

Dosse, G. and Langenscheidt, M., 1964. Morphologische, biologische und histologische Untersuchungen an Hybriden aus dem *Tetranychus urticae—cinnabarinus* Komplex (Acari, Tetranychidae). Z. Angew. Entomol., 54: 349—359.

Ehara, S., 1960. Comparative studies on the internal anatomy of three Japanese trombidiform acarinids. J. Fac. Sci. Hokkaido Univ., Ser. 6, 14: 410—434.

Feiertag-Koppen, C.C.M., 1976. Cytological studies of the two-spotted spider mite *Tetranychus urticae* Koch (Tetranychidae, Trombidiformes). I. Meiosis in eggs. Genetica, 46: 445—456.

Feiertag-Koppen, C.C.M., 1980. Cytological studies of the two-spotted spider mite *Tetranychus urticae* Koch (Tetranychidae, Trombidiformes). II. Meiosis in growing oocytes. Genetica, 54: 173—180.

Feiertag-Koppen, C.C.M. and Pijnacker, L.P., 1982. Development of the female germ cells and process of internal fertilization in the two-spotted spider mite *Tetranychus urticae* Koch (Acariformes: Tetranychidae). Int. J. Insect Morphol. Embryol., 11: 271—284.

Gasser, R., 1951. Zur Kenntnis der gemeinen Spinnmilbe *Tetranychus urticae*. 1. Mitt.: Morphologie, Anatomie, Biologie, Ökologie. Mitt. Schweiz. Entomol. Ges., 24: 217—262.

Helle, W. and Van Zon, A.Q., 1970. Linkage studies in the pacific spider mite *Tetranychus pacificus* II. Genes for white eye, lemon and flamingo. Entomol. Exp. Appl., 13: 300—306.

Jalil, M., 1969. Internal anatomy of the developmental stages of the two-spotted spider mite. Ann. Entomol. Soc. Am., 62: 247—249.

Jalil, M. and Morrison, P.E., 1969. Chemosterilization of the two-spotted spider mite. II. Histopathological effect of Apholate and 5-Fluorouracil on the reproductive organs. J. Econ. Entomol., 62: 400—403.

Jeppson, L.R., Keifer, H.H. and Baker, E.W., 1975. Mites Injurious to Economic Plants. University of California Press, Berkeley, Los Angeles, CA, London, 614 pp.

Koppen, C.C.M., 1982. Oogenesis of the Two-spotted Spider Mite *Tetranychus urticae* Koch, with Special Reference to the Behaviour of the Chromosomes. Ph.D. Thesis, University of Amsterdam.

Langenscheidt, M., 1973a. Zur Wirkungsweise von Sterilität erzeugenden Stoffen bei *Tetranychus urticae* Koch (Acari, Tetranychidae). I. Entwicklung der weiblichen Genitalorgane und Oogenese bei unbehandelten Spinnmilben. Z. Angew. Entomol., 73: 103—106.

Langenscheidt, M., 1973b. Zur Wirkungsweise von Sterilität erzeugenden Stoffen bei *Tetranychus urticae* Koch (Acari, Tetranychidae). II. Wirkungsweise von Apholate bei Spinnmilben. Z. Angew. Entomol., 74: 142—151.

Mothes, U. and Seitz, K.-A., 1981a. The transformation of male sex cells of *Tetranychus urticae* (Acari, Tetranychidae) during passage from the testis to the oocytes: an electron microscopic study. Int. J. Invertebr. Reprod., 4: 81—94.

Mothes, U. and Seitz, K.-A., 1981b. Licht- und elektronenmikroskopische Untersuchungen zur Funktionsmorphologie von *Tetranychus urticae* (Acari, Tetranychidae). II. Weibliches Geschlechtssystem und Oogenese. Zool. Jahrb., Abt. Anat., Ontog. Tiere, 105: 106—134.

Oliver, J.H., Jr., 1971. Parthenogenesis in mites and ticks (Arachnida; Acari). Am. Zool., 11: 283—299.

Pijnacker, L.P. and Drenth-Diephuis, L.J., 1973. Cytological investigations on the male reproductive system and of the sperm track in the spider mite *Tetranychus urticae* Koch (Tetranychidae, Acarina). Neth. J. Zool., 23: 446—464.

Schrader, F., 1923. Haploidie bei einer Spinnmilbe. Arch. Mikrosk. Anat. Entwicklung-smech., 97: 610—622.

Seifert, G., 1961. Der Einfluss von DDT auf die Eiproduktion von *Metatetranychus ulmi* Koch (Acari, Tetranychidae). Z. Angew. Zool., 48: 441—452.

Seitz, K.-A., 1971. Licht- und elektrononmikroskopische Untersuchungen zur Ovarentwicklung und Oogenese bei *Cupiennius salei* Keys (Araneae, Ctenidae). Z. Morphol. Tiere, 69: 283—317.

Sirsikar, A.N. and Nagbhushanam, R., 1983. Female reproductive system and oogenesis in *Oligonychus mangiferus* (R & S). Seen as manuscript.

Smith, J.W. and Boudreaux, H.B., 1972. An autoradiographic search for the site of fertilization in spider mites. Ann. Entomol. Soc. Am., 65: 69—74.

Snieder-Berkenbosch, L., 1955. Some remarks about the digestive tract and the female reproductive system of the fruit tree red spider mite, *Metatetranychus ulmi* (Koch). Proc. K. Ned. Akad. Wet., Ser. C, 58: 489—494.

Sybenga, J., 1981. Specialization in the behaviour of chromosomes on the meiotic spindle. Genetica, 57: 143—151.

Telfer, W.H., 1975. Development and physiology of the oocyte-nurse cell syncytium. Adv. Insect Physiol., 11: 223—319.

Urbani, E., 1970. A survey on some aspects of oogenesis in *Dytiscus*, *Cybister* and *Hygrobia* (Coleoptera). Acta Embryologiae Experimentalis, 3: 281—297.

Van de Lustgraaf, B., 1977. l'Ovipositeur des Tétraniques. Acarologia, 4: 642—650.

Van Eyndhoven, G.L., 1972. Some details about the genitalia of Tetranychidae (Acari). Notulae ad Tetranychidae 12. Zesz. Probl. Postepow Nauk Roln., 129: 27—42.

Weyda, F., 1980. Reproductive system and oogenesis in active females of *Tetranychus urticae* (Acari, Tetranychidae). Acta Entomol. Bohemoslov., 77: 375—377.

1.2.3 Parthenogenesis, Chromosomes and Sex

W. HELLE and L.P. PIJNACKER

PARTHENOGENESIS

Parthenogenesis is a common mode of reproduction in mites and is found in several acarine orders. Male-producing parthenogenesis (arrhenotoky) is most frequently encountered, but female-producing parthenogenesis (thelytoky) is far from rare and it occurs throughout the Acarina. Deuterotoky, the third type of parthenogenesis, producing both males and females, is rarely found in the Acarina and is not known to have evolved into a regular system.

With regard to both arrhenotoky and thelytoky, a number of phenomena have recently been described which might cause confusion in the nomenclature. Therefore a brief discussion of the terms in use is appropriate. The first confusion could arise from the term 'haploid parthenogenesis' (versus 'diploid parthenogenesis'). These terms are used with reference to the cytological phenomena in the egg (Whiting, 1945; Rieger et al., 1976). Haploid parthenogenesis has been considered as another term for arrhenotoky (a term indicating the determination of sex), since haploid parthenogenesis, as far as was known, invariably resulted in haploid males. However, it has been shown by Pijnacker et al. (1980) that haploid parthenogenesis can also give rise to haploid females. It is therefore advisable to omit the term haploid parthenogenesis and to use the terms 'male haploidy' and 'female haploidy' instead. With reference to thelytoky we have to distinguish 'haploid thelytoky' from 'diploid thelytoky'.

The second point of discussion concerns the term 'haplo-diploidy'. Haplodiploidy refers to a type of sex determination in which males are haploid and females are diploid. Arrhenotoky is a type of haplo-diploidy in which the haploid males result from parthenogenesis. In some dermanyssid mites, parthenogenetic development of the haploid eggs, giving rise to males, is initiated by pseudogamy (Oliver, 1971). This phenomenon is still in conformity with arrhenotoky, and it is acceptable to use the term 'gynogenetic arrhenotoky', or 'gynogenetic haplo-diploidy' for this phenomenon.

The third difficulty is related to haplo-diploidy in the Phytoseiidae. In phytoseiid mites, females are diploid and males are haploid, but arrhenotoky does not underlie this type of haplo-diploidy (see Chapter 2.1.2.4). Males develop from fertilized eggs in which the paternal set of chromosomes is eliminated during early embryonic development. Hoy (1979) uses the term 'parahaploidy' for this type of haplo-diploidy, referring to Hartl and Brown (1970). These authors used this word to cover *all* cases in which genetic haploidy is reached by heterochromatization, suppression, or elimination of 1 chromosome set. In our opinion the term 'parahaploidy' is confusing. In the male phytoseiids, genuine haploidy is achieved after complete elimination

Chapter 1.2.3. references, p. 138

of 1 chromosome set. Furthermore, the term 'parahaploidy' is confusing now that haploidy is being used in so many combinations. De Jong et al. (1981) use the term 'pseudo-arrhenotoky', thus denoting the absence of parthenogenesis. Absence of arrhenotoky is crucial in the peculiar haplo-diploid situation in phytoseiid mites, and we adopt the term 'pseudo-arrhenotoky' for all the cases in which male haploidy is achieved by elimination of a set of chromosomes. In pseudo-arrhenotoky, the haploid males are biparental and not impaternate, as in the case of arrhenotoky.

Difficulties often arise in demonstrating the kind of haplo-diploidy. Haplo-diploidy is easily demonstrated by studying the mitotic chromosomes in embryonic tissue. One is left with the question of whether arrhenotoky, gynogenetic arrhenotoky, or pseudo-arrhenotoky underlies the haplo-diploidy. Arrhenotoky may appear from rearing experiments, in which isolated, unmated females produce haploid eggs from which exclusively males develop. Radiation experiments, as conducted by Helle et al. (1978) and Hoy (1979), or analytical cytological studies, such as those undertaken by Nelson-Rees et al. (1980), are necessary in order to distinguish between pseudo-arrhenotoky and gynogenetic arrhenotoky. Genetic data are not conclusive for this distinction.

There is also a methodical problem in distinguishing between haploid and diploid thelytoky. If an odd number of chromosomes is found in the eggs of a thelytokous species, the thelytoky will presumably be of the haploid type. In the case of an even number of chromosomes, it has to be concluded from the shape and staining pattern of the chromosomes whether pairs of homologous chromosomes can be recognized, pointing to diploidy. However, very often it will be impossible to ascertain this by squashes of cleavage divisions and a study of the meiotic events will have to provide an answer to the question of whether the thelytoky is haploid or diploid. The occurrence of polyploidy would make classification still more difficult, but as yet polyploidy has not been demonstrated in mites. In the 3 *Brevipalpus* species which exhibit female haploidy, a characteristic was observed which possibly is indicative for female haploidy: the occurrence of spanandric males with the female chromosome number. The frequency of these spanandric males (and intersexes) increases considerably in the offspring of irradiated females (Pijnacker et al., 1980).

Haplo-diploidy has been demonstrated by karyotyping studies involving more than 100 bisexual species (Table 1.2.3.1). In more than 30 species of the genera *Bryobia, Porcupinychus, Tetranychina, Schizonobia, Eurytetranychus, Eutetranychus, Aponychus, Duplanychus, Eonychus, Eotetranychus, Neotetranychus, Panonychus, Schizotetranychus, Oligonychus* and *Tetranychus*, it has been demonstrated by rearing and karyotype studies that the offspring of virgin females is haploid and male (Gutierrez et al., 1979a). Therefore, the conclusion that haplo-diploidy in the Tetranychidae is always based on arrhenotoky is well documented. Gynogenetic arrhenotoky seems to be absent in the Tetranychidae, although it should be borne in mind that very often differences in fecundity have been found between mated and unmated females, the latter generally producing fewer eggs during the first part of reproductive life (see Chapter 1.2.6).

Thelytoky is widespread in the sub-family Bryobiinae Berlese, which is generally considered the most primitive sub-family within the Tetranychidae (Chapter 1.1.5). Besides arrhenotokous species several genera of the tribi Bryobiini, Hystrichonychini and Petrobiini also contain thelytokous species (Helle et al., 1981). It is remarkable that obligate thelytoky in the Tetranychinae Berlese hardly occurs. Only in the rather primitive genus *Oligonychus* has an obligate thelytokous species recently been described: *O. thelytokus* Gut. (Gutierrez, 1977).

TABLE 1.2.3.1

Chromosome number and parthenogenesis in the Tetranychidae

Species	Chromosome number (*n*), type of partheno-genesis (T = thelytoky; A = arrhenotoky)	Ref.
Sub-family Bryobiinae Berlese		
Bryobia		
geigeriae Meyer	*n* = 2 A	Helle et al. (1981)
kissophila van Eyndh.	*n* = 4 T	Gutierrez et al. (1970)
praetiosa Koch	*n* = 4 T	Helle et al. (1970)
rubrioculus (Scheut.)	*n* = 4 T	Helle et al. (1970)
lagodechiana Reck	*n* = 4 T	Bolland (1983)
sarothamni Geysk.	*n* = 4 A	Helle and Bolland (1967)
Pseudobryobia		
drummondi (Ewing)	*n* = 3 A	Helle et al. (1981)
Tetranycopsis		
horridus (Can. & Fanz.)	*n* = 2 T	Helle and Bolland (1967)
Hystrichonychus		
gracilipes (Banks)	*n* = 4 A	Helle et al. (1981)
sidae P. & B.	*n* = 4 A	Helle et al. (1981)
Paraplonobia		
coldeniae T. & B.	*n* = 3 A	Helle et al. (1981)
Porcupinychus		
insularis (Gut.)	*n* = 4 A	Helle et al. (1981)
Neotrichobia		
arizonensis T. & B.	*n* = 5 A	Helle et al. (1981)
Schizonobia		
sycophanta Wom.	*n* = 4 A	Gutierrez and Bolland (1973a)
Petrobia		
harti (Ewing)	*n* = 2 A	Helle et al. (1970)
moutiai B. & P.	*n* = 4 A	Bolland et al. (1981)
latens (Müller)	*n* = 4 T	Helle et al. (1970)
Sub-family Tetranychinae Berlese		
Eonychus		
curtisetosus Gut.	*n* = 2 A	Helle et al. (1970)
grewiae Gut.	*n* = 2 A	Helle et al. (1970)
Eurytetranychus		
madagascariensis Gut	*n* = 3 A	Helle et al. (1970)
buxi (Garman)	*n* = 5 A	Helle and Bolland (1967)
Eutetranychus		
africanus (Tucker)	*n* = 2 A	Gutierrez and Helle (1971)
enodes B. & P.	*n* = 2 A	Smith Meyer and Bolland (1984)
banksi (McG.)	*n* = 3 A	Helle et al. (1970)
orientalis (Klein)	*n* = 3 A	Helle et al. (1970)
eliei Gut. & Helle	*n* = 4 A	Helle et al. (1970)
Aponychus		
grandidieri (Gut.)	*n* = 2 A	Gutierrez et al. (1970)
Anatetranychus		
tephrosiae (Gut.)	*n* = 3 A	Gutierrez et al. (1970)
Duplanychus		
ranjatoi (Gut.)	*n* = 3 A	Gutierrez et al. (1970)
Mixonychus		
dulcis Meyer	*n* = 3 A	Helle et al. (1981)
tabebuiae (Flechtm.)	*n* = 3 A	G.H.W. Flechtmann (personal communication, 1983)
Panonychus		
ulmi (Koch)	*n* = 3 A	Helle and Bolland (1967)
citri (McG.)	*n* = 3 A	Gutierrez et al. (1970)
Neotetranychus		
decorus Mey. & Boll.	*n* = 2 A	Smith Meyer and Bolland (1984)
rubi Trägårdh	*n* = 7 A	Helle and Bolland (1967)

Chapter 1.2.3. references, p. 138

(*continued*)

TABLE 1.2.3.1 (continued)

Species	Chromosome number (*n*), type of partheno-genesis (T = thelytoky; A = arrhenotoky)	Ref.
Schizotetranychus		
eremophilus McG.	*n* = 3 A	Helle et al. (1981)
reticulatus P. & B.	*n* = 3 A	Bolland et al. (1981)
sacchari Flechtm. & B.	*n* = 3 A	C.H.W. Flechtmann (personal communication, 1983)
schizopus (Zacher)	*n* = 3 A	Helle and Bolland (1967)
australis Gut.	*n* = 6 A	Helle et al. (1970)
fauveli Gut.	*n* = 6 A	Helle and Gutierrez (1983)
Mononychellus		
lippiae Meyer	*n* = 3 A	Helle et al. (1981)
planki (McG.)	*n* = 3 A	C.H.W. Flechtmann (personal communication, 1983)
siccus (P. & B.)	*n* = 3 A	Helle et al. (1981)
tanajoa (Bondar)	*n* = 3 A	C.H.W. Flechtmann (personal communication, 1983)
Eotetranychus		
lewisi (McG.)	*n* = 2 A	Helle et al. (1981)
befandrianae Gut.	*n* = 2 A	Helle et al. (1970)
sakalavensis Gut.	*n* = 2 A	Helle et al. (1970)
tulearensis Gut.	*n* = 2 A	Helle et al. (1970)
malvastris (McG.)	*n* = 2 A	Helle et al. (1981)
yumensis (McG.)	*n* = 2 A	Helle et al. (1981)
garnieri Gut.	*n* = 2 A	Helle and Gutierrez (1983)
robini Gut.	*n* = 2 A	Helle and Gutierrez (1983)
cyphus B. & P.	*n* = 2 A	Smith Meyer and Bolland (1984)
falcatus Mey. & Rod.	*n* = 3 A	Helle et al. (1981)
friedmanni Gut.	*n* = 3 A	Helle et al. (1970)
grandis Gut.	*n* = 3 A	Helle et al. (1970)
imerinae Gut.	*n* = 3 A	Helle et al. (1970)
paracybelus Gut.	*n* = 3 A	Helle et al. (1970)
rinoreae Gut.	*n* = 3 A	Helle et al. (1970)
roedereri Gut.	*n* = 3 A	Helle et al. (1970)
uncatus Garman	*n* = 3 A	Bolland et al. (1981)
cameroonensis Mey. & Boll.	*n* = 3 A	Smith Meyer and Bolland (1984)
carpini (Oud.)	*n* = 4 A	Helle and Bolland (1967)
tiliarium (Herm.)	*n* = 4 A	Helle and Bolland (1967)
hellei Mey. & Boll.	*n* = 4 A	Smith Meyer and Bolland (1984)
smithi P. & B.	*n* = 5 A	Helle et al. (1970)
fremonti T. & B.	*n* = 5 A	Helle et al. (1981)
gambelii T. & B.	*n* = 5 A	Helle et al. (1981)
rubiphilus (Reck)	*n* = 5 A	Gutierrez and Helle (1983)
Tetranychus		
desertorum Banks	*n* = 3 A	Helle et al. (1981)
evansi B. & P.	*n* = 3 A	Helle et al. (1981)
urticae Koch	*n* = 3 A	Helle and Bolland (1967)
kanzawai Kish.	*n* = 3 A	Helle and Bolland (1967)
kaliphorae Gut.	*n* = 3 A	Helle et al. (1970)
lambi P. & B.	*n* = 3 A	Bolland et al. (1981)
lombardinii B. & P.	*n* = 3 A	Bolland et al. (1981)
ludeni Zacher	*n* = 3 A	Helle et al. (1970)
macfarlanei B. & P.	*n* = 3 A	Bolland et al. (1981)
neocaledonicus André	*n* = 3 A	Helle et al. (1970)
piercei McG.	*n* = 3 A	Gutierrez et al. (1979b)
turkestani Ugar. & Nik.	*n* = 3 A	Helle et al. (1970)
yusti McG.	*n* = 3 A	Bolland et al. (1981)
lintearius Duf.	*n* = 3 A	Bolland et al. (1967)
pacificus McG.	*n* = 3 A	Helle and Bolland (1967)

TABLE 1.2.3.1 (continued)

Species	Chromosome number (*n*), type of parthenogenesis (T = thelytoky; A = arrhenotoky)	Ref.
viennensis Zacher	*n* = 3 A	Helle et al. (1970)
bastosi T.B. & S.	*n* = 3 A	C.H.W. Flechtmann (personal communication, 1983)
polys P. & B.	*n* = 3 A	Helle et al. (1981)
marianae McG.	*n* = 4 A	Bolland et al. (1981)
panici Gut.	*n* = 4 A	Helle et al. (1970)
roseus Gut.	*n* = 4 A	Helle et al. (1970)
tchadi Gut. & Boll.	*n* = 4 A	Gutierrez and Bolland (1973b)
fijiensis Hirst	*n* = 4 A	Bolland et al. (1981)
tumidus Banks	*n* = 6 A	Helle et al. (1970)
Oligonychus		
andrei Gut.	*n* = 2 A	Helle et al. (1970)
gossypii (Zach.)	*n* = 2 A	Helle et al. (1970)
sylvestris Gut.	*n* = 2 A	Helle et al. (1970)
punicae (Hirst)	*n* = 2 A	Helle et al. (1981)
randriamasii Gut.	*n* = 2 A	Helle et al. (1970)
intermedius Meyer	*n* = 2 A	Helle et al. (1981)
biharensis (Hirst)	*n* = 2 A	Helle and Gutierrez (1983)
propetes P. & B.	*n* = 2 A	Helle et al. (1981)
coffeae (Nietn.)	*n* = 3 A	Helle et al. (1970)
quercinus Hirst	*n* = 3 A	Helle et al. (1970)
thelytokus Gut.	*n* = 3 T	Bolland et al. (1981)
coniferarum McG.	*n* = 3 A	Helle et al. (1981)
ununguis (Jac.)	*n* = 3 A	Helle and Bolland (1967)
pemphisi Gut.	*n* = 3 A	Gutierrez et al. (1970)
subnudus (McG.)	*n* = 3 A	Helle et al. (1981)
gambelii T. & B.	*n* = 3 A	Helle et al. (1981)
ilicis (McG.)	*n* = 3 AT	C.H.W. Flechtmann (personal communication, 1983)
mangiferus Rah. & Punj.	*n* = 3 AT	C.H.W. Flechtmann (personal communication, 1983)
bessardi Gut.	*n* = 4 A	Helle et al. (1970)
stickneyi (McG.)	*n* = 4 A	Helle et al. (1981)
chazeaui Gut.	*n* = 4 A	Helle et al. (1970)
gramineus (McG.)	*n* = 4 A	Bolland et al. (1981)
grypus B. & P.	*n* = 4 A	Helle et al. (1970)
leandrianae Gut.	*n* = 4 A	Bolland et al. (1981)
plegas B. & P.	*n* = 4 A	Bolland et al. (1981)
pratensis (Banks)	*n* = 4 A	Helle et al. (1970)
virens Gut.	*n* = 4 A	Helle et al. (1970)
neoplegas Meyer	*n* = 4 A	Helle et al. (1981)
grewiae Meyer	*n* = 4 A	Helle et al. (1981)
comptus Mey. & Boll.	*n* = 4 A	Smith Meyer and Bolland (1984)
pennisetum Meyer	*n* = 4 A	Smith Meyer and Bolland (1984)
tiwakae Gut.	*n* = 4 A	Helle and Gutierrez (1983)

The question arises as to what extent thelytoky in Tetranychidae is haploid or diploid. In the case of *O. thelytokus* it is likely that this species is diploid with $2n = 6$, particularly as the closely related arrhenotokous species *O. coffeae* has the numbers $2n = 6$ and $n = 3$. Furthermore, no males have been found, nor did irradiation of females induce masculinity in the offspring (Gutierrez, 1977). In the thelytokous *Bryobia* species, i.e. *B. kissophila* v. Eynd., *B. praetiosa* Koch and *B. rubrioculus* (Scheut.), diploidy is conceivable, since the related bisexual species *B. sarothamni* Geysk. have $2n = 8$ and $n = 4$ chromosomes, and all the thelytokous species mentioned have 8 chromosomes (Table 1.2.3.1). *Tetranycopsis horridus* (Can. & Fanz.) is a thelytokous species with 4 chromosomes in which 2 pairs of homologues are apparent. Here also, diploid thelytoky is assumed (Helle and Bolland, 1967). As yet, haploid thelytoky has only been shown to exist in the related family of the false spider mites (Helle et al., 1980), but not in the tetranychids.

A reduced proportion of males in the red spider mite species *Oligonychus ilicis* (McG) was observed in a field population (Flechtmann and Flechtmann, 1982). From laboratory rearings, it was shown that in this species unmated females give rise to either sons or daughters. The eggs from progenies developing into males had $n = 3$ chromosomes, whereas the eggs from progenies developing into females had $2n = 6$ chromosomes. This presumably is a case of tychothelytoky, since normally *O. ilicis* is an arrhenotokous species. It is of interest to mention that in such a case an unmated female has either exclusively haploid (male), or exclusively diploid (female) offspring. This is in contrast to the report of deuterotoky in a laboratory population of *Tetranychus urticae* by Jesiotr and Suski (1970). They found that only 1 out of 100 virgin females gave entirely male offspring, and another 1 produced entirely daughters. The other females produced mixed offspring of both sons and daughters. The deuterotoky appeared to be ephemeral and the population switched to arrhenotoky with $2n = 6$ and $n = 3$ chromosomes (H.R. Bolland, personal communication, 1970).

CHROMOSOMES AND POLYPLOIDY

In contrast to flowering plants, for which about 40% of the species are polyploid, polyploidy seems to occur only rarely in bisexually reproducing animals. The sex-determining mechanism constitutes an important barrier to polyploidy in animals, particularly when a balance system is operating. An obstacle to the origin of a polyploid race (species) is the bisexuality itself. For initiation of a polyploid progeny with the parental chromosome number, both parents have to possess the same degree of polyploidy. The occurrence of a polyploid animal, however, is extremely rare and this animal will probably mate with a normal diploid of the opposite sex. In the case of tetraploidy, for instance, sterile triploid progeny will arise, which interferes with the establishment of a tetraploid population. Polyploidy in animals, therefore, is usually associated with parthenogenesis (White, 1973).

Arrhenotoky possibly facilitates the origin of polyploid races (species). An occasional tetraploid female may mate with her partheno-produced sons. The latter have the diploid genotypic constitution required for the establishment of a diplo-tetraploid race. The following facts, however, will make clear that polyploidy does not play an important role in the speciation process of tetranychid mites.

In different genera (*Bryobia, Schizotetranychus, Tetranychus*) chromosome numbers were found which gave some authors cause to suppose that polyploidy was involved. Results on the tumid spider mite (*Tetranychus tumidus* Banks) in particular encouraged this supposition. This haplo-diploid species has a chromosome number of $n = 6$ and is larger than the other *Tetranychus* species with $n = 3$. Recent analysis, however, could not confirm the occurrence of polyploidy. In the genus *Tetranychus*, haplo-diploid species were found with $n = 5$ and $n = 4$, so that the chromosome numbers in this genus present a continuous series from $n = 3$ to $n = 6$. Moreover, it has been shown that the genomes of *T. tumidus* ($n = 6$) and *T. urticae* ($n = 3$) have the same DNA content, so that diplo-tetraploidy is inconceivable for *T. tumidus* (Helle et al., 1983).

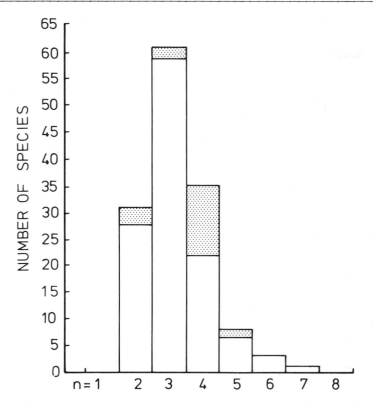

Fig. 1.2.3.1. Histogram of chromosome numbers in Tetranychidae. Stippled: sub-family Bryobiinae. Unstippled: sub-family Tetranychinae.

The species *Oligonychus grewiae* Meyer from South Africa has $n = 4$, and *O. randriamasii* Gutierrez from Madagascar has $n = 2$ (Helle et al., 1981). According to Smith Meyer (1974), these species exhibit the same diagnostic features. Since closely related tetranychid species generally have the same chromosome number, polyploidy may apply to *O. grewiae*. Determinations of DNA content are badly needed if polyploidy is to be proved.

The facts that the frequencies of chromosome numbers have a normal distribution (Fig. 1.2.3.1), and that a second peak at $n = 6$ does not occur in addition to the peak at $n = 3$, indicate that the numbers have changed by such processes as dissociations and fusions of chromosomes rather than by polyploidy. In addition, some general remarks about the chromosome numbers are appropriate. First, the numbers are always low. The number $n = 2$ is rather common and is found in many actinedid taxa (Helle and Wysoki, 1983); the highest number found so far is $n = 11$ for a camerobiid species (Bolland, 1983). As far as is known, the chromosomes of actinedid species are always holokinetic. It may thus be expected that an increase in chromosome number takes place through fragmentation because chromosome fragments enjoy a certain autonomy. It should be noted that no dramatic increase in chromosome number has taken place in the course of phylogeny of the Tetranychidae, as it has in Lepidoptera (White, 1973). Secondly, Fig. 1.2.3.1 shows that the chromosome number $n = 3$ is the modal number of the family, found in 61 species. The distribution of chromosome numbers bears some resemblance to a normal distribution. The distribution to higher numbers is $n = 4$ in 35, $n = 5$ in 8, $n = 6$ in 3 and $n = 7$ in 1 species. The distribution to the other side of the modal number gives 31 species with $n = 2$, but species with the chromosome number $n = 1$ are lacking. The reduction to a chromosome complement of a single chromosome seems to be prevented by unknown reasons.

Chapter 1.2.3. references, p. 138

SEX DETERMINATION

In most diploid animals, sex is determined by a balance between the male-determining ($= m$) and female-determining ($= f$) genes residing on sex chromosomes and/or autosomes. In the case of haplo-diploidy, a single set of chromosomes produces maleness, and a double set femaleness. The ratio between m and f is the same for haploids and for diploids and thus it is unlikely that sex is determined by a simple genic balance.

In different species of the haplo-diploid wasp *Habrobracon*, sex is determined by a series of multiple alleles (Whiting, 1945). The hemizygotes are normal males since m outweighs f. Females result from the complementary action of the female determiners in the sex alleles, when in heterozygous condition ($ff > mm$). In homozygotes, complementation is lacking; the balance, being in favour of maleness ($mm > ff$), gives rise to a diploid male (if viable). This system is easily demonstrated by an inbreeding regime, which is expected to lead either to unviable eggs or to diploid males (for further discussion, see Crozier, 1975). It has been shown by inbreeding studies that this system of complementary sex alleles does not apply to spider mites. Mother—son mating does not result in sex aberrants, nor in diploid males, nor in a drastic increase of egg lethality (Helle and Overmeer, 1973).

Sex aberrations were observed in an inbred line (line 23) of *T. urticae* (Van Eyndhoven and Helle, 1966). In this line normal females and males were found, but also sterile sex mosaics, which were described as 'giant males' and 'intersexes' (see plate IA). Further investigations (Helle, unpublished work) showed that the genetics of the sex aberrants was rather complex. In certain crosses between individuals of this line, the percentage of sex aberrants increased in subsequent offspring generations (Table 1.2.3.2). During this 'drive' the percentage of normal males did not change, but that of the females decreased. In other crosses the percentage of sex aberrants decreased in subsequent generations, while that of normal females increased. Out-crossing of females of line 23 with males of a reference strain gave rise to a normal F_1. In the next (F_2) generation, however, a varying number of sex aberrants sometimes appeared. In out-crosses of the reverse type, using males of line 23, sex aberrants never emerged. The phenomenon thus could not be transmitted by males only. In out-crosses of females of line 23 with males carrying the recessive marker *albino*, an 'albino-23' strain was established. With the aid of the marker strain it could be proven that all sex aberrants were biparental. Normal males were always partheno-produced and had the genotype of the mother. Genetic studies with this strain did not provide information on the genetics of sex determination in *T. urticae*. However, the cytogenetic analysis yielded more concrete facts. In squash preparations of embryonic tissues of a normal strain, the cells show either 6 (female tissue) or 3 (male tissue) chromosomes. In embryonic tissues of line 23, either these normal chromosome numbers or polyploid numbers were found. Tissues of 1 egg could have 6, 9, 12, 15, 18, and 24 chromosomes, indicating the basic number of $n = 3$. It is tempting to suppose that these embryos with a mixture of polyploid cells are the sex aberrants. If so, the higher degrees of ploidy favour masculinization. In terms of sex balance, degrees of ploidy higher than diploidy seem to increase the effect of the male-determining factors.

Androgenesis in *T. urticae* was studied by Overmeer et al. (1972). If unmated females were X-irradiated at a dosage of 40 kR or higher, the eggs produced by these females did not hatch. However, if these females were, after irradiation, mated with untreated males, descendants were occasionally

TABLE 1.2.3.2

Incidence of sex aberrants in offspring generations obtained from crosses with inbred line 23 of *Tetranychus urticae* Koch. The figures given represent the number of females ($\female$), males ($\male$) and sex aberrants ($\female\!\!\!^{\nearrow}$) in a sample randomly taken from a generation, and do not reflect the fecundity of the preceding generation. Crosses 1, 2, 3, 4, 5 and 6 indicate crosses between a randomly taken male and female of line 23. Crosses 7 and 8 refer to out-crosses of a female of line 23 and a male of strain *Sambucus*

Cross 1		$\female$	$\male$	$\female\!\!\!^{\nearrow}$	Cross 2		$\female$	$\male$	$\female\!\!\!^{\nearrow}$
	P_1	1	1	—		P_1	1	1	—
	F_1	26	12	0		F_1	32	16	0
	F_2	46	20	0		F_2	86	35	6
	F_3	62	24	0		F_3	59	28	0
	F_4	70	22	0		F_4	112	33	0
Cross 3		$\female$	$\male$	$\female\!\!\!^{\nearrow}$	**Cross 4**		$\female$	$\male$	$\female\!\!\!^{\nearrow}$
	P_1	1	1	—		P_1	1	1	—
	F_1	16	8	4		F_1	31	5	10
	F_2	28	8	1		F_2	18	31	35
	F_3	56	22	0		F_3	2	19	39
	F_4	50	15	0		F_4	0	26	40
Cross 5		$\female$	$\male$	$\female\!\!\!^{\nearrow}$	**Cross 6**		$\female$	$\male$	$\female\!\!\!^{\nearrow}$
	P_1	1	1	—		P_1	1	1	—
	F_1	1	28	25		F_1	0	18	35
	F_2	0	16	22					
Cross 7		$\female$	$\male$	$\female\!\!\!^{\nearrow}$	**Cross 8**		$\female$	$\male$	$\female\!\!\!^{\nearrow}$
	P_1	1	1	—		P_1	1	1	—
	F_1	22	7	0		F_1	12	8	0
	F_2	35	29	8		F_2	36	12	8
	F_3	18	44	20		F_3	121	45	8
	F_4	1	18	20		F_4	210	87	0
	F_5	0	6	2		F_5	124	47	0

obtained which were exclusively female. Using males with a marker gene for *white-eye*, the authors could prove that all of the 27 androgenetic females, obtained from 78 000 eggs, were of paternal genotype. Androgenetic males did not appear, as was expected from similar experiments on *Habrobracon juglandis* (Whiting, 1955). This probably means that a haploid male nucleus cannot survive in (irradiated) maternally inherited cytoplasm and that a fertilized egg can develop only at the diploid level, no matter how diploidy arises; however, when the male nucleus doubles, in one way or another, a diploid female (and not a male!) develops. Polyspermy does not seem to be regular in spider mites (Chapter 1.2.2), but it cannot be excluded as an explanation.

SEX RATIO

Sex ratio is primarily dependent on the amount of sperm transferred to a female. If during an experiment copulation is disturbed, the sex ratio of the offspring produced by the female changes because the number of daughters is lowered (Overmeer, 1972). If such a female is allowed to mate again with another male, the proportion of daughters will increase. Although the average tertiary sex ratio can vary a lot within offspring of individual mated females, and can also vary between different strains of a species, a ratio of 1

Chapter 1.2.3. references, p. 138

male to approximately 3 females is very often found and may be considered as 'normal'. This applies not only to *T. urticae* but to many other bisexual tetranychid species. The sex ratio is genetically controlled (Overmeer and Harrison, 1969). In fact, it is normal that in most arrhenotokous species females outnumber males.

The question arises of how a certain proportion of the eggs escapes fertilization and, consequently, provides for the origin of an adequate proportion of males in the progeny. In *T. urticae*, at least the first egg produced by a mated female is always male determined (Helle, 1967). This egg is not fertilized because sperm cannot reach it in time for the internal fertilization (Chapter 1.2.2). During the remaining oviposition period, eggs escape fertilization, notwithstanding the presence of sufficient sperm. The movement of these sperm towards the oocytes does not seem to be under female control. It is not known which eggs are fertilized and which are not. It has been observed (Feiertag-Koppen and Pijnacker, 1982) that the ventral region of the ovary receives less sperm than the dorsal region and that oocytes in the medio-ventral region possibly escape sperm penetration. If so, it is not known which mechanism is involved.

REFERENCES

Bolland, H.R., 1983. A description of *Neophyllobius aesculi* n.sp. and its developmental stages (Acari: Camerobiidae). Entomol. Ber. (Amsterdam), 43: 42—47.

Bolland, H.R., Gutierrez, J. and Helle, W., 1981. Chromosomes in spider mites (Tetranychidae — Acari). Acarologia, 22: 271—275.

Crozier, R.H., 1975. Hymenoptera. In: B. John (Editor), Animal Cytogenetics, Vol. 3, Insecta 7. Gebrüder Bornträger, Berlin, 95 pp.

De Jong, J.H., Lobbes, P.V. and Bolland, H.R., 1981. Karyotypes and sex determination in two species of laelapid mites (Acari: Gamasida). Genetica, 55: 187—190.

Feiertag-Koppen, C.C.M. and Pijnacker, L.P., 1982. Development of the female germ cells and process of internal fertilization in the two-spotted spider mite *Tetranychus urticae* Koch (Acariformes: Tetranychidae). Int. J. Insect Morphol. Embryol., 11: 271—284.

Flechtmann, B.N.N. and Flechtmann, C.H.W., 1982. Observações sobre a reprodução do ácaro vermelho do cafeeiro. Anais II Congresso Brasileiro de Imciação Cientifica em Ciências Agrárias, Piracicaba, SP, 1982. pp. 109—111.

Gutierrez, J., 1977. Un tétranyque polyphage de la zone intertropicale: *Oligonychus thelytokus* n.sp. (Acariens, Tetranychidae). Description et premières données biologiques. Cah. ORSTOM, Sér. Biol., 12: 65—72.

Gutierrez, J. and Bolland, H.R., 1973a. *Schizonobia sycophanta* Womersley (Acariens: Tetranychidae) décrit de Tasmanie est probablement originaire d'Europe. Complément de la description et étude cytogénétique. Entomol. Ber. (Amsterdam), 33: 54—60.

Gutierrez, J. and Bolland, H.R., 1973b. Description et caryotype d'une nouvelle espèce du genre Tetranychus Dufour (Acariens: Tetranychidae) récoltée au Tchad sur *Dolichos lablab* L. (Papilionaceae). Entomol. Ber. (Amsterdam), 39: 88—94.

Gutierrez, J. and Helle, W., 1971. Deux nouvelles espèces du genre Eutetranychus Banks (Acariens: Tetranychidae) vivant sur plantes cultivées à Madagascar. Entomol. Ber. (Amsterdam), 31: 45—60.

Gutierrez, J. and Helle, W., 1983. *Eotetranychus rubiphilus* espèce nouvelle pour la France et l'Europe occidentale: redescription et caryotype (Acari: Tetranychidae). Entomol. Ber. (Amsterdam), 43: 109—112.

Gutierrez, J., Bolland, H.R. and Helle, W., 1979a. Karyotypes of the Tetranychidae and the significance for taxonomy. Rec. Adv. Acarol., 2: 399—404.

Gutierrez, J., Helle, W. and Bolland, H.R., 1970. Etude cytogénétique et reflexions phylogénétiques sur la famille des Tetranychidae Donnadieu. Acarologia, 12: 732—751.

Gutierrez, J., Helle, W. and Bolland, H.R., 1979b. Etude d'une souche de *Tetranychus piercei* (Acariens: Tetranychidae) d'Indonésie: redescription, caryotype et reproduction. Entomol. Ber. (Amsterdam), 33: 155—158.

Hartl, D.L. and Brown, S.W., 1970. The origin of male haploid genetic systems and their expected sex ratio. Theor. Popul. Biol., 1: 165—190.

Helle, W., 1967. Fertilization in the two-spotted spider mite (*Tetranychus urticae*: Acari). Entomol. Exp. Appl., 10: 103—110.

Helle, W. and Bolland, H.R., 1967. Karyotypes and sex-determination in spider mites (Tetranychidae). Genetica, 38: 43—53.

Helle, W. and Gutierrez, J., 1983. Karyotypes of tetranychid species from New Caledonia. Int. J. Acarol., 9: 123—126.

Helle, W. and Overmeer, W.P.J., 1973. Variability in tetranychid mites. Annu. Rev. Entomol., 18: 97—120.

Helle, W. and Wysoki, M., 1983. The chromosomes and sex-determination of some actinotrichid taxa (Acari) with special reference to Eriophyidae. Int. J. Acarol., 9: 67—71.

Helle, W., Bolland, H.R. and Heitmans, W.R.B., 1980. Chromosomes and types of parthenogenesis in the false spider mites (Acari, Tenuipalpidae). Genetica, 54: 45—50.

Helle, W., Bolland, H.R. and Heitmans, W.R.B., 1981. A survey of the chromosome complements in the Tetranychidae. Int. J. Acarol., 7: 147—155.

Helle, W., Gutierrez, J. and Bolland, H.R., 1970. A study on sex-determination and karyotypic evolution in Tetranychidae. Genetica, 41: 21—32.

Helle, W., Tempelaar, M.J. and Drenth-Diephuis, L.J., 1983. DNA-contents in spider mites (Tetranychidae) in relation to karyotype evolution. Int. J. Acarol., 9: 127—129.

Helle, W., Bolland, H.R., Van Arendonk, R., De Boer, R., Schulten, G.G.M. and Russell, V.M., 1978. Genetic evidence for biparental males in haplo-diploid predator mites (Acarina: Phytoseiidae). Genetica, 48: 165—171.

Hoy, M.A., 1979. Parahaploidy of the 'arrhenotokous' predator *Metaseiulus occidentalis* (Acarina: Phytoseiidae) demonstrated by X-irradiation of males. Entomol. Exp. Appl., 26: 97—104.

Jesiotr, L. and Suski, Z.W., 1970. A case of deuterotoky in the two-spotted spider mite *Tetranychus urticae* Koch (Acarina, Tetranychidae). Bull. Acad. Polon. Sci., 18: 33—35.

Nelson-Rees, W.A., Hoy, M.A. and Roush, R.T., 1980. Heterochromatinization, chromatin elimination and haploidization in the parahaploid mite *Metaseiulus occidentalis* (Nesbitt) (Acarina: Phytoseiidae). Chromosoma, 77: 262—276.

Oliver, J.H., 1971. Parthenogenesis in mites and ticks (Arachnida: Acari). Am. Zool., 11: 283—299.

Overmeer, W.P.J., 1972. Notes on mating behaviour and sex ratio control of *Tetranychus urticae* Koch (Acarina: Tetranychidae). Entomol. Ber. (Amsterdam), 32: 240—244.

Overmeer, W.P.J. and Harrison, R.A., 1969. Notes on the control of the sex ratio in populations of the two-spotted spider mite, *Tetranychus urticae* Koch (Acarina: Tetranychidae). N. Z. J. Sci., 12: 920—928.

Overmeer, W.P.J., Van Zon, A.Q. and Helle, W., 1972. Androgenesis in spider mites. Entomol. Exp. Appl., 15: 256—257.

Pijnacker, L.P., Ferwerda, M.A., Bolland, H.R. and Helle, W., 1980. Haploid female parthenogenesis in the false spider mite *Brevipalpus obovatus* (Acari: Tenuipalpidae). Genetica, 51: 211—214.

Rieger, R., Michaelis, A. and Green, M.M., 1976. Glossary of Genetics and Cytogenetics. Springer-Verlag, Berlin, Heidelberg, New York, 647 pp.

Smith Meyer, M.K.P., 1974. A Revision of the Tetranychidae of Africa (Acari) with a key to the genera of the world. Repub. S. Afr., Dep. Agric. Tech. Serv., Entomol. Mem., 36.

Smith Meyer, M.K.P. and Bolland, H.R., 1984. Tetranychoid mites (Acari: Prostigmata) from the Cameroon and a survey of their chromosome complements. Phytolactica, in press.

Van Eyndhoven, G.L. and Helle, W., 1966. Sex abnormalities in the common spider mite (*Tetranychus urticae*) (Acari, Tetranychidae). Entomol. Ber. (Amsterdam), 26: 204—208.

White, M.J.D., 1973. Animal Cytology and Evolution, third edition. Cambridge University Press, Cambridge, 961 pp.

Whiting, A.R., 1955. Androgenesis as evidence for the nature of X-ray induced injury. Radiat. Res., 2: 71—78.

Whiting, P.W., 1945. The evolution of male haploidy. Q. Rev. Biol., 20: 231—260.

1.2.4 Fine Structure and Related Properties of the Chromosomes

M.J. TEMPELAAR

INTRODUCTION

The behaviour of the chromosomes of *Tetranychus urticae* Koch (Pijnacker and Ferwerda, 1976; Feiertag-Koppen, 1976) suggests that these chromosomes are probably of the holokinetic type, i.e. the kinetic faculty is spread over the entire length of the chromosome. The designation 'holokinetic' covers a variety of arrangements of association between chromosomes and microtubules (spindle fibres), which indicates that this kind of chromosome is by no means unique to *T. urticae*.

Electron and light microscopic studies were carried out at the Department of Genetics of Groningen University in order to elucidate the fine structure of the chromosomes of *T. urticae*. No other acarine species have been analysed in this respect. This chapter confirms and illustrates the holokinetic structure of the chromosomes. In addition, it reports on the behaviour of induced aberrations, especially fragments and concludes that ultimately the difference in structure with the well-known monokinetic chromosomes does not lead to such drastic effects on the karyotype of adults and progeny after induction of aberrations in embryonic stages as might be anticipated. First, the structure of the chromosomes is dealt with.

The second part is a summary of radiation experiments, carried out in order to examine the fate and properties of induced chromosome aberrations. It correlates the frequency of fragments on one hand, and the appearance of lethality and DNA deficiencies on the other. In addition, facts are given on reciprocal translocations, which also are an important type of aberration in this type of chromosome.

The main aim of the investigations described in this chapter was to obtain a coherent picture concerning the structure and associated properties of these particular holokinetic chromosomes, comparable to the detailed concepts described for the ubiquitous, well-known 'monokinetic' chromosomes (those with kinetic facilities limited to a restricted area) of most eukaryotic subjects of genetic studies.

The findings may also be compared with the few data available from other organisms with holokinetic chromosomes and some may be pertinent to radiation research on methods of genetic control.

For a review of the techniques involved, the reader is referred to Chapters 1.5.4 and 1.5.5 in this book; for more information on the subject matter, the original papers by the author, mentioned in the references section, should be consulted.

Chapter 1.2.4. references, p. 147

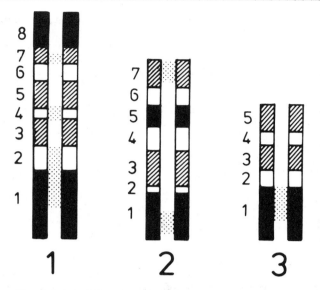

Fig. 1.2.4.1. Idiogram of Giemsa-banded metaphase chromosomes in cleavage stages of haploid (male) eggs. (Reproduced with permission from Pijnacker and Ferwerda (1976).)

CHROMOSOME STRUCTURE

Various aspects of the *T. urticae* genome may be briefly summarized. The chromosome set of females contains the diploid number, 6; that of the males the haploid number, 3. The chromosomes differ slightly in length, and they can be characterized by Giemsa-banding (Fig. 1.2.4.1) (Pijnacker and Ferwerda, 1976).

The DNA content, determined by cytophotometric measurement of the absorption of Feulgen-stained nuclei, is quite low, 0.1 pg in the haploid unreplicated genome (Tempelaar, 1980), which is somewhat lower than that of much-used research subjects like *Drosophila melanogaster* Meigen (0.18 pg; Rasch et al., 1971) and *Habrobracon hebetor* (Say) = *H. juglandis* (0.15 pg; Rasch et al., 1977). By polyploidization, the genome may increase to 8 or 256 times the initial amount of DNA in the cells of some organs.

In the early phases of development, chromosomes tend to form individual nuclei (karyomeres or micronuclei), rather than one common nucleus. The implication is that each chromosome is able to produce a nucleolus, because there are indications that without sufficient nucleolar material, chromosomes in micronuclei will not replicate (Das, 1962). Indeed, electron microscopic (EM) observations show nucleoli in karyomeres (Fig. 1.2.4.2a), while silver staining (Fig. 1.2.4.2b) and light microscopic (LM) reconstructions suggest at least as many nucleoli as chromosomes (Tempelaar and Drenth-Diephuis, 1984).

In the following paragraphs, LM and EM evidence is restricted mainly to structures and properties, related to the holokinetic nature of the chromosomes, because this is what sets the *T. urticae* chromosomes apart from their monokinetic counterparts.

With the light microscope, direct evidence cannot be obtained about the extent of the association of microtubules and mitotic chromosomes. In monokinetic chromosomes, microtubules converge onto a restricted area of the chromosome. This is called the 'primary constriction' by light microscopists and it is an easily recognizable feature of many chromosomes. The kinetochore itself and the number of associated microtubules, however, cannot be determined in this way. The orientation of the chromosomes gives

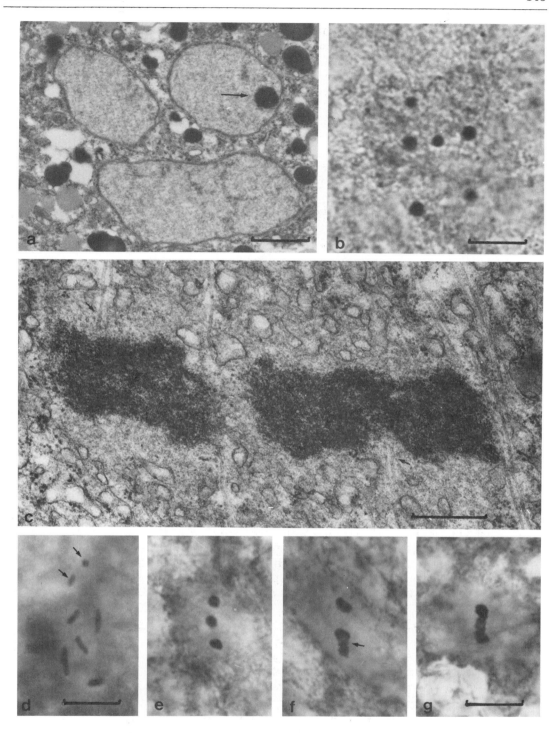

Fig. 1.2.4.2. Photographs of mitosis, interphase and meiosis. (a) Ultrastructure of karyomeres in early (5th) cleavage stage: Chromosomes have transformed into separate
karyomeres (also called 'micronuclei'). Arrow indicates nucleolar material. Bar = 2 μm.
(b) In light microscopic preparations, silver staining indicates the presence of at least as
many nucleoli (dark particles) as chromosomes: here, the 6 nucleoli originate from 3
chromosomes. Bar = 8 μm. (c) Ultrastructure of a cell in mitosis. Microtubules (arrows)
run towards the darkly stained metaphase chromosomes. Bar = 0.7 μm. (d) Induced
chromosomal aberrations in mitosis. Diploid mitotic metaphase carrying 2 chromosome
fragments (arrowed) after breakage of a chromosome. Bar = 8 μm. (e) Normal meiotic
metaphase I, showing 3 bivalents. (Cf. Fig. 1.2.4.2f, g.) (f) Induced chromosomal aberrations in meiosis. Arrow indicates association of 2 bivalents, caused by a translocation
in meiotic metaphase I. (g) Association of 3 bivalents in meiotic metaphase I, caused by
2 translocations. Bar = 8 μm.

additional clues for the position of the kinetochores: in the 'monokinetic' case, the part with the kinetochore leads the way when segregating.

In *T. urticae*, none of the set of 3 chromosomes shows a primary constriction; moreover, chromosomes move in a parallel fashion in anaphase of mitosis of the early embryonic stages (Pijnacker and Ferwerda, 1972) and meiosis (Chapter 1.2.2). This is because association between microtubules and chromosomes is not confined to one spot. Further indications to this end were obtained from LM studies of the fate of fragments in consecutive division (cf. the following section on radiobiological properties).

The number of microtubules involved and their dispersal pattern is a more elaborate matter to establish, as this requires electron microscopy and serial ultra-thin sectioning techniques (Tempelaar and Drenth-Diephuis, 1983). From individual photographs of a series, the pertinent features are traced; these drawings may be combined into reconstructions, either manually to yield three-dimensional models, or by computer, which can generate views of different angles from the same reconstruction.

Figure 1.2.4.2c is a picture of a single section through the chromosomes. Around the dark mass of chromatin, no kinetochore plate can be seen. The microtubules seem to end in or very near the chromosome body. In Figs. 1.2.4.3 and 1.2.4.4, reconstructions indicate the rather uniform dispersal of chromosome-associated microtubules over the poleward surfaces of the entire chromosome. From a number of such reconstructions, a cautious estimate puts the number of microtubules at each polar surface at 60—100

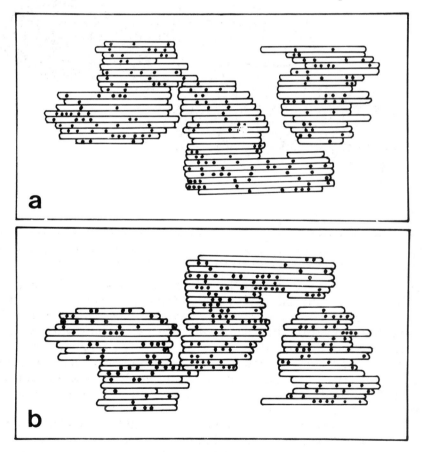

Fig. 1.2.4.3. Drawing from a plastic model of a part of a chromosome set. Images (a) and (b) depict the view from opposite sides. Microtubule-termination points are marked in black, and are seen to be scattered all over the length of the chromosomes.

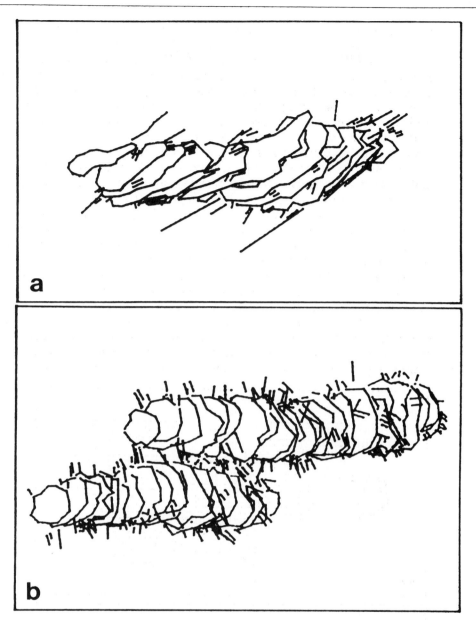

Fig. 1.2.4.4. Computer-drawn reconstructions of chromosomes, showing the shape of the sectioned chromosomes and the position of the attached microtubules.

for each of the chromosomes. This clearly confirms the holokinetic structure of the *T. urticae* chromosomes inferred from light microscopic data.

The kinetic structure of the chromosomes of *T. urticae* differs from that described in ultrastructure studies for other arthropods with diffuse kinetochores; layered kinetochores are present over some patches or over the entire length of the mitotic chromosomes of some insects (Buck, 1967; Comings and Okada, 1972; Ruthmann and Permantier, 1973) and arachnids (Riess et al., 1978; Benavente, 1982). In meiosis, some of these species have the microtubules directly inserted into the chromosomes.

Thus, *T. urticae* mitotic chromosomes resemble the meiotic chromosomes of some arthropod species rather than mitotic ones, as far as association with microtubules is concerned. A more detailed comparison of the number of microtubules and their dispersal is not possible, as the studies referred to do not provide this kind of information.

Chapter 1.2.4. references, p. 147

RADIOBIOLOGICAL PROPERTIES

When breaks or exchanges are induced in holokinetic chromosomes, the fate of these aberrations to a certain extent is determined by their kinetic abilities.

As expected from the chromosome structure described in the foregoing section, the probability of aberrations meeting with problems in mitosis is much less than for monokinetic chromosomes. This does not imply, however, that an unlimited number of aberrations can be maintained and transmitted to the subsequent generations. The following paragraphs will deal with these aspects for fragments and translocations; for more details see Tempelaar (1979a, b).

Fragments

In a typical experiment, breaks were induced by Röntgen irradiation in sperm, and the fragments of the paternal set were counted in the first 4—5 divisions of fertilized eggs (Fig. 1.2.4.2d). Less than 1% of the fragments ends up in the wrong daughter cell; this low frequency is quite in line with expectations based on the holokinetic nature of the chromosomes.

Loss of fragments is a much more common occurrence, however, amounting to 4.8% per fragment. Events leading to loss may occur not only in mitosis, but also in interphase: if a piece of chromosome does not possess sufficient nucleolar material at the onset of interphase, it may not be able to replicate (see, e.g., Das, 1962).

From correlations of cytological data and figures about lethality, it follows that loss or mis-segregation of parts of chromosomes always leads to death in the embryonic stages, even if the intact maternal genome is present. This is confirmed by cytophotometric determinations of the DNA content in larval stages, which do not indicate the presence of nuclei with a subnormal DNA content. Apparently, the presence in eggs in the early stages of development of ± 5% cells containing an incomplete or unbalanced chromosome set is enough to offset normal development.

Thus, fragments of holokinetic chromosomes are inviable aberrations in *T. urticae*, despite their large measure of stability in mitotic divisions.

Translocations

By radiation-induced exchange of parts from monokinetic chromosomes, mitotically inviable combinations arise. These are dicentrics, with 2 kineto-chores, and acentrics, without kinetochores. Another type of exchange is symmetrical, yielding mitotically viable chromosomes.

In holokinetic chromosomes, there is no distinction between these types of exchange: all combinations should be viable in mitosis. In *T. urticae*, translocations were scored in the female progeny of males which had their sperm irradiated. These females show the characteristic association of 2 or 3 bivalents in meiosis, indicating the presence of 1 or 2 translocations in the heterozygous condition (Fig. 1.2.4.2e—g). The frequency of translocation-carrying females can be very high in an experiment, e.g. 70% at 2000 R. As there is a low degree of lethality at this dose, the percentage quoted is a reliable reflection of the true frequency of induced translocations.

Not all of the translocations induced and present in adult females can be transmitted to the progeny. In some cases, such females are not able to produce the eggs which are already present in their ovaries. In other cases, embryonic lethality in the progeny of the females is nearly 100%.

In addition, there is a marked difference between the sexes of the surviving fraction: the females, which are diploids, are carriers of translocations for the expected 50%, while in males, which are haploid, only one-third of this amount are carriers. This means that translocations are often associated with recessive lethality, which is an obstacle to transmission through the haploid generation. In conclusion, a large part of the induced translocations, which meet no difficulties in mitosis of the diploid first generation, cannot be transmitted to the offspring, especially in the case of haploids.

In concluding this section, it may be stated that radiation experiments in *T. urticae* yielded detailed evidence for the kinetic properties of broken chromosomes. Earlier studies in meiotic and mitotic divisions of the actinedid mites (Feiertag-Koppen, 1976; Cooper, 1972; Keyl, 1957) gave some indications of a certain persistence of fragments over a few divisions. The experiments described in this chapter indicate that the fragments induced in radiation experiments evidently have a low probability, less than 5%, of not being transmitted to the next division: in contrast, 43% of the acentric fragments of monokinetic chromosomes are lost as paired structures, according to Carrano (1973).

Translocations, like fragments, of the holokinetic chromosomes do not meet with many problems in mitotic divisions.

Despite the ease with which aberrant holokinetic chromosomes pass through mitotic divisions, selection by lethality and sterility eliminates fragments and a portion of the translocations within 1 generation. In this respect, the difference between the holokinetic *T. urticae* chromosome and the monokinetic chromosome is not as large as would be suspected on the basis of the kinetic organization.

This is an important fact to note when considering aspects of karyotype evolution in the Tetranychidae: a large range in chromosome number, due to fragmentation, is not expected (see Chapter 1.1.5). The same applies to radiological research into methods of genetic control (see, e.g., Feldmann, 1975, 1978, where the chromosomal basis of the radiation effects is inferred from lethality and sterility figures). Knowledge of chromosome structure and properties of aberrations may provide a sound framework for such extrapolations.

ACKNOWLEDGEMENTS

The author wishes to thank L.J. Drenth-Diephuis for her part in the experiments, Dr L.P. Pijnacker for comments, H. Mulder for printing the figures and S.I. Walburgh Schmidt for typing the final version.

REFERENCES

Benavente, R., 1982. Holocentric chromosomes: Presence of kinetochore plates during meiotic divisions. Genetica, 59: 23—27.

Buck, R.C., 1967. Mitosis and meiosis in *Rhodnius prolixus*: The fine structure of the spindle and diffuse kinetochore. J. Ultrastruct. Res., 18: 489—501.

Carrano, A.V., 1973. Chromosome aberrations and radiation-induced cell-death, I. Transmission and survival parameters of aberrations. Mutat. Res., 17: 341—353.

Comings, D.E. and Okada, T.A., 1972. Holocentric chromosomes in *Oncopeltus*: Kinetochore plates are present in mitosis but absent in meiosis. Chromosoma, 37: 177—192.

Cooper, R., 1972. Experimental demonstration of holokinetic chromosomes, and of differential 'radiosensitivity' during oogenesis, in the grass mite, *Siteroptes graminum* (Reuter). J. Exp. Zool., 182: 69—94.

Das, N.K., 1962. Synthetic capacities of chromosome fragments correlated with their ability to maintain nucleolar material. J. Cell Biol., 15: 121—130.

Feiertag-Koppen, C.C.M., 1976. Cytological studies of the two-spotted spider mite *Tetranychus urticae* Koch (Tetranychidae, Trombidiniformes), Meiosis in eggs. Genetica, 46: 445-456.

Feldmann, A.M., 1975. Induction of structural chromosome mutations in males and females of *Tetranychus urticae* Koch (Acari, Tetranychidae). In: Sterility Principle for Insect Control 1974. STI/PUB/377, IAEA, Vienna, pp. 437—445.

Feldmann, A.M., 1978. Dose—response relationships and RBE values of dominant lethals induced by X-rays and 1.5-MeV Neutrons in prophase-1 oocytes and in mature sperm of the two-spotted spider mite *Tetranychus urticae* Koch (Acari, Tetranychidae). Mutat. Res., 51: 361—376.

Keyl, H.G., 1957. Zur Karyologie der Hydrachnellen (Acarina). Chromosoma, 8: 719—729.

Pijnacker, L.P. and Ferwerda, M.A., 1972. Diffuse kinetochores in the chromosomes of the arrhenotokous spider mite *Tetranychus urticae* Koch. Experientia (Basel), 28: 354.

Pijnacker, L.P. and Ferwerda, M.A., 1976. Differential Giemsa staining of the holokinetic chromosomes of the two-spotted spider mite, *Tetranychus urticae* Koch (Acari, Tetranychidae). Experientia, 32: 158.

Rasch, E.M., Barr, H.J. and Rasch, R.W., 1971. The DNA content of sperm of *Drosophila melanogaster*. Chromosoma (Berlin), 33: 1—18.

Rasch, E.M., Cassidy, J.D. and King, R.C., 1977. Evidence for dosage compensation in parthenogenetic Hymenoptera. Chromosoma (Berlin), 59: 323—340.

Riess, R.W., Barker, K.R. and Biesele, J.J., 1978. Nuclear and chromosomal changes during sperm formation in the scorpion, *Centruroides vittatus* (Say). Caryologia, 31: 147—160.

Ruthman, A. and Permantier, Y., 1973. Spindel und Kinetochoren in der Mitose und Meiose der Baumwollwanze *Dysdercus intermedius* (Heteroptera). Chromosoma, 41: 271—288.

Tempelaar, M.J., 1979a. Aberrations of holokinetic chromosomes and associated lethality after X-irradiation of meiotic stages in *Tetranychus urticae* Koch (Acari, Tetranychidae). Mutat. Res., 61: 259—274.

Tempelaar, M.J., 1979b. Fate of fragments and properties of translocations of holokinetic chromosomes after X-irradiation of mature sperm of *Tetranychus urticae* Koch (Acari, Tetranychidae). Mutat. Res., 63: 301—316.

Tempelaar, M.J., 1980. DNA-content in isolated nuclei of postembryonic stages of progeny from normal and irradiated males of *Tetranychus urticae* Koch (Acari, Tetranychidae). Chromosoma, 77: 359—371.

Tempelaar, M.J. and Drenth-Diephuis, L.J., 1983. Ultrastructure of holokinetic mitotic chromosomes and interphase nuclei of *Tetranychus urticae* Koch (Acari, Tetranychidae), in relation to loss and mis-segregation of induced fragments. Chromosoma, (Berlin), 88: 98—103.

Tempelaar, M.J. and Drenth-Diephuis, L.J., 1984. The nucleolar cycle in karyomeres of *Tetranychus urticae* Koch (Acari, Tetranychidae) in relation to replication of fragments of holokinetic chromosomes. Protoplasma, 123: 78—82.

1.2.5 Embryonic and Juvenile Development

A. CROOKER

EMBRYOLOGY

Introduction

The earliest research on spider mite embryology dates back to over 100 years ago when Claparede (1868) examined germ band formation in *Tetranychus urticae* Koch, the two-spotted spider mite. His work was one of the pioneer studies in arachnid embryology and was the first embryological study of a mite. Little additional work was done on spider mite embryos until over 80 years later when Gasser (1951) reported on *Panonychus ulmi* (Koch), the European red mite, in a paper on *T. urticae*, and Fukuda and Shinkaji (1954) published a report on the embryology of *Panonychus citri* (McGregor), the citrus red mite. In the mid-1960s to 1971, Dittrich (1965, 1968, 1969, 1971) and Dittrich and Streibert (1969) provided more detailed information on embryonic development. The current knowledge of spider mite embryology is derived mainly from the thorough work of Dittrich.

Embryonic development takes place in eggs deposited on the substrate (often vegetation, but in some species, soil or soil objects near plants) by the primitive ovipositor of the adult female (Beament, 1951; Lees, 1961; Van de Lustgraaf, 1977). During oviposition, the tube-like ovipositor is formed as a continuation of the posteroventral body surface and consists of the genital flap, folded integument of the genital region, and copulatory pore (Fig. 1.2.5.1). The egg is deposited when a portion of the genital canal is evaginated. Oviposition follows a brief (1—2 days) preoviposition period; it increases rapidly to a peak in a few days, and then declines. Deposition of 5 or 6 eggs per day appears to be average for many spider mite species of economic importance. The duration of the oviposition period varies widely depending on species and environmental conditions, but 10—15 days appears to be average. Total egg production also varies greatly from as few as 10 to 150 or more, depending on species, humidity, temperature, host and other factors (see Chapter 1.2.6).

Egg structure

The diameter of spider mite eggs varies from 110 to 150 μm depending on species (Dittrich and Streibert, 1969). This is large compared with the size of the adult female who may be only 2—3 times as wide as a fully developed egg. Egg shape varies from round to onion-like. The colour of the translucent to opaque eggs ranges from white to pale green, dark green, brown, orange and red, depending on species and physiological condition. The egg surface may be smooth (Fig. 1.2.5.2), lightly sculptured, striated to minutely striated

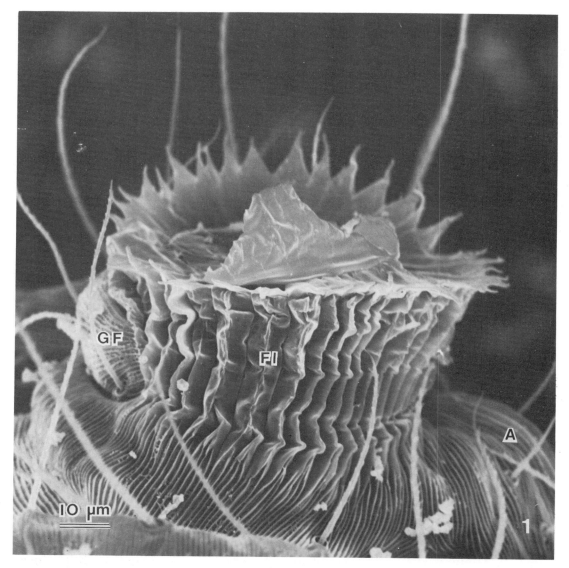

Fig. 1.2.5.1. Ovipositor of *Tetranychus urticae*. (A) anal region; (FI) folded integument of genital region; (GF) genital flap. Crooker, unpublished scanning electron micrograph.

or ribbed as a result of contact with the everted genital canal. A character-istic spike surmounts some eggs, such as those of *P. ulmi* (see Fig. 1.4.1.1) and the non-diapause eggs of *Petrobia latens* Muller, the brown wheat mite, or the eggs may be topped with a thick wax cap, as in the summer diapause eggs of *P. latens*. If embryonic development is sufficiently advanced, the respiratory system cones of *T. urticae* project above the egg surface.

The shell of *P. ulmi* consists of an outer, thick wax layer and a cement layer of oil and protein that attaches the wax to an underlying inner 'shell' layer (Beament, 1951). Beament determined that the egg received its inner 'shell' layer in the ovary and that the cement and wax layers were secreted by the glandular ovipositing pouch after the 'shell' layer contacted the sub-strate during oviposition. No wax was present at the base of the egg where it was in contact with the substrate. Hopp (1954) observed similar cement, wax and shell layers in *T. urticae*. However, cement and wax layers were not observed for this species by Dittrich (1971) who assumed these layers were dissolved during preparation for microscopy. He found the 'structureless'

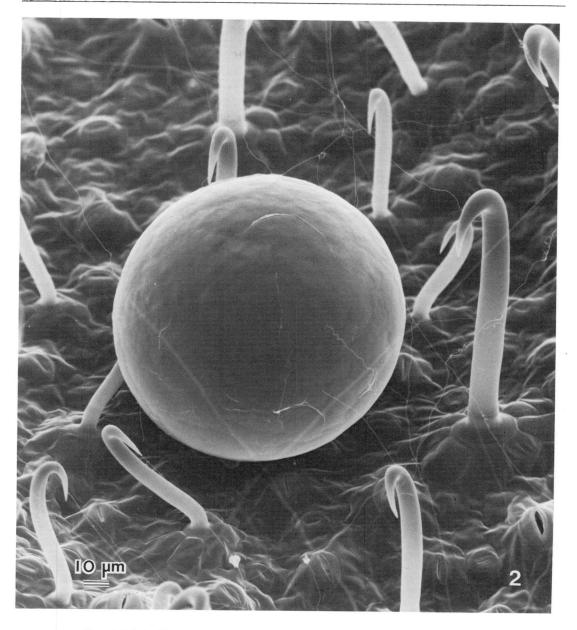

Fig. 1.2.5.2. Egg of *Tetranychus urticae* on bean leaf. Crooker, unpublished scanning electron micrograph.

shell layer to be approximately 0.1 μm thick. Mothes and Seitz (1981) studied the chitosomal (vesicular or particulate) chitin synthesizing system of the developing eggs of *T. urticae*. They determined that the shell consisted of an outer granular layer, a middle dense layer, and an inner electron-transparent layer. The chitin of the shell did not appear to be deposited in the form of chitin microfibrils. Lees (1961) found an outer wax layer and an inner 'shell' layer enclosing the ovum of *P. latens*. The presence of a cement layer could not be verified in the non-diapause eggs. A wrinkled membrane was found between the inner 'shell' layer and outer wax covering of the winter egg, but it could not be determined whether this membrane was the homologue of Beament's cement layer in *Panonychus*.

Chapter 1.2.5. references, p. 160

Embryonic development

A newly deposited egg is in the first metaphase of meiosis (Feiertag-Koppen, 1976). The metaphase bivalents can be found in an island of cytoplasm in the egg cortex. If the egg has been fertilized, a single, elongate sperm is present in the cortex: syngamy occurs later in the egg center. In spider mites, fertilized eggs produce only females; unfertilized eggs produce males (Chapter 1.2.3). The second meiotic division is completed approximately 2 h after the egg is laid. Meiosis is discussed in Chapter 1.2.2 (see also Feiertag-Koppen, 1976, 1980).

Initial development of *T. urticae* and *Eotetranychus tiliarium* (Hermann), the lime spider mite, is characterized by 2 divisions of the total, equal type (Dittrich, 1965, 1968). The first cleavage, in fertilized as well as unfertilized eggs, produces 2 blastomeres of equal size and occurs approximately 2.5 h after oviposition. The cleavage plane goes through the egg center and the site of the second polar body. The second division, at right angles to the first, occurs 30 min later, and results in a four-celled stage. Prior to the third cleavage, the cytoplasm of the blastomeres assumes a peripheral position; the ensuing cleavages become superficial. Two initial, total, equal divisions may be the rule in Tetranychidae, since Gasser (1951) and Fukuda and Shinkaji (1954) reported similar observations for *P. ulmi* and for *P. citri*, respectively.

Within the Acarina, the specialized sequence of total, equal cleavages preceding blastoderm formation, as seen in spider mite embryos, is known only for certain other acariform mites, e.g. *Caloglyphus* (Anderson, 1973). In contrast to the yolk-rich eggs characteristic of many chelicerates, these mites have small eggs containing little yolk; cleavage is considerably modified in relation to this secondary reduction of yolk. Among other arachnids, preliminary phases of total, equal cleavage can be found in the pseudoscorpions (Weygoldt, 1969) and the small, yolkless eggs of viviparous scorpions (Pflug-felder, 1930). In insects, the eggs of Collembola contain relatively little yolk and early cleavage is complete (Chapman, 1982).

Embryonic development occurs rapidly. At room temperature (22°C) all divisions occur synchronously at 30-min intervals so that in *T. urticae* the 1024 cell blastoderm is completed 7 h after egg deposition (Dittrich, 1968). Germ band formation occurs next. Primordial extremities become visible along the early germ band; a median furrow can be observed between the apical parts of the extremities. At this time, the head lobe is not strongly developed. With further development, the extremities become elongate and come to lie with their longitudinal axes parallel to each other. Leg segmentation becomes visible and the median furrow disappears. The germ band then grows in all directions until the head and tail end almost touch each other; the body becomes broader. The head lobe enlarges and the germ band contracts so that the embryo fills about one-third of the egg. At this developmental stage, the respiratory system of the egg (to be described later) is formed. The system, comprised of air duct system and perforation organs in *T. urticae*, permits gas exchange between the embryo and the atmosphere. Dorsal closure occurs and the eye spots develop. When the embryo has completed its internal organization, air enters between the shell and the space around the extremities, rendering the egg non-transparent. The surface of the shell becomes wrinkled, indicating that eclosion is imminent. Soon, the six-legged larva hatches, leaving behind the eggshell which is ruptured for more than one-half of its diameter. The perforation organs and air duct system remain behind with the shell.

The duration of the egg stage is, on average, 3—10 days for most species, depending on temperature, humidity, host and other factors (see Chapters 1.2.6 and 1.4.5). For a given species on a particular host, the main variable is temperature. Cagle (1949) found the embryonic period of *T. urticae* to range from 3 days at 23.9°C to 21 days at 11.1°C. Harrison and Smith (1961) found the incubation period of the same species to vary from 2.38 days at 32.5°C to 33.19 days at 11.5°C. Similar figures have been reported for other tetranychine species. Mathys (1957) found that the average incubation time for *Bryobia rubrioculus* Scheuten at the optimum temperature of 25°C is 9.96 days.

Embryonic adaptation

Protection against water loss is necessary because of the relatively large surface/volume ratio of the eggs. Some eggs must survive adverse environmental conditions for long periods of time. The overwintering eggs of *P. ulmi*, for example, withstand adverse conditions of low humidity and temperature throughout 6—9 months of diapause, and the summer diapause eggs of *P. latens* must withstand hot, desiccating conditions. In the case of *P. ulmi*, Beament (1951) demonstrated that resistance to water loss was due primarily to the secretion of a very thin wax layer around the inner surface of the shell. Hopp (1954) demonstrated that the shell of *T. urticae* was similar to that of *P. ulmi* in this respect. Beament believed that the water-resistant wax layer was produced by the embryo; however, Dittrich and Streibert (1969) suggested that this layer may arise from clear yolk. Lees (1961) reported that the diapause egg of *P. latens* was made resistant to water loss by the outer wax layer. Thurling (1980) demonstrated for *Tetranychus cinnabarinus* (Boisdv.) that the egg shell was not totally impermeable to gas exchange and that metabolism increased during incubation. Some spider mite eggs, such as those of *Eotetranychus sexmaculatus* (Riley), the six-spotted mite, become desiccated rapidly in dry air (Huffaker, 1958; Huffaker et al., 1963).

The developing embryo must be protected from desiccation, but respiration must also be possible. During early embryonic stages, gas exchange is possible through the water-resistant egg shell. For *T. cinnabarinus*, respiration rates (oxygen uptake per individual per hour) are 0.36 ± 0.02 nl in early embryos (Thurling, 1980). In later embryos of *T. urticae*, gas exchange occurs by way of a unique respiratory system (Dittrich and Streibert, 1969; Dittrich, 1971). The main components are the air duct system and the perforation organs (Fig. 1.2.5.3). The intermediate lamella is a major feature of the air duct system (Dittrich, 1971). It is present at oviposition as a layer immediately internal to the broad belt of superficial yolk underlying the shell. The thickness of the lamella, $0.2 \mu m$, is approximately twice that of the shell layer of a newly deposited egg; its structure consists of parabolically bent chitin fibrils. In *T. urticae* at 28°C, $0.1—0.45 \mu m$ long conic protrusions ('micropillars') form along zones of the intermediate lamella 44 h after egg deposition (Dittrich, 1971). These structures form air ducts, 1 duct on each side of the embryo, and a third duct connecting both lateral branches at the anterior end of the embryo. Micropillars appear to maintain an open space between the intermediate lamella and the shell when the embryo presses against the latter during growth.

During formation of the air duct system, primordia of the perforation organs can be seen in the interior of the egg. Approximately 68 h after oviposition, the organs penetrate the egg shell. They are located on 2 curved branches of the frontal part of the air duct system. After penetration, the shell admits air and the air duct system becomes brightly contrasted when viewed in light with a low angle of incidence. For this reason, Dittrich (1968) called the air duct system the system of 'brightly contrasted lines'. Establishment of the respiratory system makes the egg more susceptible to poisoning by certain acaricides (Dittrich and Streibert, 1969).

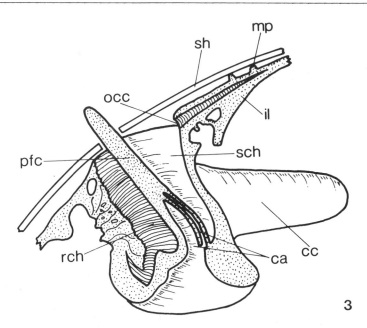

Fig. 1.2.5.3. Respiratory system of egg of *Tetranychus urticae*. After Dittrich (1971). The velum formed where rough and smooth chambers come together is not shown. (ca) canal; (cc) centripetal cone; (il) intermediate lamella; (mp) micropillar; (occ) opening to connecting canal; (pfc) perforation cone; (rch) rough chamber; (sch) smooth chamber; (sh) shell.

The perforation organs (Fig. 1.2.5.3) have the appearance of a bag, the bottom of which has been pushed upward by a finger-like object to form a slender cone (Dittrich, 1971). The cone is approximately 5 μm long and 1.8 μm wide at its base with a median groove along its longitudinal axis; internal cavities are present in the cone. After penetration, it extends as a tube above the shell surface. The cone has the important function of penetrating the shell, but the exact mechanism by which this occurs is not clear. Dittrich (1971) provided morphological evidence that internal cavities of the cone may conduct a lytic substance or plasticizing enzyme to the shell to soften it and allow penetration.

A rough chamber and a smooth chamber surround the cone (Fig. 1.2.5.3). A long narrow extension or velum is formed where the smooth and rough chambers come together. The velum is located opposite the median groove of the cone and can overlap the groove completely in the upper portions of the cone, but not at its base. A tube is thus formed which might make the respiratory system less sensitive to lateral pressure which could cause a blockage of air passages. A 'centripetal cone' extends laterally from the wall of the perforation organ to the egg center to connect with the embryonic tissues. To complete the system, 2 v-shaped superficial grooves, connecting canals in the intermediate lamella, connect the perforation organs to the zones of micropillars in the semicircular extensions of the frontal part of the air duct system. During respiration the perforation organs serve to conduct the respiratory gases from the orifices in the shell through their chambers, and into the connecting canal and air duct system.

Diapause eggs of *P. latens*, which are often laid on stones, are exposed to a relatively dry microclimate. An opening in the middle of the egg cap communicates with air spaces in an underlying porous component ('aeroscopic sponge') surrounding the ovum (Lees, 1961). The outer wax covering of the egg resists water loss, while the respiratory opening and aeroscopic sponge permit oxygen uptake and release of metabolic gases (see Fig. 1.4.6.1).

In diapause eggs (hibernating or aestivating) development is arrested at an early embryonic stage. Metabolic requirements are minimal and a respiratory system for gas exchange is not necessary. The respiratory system develops during morphogenesis after diapause termination.

JUVENILE DEVELOPMENT

Introduction

Tetranychid mites develop through egg, larva, protonymph, deutonymph and adult stages. The three active, feeding immature stages are each followed by intervening periods of quiescence called the protochrysalis (or nymphochrysalis), deutochrysalis and teleiochrysalis, respectively. During these periods of inactivity, the mite anchors itself to the substrate (Fig. 1.2.5.4) and a new cuticle is formed before the exuvium is cast off. At eclosion, the

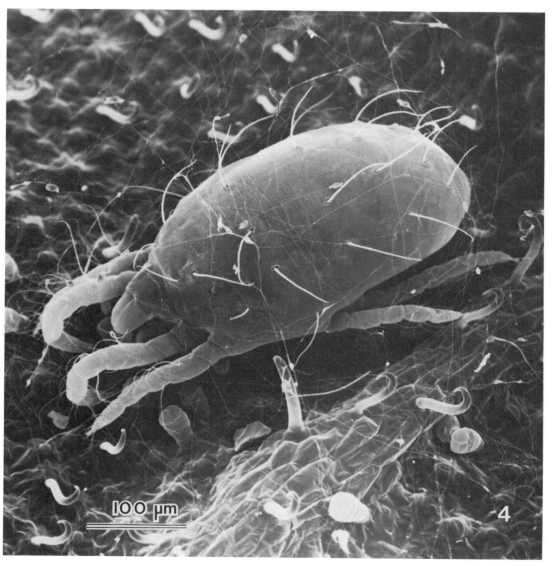

Fig. 1.2.5.4. Quiescent deutonymph of *Tetranychus urticae* on bean leaf. Crooker, unpublished scanning electron micrograph.

Chapter 1.2.5. references, p. 160

integument is split dorsally and the mite frees itself from the exuvium while the old skin is still attached to the substrate. Males mature before females and locate and remain near the female teleiochrysalis until the females emerge; copulation takes place almost immediately after emergence of the young female. As with most other acarines, the larval mite possesses 3 pairs of legs; subsequent stages have 4 pairs of legs.

Development from egg to adult may vary from 6 to 10 days or more, depending on species, temperature, host plant, humidity and other factors; development from egg to egg is about 2—5 days longer because of the pre-oviposition period. Development is primarily correlated with temperature, except where extreme temperatures and/or humidity cause the mites to enter a diapause or aestivation phase.

The development of spider mites has been investigated on numerous occasions and many references can be found in Van de Vrie et al. (1972). The following titles may serve the reader with regard to the different tetranychine species. *Tetranychus urticae*: Shih et al. (1976), Saito (1979), Herbert (1981a), Sabelis (1981), Carey and Bradley (1982). *T. cinnabarinus*: Hazan et al. (1973). *T. neocaledonicus* André: Gutierrez (1976). *T. turkestani* Ugar and Nik.: Carey and Bradley (1982). *T. pacificus* McG.: Carey and Bradley (1982). *T. evansi* B. and P.: Qureshi et al. (1969). *T. schoenei* McG.: Cagle (1943). *T. desertorum* Banks: Nickel (1960). *T. mcdanieli* McG.: Tanigoshi et al. (1975). *T. kanzawai* Kishida: Osakabe (1967). *Schizotetranychus schizopus* (Zacher): Gotoh (1983). *S. leguminosus* Ehara: Gotoh (1983). *S. cercidiphylli* Ehara: Gotoh (1983). *S. celarius* (Banks): Saito and Ueno (1979). *Eotetranychus uncatus* Garm.: Ubertalli (1955). *E. hicoriae* (McG.): Jackson et al. (1983). *Oligonychus platani* (McG.): Butler and Abid (1965). *O. ununguis* (Jacobi): Saito (1979). *O. pratensis* (Banks): Congdon and Logan (1983). *Panonychus ulmi*: Rabbinge (1976), Herbert (1981b). *P. citri*: Saito (1979). *Eutetranychus monodi* André: Coudin and Galvez (1976). *Aponychus corpuzae* Rim.: Saito and Ueno (1979). *Tenuipalpoides dorychaeta* P. and B.: Singer (1966).

Juvenile development

Although variation is evident, the biological studies of Laing (1969), Tanigoshi et al. (1975), Shih et al. (1976), Herbert (1981a,b) and others have demonstrated some apparently general features of spider mite development when mites reared under similar conditions are compared. The duration of the egg stage is greater, by a factor of approximately 2, than the individual larval and nymphal stages. For the male, developmental times for the larva, protonymph and deutonymph are roughly equal; almost one-half of each of these stages is spent in the quiescent state. For the female, developmental times for the larva and protonymph are similar; duration of the active and quiescent phases of these stages shows little difference. The duration of the female deutonymph stage is similar or somewhat longer than larval or protonymphal stages. A comparison between sexes shows that the development times for active and quiescent periods are similar for larva and protonymph. The deutonymph stage may be significantly longer in the female than in the male (see Table 1.2.5.1). That females take longer to develop than males may be a general feature of spider mite development (Herbert, 1981a,b).

Effects of temperature

The empirical relationship between temperature and embryonic and post-embryonic developmental rate is curvilinear. Tanigoshi et al. (1975) placed

TABLE 1.2.5.1

Developmental time in days for *Tetranychus urticae* Koch at 21°C (Herbert, 1981a)

	Active	Quiescent	Total
Larva			
Male	1.5	1.3	2.8
Female	1.5	1.2	2.7
Protonymph			
Male	1.0	1.3	2.3
Female	1.3	1.2	2.4
Deutonymph			
Male	1.0	1.4	2.5
Female	1.5	1.4	2.9

the immature stages of *T. mcdanieli* McGregor, the McDaniel spider mite, into 3 distinct groups, as categorized by the form of the developmental curve. Over the range 10—38°C, curves of the active stages were quartic in form and asymmetrical about the minimal developmental time. Developmental times decreased sharply as the temperature increased from 10 to 20°C, decreased gradually from 20 to 33°C and increased sharply between 33 and 38°C. Decreases in developmental time with increasing temperature followed by an increase in developmental time with further temperature increase have also been reported by Parent (1965) and Hussey et al. (1957). The egg stage curve was also quartic, dropping rapidly from 10 to 20°C, but unlike curves of the active stages, was nearly linear over a broad (20—38°C) temperature range. The quiescent stages and preoviposition period had developmental curves that were cubic in form, differing from the quartic curves in that the developmental period decreased beyond 33°C rather than increased. The curvilinear temperature—development relationship demonstrated by Tanigoshi et al. (1975) agrees with data reported earlier by other authors, and also applies to several other spider mite species.

Relationships between developmental time and temperature demonstrate the considerable variation possible in the duration of tetranychid life stages. Cagle (1949) observed that the larval stage of *T. urticae* reared in an insectary lasted 1 day at 22.8°C and 11 days at 12.5°C. The minimum time required for development of the first nymphal stage was 1 day (23.3°C) and the maximum was 13 days (9.0°C). Developmental time for the deutonymph ranged from 1 day (23.4°C) to 45 days (4.3°C). Tanigoshi et al. (1975) presented temperature—development data for the life stages of *T. mcdanieli* in the form of developmental curves; these authors found the egg-to-egg developmental time to range from 8.12 days at 35°C to 61.96 days at 10°C. Information contained in the tables of Herbert (1981a,b) indicates that developmental times at 15°C were approximately twice those at 21°C for all life stages of both sexes.

The optimum temperature for the most rapid development of many tetranychids is from 24 to 29°C (Boudreaux, 1963). Examples are given by Mori (1961), who reported the temperature preferences of 4 species: *P. ulmi*, 25—28°C; *Bryobia praetiosa* Koch (clover mite), 21—24°C; *Tetranychus viennensis* Zacher (hawthorn spider mite), 25—30°C; and *T. urticae*, 13—35°C. Some species, such as *T. mcdanieli*, develop most rapidly at 35°C (Tanigoshi et al., 1975).

The influence of varying temperatures on the development of the egg and of the immature stages is difficult to assess. Fluctuating temperature may accelerate or decelerate spider mite development, or development may remain

Chapter 1.2.5. references, p. 160

unaffected depending on the magnitude and duration of the fluctuation. Miller (1952) studied the effect of temperature on the development of *P. ulmi* and found a slightly faster development under fluctuating temperatures in contrast to constant temperatures of approximately the same value. Tanigoshi et al. (1976) found no evidence of stimulation or retardation by cyclic temperature alterations when the observed development of *T. mcdanieli* was compared with a variable temperature model derived from constant temperature life stage studies. These authors also examined the life stage developmental rates of *T. mcdanieli* at various alternating temperatures (low temperatures at night, high temperatures during the day). Mites developed slowly at an alternating cycle of 5—20°C, but progressively more rapidly at the regimes of 10—25, 15—30 and 15—35°C.

The threshold temperature for development has been studied in several mites. Parent (1965) found the embryonic and postembryonic development of *P. ulmi* in Quebec to begin at 8.9 and 8.4°C, respectively. Putman (1970) determined, from rearing *P. ulmi* on peach in Ontario, that the threshold for complete postovarial development was between 9.2 and 11.7°C. A similar temperature was reported by Herbert (1981b) in Nova Scotia who found that the threshold temperature of development varied from 10.0 to 11.3°C and 9.3 to 12.5°C for egg and active and quiescent immature stages of the female and male, respectively. In Japan, Tsugawa et al. (1961) found no development of winter eggs of *P. ulmi* at 7°C or below. Osakabe (1967) found that the theoretical zero points for development of eggs, larvae, protonymphs, and deutonymphs of *T. kanzawai* Kishida were 8.7, 14.6, 13.4 and 13.2°C, respectively. Recently, Herbert (1981a) determined that the threshold for development for *T. urticae* was 10.0°C. Threshold temperatures for development may vary among species. There is evidence for selection of cold-tolerant strains within species (Van de Vrie et al., 1972).

The upper limits of temperature tolerance have been reported for some adult mites, e.g. Roesler (1953), Mori (1961) and Shinkaji (1962), but very little is known about the upper limits for tetranychid development. Immature stages of *T. urticae* may tolerate higher temperatures better than adults (Herbert, 1981a). The upper lethal limit for some immature mites may be near 38—40°C, but much variation can be expected depending on species, life stage, humidity and duration of exposure. Tanigoshi et al. (1975) observed high mortality rates in active immature stages of *T. mcdanieli* at 38°C. Interestingly, these authors observed that the developmental times for the quiescent stages were still decreasing at 38°C, the highest temperature used in the study. Therefore, the quiescent stages may possess higher lethal threshold limits than the active stages. Adult longevity is decreased at higher temperatures.

Effects of humidity

Humidity, and humidity and temperature, exert strong influences on spider mite development (Van de Vrie et al., 1972). Loss or gain of water from the atmosphere is important for these small organisms because of the large surface/volume ratio. Except for the egg stage, water lost to the atmosphere in a dry environment can be replaced during feeding. Under low humidities, intake of nutrients is greater to compensate for water loss, and reproduction and developmental rates may also be greater. High relative humidity may interfere with mite feeding by preventing loss of moisture from the body by evaporation through the cuticle (Boudreaux, 1958), although excretion of excess water from the digestive system as a fluid helps offset this difficulty (McEnroe, 1963).

Hot, dry weather is conducive to rapid development and high population levels of many spider mite species, however cooler, more humid conditions are favourable for the development of other species. Nickel (1960) found that *T. urticae* developed faster, with higher egg production, under low humidity (25—30%) than high humidity (85—90%). He found the reverse to be true for *T. desertorum*. Boudreaux (1958) observed greater egg production and longevity in adult females and increased mortality of newly-hatched larvae under high humidity. Harrison and Smith (1961) found all relative humidities, except 100%, to have little influence on egg incubation time, but to have a major effect on determining the number of eggs which hatch. Both low and high humidities reduced the natural hatching percentage of eggs, and 100% relative humidity delayed hatching time so long that the embryo usually died. Osakabe (1967) found the ranges of temperature and humidity for successful hatching of *T. kanzawai* to be 15—30°C and 46—93% relative humidity. Relative humidities of 20 and 100% had adverse effects on egg hatching and inhibited development of immature stages. Mori (1957) indicated that for *P. ulmi*, 43—100% relative humidity at 15—32°C was most suitable for embryonic development, whereas 65—100% relative humidity at 15—31°C was most suitable for hatching. Das and Das (1967) found the optimum conditions for development and egg hatch of *Oligonychus coffeae* (Nietner), the tea red spider mite, to be 20—30°C and 49—94% relative humidity. No eggs hatched at 34°C at any humidity, or at 17% humidity irrespective of temperature. Boyne and Hain (1983) reported that *O. ununguis*, the spruce spider mite, responds most favourably to temperatures of about 26°C and relative humidities of 50—60%. Developmental times at 26°C for all stages showed significant differences when low (30—40%) and moderate (50—60%) relative humidity levels were compared with high (88—98%) relative humidity levels. Survival was significantly less at high humidity levels.

Effect of host plant

Differences in development, reproduction, longevity and population development of mites on different host plants are common. These differences may be associated with impediments to feeding such as host plant texture and vestiture, nutritional value of the host plant, host physiology, or the favourability of the microenvironment. Agricultural chemicals (not discussed here) may also affect development (see Van de Vrie et al., 1972). Puttaswamy (1980) observed that the egg-to-adult developmental time of *Tetranychus ludeni* Zacher ranged from 9.24 days on brinjal to 9.91 days on South American cucurbit. Puttaswamy (1981) also determined that the egg-to-adult developmental time of *T. neocaledonicus* was significantly longer on tapioca (12.11 days) than on castor (10.48 days), amaranthus (10.20 days) and mulberry (10.14 days). Other examples of the effects of host plants on spider mite biology can be found in Van de Vrie et al. (1972) and Jeppson et al. (1975).

The chemical constitution of the host plant may influence fecundity and mortality and development of the immature stages. Numerous reports indicate that host plant nitrogen content or nitrogen fertilization is associated with increases in mite populations, often through effects on fecundity. Breukel and Post (1959) found that *P. ulmi* on apple leaves with a higher nitrogen content exhibited higher egg production and faster development. Many additional examples can be found in Van de Vrie et al. (1972), together with conflicting results of the influence of nitrogen on development and reproduction. The effects of potassium, phosphorus, and other elements on

Chapter 1.2.5. references, p. 160

development and reproduction are also discussed by Van de Vrie et al. (1972).

Structural changes

Little is known about structural changes which occur during development. Externally, setae are added during successive life stages (see Chapter 1.1.2). Jalil (1969) examined briefly the internal anatomy of developmental stages of *T. urticae*. The ovary and central nervous mass (synganglion) were found to be more closely juxtaposed during larval and protonymphal stages than in later stages. Oogenesis began in the quiescent deutonymph. For the male, the testes first appeared in the late active deutonymphal stage and developed during the following quiescent phase. Salivary and silk glands were rudimentary in the larval stage and well-developed during the ensuing stages.

Feiertag-Koppen and Pijnacker (1982) studied the development of female germ cells of *T. urticae*. These authors describe the descent of major ovarian cell types from mitotic oogonial cells present in the larva; differentiation of new cell types occurred at each successive life stage of the mite. The details are presented and discussed by these authors in Chapter 1.2.2.

CONCLUDING REMARKS

The embryonic and juvenile development of spider mites is not well known. Only very general features of embryonic development and adaptation have been examined. These features include certain aspects of early cleavage, blastoderm and germ band formation, development of the respiratory system, and further development of the major body regions. Fundamental topics, such as the origin and development of the mesoderm and endoderm, have not yet been investigated. Neither have many other general aspects of embryonic development including gastrulation, organogenesis and organ system development, segmentation, limb formation, dorsal closure, morphogenetic movements and extraembryonic membranes. Metabolism, growth rates, developmental order, gene activity, and control of differentiation also require study.

Juvenile development has been studied primarily from a life cycle and life table approach, e.g. duration of life stages, rates of increase, and effects of factors such as temperature on rate of development. Additional information of this nature is needed, particularly as it relates to a better understanding of population dynamics and factors regulating field populations of these pests. Very little is known about structural and physiological changes which occur during development.

Study of the embryonic and juvenile development of the fascinating and highly successful spider mites has not yet received the attention it merits. Field and laboratory examinations of developmental parameters and population dynamics, and morphological, physiological, biochemical, molecular and genetic studies will be required to further clarify the many unanswered questions.

REFERENCES

Anderson, D.T., 1973. Embryology and Phylogeny in Annelids and Arthropods. Pergamon Press, Oxford, 495 pp.
Beament, J.W.L., 1951. The structure and formation of the egg of the fruit tree red spider mite, *Metatetranychus ulmi* Koch. Ann. Appl. Biol., 38: 1—24.

Boudreaux, H.B., 1958. The effect of relative humidity on egg-laying, hatching, and survival in various spider mites. J. Insect Physiol., 2: 65—72.

Boudreaux, H.B., 1963. Biological aspects of some phytophagous mites. Annu. Rev. Entomol., 8: 137—154.

Boyne, J.V. and Hain, F.P., 1983. Effects of constant temperature, relative humidity, and simulated rainfall on development and survival of the spruce spider mite (*Oligonychus ununguis*). Can. Entomol., 115: 93—105.

Breukel, L.M. and Post, A., 1959. The influence of manurial treatment on the population density of *Metatetranychus ulmi* (Koch) (Acari, Tetranychidae). Entomol. Exp. Appl., 2: 38—47.

Butler, G.D. and Abid, M.K., 1965. The biology of *Oligonychus platani* on pyracantha. J. Econ. Entomol., 58: 687—688.

Cagle, L.R., 1943. Life history of the spider mite, *Tetranychus schoenei* McG. Va. Agric. Exp. Stn., Tech. Bull., 87: 1—16.

Cagle, L.R., 1949. Life history of the two-spotted spider mite. Va. Agric. Exp. Stn., Tech. Bull., 113: 1—31.

Carey, J.R. and Bradley, J.W., 1982. Developmental rates, vital schedules, sex ratios, and life tables for *Tetranychus urticae, T. turkestani* and *T. pacificus* (Acarina: Tetranychidae) on cotton. Acarologia, 23: 333—345.

Chapman, R.F., 1982. The Insects. Structure and Function. Harvard University Press, 3rd edn., 992 pp.

Claparede, E., 1868. Studien an acariden. Z. Wiss. Zool., 18: 445—546.

Congdon, B.D. and Logan, J.A., 1983. Temperature effects on development and fecundity of *Oligonychus pratensis* (Acari: Tetranychidae). Environ. Entomol., 12: 359—362.

Coudin, B. and Galvez, F., 1976. Etude de facteurs climatiques sur le développement et la multiplication de *Eutetranychus monodi* André. Fruits, 31(10): 623—630.

Das, G.M. and Das, S.C., 1967. Effect of temperature and humidity on the development of tea red spider mite, *Oligonychus coffeae* (Nietner). Bull. Entomol. Res., 57: 433—436.

Dittrich, V., 1965. Embryonic development of tetranychids. Boll. Zool. Agrar. Bachic., Ser. 2, 7: 101—104 (Proceedings 5th European Mite Symposium, Milano).

Dittrich, V., 1968. Die Embryonalentwicklung von *Tetranychus urticae* Koch in der Auflichtmikroskopie. Z. Angew. Entomol., 61: 142—153.

Dittrich, V., 1969. Recent investigations of the embryonic development of *Tetranychus urticae* Koch. Proc. 2nd. Intl. Congr. Acarology, pp. 477—478.

Dittrich, V., 1971. Electron-microscopic studies of the respiratory mechanism of spider mite eggs. Ann. Entomol. Soc. Am., 64: 1134—1143.

Dittrich, V. and Streibert, P., 1969. The respiratory mechanism of spider mite eggs. Z. Angew. Entomol., 63: 200—211.

Feiertag-Koppen, C.C.M., 1976. Cytological studies of the two-spotted spider mite *Tetranychus urticae* Koch (Tetranychidae, Trombidiformes). I. Meiosis in eggs. Genetica, 46: 445—456.

Feiertag-Koppen, C.C.M., 1980. Cytological studies of the two-spotted spider mite *Tetranychus urticae* Koch (Tetranychidae, Trombidiformes). II. Meiosis in growing oocytes. Genetica, 54: 173—180.

Feiertag-Koppen, C.C.M. and Pijnacker, L.P., 1982. Development of the female germ cells and process of internal fertilization in the two-spotted spider mite *Tetranychus urticae* Koch (Acariformes: Tetranychidae). Int. J. Insect Morphol. Embryol., 11: 271—284.

Fukuda, J. and Shinkaji, N., 1954. Experimental studies on the influence of temperature and relative humidity upon the development of citrus red mite (*Metatetranychus citri* McGregor). I. On the influence of temperature and relative humidity upon the development of the eggs. Tokai—Kinki Agric. Exp. Stn., Bull. 2 (Horticulture): 160—171 (In Japanese; English summary).

Gasser, R., 1951. Zur Kenntnis der gemeinen Spinnmilbe *Tetranychus urticae* Koch. I. Mitteilung: Morphologie, Anatomie, Biologie und Oekologie. Mitt. Schweiz. Entomol. Ges., 24: 217—262.

Gotoh, T., 1983. Life history parameters of three species of *Schizotetranychus* on deciduous trees. Appl. Entomol. Zool., 18: 122—128.

Gutierrez, J., 1976. Etude biologique et écologique de *Tetranychus neocaledonicus* André (Acariens, Tetranychidae). Trav. Doc. ORSTOM, 57: 1—173.

Harrison, R.A. and Smith, A.G., 1961. The influence of temperature and relative humidity on the development of eggs and on the effectiveness of ovicides against *Tetranychus telarius* (L.) (Acarina: Tetranychidae). N. Z. J. Sci., 4: 540—549.

Hazan, A., Gerson, U. and Tahori, S., 1973. Life history and life tables of the carmine spider mite. Acarologia, 15: 414—440.

Herbert, H.J., 1981a. Biology, life tables, and innate capacity for increase of the two-spotted spider mite, *Tetranychus urticae* (Acarina: Tetranychidae). Can. Entomol., 113: 371—378.

Herbert, H.J., 1981b. Biology, life tables, and intrinsic rate of increase of the European red mite, *Panonychus ulmi* (Acarina: Tetranychidae). Can. Entomol., 113: 65—71.

Hopp, H.H., 1954. Beiträge der Wirkungsweise einiger Akarizide auf Eier der Spinnmilben *Paratetranychus pilosus* Can. et Fanz. und *Tetranychus urticae* Koch. Z. Angew. Zool., 41: 269—286.

Huffaker, C.B., 1958. Experimental studies on predation: dispersion factors and predator— prey oscillations. Hilgardia, 27: 343—383.

Huffaker, C.B., Shea, K.P. and Herman, S.G., 1963. Experimental studies on predation: complex dispersion and levels of food in an acarine predator—prey interaction. Hilgardia, 34: 305—330.

Hussey, N.W., Parr, W.J. and Crocker, C.D., 1957. Effect of temperature on the development of *Tetranychus telarius* L. Nature (London), 179: 739—740.

Jackson, P.R., Hunter, P.E. and Payne, J.A., 1983. Biology of the pecan leaf scorch mite (Acari: Tetranychidae). Environ. Entomol., 12: 55—59.

Jalil, M., 1969. Internal anatomy of the developmental stages of the two-spotted spider mite. Ann. Entomol. Soc. Am., 62: 247—249.

Jeppson, L.R., Keifer, H.H. and Baker, E.W., 1975. Mites Injurious to Economic Plants. University of California Press, 614 pp.

Laing, J.E., 1969. Life history and life table of *Tetranychus urticae* Koch. Acarologia, 9: 32—42.

Lees, A.D., 1961. On the structure of the egg shell in the mite *Petrobia latens* Muller (Acarina: Tetranychidae). J. Insect Physiol., 6: 146—151.

Mathys, G., 1957. Contribution à la connaissance de la systématique et de la biologie du genre *Bryobia* en Suisse romande. Bull. Soc. Entomol. Suisse, 30(3): 189—284.

McEnroe, W.D., 1963. The role of the digestive system in the water balance of the two-spotted spider mite. In: Advances in Acarology, Vol. I. Cornell University Press, pp. 225—231.

Miller, L.W., 1952. The hatching of the overwintering eggs of the European red mite. Tasmanian J. Agric., 23: 102—116.

Mori, H., 1957. The influence of temperature and relative humidity upon the development of the eggs of the fruit tree red spider mite *Metatetranychus ulmi* (Koch) (Acarina: Tetranychidae). J. Fac. Agric. Hokkaido Univ., 50: 363—370.

Mori, H., 1961. Comparative studies on thermal reaction in four species of spider mites. J. Fac. Agric., Hokkaido Univ., 51: 574—591.

Mothes, U. and Seitz, K.A., 1981. A possible pathway of chitin synthesis as revealed by electron microscopy in *Tetranychus urticae* (Acari, Tetranychidae). Cell Tissue Res., 214: 443—448.

Nickel, J.L., 1960. Temperature and humidity relationships of *Tetranychus desertorum* Banks with special reference to distribution. Hilgardia, 30: 41—100.

Osakabe, M., 1967. Biological studies on the tea red spider mite, *Tetranychus kanzawai* Kishida, in tea plantation. Bull. Tea Res. Stn., Min. Agric. For., 4: 35—156.

Parent, B., 1965. Influence de la température sur le developpement embryonnaire et post-embryonnaire de tetranyque rouge du pommier, *Panonychus ulmi* (Koch) (Acariens: Tetranychidae). Ann. Soc. Entomol. Que., 10: 3—10.

Pflugfelder, O., 1930. Zur Embryologie des Skorpions *Homurus australasiae* (F). Z. Wiss. Zool., 137: 1—29.

Putman, W.L., 1970. Threshold temperatures for the European red mite, *Panonychus ulmi* (Acarina: Tetranychidae). Can. Entomol., 102—421—425.

Puttaswamy, C.G.P., 1980. Influence of host plants on the development, fecundity and longevity of *Tetranychus ludeni* (Acari: Tetranychidae). Indian J. Acarology, 5: 80—84.

Puttaswamy, C.G.P., 1981. Influence of host plants on the reproductive biology of *Tetranychus neocaledonicus* (Acari: Tetranychidae). Indian J. Acarology, 6: 72—76.

Qureshi, A.H., Oatman, E.R. and Fleschner, C.A., 1969. Biology of the spider mite *Tetranychus evansi* Pritchard and Baker. Ann. Entomol. Soc. Am., 62(4): 898—903.

Rabbinge, R., 1976. Biological Control of the Fruit Tree Red Spider Mite. Simulation Monographs. Pudoc, Wageningen, 228 pp.

Roesler, R., 1953. Rote Spinne und Witterung. Z. Angew. Entomol., 36: 197—200.

Sabelis, M.W., 1981. Biological Control of Two-spotted Spider Mites using Phytoseiid Predators. Part 1. Modelling the Predator—prey Interaction at the Individual Level. Pudoc, Wageningen, 242 pp.

Saito, Y., 1979. Comparative studies on life histories of three species of spider mites (Acarina: Tetranychidae). Appl. Entomol. Zool., 14(1): 83—94.

Saito, Y. and Ueno, J., 1979. Life history studies on *Schizotetranychus celarius* (Banks) and *Aponychus corpuzae* Rimando as compared with other tetranychid mite species (Acarina: Tetranychidae). Appl. Entomol. Zool., 14(4): 445—452.

Shih, C.T., Poe, S.L. and Cromroy, H.L., 1976. Biology, life table, and intrinsic rate of increase of *Tetranychus urticae*. Ann. Entomol. Soc. Am., 69: 362—364.

Shinkaji, N., 1962. Some investigations on the behavior of citrus red mite (*Panonychus citri* McGregor). Tokai—Kinki Agric. Exp. Stn., Bull. (Horticulture, Ser. B), I: 192—205. (In Japanese; English summary).

Singer, G., 1966. The bionomics of *Tenuipalpoides dorychaeta* Pritchard and Baker (1955) (Acarina, Trombidiformes, Tetranychidae). Univ. Kans. Sci. Bull., 46(17): 625—645.

Tanigoshi, L.K., Hoyt, S.C., Browne, R.W. and Logan, J.A., 1975. Influence of temperature on population increase of *Tetranychus mcdanieli* (Acarina: Tetranychidae). Ann. Entomol. Soc. Am., 68: 972—978.

Tanigoshi, L.K., Browne, R.W., Hoyt, S.C. and Lagier, R.F., 1976. Empirical analysis of variable temperature regimes on life stage development and population growth of *Tetranychus mcdanieli* (Acarina: Tetranychidae). Ann. Entomol. Soc. Am., 69: 712—716.

Thurling, D.J., 1980. Metabolic rate and life stage of the mites *Tetranychus cinnabarinus* Boisd. (Prostigmata) and *Phytoseiulus persimilis* A—H. (Mesostigmata). Oecologia (Berlin), 46: 391—396.

Tsugawa, C., Yamada, H. and Shirasaki, S., 1961. Forecasting the outbreak of destructive insects in apple orchards. III. Forecasting the initial date of hatch in respect to the overwintering eggs of the European red mite, *Panonychus ulmi* (Koch), in Aomori Prefecture. Jpn. J. Appl. Entomol. Zool., 5: 167—173.

Ubertalli, J.A., 1955. Life history of *Eutetranychus uncatus* Garman. J. Econ. Entomol., 48: 47—49.

Van de Lustgraaf, B., 1977. L'ovipositeur des Tetraniques. Acarologia, 18: 642—650.

Van de Vrie, M., McMurtry, J.A. and Huffaker, C.B., 1972. Ecology of Tetranychid Mites and their Natural Enemies: A Review. III. Biology, Ecology and Pest Status and Host—Plant Relations of Tetranychids. Hilgardia, 41: 343—432.

Weygoldt, P., 1969. The Biology of Pseudoscorpions. Harvard University Press, 145 pp.

1.2.6 Reproductive Parameters

D.L. WRENSCH

INTRODUCTION

Reproduction in spider mites is extremely sensitive to a wide variety of intrinsic and extrinsic conditions. Individual reproductive parameters, or fitness traits, which determine to a greater or lesser degree the magnitude of r_m, the intrinsic rate of increase, include fecundity, hatchability, lengths of period of oviposition and longevity, rate of development, survivorship and certain aspects of the sex ratio (see also Chapter 1.2.3). Extrinsic factors which influence these parameters include temperature, humidity, light, level of predation, intra- and interspecific competition, quantity, quality and timing of pesticides, and various features of host plants, such as strain, plant and soil nutrition and plant age. Among intrinsic factors affecting reproductive potential are mite strain and level of inbreeding, colony density, age of female and of population, female's fertilization status, quality of mate, duration of insemination, and various aspects of behaviour. Many of these factors have been discussed elsewhere, for example by Huffaker et al. (1969), Van de Vrie et al. (1972) and Wrensch (1979). Limitations on the length of this chapter do not permit a full exploration of all the literature pertaining to these factors but key articles, recent experimental work and some new avenues of research will be examined.

The goal of many life history studies of spider mites has been an assessment of the intrinsic rate of increase in order to quantify the now well-documented ability of spider mites to undergo population explosions or outbreaks (Laing, 1969; Hazan et al., 1973; Wrensch and Young, 1975; Shih et al., 1976). It seems that every chemical manipulation favouring agricultural production, whether it be water, fertilizers, or pesticides, is a treatment that can also favour mite reproduction. Some pesticides, for example, paradoxically appear to enhance population growth via direct stimulation of fecundity or indirectly by virtue of killing susceptible predators. Direct stimulation may occur as trophobiosis (Chabousseau, 1966) in which the pesticide improves host plant favourability to tetranychids or by hormoligosis (Luckey, 1968) wherein pesticides directly stimulate the rate of development and fecundity. Van de Vrie et al. (1972) state that stimulation caused by pesticides is a generally occurring phenomenon, although there are contradictory instances in the literature (Boykin and Campbell, 1982).

A great deal of research has focussed on the inherent resistance or susceptibility of host plants to mite infestation. Huffaker et al. (1969) concluded that

'condition of host plant seems ... to be a major factor in abundance of spider mites. Factors such as weather, seasonal growth cycle, soil, water and pesticides can alter physiology of the plant affecting its favorability to mites.'

Chapter 1.2.6. references, p. 168

Seasonal fluctuations in r_m were noted by Jesiotr and Suski (1976). Post (1962) showed that soil nitrogen correlates directly with population increase and that plant nutrition was more important to increases of the pest populations than the killing of predators by insecticides. Rodriguez et al. (1970) implicated foliage nitrogen with mite increase. Most results indicate advantages to spider mite populations from nitrogen but results for other minerals are ambiguous.

Jesiotr et al. (1979) point out that there are many ways in which a plant might affect a mite population and list

> 'structure of the leaf surface, thickness of the cuticle, chemical composition of the saps, osmotic pressure within a cell, microclimate in the canopy, and most probably many others.'

Neiswander et al. (1950) found that two-spotted spider mites reared on tomatoes were more susceptible to acaricides than those grown on beans. Patterson et al. (1974) showed that resistance in *Nicotiana* species to *Tetranychus urticae* Koch was due to a combination of non-preference and antibiosis, probably from alkaloids. Resistance was associated with a viscid secretion from trichomes that was toxic and also entrapped the mites. Patterson et al. (1975) demonstrated that resistance in tomatoes to *T. urticae* was due to avoidance and toxicity caused by sesquiterpenoids. The chemical basis of resistance either by avoidance or antibiosis has been and is the subject of much research interest.

The potential of spider mites for increase can vary widely between species of hosts. Gerson and Aronowitz (1980) found r_m values ranging from a low of 0.091 per day on Algerian ivy to a high of 0.270 per day on beans under similar environmental conditions. Different cultivars have been shown repeatedly to affect spider mite success (Dabrowski et al., 1971; Dabrowski and Bielak, 1978; Gould, 1978; Al-Abbasi and Weigle, 1982). Leigh and Hyer (1963) found that wild cotton biotypes or crosses to them were more resistant. However, spider mites exhibit the capacity to adapt rapidly to new hosts (Jesiotr and Suski, 1976; Jesiotr, 1980), apparently through selection for fittest genotypes. Thus they are implicitly adapting to host defences. Considering the well-documented propensity for evolution of resistance to pesticides, the fact that spider mites have the potential for evolving escapes to plant defences is not surprising.

Another rather curious finding is that water stress tends to enhance susceptibility to spider mites (Dabrowski et al., 1971; Hollingsworth and Berry, 1982; A. Colijn, personal communication, 1984). In a series of papers, DeAngelis et al. (1982, 1983a, 1983b) demonstrated that feeding injury actually increases night-time transpiration rates and damages leaf epidermis and cuticle, leading to water stress and accumulation of soluble leaf carbohydrates. Apparently this concentration or enrichment of the food source affects spider mite reproduction positively.

The 2 life history parameters of paramount importance in determining the intrinsic rate of increase are developmental time and fecundity (Snell, 1978). Cole (1954) demonstrated that the age of first reproduction has a far greater effect on r_m than does fecundity. Lewontin (1965) proved that small decreases in the rate of development were equivalent to very large increases in fecundity, in terms of increasing r_m. For example, a decrease in development time of the order of 10% might have the same effect as an increase in fecundity of the order of 100%. Lewontin's methodology was applied to spider mites by Wrensch and Young (1975), who confirmed his principal conclusions. A limitation of the methodology was examined by Meats (1971) who showed that when r_m approached zero, fecundity greatly surpassed developmental rate in relative importance. Further refinements for cases when r_m is very small or when fecundity and developmental time are correlated were examined by Snell (1978) who found that decreases in developmental time had the largest effect on r_m when fecundity was large.

Many studies that included the 2 primary spider mite reproductive parameters have also incorporated temperature as a variable, often in combination with a variety of humidities. Boudreaux (1958) first pointed to the negative effects of high humidity on fecundity and survivorship. Mori and Chant (1966) demonstrated hygrokinetic behaviour in *T. urticae* with avoidance movement away from higher humidity. Nickel (1960) found for *Tetranychus desertorum* Banks that r_m at 30°C was 0.46 per day at 25—30% RH and 0.36 per day at 85—90% RH, reflecting the promotion of egg-laying by lower humidities. Although Hazan et al. (1973), working with various combinations of temperature and humidity on *T. cinnabarinus* (Boisduval), found highest fecundity at 24°C and 38% RH, the highest r_m, 0.355 per day, occurred at 35°C and 38% RH. No doubt this high r_m reflects the faster development associated with increasing temperature found in this and other studies. The shorter time for development overrides the impact of reduced survivorship and lower fecundity. Herbert (1981), working with *T. urticae* on apple, found r_m values ranging from 0.069 per day at 15°C to 0.372 per day at 21°C, at a relative humidity of 80%. Development time was studied and shown to decrease with increasing temperature. Similar results for *T. pacificus* McGregor and *T. turkestani* (Ugar. & Mik.) were found by Carey and Bradley (1982). Tanigoshi et al. (1975), working with *T. mcdanieli* McGregor, found r_m values ranging from 0.115 per day at 18°C to 0.431 per day at 35°C, at 40—65% RH. Their highest r_m occurred in spite of survivorship decreasing with increasing temperature and a lower fecundity than was found at 32°C. The time of development was shortest at 35°C. They state that this highest r_m was due to shortest development time and generation time. Carey and Bradley (1982) found the development in *T. urticae* of egg to adult to take only 6.2 days at 29.4°C, resulting in an r_m of 0.293 per day. More recent work investigating the relationship between temperature and developmental rate indicates a complexity beyond simple linearity (Tanigoshi and Logan, 1979). From these few examples, the importance of the environmentally sensitive developmental time emerges as virtually transcendant in influencing the innate capacity for increase. The developmental rate also varies between strains (Nickel, 1960) and species (Murtaugh and Wrensch, 1978; Saito, 1979), but these differences are not as pronounced as those due to temperature. They are sufficient, however, to evoke differential success in competition within and among species.

Although an interesting and difficult aspect to study, population density has been shown to influence reproductive parameters. Davis (1952) found that dense populations resulted in lower individual female egg production and a higher percentage of non-viable eggs. Mortality of immatures was also associated with crowding. Attiah and Boudreaux (1964) found that density tended to reduce longevity. Marked reductions in developmental rate, survivorship and sex ratio were found at high, rather than low density by Wrensch and Young (1978). They concluded that the major elements influencing the rate of population increase are density dependent.

As indicated earlier, next to developmental rate, fecundity is of greatest importance to the innate capacity for increase. The sensitivity of fecundity to the various environmental factors has been fairly well documented. Young and Wrensch (1981) showed that variations in the adult offspring number, or net fitness, are determined mainly by variations in fecundity and modified in minor ways by survivorship. They concluded that the ability to reproduce a large number of offspring under marginal conditions is an important reason

Chapter 1.2.6. references, p. 168

why spider mites are highly successful colonizing organisms. Williams (1954) studied one *T. urticae* female that produced 312 eggs and Shih et al. (1976) observed one female, also *T. urticae*, that produced a total of 304 eggs. Young and Wrensch (1981) observed groups of *T. urticae* females that on average produced almost 200 eggs, resulting in an average of 176 adult offspring. However, fecundity alone is actually inadequate to use as a measure or correlation of capacity for increase, because spider mites ordinarily do not have equal frequencies of the sexes produced. Thus, authors who have developed life tables assuming 1:1 sex ratios are underestimating r_m whenever the frequency of female offspring exceeds that of males. Hazan et al. (1973) compared r_m values with and without a correction for sex ratio and Wrensch and Young (1975) and the present author (Wrensch, 1979) include adjustments of the sex ratio in their estimates of r_m. Carey (1983) and Feldmann (1981) also adjust for sex ratio in their demographic studies.

Although Boudreaux (1963) stated that there is no normal sex ratio, the abundance of observations available now suggests that for *T. urticae* and related species, and also for other tetranychines, the sex ratio is usually female biased and life-time progeny sex ratios average around 3 times as many daughters as sons. Factors determining the sex ratio are dealt with elsewhere (see Chapter 1.2.3). The sex ratio is sensitive to leaf quality (Wrensch and Young, 1983), density (Wrensch and Young, 1978), and temperature (Hazan et al., 1973), as well as other factors. Also, many intrinsic factors influence the sex ratio. Overmeer (1972) found that the sex ratio depends on the amount of sperm introduced, which depends in turn on the length of copulation and sperm supply. Both Shih (1979) and Wrensch (1979) point out that very old females produce more, and eventually exclusively, sons and this has been interpreted as females depleting their supply of sperm. Kongchuensin (1983) confirmed the results of Young and Wrensch (1984) and found that females fertilize a fixed fraction of eggs daily during the majority of their period of oviposition. A maternal effect has long been noted (Overmeer and Harrison, 1969). Mitchell (1972) found evidence for an hereditary basis. Helle (1967) presented evidence that the first egg produced is almost always unfertilized. Sex ratio is a fitness trait that needs to be regularly included in evaluations of spider mite capacity for increase.

The wealth of experimental observations on various reproductive parameters of spider mites clearly reinforces the prevalent theme that these are serious pests with biological attributes that are not readily subdued. The fitness traits examined here, and others, are so sensitive that great variability in potential can be seen in the literature. In crucial fitness traits, especially development rate, spider mites are extremely flexible. They are physiologically as well as evolutionarily labile, which ensures them timely adaptive responsiveness. Finally, as species characterized by selection among demes, they possess a great capacity to circumvent even the most imaginative and complex man-made attempts to curtail their devastating economic impact.

REFERENCES

Al-Abbasi, S.H. and Weigle, J.L., 1982. Resistance in New Guinea *Impatiens* species and hybrids to the two-spotted spider mite. HortScience, 17: 47—48.
Attiah, H.H. and Boudreaux, H.B., 1964. Influence of DDT on egglaying in spider mites. J. Econ. Entomol., 57: 50—53.
Boudreaux, H.B., 1958. The effect of relative humidity on egg-laying, hatching, and survival in various spider mites. J. Insect Physiol., 2: 65—72.
Boudreaux, H.B., 1963. Biological aspects of some phytophagous mites. Annu. Rev. Entomol., 8: 137—154.

Boykin, L.S. and Campbell, W.V., 1982. Rate of population increase of the two-spotted spider mite (Acari: Tetranychidae) on peanut leaves treated with pesticides. Ann. Entomol. Soc. Am., 75: 966—971.

Carey, J.R., 1983. Demography of the two-spotted spider mite. Oecologia, 52: 389—395.

Carey, J.R. and Bradley, J.W., 1982. Developmental rates, vital schedules, sex ratios, and life tables for *Tetranychus urticae*, *T. turkestani* and *T. pacificus* (Acarina: Tetranychidae) on cotton. Acarologia, 23: 333—345.

Chaboussou, F., 1966. Nouveaux Aspects de la Phytiatrie et de la Phytopharmacie. Le Phenomène de la Trophobiose. Proc. FAO Symp. Integrated Pest Control, Rome, 1965, Vol. 1, pp. 33—61.

Cole, L., 1954. The population consequences of life history phenomena. Q. Rev. Biol., 29: 103—137.

Dabrowski, Z.T. and Bielak, B., 1978. Effect of some plant chemical compounds on the behaviour and reproduction of spider mites (Acarina: Tetranychidae). Entomol. Exp. Appl., 24: 117—126.

Dabrowski, Z.T. and Rodriguez, J.G., 1971. Studies on resistance of strawberries to mites. 3. Preference and non-preference responses of *Tetranychus urticae* and *T. turkestani* to essential oils of foliage. J. Econ. Entomol., 64: 387—391.

Dabrowski, Z.T., Rodriguez, J.G. and Chaplin, C.E., 1971. Studies in the resistance of strawberries to mites. IV. Effect of season on preference and non-preference of strawberries to *Tetranychus urticae*. J. Econ. Entomol., 64: 806-809.

Davis, D.W., 1952. Influence of population density on *Tetranychus multisetis*. J. Econ. Entomol., 45: 652—654.

DeAngelis, J.D., Berry, R.E. and Krantz, G.W., 1983a. Evidence for spider mite (Acari: Tetranychidae) injury-induced leaf water deficits and osmotic adjustment in peppermint. Environ. Entomol., 12: 336—339.

DeAngelis, J.D., Berry, R.E. and Krantz, G.W., 1983b. Photosynthesis, leaf conductance, and leaf chlorophyll content in spider mite (Acari: Tetranychidae)-injured peppermint leaves. Environ. Entomol., 12: 345—348.

DeAngelis, J.D., Larson, K.C., Berry, R.E. and Krantz, G.W., 1982. Effects of spider mite injury on transpiration and leaf water status in peppermint. Environ. Entomol., 11: 975—978.

Feldmann, A.M., 1981. Life table and male mating competitiveness of wild-type and of a chromosome mutation strain of *Tetranychus urticae* in relation to genetic pest control. Entomol. Exp. Appl., 29: 125—137.

Gerson, U. and Aronowitz, A., 1980. Feeding of the carmine spider mite on seven host plant species. Entomol. Exp. Appl., 28: 109—115.

Gould, F., 1978. Resistance of cucumber varieties to *Tetranychus urticae*: genetic and environmental determinants. J. Econ. Entomol., 71: 680—683.

Hazan, A., Gerson, U. and Tahori, A.S., 1973. Life history and life tables of the carmine spider mite. Acarologia, 15: 414—440.

Helle, W., 1967. Fertilization in the two-spotted spider mite (*Tetranychus urticae*: Acari). Entomol. Exp. Appl., 10: 103—110.

Herbert, H.J., 1981. Biology, life tables, and innate capacity for increase of the two-spotted spider mite, *Tetranychus urticae* (Acarina: Tetranychidae). Can. Entomol., 113: 371—378.

Hollingsworth, C.S. and Berry, R.E., 1982. Two-spotted spider mite (Acari: Tetranychidae) in peppermint: population dynamics and influence of cultural practices. Environ. Entomol., 11: 1280—1284.

Huffaker, C.B., Van de Vrie, M. and McMurtry, J.A., 1969. The ecology of tetranychid mites and their natural enemies. Annu. Rev. Entomol., 14: 12ɔ— ɪ ɪ ɔ.

Jesiotr, J., 1980. The influence of host plants on the reproduction potential of the two-spotted spider mite, *Tetranychus urticae* Koch (Acarina: Tetranychidae) IV. Changes within different populations affected by a new species of host plant. Ekol. Pol., 28: 633—647.

Jesiotr, J. and Suski, W., 1976. The influence of the host plants on the reproduction potential of the two-spotted spider mite, *Tetranychus urticae* Koch (Acarina: Tetranychidae). Ekol. Pol., 24: 407—411.

Jesiotr, J., Suski, Z.W. and Badowska-Czubik, T., 1979. Food quality influences on a spider mite population. Rec. Adv. Acarol., 1: 189—196.

Kongchuensin, M., 1983. Geographic variation and combining abilities of sex ratio and offspring number in the two-spotted spider mite, *Tetranychus urticae* Koch. Thesis, The Ohio State University, viii, 166 pp.

Laing, J.E., 1969. Life history and life table of *Tetranychus urticae* Koch. Acarologia, 11: 32—42.

Leigh, T.F. and Hyer, A., 1963. Spider mite-resistant cotton. Calif. Agric., Feb.: 6—7.

Lewontin, R.C., 1965. Selection for colonizing ability. In: H.G. Baker and G. Ledyard Stebbins (Editors), The Genetics of Colonizing Species. Academic Press, New York, NY, pp. 77—91.

Luckey, T.D., 1968. Insecticide hormoligosis. J. Econ. Entomol., 61: 7—12.

Meats, A., 1971. The relative importance to population increase of fluctuations in mortality, fecundity and the time variables of the reproductive schedule. Oecologia, 6: 223—237.

Mitchell, R., 1972. The sex ratio of the spider mite *Tetranychus urticae*. Entomol. Exp. Appl., 15: 299—304.

Mori, H. and Chant, D.A., 1966. The influence of humidity on the activity of *Phytoseiulus persimilis* Athias-Henriot and its prey, *Tetranychus urticae* (C.L. Koch) (Acarina: Phytoseiidae, Tetranychidae). Can. J. Zool., 44: 863—871.

Murtaugh, M.P. and Wrensch, D.L., 1978. Interspecific competition and hybridization between two-spotted and carmine spider mites. Ann. Entomol. Soc. Am., 71: 862—864.

Neiswander, C.R., Rodriguez, J.G. and Neiswander, R.B., 1950. Natural and induced variations in two-spotted spider mite populations. J. Econ. Entomol., 43: 633—636.

Nickel, J.L., 1960. Temperature and humidity relationships of *Tetranychus desertorum* Banks with special reference to distribution. Hilgardia, 30: 41—100.

Overmeer, W.P.J., 1972. Notes on mating behaviour and sex ratio control of *Tetranychus urticae* Koch (Acarina: Tetranychidae). Entomol. Ber., 32: 240—244.

Overmeer, W.P.J. and Harrison, R.A., 1969. Notes on the control of the sex ratio in populations of the two-spotted spider mite, *Tetranychus urticae* Koch (Acarina: Tetranychidae). N.Z. J. Sci., 12: 920—928.

Patterson, C.G., Knavel, D.E., Kemp, T.R. and Rodriguez, J.G., 1975. Chemical basis for resistance to *Tetranychus urticae* Koch in tomatoes. Environ. Entomol., 4: 670—674.

Patterson, C.G., Thurston, R. and Rodriguez, J.G., 1974. Two-spotted spider mite resistance in *Nicotania* species. J. Econ. Entomol., 67: 341—343.

Post, A., 1962. Effect of cultural measures on the population density of the fruit tree red spider mite, *Metatetranychus ulmi* Koch (Acari/Tetranychidae). Dissertation, University of Leiden, 100 pp.

Rodriguez, J.G., Chaplin, C.E., Stoltz, L.P. and Lasheen, A.M., 1970. Studies on resistance of strawberries to mites. I. Effects of plant nitrogen. J. Econ. Entomol., 63: 1855—1858.

Saito, Y., 1979. Comparative studies on life histories of three species of spider mites (Acarina: Tetranychidae). Appl. Entomol. Zool., 14: 83—94.

Shih, C.T., 1979. The influences of age of female *Tetranychus kanzawai* on sex ratio and life cycle of its progeny. Rec. Adv. Acarol., 1: 511—517.

Shih, C.T., Poe, S.L. and Cromroy, H.L., 1976. Biology, life table, and intrinsic rate of increase of *Tetranychus urticae*. Ann. Entomol. Soc. Am., 69: 362—364.

Snell, T.W., 1978. Fecundity, developmental time, and population growth rate. Oecologia, 32: 119—125.

Tanigoshi, L.K., Hoyt, S.C., Browne, R.W. and Logan, J.A., 1975. Influence of temperature on population increase of *Tetranychus mcdanieli* (Acarina: Tetranychidae). Ann. Entomol. Soc. Am., 68: 972—978.

Tanigoshi, L.K. and Logan, J.A., 1979. Tetranychid development under variable temperature regimes. Rec. Adv. Acarol., 1: 165—175.

Van de Vrie, M., McMurtry, J.A. and Huffaker, C.B., 1972. Ecology of tetranychid mites and their natural enemies: A review. Hilgardia, 41: 343—432.

Williams, A.J., 1954. Biology of the common red spider. J. Kans. Entomol. Soc., 27: 97—99.

Wrensch, D.L., 1979. Components of reproductive success in spider mites. Rec. Adv. Acarol., 1: 155—164.

Wrensch, D.L. and Young, S.S.Y., 1975. Effects of quality of resource and fertilization status on some fitness traits in the two-spotted spider mite, *Tetranychus urticae* Koch. Oecologia, 18: 259—267.

Wrensch, D.L. and Young, S.S.Y., 1978. Effects of density and host quality on rate of development, survivorship, and sex ratio in the carmine spider mite. Environ. Entomol., 7: 499—501.

Wrensch, D.L. and Young, S.S.Y., 1983. Relationship between primary and tertiary sex ratio in the two-spotted spider mite (Acarina: Tetranychidae). Ann. Entomol. Soc. Am., 76: 786—789.

Young, S.S.Y. and Wrensch, D.L., 1981. Relative influence of fitness components on total fitness of the two-spotted spider mite in different environments. Environ. Entomol., 10: 1—5.

Young, S.S.Y. and Wrensch, D.L., 1984. Control of sex ratio by female spider mites. Seen as manuscript.

Chapter 1.3 Physiology and Genetics

1.3.1 Aspects of Physiology

L.P.S. VAN DER GEEST

INTRODUCTION

Spider mites have not been the subject of extensive physiological studies: their small size makes them unsuited for many investigations, especially when microsurgical methods are needed to aid the student in unravelling physiological processes. Large areas in the field of spider mite physiology, therefore, show many gaps, but other areas have been studied in detail, often because of the great economic importance of these organisms and because of the ease of growing spider mites in large quantities.

In this chapter, only a few aspects of spider mite physiology will be reviewed. Some well-studied subjects are treated elsewhere, e.g. diapause by Veerman (Chapter 1.4.6), pheromones by Cone (Chapter 1.4.3) and webbing by Gerson (Chapter 1.4.1).

DIGESTION

The morphology of the digestive system has been studied by several authors and will be reviewed in this book in Chapter 1.1.2. However, little information is available about the functional characteristics of the different parts of the digestive system. Tetranychidae are parenchym-sucking mites. They acquire the food by piercing the cells with their stylets and by sucking out the contents. These long, slender, grooved stylets — the movable digits of the chelicerae — measure approximately one-third of the body length of the mites (Akimov and Barabanova, 1977). The other basal segments of the chelicerae have been fused to the unpaired stylophore, a fleshy and well-developed body part occupying a large portion of the gnathosomal region. The stylets fit together to form a food channel that ends distally in a hole through which the plant juices are sucked. The channel is connected proximally to the pharynx, which seems to act as a pump (André and Remacle, 1984). These authors found no evidence for the hypothesis of Hislop and Jeppson (1976), that salivary excretions may pass through the chelicerae into the plant tissues. For a more detailed description of the structure of the mouth parts of spider mites, the reader is referred to Chapter 1.1.1.

Large quantities of plant juices pass through the digestive system of spider mites, amounting to $6 \times 10^{-3} \mu l$ per 30 min and a simultaneous defaecation rate of $5 \times 10^{-3} \mu l$ (McEnroe, 1961a, 1963). This implies that 20—25% of the female body weight passes through the gut every 30 min. Liesering (1960) estimated the number of parenchym cells punctured and emptied to be 100 per 5 min, which corresponds to an ingestion rate of $2.5 \times 10^{-3} \mu l$ per 30 min

Chapter 1.3.1. references, p. 182

(M. Sabelis, personal communication, 1983). Spider mites must thus be able to handle a large volume of fluid, which they ingest during feeding. McEnroe (1963) discussed adaptations in the digestive system necessary for normal functions under conditions of high fluid intake. When mites feed on leaves, particulate material, such as chloroplasts, accumulates in the midgut, while the fluid cell sap is immediately transported to the hindgut. The fluid is eliminated in the urine: actively feeding mites have been observed to urinate at intervals of 30 min. The connection between hindgut and oesophagus makes it possible for fluid cell sap to flow directly towards the hindgut, while the chloroplasts and other particles are shunted to the midgut caeca. The entrance to the hindgut consists of a slit with the lips normally pressed together. However, during feeding, a pair of dorsoventral muscles contracts, resulting in a compression of the anterior midgut caecum. The fluid part of the cell sap is then forced into the hindgut through the partially opened lips, while the particulate components remain in the anterior midgut caecum.

Mothes and Seitz (1981a) studied the structure of the digestive system of *Tetranychus urticae* Koch by means of light and electron microscopical techniques. They assumed that salivary excretion destroys the outer membrane of the chloroplasts, since they observed only thylakoid granules inside the oesophagus and the ventriculus. The midgut epithelium may be involved in the production of digestive excretions, as the fine structure of these cells indicates the occurrence of protein synthesis. They believe that fatty acids, amino acids and carbohydrates are absorbed by certain midgut epithelial cells, in contrast to McEnroe (1961b) who demonstrated the elimination of low-molecular-weight material by a direct route from the oesophagus to the hindgut. Certain cells of the midgut become protruding and are later released inside the lumen; they seem to deposit excretory material in the gut (Blauvelt, 1945), but may also become phagocytes inside the gut lumen. Mothes and Seitz (1981a) assumed that these cells phagocytize thylakoid granules and starch and that after degradation of these products, low-molecular-weight components are released by these cells and used by other cells for the formation of storage material such as glycogen and protein. At the end of their life span, the phagocytes become excretory cells and pass through the hindgut. They become condensed in the hindgut into excretory balls and are eliminated through the anus (Wiesmann, 1968), glued together by a secretion, probably of a glycoprotein nature (Weyda, 1981). These strawberry-shaped faecal pellets are blackish and contain a large amount of plant pigment waste products (Fig. 1.3.1.1a, b). In addition, the so-called white pellets are excreted by spider mites. These pellets, which are formed inside the hindgut, contain mainly guanine (McEnroe, 1961). Weyda (1981 and personal communication, 1983) compared the surface of the black and white faecal pellets (cf. next section) by means of X-ray microanalysis together with scanning electron microscopy (SEM) (Fig. 1.3.1.2). Similar element patterns and the presence of sulphur indicated a probable glyco-proteinaceous nature of the surface excretion of both kinds of pellets. X-ray microanalysis of sections through the black faecal pellet revealed a high content of magnesium (F. Weyda, personal communication, 1983) which is in accordance with knowledge obtained from older reports on the presence of chlorophyll derivatives inside the pellets.

The metabolism of carbohydrates in spider mites is similar to that in other organisms. Ehrhardt and Voss (1961) identified a number of enzymes capable of hydrolyzing the following substrates in homogenates of whole spider mites (*T. urticae*): starch, maltose, sucrose, trehalose, melibiose, lactose, melezitose and raffinose. They assumed the presence of amylase, α-glucosidase, β-h-fructosidase and α- and β-galactosidase. No indications were found for the

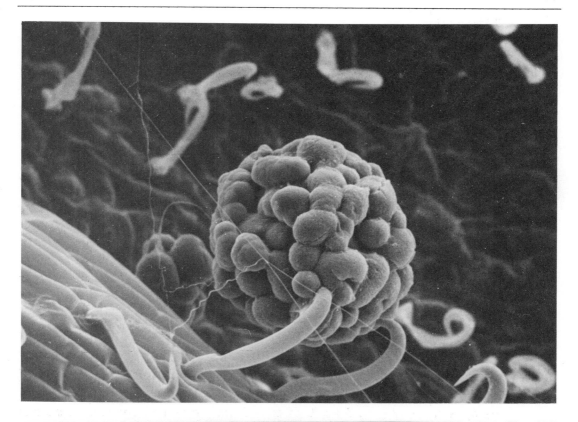

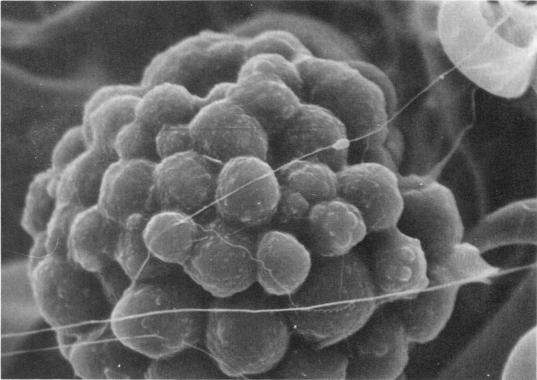

Fig. 1.3.1.1. (a) Black faecal pellet of *Tetranychus urticae* Koch, anchored to the leaf surface by silken threads (600×). (b) Surface of the black faecal pellet covered by a thin secretion, probably of a glycoprotein nature (1300×). Photographs courtesy of F. Weyda and P. Berkowský.

Chapter 1.3.1. references, p. 182

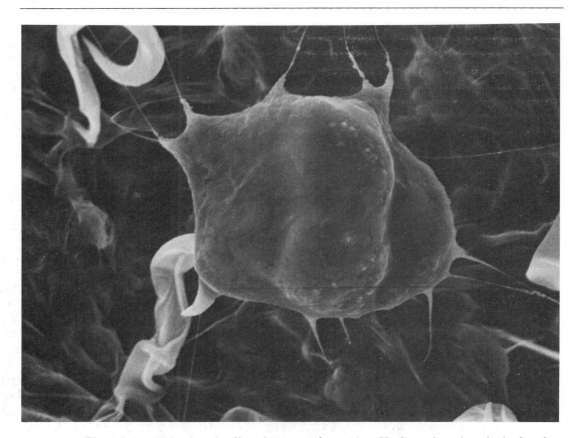

Fig. 1.3.1.2. White faecal pellet of *Tetranychus urticae* Koch, anchored to the leaf surface by silken threads. Surface of the pellet covered by a thin secretion, probably of the same origin as that found on black faecal pellets (2000×). Photograph courtesy of F. Weyda and P. Berkowský.

presence of β-glucosidase, cellulase, pectinase and polygalacturonase. Mehrotra (1961, 1963a) and later, Kötter (1978), demonstrated the presence of a number of enzymes in *T. urticae* necessary for the glycolytic pathway and for the hexose monophosphate cycle (pentose cycle). For the tricarboxy-acid cycle, the presence of malate dehydrogenase (Mehrotra, 1962; Kötter, 1978) was shown. The glycolytic pathway in spider mites, however, differs from that in various other invertebrate tissues, but resembles that in the flight muscle of insects, as the carbohydrates appear to be glycolyzed by the pyruvate-α-glycerophosphate dismutation system (Mehrotra, 1963a).

Food preference seems to affect the activity of digestive enzymes in spider mite species (Akimov and Barabanova, 1977). In polyphagous species, as *T. cinnabarinus* Boisduval (a red form in the *T. urticae* complex) and *T. turkestani* Ugarov & Nikolski, high activity was found for carbohydrase hydrolyzing poly- and oligosaccharides, as well as for protease. In the oligophagous species *Panonychus ulmi* (Koch), *Tetranychus przhevalskii* Reck, *Bryobia reditkorsevi* Reck and *Oligonychus biotae* (Reck) and in the monophagous species *Eotetranychus tiliarium* (Hermann) only traces of proteases were found, while the most prevalent carbohydrase appeared to be invertase. The authors, however, were unable to explain the mechanism by which host plant specificity affects the activity of the various digestive enzymes. Akimov and Barabanova (1977) also showed that the feeding of mites on host plants may be influenced by the presence of certain accompanying substances: NaCl, KNO_3, and glutamine solutions were found to activate digestive enzymes in *T. cinnabarinus*.

Metabolic rates (oxygen consumption per unit live weight) of different stages of *T. cinnabarinus* vary from 0.27 to 2.32 nl O_2 per microgram live weight per hour, which is about half that of the predaceous mite *Phytoseiulus persimilis* A.-H. (Thurling, 1980). This difference is attributed to the different modes of life of these 2 mite species. The respiration rate (oxygen consumption per individual) of *T. cinnabarinus* is a function of live weight for all mobile stages: the respiration—body weight relation can be expressed by the following equation: $\log_{10}R = -0.091 + 1.213 \log_{10}W$, where R is the respiration per individual and W is the weight.

EXCRETION

Several organs in spider mites participate in the excretion of metabolic waste products and in osmoregulation. Blauvelt (1945) has already described the hindgut as an organ which serves for the evacuation of food residue and as a nephritic organ for the elimination of waste products. He suggested that the anterior part of the hindgut is formed from Malpighian tubules which are stretched out between mid- and hindgut. According to Mothes and Seitz (1980), not only the hindgut, but also part of the midgut is involved in the elimination of excretory products. The lateral walls of the hindgut consist of a single layer of cells with long branches extending towards the gut lumen. The dorsal wall is composed of highly vacuolated cells in the form of a syncytium. The midgut epithelium forms a dense layer of long and branched microvilli around the lumen. These authors observed the concentration of excretions inside the mitochondria of the midgut epithelium. These organelles swell, owing to the presence of these waste products, but do not lose their structure. The excretions are later released into the cytoplasm. Excretory material was also observed in the hindgut epithelial cells. The nature of this crystal-like material has been analysed by McEnroe (1961a). He found that it is mainly composed of guanine, which is excreted as the so-called white pellets. The black faecal pellets, formed by the phagocytes and consisting mainly of food residues, also contain a certain amount of guanine.

In contrast to the observations of several others (see, e.g., Blauvelt, 1945; McEnroe, 1961b), a Malpighian complex was found to exist in *T. urticae* by Mothes and Seitz (1980). It seems to occur only in sexually mature adult females, in which it can easily be observed under the light microscope around the female sexual organ. Two different parts can be distinguished: an initial part characterized by basal membrane foldings and an end part with a wide lumen that is lined with small epithelial cells. The authors did not observe any compact excretory material in this organ. They assumed that the Malpighian complex has a function in osmoregulation. F. Weyda (personal communication, 1983) does not agree with the results of Mothes and Seitz (1980) and considers this organ to be the second part of the oviduct (Weyda, 1980, 1981). The ultrastructure of this so-called Malpighian complex points to a secretory and not an excretory function.

Coxal organs are also claimed to have a function in osmoregulation (Mothes and Seitz, 1980). These paired coxal glands, typical for arthropods, consist of a proximal tube, a middle region and a distal tube. The proximal tube has a lumen which is formed by 3 cells with a cytoplasm in which intraplasmatic tubules can be discerned. These seem to be surrounded by a double membrane. This proximal part of the coxal organ is connected to the cells of the midgut. The distal tube, with an opening near coxa I, does not show this tubular structure, but possesses infoldings in the basal membrane.

Chapter 1.3.1. references, p. 182

Mothes and Seitz (1980) observed granular material inside the lumen of the coxal organ, but no excretion crystals. The coxal organs probably have a function in the excretion of certain ions and therefore a role in osmoregulation.

WATER BALANCE

Organisms of the size of a spider mite have a ratio of surface area to volume which imposes severe stress when they encounter a vapour deficit. Under normal feeding conditions, however, spider mites have to handle a vast amount of fluid which is ingested and eliminated by the direct pathway from the oesophagus to the hindgut. Evaporation through the cuticle will be limited under feeding conditions, as the mites feed on the leaf surface where relative humidity is in general high, particularly when webbing is present. Under non-feeding conditions (e.g. diapause) and premoult conditions, however, the mites must conserve water. Control of water balance can be effected by means of the respiratory system. Blauvelt (1945) demonstrated the direct communication of the peritremes with the open air through the slit-like stigmata. Closing of the peritremes can be effected by retraction of the mandibular plate, as the peritremes are then inverted and enclosed between the epistome and the mandibular plate. Especially under diapause conditions, when no feeding occurs, loss of water will be an important problem, unless evaporation is effectively limited. McEnroe (1961b) compared the water loss of diapausing and summer forms of *T. urticae*. By sealing the tracheal openings and by cyanide treatments, he was able to demonstrate the role of the respiratory system in evaporation. Moreover, he could show that transpiration through the cuticle is of hardly any importance. Diapausing females are very resistant to desiccation. The stigmata are apparently closed, which was also learned from the fact that such females are insensitive to cyanide treatment.

Winston (1964) studied the mechanism by which water balance is regulated in *Bryobia praetiosa* Koch. This mite is active at high temperature under very dry conditions. He showed that the mite is able to actively regulate water loss at relative humidities between 53 and 85%. At humidities below 53%, water loss is still restricted, but not regulated. He also showed that well-fed mites prefer conditions of low humidity when they are offered a choice. However, mites that are somewhat desiccated prefer conditions with a high relative humidity.

Eggs of spider mites, in particular hibernating eggs, are subject to extreme desiccation, unless special adaptations safeguard the eggs while gas exchange (CO_2 and O_2) can still proceed. In spider mite eggs, 2 embryonic stigmata of complicated structure are found, which pierce the egg shell during the contractive phase of the germ band (Dittrich and Streibert, 1969). The stigmata are connected with a certain part of the intermediate membrane that surrounds the embryo. This part of the membrane contains a large number of perforations and ridges, thus allowing the presence of an air cushion of thickness 0.2—0.3 μm between egg shell and embryo. The authors assume that the perforations are the sites of gas diffusion out of and into the embryo. Winter eggs of *P. ulmi* remain undeveloped until the second half of April, after which development to the germ band can proceed. This late development probably ensures protection of the egg against desiccation.

ESTERASES AND CHOLINEACETYLASES

Several studies have been made on enzymes with esterase activity, mainly because of the role they may play in resistance of spider mites to acaricides. In this chapter, however, no attention will be paid to esterase studies dealing with resistance to pesticides, which will be dealt with in Chapter 3.4.

Smissaert (1965) demonstrated the presence of α-naphthyl acetate-hydrolyzing esterases in the two-spotted spider mite *T. urticae*. At least three different enzymes were detected, which differed in heat sensitivity and rate of inhibition by diazoxon. Blank (1979) studied esterases in *T. urticae* by means of electrophoresis. Seven distinct non-specific esterase bands were detected on electropherograms from soluble mite extracts, while additional faint bands were sometimes visible when freshly prepared extracts were used. It was not possible to fit all esterases into a current classification system on the basis of substrate and inhibition characteristics. For example, one of the esterases was insensitive to eserine and organophosphate inhibitors and would, on this basis, be classified as an aromatic esterase, but such a classification conflicts with the insensitivity to EDTA. Other esterases were inhibited by eserine and showed differential inhibition by organophosphates. One of these was sensitive to malaoxon and mevinphos and would, by these criteria, be classified as a cholinesterase, but its insensitivity to dicrotophos was in contradiction to such a classification. It was therefore necessary to determine esterases having cholinesterase activity by using specific staining techniques with acetyl thiocholine as substrate. A considerable amount of material having cholinesterase activity failed to penetrate the electrophoresis gel, an indication that in spider mites these enzymes are strongly bound to membranes, or other proteinaceous material.

The toxic effect of anticholinesterase compounds suggests that the cholinergic system is responsible for stimulus transmission in the nervous system of spider mites. All components of the cholinergic system have been demonstrated in spider mites: acetylcholine, cholinesterase and cholineacetylase. Other types of transmitter substance, such as epinephrine, γ-amino butyric acid and 5-hydroxytryptamine have not yet been detected in spider mites (Mehrotra, 1963b).

The substrate specificity of choline acetylase in spider mites was studied by Mehrotra and Dauterman (1963). As substrate, a number of *N*-alkyl substituted analogues of choline and analogues of 2-dialkylaminoethanol were studied. The enzyme only acetylated analogues having dimethyl or diethyl groups, and was not able to use analogues having more than 2 carbon atoms between the quarternary nitrogen and the hydroxyl group, or to use tertiary alcohol. Similar results were obtained for housefly enzymes. Choline acetylase from rat brain, however, can only utilize dimethyl derivatives of choline. On the basis of their results, the authors concluded that the enzyme possesses 2 active sites, an 'acetylatic' and an 'anionic' site, analogous to 'esteratic' and 'anionic' sites on cholinesterase. The distance between these sites is such that only choline is effectively used by the enzyme. The distance between the 2 sites was estimated to be approximately 4.6 Å.

CHITIN SYNTHESIS

Chitin is an important component of the cuticle of arthropods and of the cell wall of fungi. It is a high-molecular-weight polysaccharide, consisting predominantly of *N*-acetyl-D-glucosamine. There is evidence that chitin synthesis in arthropods and in fungi is regulated by similar enzyme systems (Mothes and Seitz, 1981b).

Interest in the pathway of chitin synthesis has increased in the last decade, since the introduction of pesticides such as diflubenzuron, and the discovery of antibiotics of the Nikkomycin group which impede the action of chitin synthesis.

Two different methods of chitin synthesis in *T. urticae* have been described (Mothes and Seitz, 1981b). One system is situated inside the egg, the other in the hypodermis below the cuticle. Small vesicles (20—40 nm) can be observed during the formation of the egg shell in the periphery of the previtellogenic oocyte. These vesicles (chitosomes), derived from a cytoplasmatic condensation of numerous ribosomes, move towards the periphery of the oocyte and obtain an electron-dense periphery and an electron-transparent centre. Other vesicles which are in contact with those parts of the egg shell which are already formed have a shell composed of an outer granular layer, a dense middle layer and an electron-transparent layer around an electron-dense matrix. After establishment of contact with the egg shell, the contents of the vesicles are released onto the partially formed egg shell and condensed.

No vesicles appear in the hypodermis during the moulting process. The non-chitinous epicuticle is deposited on top of small hypodermal microvilli, while the chitinous procuticle deposition occurs in spaces between the epicuticle and the cell surfaces of the hypodermis.

Treatment of *T. urticae* mites with Nikkomycins results in inhibition of cuticle synthesis, apparently because of inhibition of chitin synthetase. Also, egg shell formation is disturbed: abnormally thickened or plaque-like egg shells are produced. No satisfactory explanation for the effect of Nikkomycin on egg shell formation could hitherto be given (Mothes and Seitz, 1982).

CAROTENOID PIGMENTS

Spider mites are often characterized by an intense colouration, which is caused mainly by the presence of pigments in the viscera, haemolymph and tissues. The cuticle is usually colourless and devoid of these biochromes. The first investigations on the characterization of the nature of these pigments were carried out by Metcalf and Newell (1962). They found that the pink, rose and deep orange pigments are all carotenoids, as is apparent from their adsorption spectra. In addition, chlorophyll and chlorophyll-derived pigments were found, but these are only present in the gut in food residue.

The discovery of a number of pigment mutations in the species *Tetranychus pacificus* McGregor and *T. urticae* in the Laboratory of Experimental Entomology of the University of Amsterdam initiated a study on the presence of carotenoids in these species, in *T. cinnabarinus*, which is characterized by a red colour, and in the primitive species *Schizonobia sycophanta* Womersley (Veerman, 1970, 1972, 1974a, b). The normally pigmented wild types of these species showed a striking similarity with respect to the relative abundance of the different carotenoids present. All species contain 3-hydroxy-4-keto-β-carotene, astaxanthin and traces of isocryptoxanthin, echinone and 2 unidentified carotenoid epoxides, often in esterified form, in addition to the carotenoids which are present in their food (*Phaseolus vulgaris* L. leaves): α-carotene, β-carotene, lutein, lutein-5,6-epoxide, violaxanthin and neoxanthin. These plant carotenoids probably occur only in the gut, with the exception of β-carotene, which is also present in the body of the mites. *Albino* mutants of *T. urticae* and *T. pacificus* are devoid of any trace of keto-carotenoids: only plant carotenoids could be detected, probably because they are present in the intestines of the mites. Veerman (1974a) assumed that the complex locus for albinism in *T. urticae* is concerned with the uptake of carotenoids by the gut from the food. The mutant *lemon*, a yellow-coloured mutant, also contains only plant-derived carotenoids, as does *albino*, but β-carotene is present in a considerably higher concentration. This accumulation of β-carotene is an indication that β-carotene is the precursor for the synthesis of keto-carotenoids in the mites and that the first oxidative step is blocked in *albino*. The *white eye* mutations lack the highest oxidized pigment astaxanthin. It seems probable, therefore, that in this mutant the oxidation of 3-hydroxy-4,4′-diketo-β-carotene to astaxanthin is blocked. The 2 mutants *flamingo* and *stork* both showed an aberrant distribution of pigments in the body of the mites. *Flamingo* differs from *stork* in having small red eye spots which are absent in *stork*. In both these mutants all hydroxy-keto-carotenoids are present, but in an unesterified form, in contrast to the way they occur in *wild-type* mites. *Flamingo*, however, is capable of forming an ester of astaxanthin which, according to Veerman (1974a), is responsible for the presence of eye spots in this mutant. The aberrant distribution of pigments in these mutant mites seems to be caused by the fact that no esterified hydroxy-keto-carotenoids are present.

Based on these findings, the scheme in Fig. 1.3.1.3 for the biosynthesis of keto-carotenoids was proposed by Veerman (1970).

Astaxanthin in spider mites shows the 3S, 3′S chirality, which has also been found for astaxanthin from Crustaceae (Veerman et al., 1975). This configuration supports the precursor role of β-carotene.

Diapausing females of *T. urticae* are characterized by a deep orange colour. This is caused by the fact that the 3-hydroxy-keto-carotenoids are present in a concentration which is 2.5 times higher than that in summer

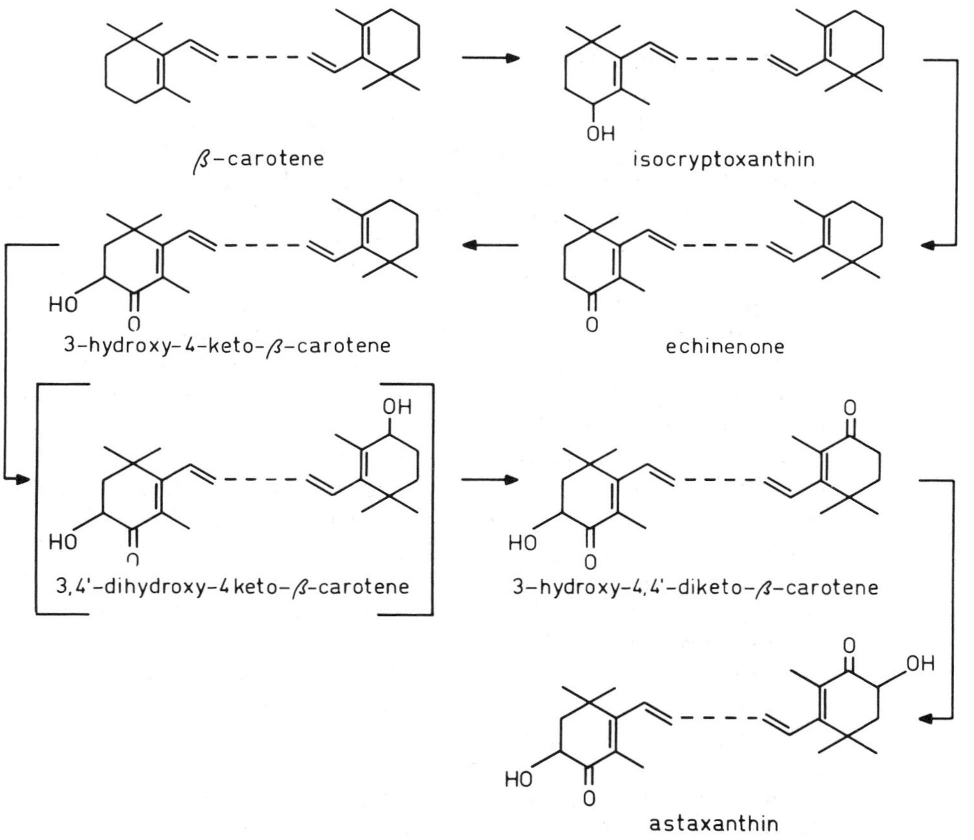

Fig. 1.3.1.3. Possible pathways for the biosynthesis of keto-carotenoids in spider mites. [ ] = not isolated (Veerman, 1970).

Chapter 1.3.1. references, p. 182

females. The possible function of this increase in pigment is still unknown. The role that carotenoids play in the induction of diapause in spider mites is discussed in Chapter 1.4.6 of this book.

RESPONSES TO EXTERNAL FACTORS

Temperature affects the behaviour and activity of spider mites to a great extent. Mori (1961) studied the reaction of 4 species of spider mites to temperatures varying from 2°C to approximately 50°C. *T. urticae* and *P. ulmi* showed similar reactions, with activity zones ranging from 5—9°C to 41—44°C, but the zones of activity of *B. praetiosa* (10.8—40.2°C) and in particular of *Tetranychus viennensis* Zacher (14.8—40.8°C) were considerably narrower. Temperature preference for most of these species was around 25°C, but *T. urticae* was somewhat deviant, as it always showed a preference for the lower temperature when given the choice between 2 temperatures in the range 13—35°C. A positive phototaxis to white light was demonstrated by Mori (1962a) in *P. ulmi*, *T. viennensis* and *T. urticae*, irrespective of sex, feeding condition and season in which they were collected. This photokinetic response even disturbs the temperature preference reaction of spider mites, except at temperatures above 35°C (Mori, 1962b). Under moist conditions, *T. urticae* becomes sluggish and is almost immobile (Mori and Chant, 1966). The species also shows a negative hygrotaxis: it avoids places with a high relative humidity, when given the choice between moist and drier places.

The photoperiodic response to white light has been studied in more detail only for the species *T. urticae*. The degree of photoresponse to light depends on the angle of incidence and on the intensity of the light (Suski and Naegele, 1963a). The response to light, however, is also affected by the physiological condition of the mite (Suski and Naegele, 1963b). A low ambient relative humidity, caused by a depletion and desiccation of the food, results in an increased activity and a stronger response to light, which may enable the mites to search more efficiently for new food sources. The sensitivity of *T. urticae* females to light of wavelengths varying from 350 to 700 nm was studied by Naegele et al. (1966). No response was observed to light with wavelengths longer than 600 nm, although McEnroe (1971a) demonstrated a slight photoeffect of light in the far red (around 700 nm). It is believed that this red light stimulates the central nervous system directly through the transparent integument. Peaks in sensitivity to light are found in the near-UV at 375 nm and at 525 nm. Barcelo (1981) observed a strong avoidance reaction in spider mites exposed to UV light in the region 280—320 nm, but no reaction when they were subjected to UV of 254 nm. *T. urticae* cannot discern the plane of polarized light (McEnroe 1971b).

McEnroe and Dronka (1966) studied the light response of *T. urticae* by exposing adult females to increasing intensities of light of a certain wavelength. For most colours, a random response was observed, irrespective of the light intensity. An increasing photoresponse, however, was observed when mites were exposed to increasing intensities of light in the near-UV (360—400 nm). This is in agreement with the observation that *T. urticae* shows a peak in sensitivity to light of 375 nm. The response to green light is particularly interesting, as the mites show either a photopositive (green +) or a photonegative (green —) reaction to light. 'Green +' mites, subjected to increasing intensities of light of 380 or 520 nm, show a steep threshold of response, while the response index of 'green—' mites to increasing intensities of light of 380 nm follows a straight line. Response of 'green—' mites to light of 530 nm is random, with an average response index of near zero. The response index of 'green +' mites to light of 530 nm decreased to zero when the mites were kept for 1 h at a relative humidity of 55%. Their response to light of 380 nm remained unchanged, which is a strong indication that responses of *T. urticae* to green (± 500 nm) and UV light are independent and require separate receptor systems. Both males and females have a pair of eyes located on either side of the propodosoma (Mills, 1974). The anterior eye has a biconvex lens and is surrounded by a red oily fluid. These eyes serve as scanning detectors (McEnroe, 1969) and have separate receptors for both near-UV and green light (McEnroe and Dronka, 1969), while the posterior eyes with simple convex lenses are non-directional receptors for ultraviolet light only. Later research showed that the reaction to green light is dependent on the feeding condition of the mites: teneral females which have not had an opportunity to feed belong to the 'green +' category, while 'green —' mites have already fed for at least 1 day (McEnroe and Dronka, 1971). 'Green —' mites held at low ambient relative humidity conditions slowly change to 'green +' mites. The decrease in volume of these mites, the actual basis for the switch in behavioural class, is apparently perceived by a pressure receptor. On the other hand, rapid loss of green photopositive response by 'green +' mites under high relative humidity is probably controlled by a humidity receptor. In addition, changes in geotaxis were observed by these authors. 'Green +' and 'green —' mites show a negative geotaxis in darkness, which, however, can be overridden by UV illumination from below. Green illumination results in a positive geotaxis in the 'green +' class, while the 'green —' mites continue to show a negative geotactic response. The shift in photobehavioural classes is apparently due to water stress and ambient relative humidity. As long as fresh food is available, 'green —' mites will predominate, but when leaves are destroyed by heavy feeding activity, 'green +' mites will develop and migration may occur.

Results obtained by Herne (1968) indicate that *P. ulmi* can withstand immersion in water over prolonged periods of up to 48 h. Active stages showed only 25% mortality after 48 h of immersion, while no mortality at all was observed among eggs. An interesting aspect is that feeding and oviposition does not occur under water, nor hatching of the eggs. On the basis of these results, one may expect that long periods of wet weather delay the build-up of mite infestation in the field. The response of spider mites to sub-zero temperatures is discussed in Chapter 1.4.6 of this book.

DAILY RHYTHMICITY

Most physiological processes in organisms do not proceed at a continuous rate, but their rate is often synchronized in some way with the daily environmental fluctuations. Daily periodicity in spider mites has been studied for a few processes, and only in the two-spotted spider mite *T. urticae*. Nowosielski et al. (1964) studied the sensitivity of this mite to ethyl ether, chloroform and carbon tetrachloride at different times of the day by determining the 25% recovery time after narcotization with these solvents. These authors considered recovery time to be a measure of the sensitivity of the mites to these solvents. The mites showed the greatest sensitivity to the narcotic effect of all solvents around dawn, while the lowest sensitivity was found around nightfall. In another study, Polcik et al. (1964) observed a daily rhythm in sensitivity to DDVP (dimethyl-2,2-dichlorovinyl phosphate) in *T. urticae*. Maximum sensitivity of the mite to the insecticide was found 2 h after dawn, after which the sensitivity gradually declined during the day to reach a minimum around 2 h after nightfall. No experiments were done to

Chapter 1.3.1. references, p. 182

investigate a possible endogenous character of this rhythm. A dial periodicity in sensitivity to dicofol was also demonstrated in *T. urticae* (Fisher, 1967). In contrast to the results of Polcik et al. (1964), a two-peak cycle of susceptibility was found, one occurring between 2:00 and 5:00 h and the second between 13:00 and 18:00 h. Experiments were carried out in total darkness under conditions of constant temperature and relative humidity. Polcik et al. (1965) observed a daily rhythm in oviposition when the two-spotted spider mite was maintained in an LD 14:10 cycle (14 h of light alternating with 10 h of darkness). They compared oviposition in 2 different mite populations, one maintained in a light—dark cycle which coincided with the normal day—night cycle and the other kept in a reverse light—dark cycle. For both populations, oviposition frequency decreased after experimental nightfall, remained at a low level during the entire night period and increased after dawn again to the higher daytime levels. No such rhythm could be demonstrated by these authors when mites were kept under constant light conditions, nor could rhythmicity in oviposition by shown under conditions of constant darkness (Fisher, 1967).

Veerman and Vaz Nunes (1980) demonstrated the involvement of circadian rhythmicity in the photoperiodic determination of diapause in *T. urticae*. They exposed mites to dark phases varying from 4 to 48 h, alternated by light phases of 8 h. From their results, it is apparent that the incidence of diapause is a rhythmic function of the duration of darkness. Peaks of high incidence recur with a cycle length (duration of light plus dark) of about 24, 44, 64 and 84 h. The circadian rhythm involved may thus have a relatively short free-running period of 20 h, which is apparently corrected by external factors under normal daylength conditions.

CONCLUDING REMARKS

In this chapter, an attempt was made to review our present knowledge of the physiology of spider mites. Large areas are still unexplored: no studies have yet been made on hormones and neuromones in spider mites, while the physiological mechanisms underlying the action of mechanical and chemical sensilla are completely unknown. The author hopes that this chapter will be a stimulus for students to fill the gaps in our knowledge on the physiology of these economically important organisms.

REFERENCES

Akimov, I.A. and Barabanova, V.V., 1977. Morphological and functional characteristics of the digestive system in tetranychid mites (Trombidiformes, Tetranychoidea). Rev. Entomol. URSS, 56: 912—922.

André, H.M. and Remacle, Cl., 1984. Comparative and functional morphology of the gnathosoma of *Tetranychus urticae* (Acari: Tetranychidae). Acarologia, 25: 179—190.

Barcelo, J.A., 1981. Photoeffects of visible and ultraviolet radiation on the two-spotted spider mite, *Tetranychus urticae*. Photochem. Photobiol., 33: 703—706.

Blank, R.H., 1979. Studies on the non-specific esterase and acetylcholinesterase isozymes electrophoretically separated from the mites *Sancassania berlesei* (Tyroglyphidae) and *Tetranychus urticae* (Tetranychidae). N. Z. J. Agric. Res., 22: 497—506.

Blauvelt, W.E., 1945. The internal morphology of the common red spider mite (*Tetranychus telarius* Linn.). Mem. Cornell Univ. Agric. Exp. Station, Ithaca, NY, 270: 1—35.

Dittrich, V. and Streibert, P., 1969. The respiratory mechanism of spider mite eggs. Z. Angew. Entomol. 63: 200—211.

Ehrhardt, P. and Voss, G., 1961. Die Carbohydrasen der Spinnmilbe *Tetranychus urticae* Koch (Acari, Trombidiformes, Tetranychidae). Experientia, 17: 307.

Fisher, R.W., 1967. Diel periodicity in sensitivity of *Tetranychus urticae* (Acarina: Tetranychidae) to dicofol. Can. Entomol., 99: 281—284.

Herne, D.H.C., 1968. Some responses of the European red mite, *Panonychus ulmi*, to immersion in water. Can. Entomol., 100: 540—541.

Hislop, R.G. and Jeppson, L.R., 1976. Morphology of the mouthparts of several species of phytophagous mites. Ann. Entomol. Soc. Am., 69: 1125—1135.

Kötter, C., 1978. Ein Beitrag zur Stoffwechselphysiologie von *Tetranychus urticae* Koch (Acari, Tetranychidae). Z. Angew. Entomol., 86: 337—348.

Liesering, R., 1960. Beitrag zum phytopathologischen Wirkungsmechanismen von *Tetranychus urticae* Koch (Acarina: Tetranychidae). Z. Pflanzenkr. Pflanzenschutz, 67: 524—543.

McEnroe, W.D., 1961a. Guanine excretion by the two-spotted spider mite (*Tetranychus telarius* (L)). Ann. Entomol. Soc. Am., 54: 925—926.

McEnroe, W.D., 1961b. The control of water loss by the two-spotted spider mite (*Tetranychus telarius*). Ann. Entomol. Soc. Am., 54: 883—887.

McEnroe, W.D., 1963. The role of the digestive system in the water balance of the two-spotted spider mite. Adv. Acarol., 1: 225—231.

McEnroe, W.D., 1969. Eyes of the female two-spotted spider mite, *Tetranychus urticae*. I. Morphology. Ann. Entomol. Soc. Am., 62: 461—466.

McEnroe, W.D., 1971a. The red photoresponse of the spider mite *Tetranychus urticae* (Acarina: Tetranychidae). Acarologia, 13: 113—118.

McEnroe, W.D., 1971b. Eyes of the two-spotted spider mite *Tetranychus urticae*. III. Analysis of polarized light. Ann. Entomol. Soc. Am., 64: 879—883.

McEnroe, W.D. and Dronka, K., 1966. Color vision in the adult female two-spotted spider mite. Science, 154: 782—784.

McEnroe, W.D. and Dronka, K., 1969. Eyes of the two-spotted spider mite, *Tetranychus urticae*. II. Behavioural analysis of the photoreceptors. Ann. Entomol. Soc. Am., 62: 466—469.

McEnroe, W.D. and Dronka, K., 1971. Photobehavioural classes of the spider mite *Tetranychus urticae* (Acarina: Tetranychidae). Entomol. Exp. Appl., 14: 420—424.

Mehrotra, K.N., 1961. Carbohydrate metabolism in the two-spotted spider mite, *Tetranychus telarius* L. I. Hexose monophosphate cycle. Comp. Biochem. Physiol., 3: 184—198.

Mehrotra, K.N., 1962. Malic dehydrogenase in the two-spotted spider mite *Tetranychus telarius* (L). Can. J. Biochem., 40: 1529—1533.

Mehrotra, K.N., 1963a. Carbohydrate metabolism in the two-spotted spider mite, *Tetranychus urticae*. II. Embden—Meyerhof pathway. Can. J. Biochem. Physiol., 41: 1592—1602.

Mehrotra, K.N., 1963b. Biochemistry of nerve function in Acarina. Adv. Acarol., 1: 209—213.

Mehrotra, K.N. and Dauterman, W.C., 1963. The *N*-alkyl group specificity of choline acetylase from the housefly *Musca domestica* L., and the two-spotted spider mite, *Tetranychus telarius* L. J. Insect Physiol., 9: 293—298.

Metcalf, R.L. and Newell, I.M., 1962. Investigation of the biochromes of mites. Ann. Entomol. Soc. Am., 55: 350—352.

Mills, L.R., 1974. Structure of the visual system of the two-spotted spider mite, *Tetranychus urticae*. J. Insect Physiol., 20: 795—808.

Mori, H., 1961. Comparative studies of thermal reaction in four species of spider mites (Acarina: Tetranychidae). J. Fac. Agric., Hokkaido Univ., 51: 574—591.

Mori, H., 1962a. Seasonal difference of phototactic response in three species of spider mites (Acarina: Tetranychidae). J. Fac. Agric. Hokkaido Univ., 52: 1—9.

Mori, H., 1962b. The effects of photostimulus on the thermal reaction in four species of spider mites (Acarina: Tetranychidae). J. Fac. Agric. Hokkaido Univ., 52: 10—19.

Mori, H. and Chant, D.A., 1966. The influence of humidity on the activity of *Phytoseiulus persimilis* Athias-Henriot and its prey, *Tetranychus urticae* (C.L. Koch) Acarina: Phytoseidae, Tetranychidae). Can. J. Zool., 144: 863—871.

Mothes, U. and Seitz, K.-A., 1980. Licht- und elektronenmikroskopische Untersuchungen zur Funktionsmorphologie von *Tetranychus urticae* (Acari: Tetranychidae). I. Exkretionssysteme. Zool. Jahrb., Abt. Anat. Ontog. Tiere, 104: 500—529.

Mothes, U. and Seitz, K.-A., 1981a. Functional microscopic anatomy of the digestive system of *Tetranychus urticae* (Acari: Tetranychidae). Acarologia, 22: 257—270.

Mothes, U. and Seitz, K.-A., 1981b. A possible pathway of chitin synthesis as revealed by electron microscopy in *Tetranychus urticae* (Acari: Tetranychidae). Cell Tissue Res., 214: 443—448.

Mothes, U. and Seitz, K.-A., 1982. Action of the microbial metabolite and chitin synthesis inhibitor Nikkomycin on the mite *Tetranychus urticae*; an electron microscope study. Pesticide Sci., 13: 426—441.

Naegele, J.A., McEnroe, W.D. and Soans, A.B., 1966. Spectral sensitivity and orientation response of the two-spotted spider mite, *Tetranychus urticae* Koch from 350 mμ to 700 mμ. J. Insect Physiol., 12: 1187—1195.

Nowosielski, J.W., Patton, R.L. and Naegele, J.A., 1964. Daily rhythm of narcotic sensitivity in the house cricket, *Gryllus domesticus* L. and the two-spotted spider mite, *Tetranychus urticae* Koch. J. Cell. Comp. Physiol., 63: 393—398.

Polcik, B., Nowosielski, J.W. and Naegele, J.A., 1964. Daily sensitivity rhythm of the two-spotted spider mite, *Tetranychus urticae*, to DDVP. Science, 145: 405—406.

Polcik, B., Nowosielski, J.W. and Naegele, J.A., 1965. Daily rhythm of oviposition in the two-spotted spider mite. J. Econ. Entomol., 58: 467—469.

Smissaert, H.R., 1965. Esterases in spider mites hydrolysing α-naphthylacetate. Nature (London), 205: 158—160.

Suski, Z.W. and Naegele, J.A., 1963a. Light response in the two-spotted spider mite. I. Analysis of behavioral response. Adv. Acarol., 1: 435—444.

Suski, Z.W. and Naegele, J.A., 1963b. Light response in the two-spotted spider mite. II. Behaviour in the sedentary and dispersal phases. Adv. Acarol., 1: 445—453.

Thurling, D.J., 1980. Metabolic rate and life stage of the mites *Tetranychus cinnabarinus* Boisd. (Prostigmata) and *Phytoseiulus persimilis* A.-H. (Mesostigmata). Oecologia, 46: 391—396.

Veerman, A., 1970. The pigments of *Tetranychus cinnabarinus* Boisd. (Acari: Tetranychidae). Comp. Biochem. Physiol., 36: 749—763.

Veerman, A., 1972. Carotenoids of wild-type and mutant strains of *Tetranychus pacificus* McGregor (Acari: Tetranychidae). Comp. Biochem. Physiol. B, 42: 329—340.

Veerman, A., 1974a. Carotenoid metabolism in *Tetranychus urticae* Koch (Acari: Tetranychidae). Comp. Biochem. Physiol. B, 47: 101—116.

Veerman, A., 1974b. The carotenoid pigments of *Schizonobia sycophanta* Womersley (Acari: Tetranychidae). Comp. Biochem. Physiol. B, 48: 321—327.

Veerman, A., Borch, G., Pederson, R. and Liaaen-Jensen, S., 1975. Animal carotenoids. 10. Chirality of astaxanthin of different biosynthetic origin. Acta Chem. Scand., Ser. B, 29: 525.

Veerman, A. and Vaz Nunes, M., 1980. Circadian rhythmicity participates in the photoperiodic determination of diapause in spider mites. Nature (London), 287: 140—141.

Weyda, F., 1980. Reproductive system and oogenesis in active females of *Tetranychus urticae* (Acari: Tetranychidae). Acta Entomol. Bohemoslov., 77: 375—377.

Weyda, F., 1981. Biotypy Svilušky Chmelové, *Tetranychus urticae* ve Vztahu k Rezistenci (Biotypes of the Two-spotted Spider Mite, *Tetranychus urticae*, in Relation to Resistance), Vol. 1 (text, 320 pp.) and Vol. 2 (tables, graphs and photographs). Ph.D. Dissertation, Institute of Entomology, Prague.

Wiesmann, R., 1968. Untersuchungen über die Verdauungsvorgänge bei der gemeine Spinnmilbe *Tetranychus urticae* Koch. Z. Angew. Entomol., 61: 457—465.

Winston, P.W., 1964. The physiology of waterbalance in Acarina. Acarologia, Facsimile Hors Série: 307—314.

1.3.2 Genetics

W. HELLE

INTRODUCTION

The rapid development of a number of resistances to various pesticides after 1950 stimulated interest into the genetic basis of resistance in spider mites. Some types of resistance were analyzed by means of conventional crossing procedures (see Chapter 3.4). These studies resulted in the identification of a mutation for resistance to organophosphates residing at the locus *OP* (Helle, 1962; Schulten, 1968), and of a mutation responsible for resistance to tetradifon, designated *T* by Overmeer (1967). Thereupon it was felt appropriate to map the genes for resistance and to look for visible mutations, which were detected in the mid-1960s in *Tetranychus pacificus* McGregor and in *T. urticae* Koch. Marker stocks were established and have been maintained since at the Laboratory of Experimental Entomology, University of Amsterdam.

The studies with marker genes in spider mites have elucidated a number of obvious questions dealing with the pathway of inheritance within the haplo—diploid system. An account of the results of these studies can be found in a review by Helle and Overmeer (1973).

SUITABILITY FOR GENETIC STUDIES

Spider mites offer certain advantages for genetic studies. They can be reared in great numbers on a small surface and at a very low price. They are easily manipulated in a two-dimensional plane under the dissecting microscope. They are extremely proliferous, and have a short generation time. Furthermore, the partheno-produced haploid males are convenient for gametic analysis, since in the males both dominant as well as recessive traits are expressed. A mother can be crossed with her haploid sons, which makes special inbreeding procedures possible (Helle, 1965, 1968).

Spider mites are particularly suited for the detection of rare mutants. The suitability of spider mites for such studies is related to the fact that haploid males can easily be produced by the thousands by using large numbers of unfertilized females for their production. This can be achieved by means of a rearing procedure involving the use of a mutation for resistance to tetradifon, an ovo-larvicide: tetradifon-susceptible females are crossed with tetradifon-resistant males. The offspring obtained are reared on leaves treated with tetradifon. All sons (being susceptible) are eliminated by the toxicant; thus the daughters, all heterozygous for the dominant gene of resistance, will not be fertilized on reaching adulthood. These virgins are used for mass rearing of males, which are examined for the occurrence of rare mutants.

On the other hand, spider mites also present some difficulties to the geneticist. It is difficult and laborious to keep genetic strains pure from alien immigrants. The ubiquity of the polyphagous species on all kinds of plants will result sooner or later in contamination. Most commonly, the introduction of alien mite material occurs with host plants reared in a greenhouse, which are brought into the stock room. Special precautions have to be taken in order to prevent this kind of contamination of genetically defined stock material.

Mites are not expected to offer many visible mutations to the geneticist. In insect species, the wings are the sites providing a wealth of conspicuous mutants. Spider mites, by their size and general appearance, do not possess peculiarities of conspicuous significance for visible mutation. In fact, the only visible markers in mites affect the coloration of the eyes and the body. A limited number of markers has so far been detected. Undoubtedly, the genetic variation present in spider mites will be rich enough to extend the number of visible markers, if less conspicuous mutants are sought. However, the practical significance of markers other than pigment markers is probably restricted, and for phenotype recognition it would require a light microscope instead of a dissecting microscope.

On the other hand, it seems worthwhile to study enzyme polymorphism in order to establish additional marker material. However, the small size of spider mites only allows iso-enzyme studies for a very limited number of enzyme systems (L.P.S. van der Geest, personal communication, 1984).

VISIBLE MARKERS

Marker mutation affecting body and/or eye pigmentation have been detected initially in *T. pacificus* (Van Zon and Helle, 1966a) and later in other species: *T. urticae*, *T. neocaledonicus* André and *Eutetranychus orientalis* Klein. The symbols of the markers, the descriptions of the phenotype and the references are given in Table 1.3.2.1.

Mutations affecting the gross morphology of the mite, such as a fusion of leg IV and body, have been observed in several species. However, these mutations could never be established as mutant stocks because of a severe deterioration in viability (Bolland and Helle, unpublished). All mutations affecting pigmentation appeared to be fully recessive. The viability of most mutants is good to excellent, and no special precautions have to be taken for maintenance. Several of the Amsterdam mutant stocks have been reared for nearly 20 years, i.e. for several hundreds of generations, and difficulties with inbreeding depression have never been encountered.

White-eye

Two distinct mutations (w and we) give rise to the phenotype white-eye. Hybrids obtained from crosses between w and we are completely wild-type. The female hybrids produce (haploid) males, which are 25% wild-type and 75% white-eye, indicating an absence of linkage between the two loci. Since no measurable linkage could be shown with other mutations, the identification of the mutations resulting in the white-eye phenotype in the different species offer difficulties with respect to the nomenclature. The assignments of the symbols for white-eye in Table 1.3.2.1 and Fig. 1.3.2.1 is purely arbitrary.

TABLE 1.3.2.1

Pigment mutations in 4 spider mite species

Tetranychus pacificus McGregor
    Symbol *p: pigmentless*. References: Van Zon and Helle, 1966a, b, 1967; Ballantyne, 1969; Veerman, 1972. Complete lack of body and eye pigmentation. Good expression in the eggs. Segregants obtained from a wild-type female usually have a maternally inherited eyespot, which exhibits variable expression. Nearly complete linkage with *a*; linkage with *f*.

    Symbol *a: albino* (same as *al* in Helle and Van Zon, 1967). References: Ballantyne, 1969; Veerman, 1972. Same phenotype as described for *pigmentless*. Certain mutant alleles of *a* show the 'albino-eyespot' phenotype, i.e. complete absence of body pigmentation and a small red eyespot under the anterior cornea lens (see Ballantyne, 1969: alleles a_1, a_2, a_4). The 'eyespot' phenotype often exhibits incomplete penetrance. Nearly complete linkage with *p*; linkage with *f*.

    Symbol *we: white-eye* I. References: Van Zon and Helle, 1966b, 1967; Veerman, 1972. Pigmentation of the eyes is lacking. Body coloration variable, usually increased in comparison to wild-type.

    Symbol *w: white-eye* II. References: Helle and Van Zon, 1970; Veerman, 1972. Same phenotype as *we*.

    Symbol *l: lemon* (same as *le* in Helle and Van Zon, 1967). References: Helle and Van Zon, 1970; Veerman, 1972. Absence of eye pigmentation. Body with a conspicuous bright yellow colour. Good expression in the eggs.

    Symbol *st: stork*. References: Van Zon and Helle, 1966b, 1967; Veerman, 1972. Absence of eye pigmentation. Red pigments accumulating near or in the proximal parts of the legs; distal part of the legs clearly white. Expression variable, depending often on age. Linkage with *OP* for resistance to organophosphates.

    Symbol *f: flamingo*. References: Helle and Van Zon, 1970; Veerman, 1972. The reddish pigments in the granulae in the posterior cornea lenses are partly lacking. Body pigmentation reduced. Reddish pigments present in or near the legs.

    Symbol *r: rose*. References: Ballantyne, 1969; Veerman, 1972. Pinkish colours all over the body (often more intense in the gnathosoma and in legs I and II), in addition to the normal eye and body pigmentation. The expression of *rose* is very distinct in combination with the mutations for albinism (*p* or *a*).

Tetranychus urticae Koch
    Symbol *p: pigmentless*. References: Helle, 1967; Ballantyne, 1969; Veerman, 1974. The same phenotype as in *T. pacificus*. The maternal effect in *T. urticae* is less compared with that found in *T. pacificus*. Pleiotropy for a decreased ability to diapause (Veerman and Helle, 1978). Close linkage with *a*, but separable from it by recombination; linkage with *flamingo*, *f*.

    Symbol *a: albino*. References: Helle and Van Zon, 1966; Ballantyne, 1969; Veerman, 1974. Same phenotype and characteristics as for *pigmentless p*.

    Symbol *we: white-eye* I. Reference: Veerman, 1974. Same phenotype as described for *T. pacificus*. It was shown by Dupont (1980) that this marker mutation could be transmitted to *Tetranychus cinnabarinus* (Boisdv.).

    Symbol *w: white-eye* II. Reference: Veerman, 1974. Same phenotype as described for *T. pacificus*.

    Symbol *l: lemon*. Reference: Veerman, 1974. Same phenotype as described for *T. pacificus*.

    Symbol *st: stork*. Reference: Veerman, 1974. Same phenotype as described for *T. pacificus*. Linkage with *OP* for resistance to organophosphates (Helle and Van Zon, unpublished).

Chapter 1.3.2. references, p. 191

(*Continued*)

Table 1.3.2.1 (continued)

Symbol *f: flamingo*. Reference: Veerman, 1974. Same phenotype as described for *T. pacificus*. Linkage with *a—p* (Helle and Van Zon, unpublished).

Symbol *pa: pale*. (Helle and Van Zon, unpublished). Body and eye pigmentation decreased.

Tetranychus neocaledonicus André
Symbol *we: white-eye* I. Reference: Gutierrez and Van Zon, 1973. Same phenotype as described for *T. pacificus*. The body coloration of *we*, however, is not different from wild-type.

Eutetranychus orientalis Klein
Symbol *we: white-eye* I. (Helle, unpublished). Same phenotype as described for *T. pacificus*. The body of *we*, however, is similar to wild-type.

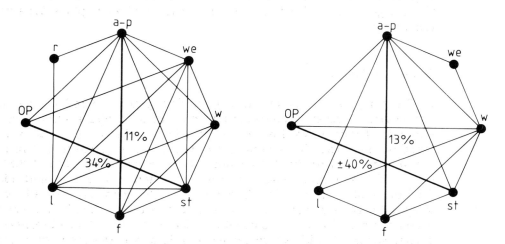

Fig. 1.3.2.1. Linkage relationships between marker loci in *Tetranychus pacificus* McGregor (left) and *T. urticae* Koch (right). The thin lines indicate that the cross-over percentage is 50% and that no linkage has been found. The thick lines represent linkage. The designation of white-eye I and II is chosen arbitrarely (Helle and Van Zon, 1970 and unpublished data).

Complex locus *a—p*

Mutations at this albino-locus, involving the paraloci *a* and *p*, interfere with the production of body and eye pigments and result in the albino phenotype. Both in *T. pacificus* and *T. urticae*, a series of mutations at the *a* and the *p* site has been isolated. This locus has been studied in depth for *T. pacificus* by Ballantyne (1969). Crosses between *a* and *p* mutants give compound hybrids which generally show a normal pigmentation of body and eyes. Most of the mutations complement fully for their mutual defects, and this suggests that *a* and *p* are independent functional units. Crossings between *a* mutants are mutually complementary to a much lower degree, while crossings between *p* mutants fail to complement one another. Scoring the degree of complementation produced by all possible combinations of mutants at the *a—p* locus permitted the construction of a linear complementation map, based on the assumption that the degree of complementation depends on the linear distance between the sites of the mutations. Attempts were also made by Ballantyne to construct a genetic map of the mutations at the *a—p* locus by scoring the frequency of the rare wild-type recombinants in the mass-reared haploid offspring obtained from compound hybrids. This was based on the assumption that a cross-over between an *a* and a *p* mutant, giving rise to the so-called double mutant, depends on the linear distance between the mutation site at the *a* and the *p* locus.

Linkage relationships

Linkage relationships have been studied in *T. urticae* and *T. pacificus* by means of two-factor crosses (Van Zon and Helle, 1966c, 1967; Helle and Van Zon, 1970), including 8 loci for *T. pacificus* (Van Zon and Helle, 1966b) and 7 for *T. urticae* (Helle and Van Zon, unpublished). For both *T. pacificus* and *T. urticae* linkage between albino (*a*—*p*) and flamingo (*f*), and between *OP* (organo-phosphate resistance) and *st* (stork) was demonstrated. For all other combinations (20 for *T. pacificus*) no other linkage relationships were found (see Fig. 1.3.2.1). There seems to be a striking similarity in genetic organization between the two species. It should be taken into account that the designation of the two symbols for the white-eye phenotype is chosen arbitrarily.

General conclusions

The study of the inheritance of marker genes and linkage relationships has answered several questions and confirmed earlier assumptions. It was ascertained that males, irrespective whether produced by mated or unmated females, are always impaternate. The females of bisexual spider mite species are of biparental origin (for exceptions, however, see Chapter 1.2.3). Segregation of the marker genes is according to mendelian ratios, and this is true for both males and females. The studies of *T. pacificus* involved 8 different loci, and it is conceivable that these are scattered over the 3 pairs ($2n = 6$) of chromosomes. There are no genetic indications that polarized segregations of chromosomes should occur: no systematic departures from expected ratios were ever observed; the distribution of alleles over the gametes was always random. The fact that only 2 out of 20 combinations of factors showed linkage in *T. pacificus* indicates that the recombination index is rather high; the chiasmata frequency may be of the order of 2—3 (cf. Chapter 1.2.2).

MUTATION RATE

Pigment mutations have been found and isolated from laboratory populations of different species. Apparently, all these mutants were regenerated by spontaneous mutation processes. Attempts to increase the number of pigment mutants in a laboratory population by means of X- or γ-irradiation have not been unequivocally successful.

It has been observed, however, that in some strains considerably more mutants occurred than in others. In a particular strain of *T. pacificus*, in which a very high number of different pigment mutants had been found during a period extending over at least 2 years, Helle and Van Zon (1967) studied the rates of spontaneous mutations at the loci *p, a, we, w, l* and *st* by examining some 25 000 males obtained from females homozygous for the wild-type alleles of the loci chosen. The mutation rate for all loci chosen appeared to range between 0.8 and 2.8×10^{-4}. This is a high figure. Since the strain studied also suffered from a high egg lethality (up to 30% of the haploid eggs were unviable), it is conceivable that the overall mutability was increased. This high mutability was temporary and disappeared slowly after many generations.

Chapter 1.3.2. references, p. 191

Attempts to measure the production of mutations rendering resistance to acaricides have been reported by Helle (1984), using marker populations of *T. urticae*. Gametes (3.1×10^6) were tested on mutations for parathion resistance, and 1.1×10^6 gametes were tested on mutations for tetradifon resistance; no mutations for the chosen resistances were found. Apparently, the mutation rate for these resistances, so commonly encountered in the control practice, is too low to be measured within a laboratory experiment.

DIAPAUSE

In *T. urticae*, resistance to pesticides is very often accompanied by a decreased ability to diapause. Helle (1961) and Saba (1961) reported that diapause-inducing conditions, such as short day cycles and moderate temperatures during development, results in 100% reproductive diapause in susceptible populations only, but not in resistant ones. In the latter, diapause incidence appeared to be considerably reduced, or even absent. This curious association of the reduced ability to diapause and resistance has been discussed by several authors: Helle, 1962; French and Ludlam, 1973; Zilbermintz et al., 1976; Gotoh and Shinkaji, 1981. It is now generally agreed upon that 'resistance' and 'non-diapause' are genetically independent phenomena. Resistance manifested itself first in greenhouses, and resistant strains of *T. urticae* very often originated from heated greenhouses. In heated greenhouses, there are two typical selective forces: (1) the high temperatures during the winter season, which eliminate the diapausing females and favour mites with a homodynamic cycle; and (2) the abundant applications of pesticides, eliminating the susceptible mites and selecting for traits rendering resistance. By backcrossing procedures it is possible to dissociate 'non-diapause' and resistance (Helle, 1962).

It has been shown several times that a non-diapausing strain can be selected and established from a population exhibiting normal diapause (Helle, 1968). It is remarkable that selection on 'non-diapause' is responded to within a few generations, even if it concerned a laboratory population of limited size reared under constant environmental conditions for hundreds of generations (Helle, unpublished). It is certainly of interest to investigate whether genetic polymorphism is maintained with respect to the ability to diapause in spider mites. In this connection, it is worth mentioning that diapause is a sex-limited trait: it is expressed in the female, but not in the male. Selection on diapause variants, therefore, is restricted to the diploid individuals, and has no effect on the haploid ones. This peculiarity favours the maintenance of genetic variation in diapause. Diapause loci can be considered as 'autosomal' (see Crozier, Chapter 1.3.4) and are more compatible with the presence of allelic polymorphism than are 'sex-linked' loci.

An attempt to elucidate the genetic basis of the reduced ability to diapause in *T. urticae* has been reported by Helle (1968). From this study, it appeared that 'diapause' is dominant over 'non-diapause'. There were indications that the suppressed ability to diapause is inherited as a single recessive gene. However, the backcrosses did show departures from a 1:1 ratio, dependent on the direction of the crossings. These departures were not understood.

Recently, Ignatowicz and Helle (1985) succeeded in demonstrating that a single recessive factor was underlying the reduced ability to diapause in a laboratory-selected line of *T. urticae*. Since the marker mutations for albinism exhibit pleiotropism with regard to a decreased ability to diapause (Veerman and Helle, 1978), the question arises whether the recessive gene for the reduced ability to diapause in the laboratory-selected line is in fact an allele of the *a—p* locus. Apparently, it is not. Ignatowicz and Helle made the crosses between *albino* and the non-diapause selected line, which resulted in a wild-type progeny exhibiting a normal ability to diapause. For further information on genetic variability with respect to factors controlling diapause see Chapter 1.4.6.

INCOMPATIBILITY BARRIERS

Reproductive barriers to various degrees are commonly found in *T. urticae* and these barriers may interfere with genetic procedures. Very often, hybrids obtained from crossings between populations of different origin exhibit some degree of infertility: some of the eggs produced by these hybrids do not hatch. This phenomenon, with its genetic aspects, is dealt with in Chapters 1.3.3 and 1.3.4. In most cases it is possible to cast down genetic incompatibility barriers by means of repeated backcrosses with one parental strain. An example is described by Overmeer (1967).

ISO-ENZYMES

Studies on enzyme polymorphism in mites have revealed that marker mutations can be regenerated by electrophoretic means. Sula and Weyda (1983) reported that intra- and inter-populational variation in *T. urticae* is considerable with respect to esterases. The drawback for the development of iso-enzyme marker stocks is obvious: the maintenance and isolation of such stocks require regular examination, which will be a rather laborious job. For the electrophoretic techniques with mites, the reader is referred to Sula and Weyda, 1983; Ogita and Kassai, 1965 and to Chapter 2.1.4.2.

REFERENCES

Ballantyne, G.H., 1969. Genetic fine structure and complementation at the albino locus in spider mites (*Tetranychus*-species: Acarina). Genetica, 40: 289—323.

Dupont, L.M., 1979. On gene flow between *Tetranychus urticae* Koch, 1836 and *Tetranychus cinnabarinus* (Boisduval) Boudreaux, 1956. (Acari: Tetranychidae). Synonymy between the two species. Entomol. Exp. Appl., 25: 297—303.

French, N. and Ludlam, F.A.B., 1973. Observations on winter survival and diapausing behaviour of red spider mite (*Tetranychus urticae*) on glasshouse roses. Plant Pathol., 22: 16—21.

Gotoh, T. and Shinkaji, N., 1981. Critical photoperiod and geographical variation of diapause induction in the two-spotted spider mite, *Tetranychus urticae* Koch (Acarina: Tetranychidae), in Japan. Jpn J. Appl. Entomol. Zool., 25: 113—118 (in Japanese).

Gutierrez, J. and Van Zon, A.Q., 1973. A comparative study of several strains of the *Tetranychus neocaledonicus* complex and sterilization tests on males by X-rays. Entomol. Exp. Appl., 16: 123—134.

Helle, W., 1961. Relation between organophosphorus-resistance and non-diapause in spider mites. Nature, 192: 1314—1315.

Helle, W., 1962. Genetics of resistance to organophosphorus compounds and its relation to diapause in *Tetranychus urticae* Koch (Acari). Tijdschr. Plantenziekten, 68: 155—195.

Helle, W., 1965. Inbreeding depression in an arrhenotokous mite (*Tetranychus urticae* Koch). Entomol. Exp. Appl., 8: 299—304.

Helle, W., 1967. Fertilization in the two-spotted spider mite (*Tetranychus urticae*: Acari). Entomol. Exp. Appl., 10: 103—110.

Helle, W., 1968. Genetic variability of photoperiodic response in an arrhenotokous mite (*Tetranychus urticae*). Entomol. Exp., Appl., 11: 101—113.

Helle, W., 1984. Aspects of pesticide resistance in mites. In: D.A. Griffith and C.E. Bowman (Eds.), Acarology VI, Vol. I, Ellis Horwood, Chichester, England, pp. 122—131.

Helle, W. and Van Zon, A.Q., 1966. Albinism in two spider mite species. Genen Phaenen, 2: 24—25.

Helle, W. and Van Zon, A.Q., 1967. Rates of spontaneous mutation in certain genes of an arrhenotokous mite, *Tetranychus pacificus* McGregor. Entomol. Exp. Appl., 10: 189—193.

Helle, W. and Van Zon, A.Q., 1970. Linkage studies in the pacific spider mite *Tetranychus pacificus* II. Genes for white eye II, lemon and flamingo. Entomol. Exp. Appl., 13: 300—306.

Helle, W. and Overmeer, W.P.J., 1973. Variability in tetranychid mites. Annu. Rev. Entomol., 18: 97—120.

Ignatowicz, S. and Helle, W., 1985. On the genetics of diapause suppression in the two-spotted spider mite, *Tetranychus urticae* Koch. Exp. Appl. Acarol. (in press).

Ogita, Z. and Kassai, T., 1965. A microtechnique for enzyme separation of individual spider mites with thin layer electrophoresis. SABCO J., 1: 117.

Overmeer, W.P.J., 1967. Genetics of resistance to tedion in *Tetranychus urticae* C.L. Koch. Arch. Neérl. Zool., 17: 295—349.

Saba, F., 1961. Ueber die Bildung der Diapauseform bei *Tetranychus urticae* Koch in Abhängigkeit von Giftresistenz. Entomol. Exp. Appl., 4: 264—272.

Schulten, G.G.M., 1968. Genetics of organophosphate resistance in the two-spotted spider mite (*Tetranychus urticae* Koch). Publ. R. Trop. Inst., Amsterdam, The Netherlands, 57:1—57.

Sula, J. and Weyda, F., 1983. Esterase polymorphism in several populations of the two-spotted spider mite, *Tetranychus urticae* Koch. Experientia, 39: 78—79.

Van Zon, A.Q. and Helle, W., 1966a. Albinism as a marker in *Tetranychus pacificus*. Entomol. Exp. Appl., 9: 205—208.

Van Zon, A.Q. and Helle, W., 1966b. Pigment mutations in *Tetranychus pacificus*. Entomol. Exp. Appl., 9: 402—403.

Van Zon, A.Q. and Helle, W., 1966c. A search for linkage between genes for albinism and parathion resistance in *Tetranychus pacificus*. McGregor. Genetica, 37: 181—185.

Van Zon, A.Q. and Helle, W., 1967. Linkage studies in the pacific spider mite *Tetranychus pacificus*. I. Genes for pigmentless, white-eye, stork and organophosphate resistance. Entomol. Exp. Appl., 10: 69—74.

Veerman, A., 1972. Carotenoids of wild-type and mutant strains of *Tetranychus pacificus* McGregor (Acari: Tetranychidae). Comp. Biochem. Physiol., 42B: 329—340.

Veerman, A., 1974. Carotenoid metabolism in *Tetranychus urticae* Koch (Acari: Tetranychidae). Comp. Biochem. Physiol., 47B: 101—116.

Veerman, A. and Helle, W., 1978. Evidence for the functional involvement of carotenoids in the photoperiodic reaction of spider mites. Nature, 275: 234.

Zilbermintz, I.V., Dubynina, T.A. and Solomatina, V.I., 1976. Photoperiodic reaction and ability for diapausing in *Tetranychus* mites resistant to acaricides. Skh. Biol., 11: 711—715 (in Russian).

1.3.3 Reproductive Barriers

R. DE BOER

CLASSIFICATION OF REPRODUCTIVE BARRIERS

It goes without saying that only the sexually reproducing species of spider mites are considered. In a bisexual species, the first reproductive barrier that has to be overcome is a *spatial barrier*; males and females have to meet. This could be a problem if mates on separate plants are involved. As unwinged arthropods, spider mites have to rely on passive dispersal. Only if suitable host plants are nearby can directed dispersal play a role. It is expected that the area occupied by a population of spider mites sufficiently isolated to preserve its own genetic peculiarities will very often be rather small. This area will be determined by the distribution pattern of suitable host plants.

Populations of the same species may differ in their host plant adaptation. For instance, populations of *Tetranychus urticae* Koch have been described living on carnations, whereas other populations could not be maintained on this host plant. Host plant adaptation itself represents a barrier to gene exchange, which should be classified as an *ecological barrier*. Between different species of spider mites, ecological barriers generally must be of great importance, especially in monophagous species.

Cross-breeding between mites from different populations may give rise to less well-adapted and thus reproductively less successful offspring. It is conceivable therefore that the development of behavioural, mechanical and gametic mating barriers can be enhanced by natural selection. This mechanism has sometimes been referred to as the 'Wallace effect' (Grant, 1966).

Isolation mechanisms other than spatial and ecological barriers can be conveniently studied in the laboratory. The mites are placed on a leaf culture (Chapter 1.5.1) and observed through a dissecting microscope. Eggs will then be deposited on the same leaf disc and can be accurately counted; hatchability can be assessed as well as the sex ratio of the surviving hatchlings. If egg mortality is negligible, the proportion of females in a progeny reflects the proportion of eggs fertilized, because the males arise from unfertilized eggs. If egg mortality is high, the degree of fertilization can be estimated by isolating a few virgin females in order to obtain a sample of unfertilized eggs. The ratio between viable and inviable males is obtained from this egg sample. This ratio is used to estimate the number of inviable male eggs in a progeny comprising viable males and females together with dead eggs of unknown sex. It has been observed (W. Helle, personal communication, 1984) that mites belonging to the genera *Oligonychus* and *Tetranychus*, when placed on the same leaf disc, may attempt to copulate. This may indicate that a *behavioural barrier* is poorly developed. On the other hand, it has been repeatedly observed that mites are reluctant to accept a partner from a

Chapter 1.3.3. references, p. 199

different population of the same species. In artificially mixed populations of *T. urticae*, allotypic matings are often under-represented, and Dieleman and Overmeer (1972) reported that such copulations are often broken off prematurely. It is not possible to express a definite opinion on the significance of behavioural barriers between tetranychid species, since nobody has investigated this subject systematically. A remarkable case is reported by Lee (1969) dealing with the aggressive behaviour of *T. urticae* males towards females of *Panonychus ulmi* (Koch). It is expected that silk may also function as a barrier between species. Silken threads, and the various webbing structures (see Chapter 1.4.4) may prevent access of alien mates, and thus operate as mating barriers. Studies dealing with these interesting aspects of webbing are still lacking.

Mechanical and *gametic barriers* cannot be easily distinguished. In both cases copulations or attempts to copulate are observed but no females appear in the offspring. Either sperm transfer has not been accomplished (mechanical isolation) or the gametes do not fuse (gametic isolation). Ghobrial et al. (1969) described a reproductive barrier, apparently based on one of these mechanisms. Mites of the red and the green type of *T. urticae* lived in mixed populations on the same plants in Egypt. Matings between different types were observed but no hybrids were obtained. Such situations are definitely rare within this spider mite complex. In contrast, an intermediate situation is often observed i.e. an abnormally low number of F_1 daughters among a preponderance of males in the progeny of an inter-strain crossing. Usually, this is accompanied with a low fertility of the F_1 females. Attempts have been made to hybridize at least 9 different *Tetranychus* species in 10 different combinations without a single F_1-individual arising. In most cases the mites were observed to copulate, although sometimes reluctantly (Newcomer, 1954; Helle and Van de Bund, 1962; Gutierrez and Van Zon, 1973; Jordaan, 1977). It is conceivable that a mechanical barrier may arise due to differences in the shape of the aedeagus. In fact, morphologically very similar species can often be distinguished only by examining the aedeagus. Presumably, the Wallace effect is operating here.

Only one well-documented case of *hybrid inviability* is known in spider mites (De Boer and Veerman, 1983). Females of an Egyptian strain of *T. urticae* inseminated by males from a Dutch strain produced a mixture of unfertilized eggs, which were normally viable, and fertilized eggs, which were largely inviable. Gene exchange is prevented here at the expense of the reproductive capacity of the female. Of course, such mechanisms are not favoured by natural selection, and must be merely a side effect of the isolation of the populations. The same is true of the next class of reproductive barriers: *hybrid sterility* and *hybrid breakdown*. Again, the available information mainly concerns *T. urticae*. Within this species the phenomenon is very widespread. F_1 females from inter-strain crosses may exhibit normal egg production, but a proportion of their eggs, ranging from a few to almost 100%, usually are inviable. The number of eggs per female may also be reduced. In extreme cases less than 10 eggs are produced on average. This may be due to degeneration of the ovaries or to the retention of eggs in the oviduct (Dosse and Langenscheidt, 1964). In the latter case the females adopt a typically swollen appearance. In the F_2 and F_3 generation, fertility is usually somewhat improved, but it takes many generations before complete restoration of fertility is achieved.

Studies by Helle and Pieterse (1965) and Overmeer and Van Zon (1976) revealed a high degree of hybrid infertility between any 2 strains collected in the same greenhouse complex in Aalsmeer, The Netherlands. De Boer (1980, 1981) crossed *T. urticae* populations which were collected from the

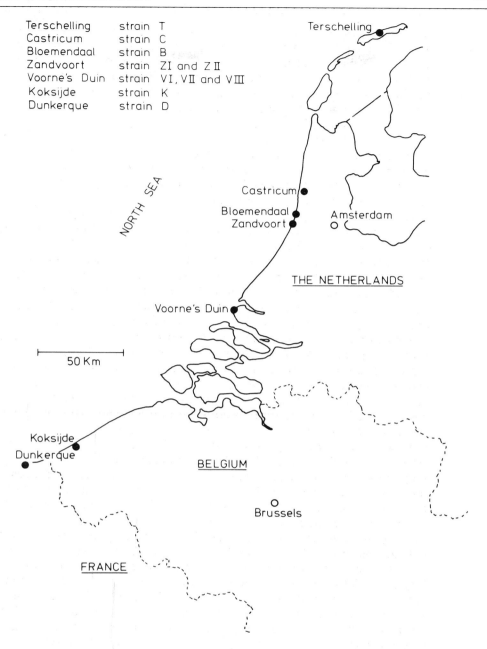

Fig. 1.3.3.1. Localities where the strains of *Tetranychus urticae* Koch were collected (cf. Fig. 1.3.3.2).

same host plant species in the dunes of the Dutch and Belgian west coast (Figs. 1.3.3.1 and 1.3.3.2). He found that the size of an area occupied by an intra-fertile population was 5—10 Km across. The evidence for clinal variation was rather weak. Since all these mites readily hybridized in the laboratory, the maintenance of genetic differences probably depends entirely on spatial isolation. The number of fertility races seems to be unlimited. It has been suspected by several acarologists that infertility barriers may develop after only a few generations of isolation in the laboratory. However, the supporting evidence is largely circumstantial and an attempt to demonstrate the phenomenon in a well-controlled experiment failed to prove it (De Boer, 1979). Also, the idea that many fertility races can arise through hybridization of

Chapter 1.3.3. references, p. 199

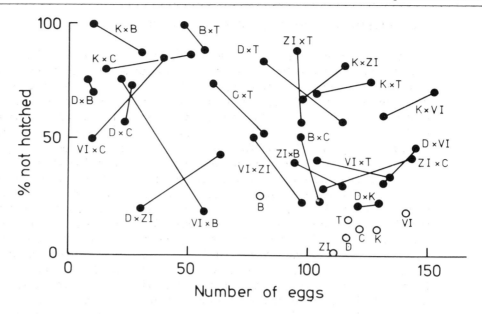

Fig. 1.3.3.2. Egg production and egg mortality of 10 pooled virgin F_1 females of *T. urticae* Koch in their 6th day of adult life. Seven strains, T, C, B, ZI, VI, K and D (cf. Fig. 1.3.3.1) were crossed in all 21 combinations. Lines connect points representing 2 corresponding reciprocal crosses. (○) Control crosses (intra-strain). The origin of the strains is given in Fig. 1.3.3.1.

just 2 extant populations could not be confirmed by experiments (De Boer, 1982a).

RACES AND SPECIES

Since *T. urticae* is a highly polytypic species, it is not surprising that morphological differences often accompany the infertility barriers. Pritchard and Baker (1955) combined no less than 58 formerly described species under the name *T. urticae*. However, Boudreaux (1963) found considerable hybrid infertility between morphologically inseparable populations.

There is still a discussion going on about the status of the so-called carmine or red spider mite as opposed to the green spider mite. Splitters consider the red form to be a separate species, *Tetranychus cinnabarinus* (Boisduval). Distinctions between the 2 forms are supposed to be at least four-fold:

(1) presence or absence of red pigmentation;

(2) the striae on the back of adult females bear lobes, which have a different shape in the 2 forms;

(3) the red forms are unable to diapause and

(4) crosses between the 2 forms reveal virtually complete hybrid sterility.

Mites bearing lobes intermediate in shape between those of *T. urticae* and *T. cinnabarinus* have been discovered. Strangely enough, these atypical forms were explained away as the results of hybridization. Now at least 1 red strain is known which is able to diapause and which can be easily hybridized with certain green strains (Dupont, 1979). Conversely, severe hybrid sterility occurs between green strains as well as between red strains (Dillon, 1958). In temperate regions, the red forms are found only inside greenhouses, indicating some adaptive significance of the pigmentation associated with climate.

THE GENETIC ANALYSIS OF HYBRID STERILITY

The genetic basis of hybrid sterility in the *T. urticae* complex is not fully understood. Presumably, spider mite populations are very often founded by a single uninseminated female, reaching a new host plant as a sexually immature colonist. The first eggs will of course be unfertilized and develop into sons, but as soon as the first son becomes an adult male, the female will be fertilized and start to produce daughters. These mother × son matings are genetically equivalent to selfings and the circumstances seem to be ideal for the fixation of chromosome mutations such as translocations. It is a very attractive idea therefore that inter-strain crossings give rise to translocation heterozygotes. The semi-sterility associated with translocation heterozygosity is well known. Unfortunately, the characteristics of partial hybrid sterility in the *T. urticae* complex do not agree with this assumption. In the first place, there is great variability in egg mortality between the F_1 females of the same progeny. Many papers report on this variability more or less explicitly, so it seems to be a general feature. If a limited number of F_1 females are studied individually they may prove to be totally sterile. Yet a mass crossing of the 2 parent strains eventually results in a vigorous strain, owing to a few exceptional females laying viable eggs. Also, less extreme but statistically significant differences in fertility between F_1 females have been described repeatedly. The involvement of modifying genes cannot be totally excluded as the strains were usually not inbred prior to hybridization. However, 2 samples of eggs obtained subsequently from the same F_1 female frequently show significantly different levels of egg mortality (De Boer, 1982b). This departure from Mendel's first law of inheritance is a first indication that extra-chromosomal factors are involved. At the same time, this lack of uniformity is an obstacle for further genetic analysis of the phenomenon. Differences between reciprocal crosses, a standard test for extra-chromosomal inheritance, cannot be reliably shown unless many replicates of the same crossings are performed. Systematically different egg mortality percentages in reciprocal crosses have now been firmly established in only 2 cases.

Egg mortality is usually reduced through fertilization. A fertilized egg has a better chance of hatching, but not quite as good a chance as an egg from a non-hybrid female (De Boer, 1982b). The source of the sperm is important. A sperm from the female parent strain has a stronger viability-restoring effect than one from the male parent strain. This is basically a difference between reciprocal crossings and it seems to be another general feature of the sterility phenomenon (Overmeer and Van Zon, 1976). The difference between reciprocals is extreme in crossings between the ZI and the ZII strain (De Boer, 1982c). These 2 strains were originally collected in the Dutch dunes at a distance of 1400 m apart. The eggs from F_1 (ZI♀ × ZII♂) females exhibit normal viability. In contrast, the F_1 (ZII♀ × ZI♂) females have on average 35% mortality in their eggs. The egg mortality characteristics are typical for the sterility phenomenon in *T. urticae*, with a great difference between F_1 females and a viability-restoring effect of fertilization. The genetic analysis can be carried somewhat further with these 2 strains. Strong evidence was found that egg mortality is the result of an interaction between a nuclear gene N and an extra-chromosomal factor E. The E factor is passed on from a mother to her daughters but there is no strict maternal inheritance. The factor disappears after repeated back-crossing with ZI males. The synthesis of E probably is controlled by 1 or more nuclear genes from the ZII strain. A gene at the N locus of the ZII strain may be responsible. This possibility has not been tested so far. The matter is of interest because it would indicate that the incompatibility resulted from a single mutation at the N locus. It is

Chapter 1.3.3. references, p. 199

significant that the viability-restoring effect of fertilization also seems to depend on a gene at the N locus (De Boer, 1983).

The actual cause of the death of an egg remains totally obscure. The possibility that a mutator gene is involved, giving rise to recessive lethals, had to be rejected (De Boer, 1983).

IS THE STERILITY PHENOMENON UNIQUE?

Because of their economic importance, spider mites are relatively well-studied organisms. In addition, as has been argued before, they are out-standingly suitable for the study of reproductive barriers. Inviable eggs can hardly escape the attention of an investigator. In a species frequently used for crossing experiments, *Drosophila melanogaster* Meigen, a very similar sterility phenomenon was not recognized until the early 1970s. In order to distinguish the inviable eggs, the medium was stained with carbon black. It seems quite possible that similar sterility phenomena are only waiting to be discovered in other organisms. In spider mites, the haplo-diploid mechanism of sex determination facilitates the recognition of the egg mortality, since it is expressed especially in haploid eggs. The question of whether the peculiar sex determination system made possible the development of these sterility barriers cannot be answered until more is understood of its nature. It would be interesting to know whether the same phenomena occur in unrelated haplo-diploid organisms such as the Hymenoptera (Chapter 1.3.4). Apart from *T. urticae*, hybrid infertility of strains has been described in 2 other species: *Tetranychus neocaledonicus* André and *Panonychus citri* (McG) (Inoue, 1972; Gutierrez and Van Zon, 1973). Inter-strain crosses with *Panonychus ulmi*, quite surprisingly, did not reveal any F_1 infertility, implying that in this species the phenomenon is less widespread or is absent (Cranham, 1982). In the phytoseiid mite *Typhlodromus occidentalis* Nesbitt F_1 infertility between strains was reported by Croft (1970).

A phenomenon called hybrid dysgenesis in *D. melanogaster* has been referred to before. It has some striking similarities with the sterility pheno-menon in *T. urticae* (Bregliano et al., 1980). Some of the eggs laid by F_1 females from inter-strain crosses in *D. melanogaster* are inviable. The pro-portion changes with the age of the F_1 female. The sterility is the result of interaction between a chromosomal factor I and a cytoplasmic factor R. The R factor is maternally inherited but in the absence of chromosomes from the maternal parent strain it gradually disappears.

In spite of these similarities, there are also differences. The phenomenon called *chromosomal contamination* associated with hybrid dysgenesis did not seem to occur in the F_1 (ZII♀ × ZI♂) females of *T. urticae*. Also, the geographic distribution of the fertility races is quite different. In *Drosophila* only 2 types of strains can be distinguished, reactive (carrying R) and inducer (carrying I). Consequently, hybrid infertility is always unidirectional i.e. only F_1 females with a reactive mother and an inducer father are semi-sterile. Natural populations are always inducers. Only some of the laboratory strains of *D. melanogaster* are reactive. This indicates that the laboratory conditions somehow promote the development of the reactive condition. Possibly the smaller population size and concomitant inbreeding is a necessary pre-requisite. It should be remembered that small, genetically impoverished populations are probably quite natural in *T. urticae*. The establishment of infertility barriers may well be a general consequence of such population structure. Fertility races in other organisms with a similar population struc-ture are not known, but the matter deserves investigation.

REFERENCES

Boudreaux, H.B., 1963. Biological aspects of some phytophagous mites. Annu. Rev. Entomol., 8: 137—154.

Bregliano, J.C., Picard, G., Bucheton, A., Pelisson, A., Lavige, J.M. and L'Heritier, P., 1980. Hybrid dysgenesis in *Drosophila melanogaster*. Science, 207: 606—611.

Cranham, J.E., 1982. Resistance to organophosphates, and the genetic background, in fruit tree red spider mite, *Panonychus ulmi*, from English apple orchards. Ann. Appl. Entomol., 100: 11—23.

Croft, B.A., 1970. Comparative studies on four strains of *Typhlodromus occidentalis* (Acarina: Phytoseiidae). I. Hybridization and reproductive isolation studies. Ann. Entomol. Soc. Am., 63: 1558—1563.

De Boer, R., 1979. Investigations concerning the development of reproductive incompatibilities between populations of the spider mite, *Tetranychus urticae*. Z. Angew. Entomol., 87: 113—121.

De Boer, R., 1980. Genetic affinities between spider mite *Tetranychus urticae* populations in a non-agricultural area. Entomol. Exp. Appl., 28: 22—28.

De Boer, R., 1981. Genetic affinities between spider mite *Tetranychus urticae* populations in a non-agricultural area II. Entomol. Exp. Appl., 30: 63—67.

De Boer, R., 1982a. Laboratory hybridization between semi-incompatible races of the arrhenotokous spider mite *Tetranychus urticae* Koch. Evolution, 36: 553—560.

De Boer, R., 1982b. Partial hybrid sterility between strains of the arrhenotokous spider mite, *Tetranychus urticae*, complex. Genetica, 58: 23—33.

De Boer, R., 1982c. Nucleo-cytoplasmic interactions causing partial female sterility in the spider mite *Tetranychus urticae* Koch (Acari: Tetranychidae). Genetica, 58: 17—22.

De Boer, R., 1983. Nucleocytoplasmic interactions causing partial female sterility in the spider mite *Tetranychus urticae* Koch II. Genetica, 61: 107—111.

De Boer, R. and Veerman, A., 1983. A case of hybrid inviability in the two-spotted spider mite, *Tetranychus urticae*. Entomol. Exp. Appl., 34: 127—128.

Dieleman, J. and Overmeer, W.P.J., 1972. Preferential mating hampering the possibility to apply a genetic control method against a population of *Tetranychus urticae* Koch. Z. Angew. Entomol., 71: 156—161.

Dillon, L.S., 1958. Reproductive isolation among certain spider mites of the *Tetranychus telarius* complex, with preliminary systematic notes. Ann. Entomol. Soc. Am., 51: 441—448.

Dosse, G. and Langenscheidt, M., 1964. Morphologische, biologische und histologische Untersuchungen an Hybriden aus dem *Tetranychus urticae—cinnabarinus*-Komplex. Z. Angew. Entomol., 54: 349—359.

Dupont, L.M., 1979. On gene flow between *Tetranychus urticae* Koch, 1836 and *Tetranychus cinnabarinus* (Boisduval) Boudreaux, 1956 (Acari: Tetranychidae): Synonomy between the two species. Entomol. Exp. Appl., 25: 297—303.

Ghobrial, A., Attiah, H., Voss, G. and Dittrich, V., 1969. The *Tetranychus telarius* complex (red and green forms) in Egyptian cotton: two separate species. J. Econ. Entomol., 62: 1304—1306.

Grant, V., 1966. The selective origin of incompatibility barriers in the plant genus *Gilia*. Am. Nat., 100: 99—118.

Gutierrez, J. and Van Zon, A.Q., 1973. A comparative study of several strains of the *Tetranychus neocaledonicus* complex and sterilization tests of males by X-rays. Entomol. Exp. Appl., 16: 123—134.

Helle, W. and Pieterse, A.H., 1965. Genetic affinities between adjacent populations of spider mites. Entomol. Exp. Appl., 8: 305—308.

Helle, W. and Van de Bund, C.F., 1962. Crossbreeding experiments with some species of the *Tetranychus urticae* group. Entomol. Exp. Appl., 5: 159—162.

Inoue, K., 1972. Sterilities, visible mutations in F_1 hybrid females obtained by crosses between different strains and mortalities of their eggs in Citrus red mites, *Panonychus citri* (McG). Bull. Hortic. Res. Stn. (Minist. Agric. Forest.), 7: 29—36.

Jordaan, L.C., 1977. Hybridization studies in the *Tetranychus cinnabarinus* complex in South Africa. J. Entomol. Soc. S. Afr., 40: 147—156.

Lee, B., 1969. Cannibalism and predation by adult males of the two-spotted mite *Tetranychus urticae* Koch (Acarina: Tetranychidae). J. Aust. Entomol. Soc., 8: 210.

Newcomer, E.J., 1954. Identity of *Tetranychus pacificus* and *mcdanieli*. J. Econ. Entomol., 47: 460—462.

Overmeer, W.P.J. and Van Zon, A.Q., 1976. Partial reproductive incompatibility between populations of spider mites. Entomol. Exp. Appl., 20: 225—236.

Pritchard, A.E. and Baker, E.W., 1955. A revision of the spider mite family Tetranychidae. Pac. Coast Entomol. Soc., Cal. Acad. Sci., San Francisco, 472 pp.

Spider Mites. Their Biology, Natural Enemies and Control. Volume 1A
Edited by W. Helle and M.W. Sabelis
© 1985 Elsevier Science Publishers B.V., Amsterdam — Printed in The Netherlands   201

1.3.4 Adaptive Consequences of Male-Haploidy

R.H. CROZIER

THE BASIC FEATURES OF MALE-HAPLOID POPULATION GENETICS

The null case: genetics of the archetypal organism

The archetypal organism of elementary population genetics theory has yet to be discovered. Such an organism lives in populations of infinite size, engages in completely random mating (or, better still, pours its gametes into a completely mixed 'gene pool'), and does not indulge in migration. Furthermore, it exhibits purely diploid autosomal inheritance, is often portrayed as hermaphroditic, and is not plagued by mutation. Such a creature is unlike any known, and certainly quite different from any spider mite. Despite these failings, a consideration of this mythical beast makes a good starting point for understanding the biology of tetranychids, which differ from our archetypal organisms in most of the characteristics mentioned so far.

Consider our archetypes pouring their gametes into the common mixing pot (in fact, random mating is fortunately the same as the random union of gametes, so that our picture does not rob us of reality too much). If we consider a single polymorphic locus with 2 alleles, A with frequency p, and a with frequency q, and then pull gametes out of the pool in pairs so as to make diploid individuals, we find that the chance of pulling out 2 As is p^2, 2 as is q^2, and that leaves the chance of 1 A and 1 a, resulting in a heterozygote, as $2pq$. We thus have the famous Hardy—Weinberg Principle, also known as the Binomial Square Law because of its ready derivation

$$(p + q)^2 = p^2 + 2pq + q^2 \qquad\qquad (1)$$
$$AA \quad Aa \quad\ aa$$

These proportions will arise in 1 generation provided that the conditions mentioned above obtain, and that the allele frequencies are the same in the 2 sexes. If there are different allele frequencies in males and females, 2 generations are required for Hardy—Weinberg proportions to arise: the first generation will see an excess number of heterozygotes but will also eliminate the difference between the sexes.

We can now see that the conditions for Hardy—Weinberg proportions to occur *exactly* are only slightly more restrictive than those for evolution (change in the genetic make-up of a lineage) *not* to occur, in that for a population to evolve requires violation of 1 or more of the first 4 of the conditions for Hardy—Weinberg proportions:

(1) no mutation;
(2) no selection;

Chapter 1.3.4. references, p. 219

(3) infinite population size;

(4) no immigration from populations with different allele frequencies;

(5) random mating.

No known population fulfils all these conditions, not even the first 4. Evolution is thus inevitable, and the many examples of populations whose genotypic proportions cannot be distinguished from those expected under the Hardy—Weinberg Principle (HWP) simply reflect the fact that the deviations are usually small, and perhaps that evolution is often slow.

However, no organism has only 1 genetic locus, although some viruses come close with only 3 or 4. We thus become concerned not only with the allele frequency changes and genotypic proportions of 1 locus, but the change in association between alleles at different loci. To put it another way, we become concerned with *gamete* frequencies, because such frequencies measure the strengths of association between alleles at different loci. Consider 2 loci, one with alleles A and a at frequencies of p and q, respectively, and the other with alleles B and b, at frequencies r and s. Let the frequencies of the 4 gametes be h, i, j and k. Now, if we waited long enough, and there is at least some recombination between the loci, the frequencies of the 4 gametes would simply reflect the frequencies of the alleles at each locus, thus

Gamete	Frequency symbol	'Ultimate' value
AB	h	pr
Ab	i	ps
aB	j	qr
ab	k	qs

The deviation of the gamete frequencies from 'ultimate' values can be measured as 'linkage disequilibrium'

$$D = hk - ij \tag{2}$$

A population reduces D slowly, at a rate according to the fraction, c, of the gametes of double heterozygotes that are recombinant, so that the value of D in the next generation, D', is

$$D' = D - cD \tag{3}$$

The attainment of linkage equilibrium ($D = 0$) thus resembles Achilles's pursuit of the tortoise as described by Aesop: the closer D is to zero, the slower is its approach to it (and did Achilles catch the tortoise?). We can intuitively understand the situation by reflecting on the fact that recombination only affects D when it occurs in double heterozygotes. Furthermore, such double heterozygotes are of 2 possible types (AB/ab and Ab/aB), with recombination in one having an opposite effect on D to recombination in the other. As $D = 0$ is approached, the 2 effects increasingly cancel out. Crow and Kimura (1970a) provide a more detailed account of this and other population genetic phenomena, and suggest that, because the number of generations to attain linkage equilibrium is infinite, it makes more sense to ask how long it takes to reduce D to half its initial value. For unlinked loci ($c = 0.5$), the mean time to this reduction is 1 generation; when $c = 0.1$, 7 generations, and for $c = 0.001$ (very tight linkage) the time required is about 693 generations.

As might be expected, further complications arise with 3 or more loci, but the principles established for 2 loci are valid enough to form the basis of our understanding.

Male-haploid population genetics is (largely) isomorphic with sex-linkage

Male-haploids lack autosomes and dosage compensation

So far, we have considered autosomal loci, those present in double dose in both males and females. Spider mites lack such loci. Instead, each female has 2 copies of each chromosome, but males only 1. All male-haploid (and hence, spider mite) loci thus follow the inheritance rules of loci on the X-chromosomes of organisms such as *Drosophila* and ourselves. Thus, for a locus with 2 alleles, there are 6 possible genotypes if it is autosomal (3 in females and 3 in males), but only 5 if it is male-haploid, because then there are no heterozygote males. While there is thus a respectable body of theory from standard population genetics directly applicable to male-haploid organisms, it is worth noting that there are 2 significant differences between male-haploid genetics and that of 'conventional' population genetic work-horses. First, *Drosophila* (and we too) have only one X-chromosome pair, whereas male-haploids have at least 2 chromosome pairs. This is not to say that organisms with multiple X-chromosomes are unknown, only that they are not known genetically apart from their cytogenetics. The effects of having several X-chromosome linkage groups are not known. Secondly, organisms such as *Drosophila* have interactions between X-linked loci and those on their autosomes, whereas such interactions are of course lacking in male-haploids. Thus, mixed autosome/X systems face the problem of 'dosage compensation', namely that of adjusting for different relative X and autosome dosages in males and females. Dosage compensation is therefore unnecessary in male-haploids, although an analogue may occur in terms of a balance between the cytoplasm and the nucleus (Crozier, 1975a).

The weighted allele frequency mean is usually appropriate for male-haploids

For autosomal genes, the allele frequency for the whole population is the simple average of the allele frequencies in females (p_F) and males (p_M). This is inappropriate for most population applications in spider mites, where the appropriate value for the allele frequency is given by

$$p_T = (p_M + 2p_F)/3 \tag{4}$$

Although it is most easily remembered in terms of females having 2 sets of chromosomes and males only 1, the real basis for eqn. (4) concerns the transmission of genetic material. Males as a group pass on only half as much genetic material as females as a group, because females have both sons and daughters whereas males have only daughters. Consequently, females make twice the contribution to remote generations as do males.

Genotypes present in male-haploids

At autosomal loci with 2 alleles there are 6 genotypes, 3 in males and 3 in females. But heterozygote males do not occur in male-haploids, so that for these organisms there are only 5 genotypes. While the 3 female genotypes in a male-haploid species resemble those of a male-diploid species in their relationship to allele frequencies, the 2 haploid male genotypes are simply equivalent to allele frequencies. It is, in fact, an important aspect of male-haploid population genetics that a male passes on exactly the same gametes to all his offspring. The genotypes of a biallelic male-haploid locus are

Females			Males		
AA	Aa	aa	A	a	
$p_F \cdot p_M$	$(p_F \cdot q_M + p_M \cdot q_F)$	$q_F \cdot q_M$	p_F	q_F	(5)

Chapter 1.3.4. references, p. 219

where p_F and q_F refer to the female and p_M and q_M to the male allele frequencies in the preceding (parental) generation.

The approach to HWP proportions for male-haploids

As noted above, an autosomal locus reaches HWP proportions after 1 generation of random mating if the sexes have the same allele frequencies, and takes only 1 more if the sexes differ in allele frequency. For the case of non-overlapping generations, however, male-haploids with an initial allele frequency difference between the sexes do not reach HWP proportions in 2 generations, but rather the frequencies of genotypes in the 2 sexes oscillate indefinitely while approaching p_T. This oscillation arises because males have the allele frequency of the females of the preceding generation (eqn. (5)), but females have the mean of the frequencies of the preceding generation. p_M always deviates twice as much from p_T as does p_F, and the deviations from p_T change in sign and halve in absolute magnitude with each generation. The oscillations become irregular if generations overlap (Cornette, 1977).

Dyson (1965) studied the approach to p_T of 2 visible markers in laboratory populations of the parasitoid wasp *Bracon hebetor* Say, and Clegg· and Cavener (1982) studied 2 sex-linked allozyme loci in discrete and overlapping-generation laboratory populations of *Drosophila melanogaster* Meigen. Allele frequency oscillations occurred in both studies, and Clegg and Cavener noted the potential usefulness of sex-linked loci in elucidating the demographic structure of populations.

The effects of allele frequency oscillations in male-haploid populations are likely to be complex and far-reaching. Thus, although the lack of male heterozygotes will slow the reduction of D because of an effective halving of c, oscillations in gamete frequencies between the sexes increase the fraction of heterozygotes in females, thus increasing the scope for recombination. This recombination further alters the gamete frequencies between sexes, helping to prolong the oscillations.

Male-haploidy reduces effective population size

An important concept for understanding the loss of variation in small populations is *effective population size*, also called *effective population number* (Crow and Kimura, 1970b; Wright, 1969a). This value, N_E, enables a comparison to be made between populations of different size, sex ratio, and genetic system in terms of the rate of loss of genetic variation, which will be the same for populations with the same N_E value, however much they differ in the actual number of individuals. If we designate as N_F the number of females and N_M the number of males in the breeding population, then for an autosomal locus

$$N_{EA} = 4N_F N_M /(N_F + N_M) \tag{6}$$

For a male-haploid locus, the corresponding value is given by

$$N_{EM} = 9N_F N_M /(4N_M + 2N_F) \tag{7}$$

The ratio of the male-haploid to the autosomal sizes, given the same numbers of breeding males and females, is

$$N_{EM}/N_{EA} = (9/8)(N_F + N_M)/(N_F + 2N_M) \tag{8}$$

Where there are equal numbers of males and females, the size ratio given by eqn. (8) is 0.75. However, this ratio increases with increasingly female-biased sex ratios, so that when there are more than 7 females per male the male-haploid size is greater than the autosomal one (Crozier, 1979).

Although it generally returns the same values as are given by the approximate formulae above, the effective population size, as determined by the breeding structure of the population, and its actual size, can differ. Wright (1969b) discusses these *inbreeding* and *variance* effective sizes, and Nagylaki (1981) considers the inbreeding effective size for male haploids.

SELECTION WITHIN POPULATIONS

Mutation and deleterious genes

Male-haploids have lower equilibrium frequencies

Let us assume that selection against a deleterious allele at an autosomal locus is the same in each sex and follows the scheme

$$AA \quad Aa \quad aa$$

Fitness   1   1 $1 - s$

Then selection will decrease the frequency of the a allele until it is balanced by mutation, so that the equilibrium frequency of a, $\hat{q}_1$, is determined purely by the selective disadvantage to aa, s, and the mutation rate, u, from A to a, according to the approximation

$$\hat{q}_1 = (u/s)^{0.5} \tag{9}$$

Whereas for recessive alleles at autosomal loci, selection can only act when they occur in a double dose as homozygotes, this is not the case for male-haploid loci, because effectively one-third of the genetic material at any moment is in the haploid males ('effectively one-third' because, even if males are rare, they are responsible for one-third of the population's genetic output). This selection in the haploid males is much more effective than selection in the females, so that the equilibrium frequency, $\hat{q}_2$, for a recessive deleterious allele at a male-haploid locus is given by

$$\hat{q}_2 = 3(u/s) \tag{10}$$

The frequency of deleterious recessives in autosomal systems will therefore be much higher than that in male-haploids, with a ratio between them of

$$\hat{q}_1/\hat{q}_2 = (1/3)(s/u)^{1/2} \tag{11}$$

Male-haploids show an illusory bias to sex limitation

Genetic loci are not necessarily expressed equally in each sex. In fact, if they all were, there could be no sexual differentiation! The possibility has occurred to various authors (see, e.g., Crozier, 1979) of a high proportion of loci in male-haploids being limited in expression to females, effectively making these loci autosomal with respect to the effects of selection (although not with respect to stochastic processes). In fact, a high proportion of the known loci in male-haploids are known from alleles with sex-limited effects (Crozier, 1975b). Noteworthy was Kerr's (1976) finding that some 14% of the mutational genetic load in a population of the honey bee *Apis mellifera* L. was due to sex-limited gene effects. However, a large sex-limited genetic load does not necessarily indicate a high proportion of sex-limited loci, because deleterious alleles (which cause the load) reach higher frequencies at sex-limited than at non-sex-limited loci in male-haploids (Crozier, 1979). Genetic load estimates generally involve inbreeding, which therefore picks up an inflated proportion of deleterious alleles

Chapter 1.3.4. references, p. 219

as being at sex-limited loci. Let us see how this result comes about. Recall the equilibrium frequencies for autosomal (eqn. (9)) and male-haploid (eqn. (10)) deleterious recessives at loci that are not sex-limited. Because alleles are 'protected' while in males for sex-limited loci, the equilibrium frequencies for such alleles are higher than those from non-sex-limited loci, for autosomal loci becoming

$$q_{1L} = (2u/s)^{1/2} \qquad (12)$$

and for male-haploids becoming

$$q_{2L} = (3u/2s)^{1/2} \qquad (13)$$

At male-haploid loci, the ratio of the frequencies of deleterious recessive alleles at a sex-limited locus to those at a non-sex-limited locus becomes, from eqns. 6 and 9

$$R_F = (s/6u)^{1/2} \qquad (14)$$

Let the frequency of sex-limited *loci* be x, so that the ratio of sex-limited to non-sex-limited *loci* is given by

$$R_L = x/(1-x) \qquad (15)$$

The product $R_F R_L$ is therefore the expected ratio of the *overall* frequencies of sex-limited deleterious *alleles* to non-sex-limited *alleles*, across the entire genome. Thus, for $x = 0.01$ (1% of the genome sex-limited), $s = 1$ (lethal alleles), $u = 10^{-5}$, and

$$z/(1-z) = R_F R_L = 1.304 \qquad (16)$$

where z is the proportion of deleterious alleles (here, lethals) that are at sex-limited loci. Thus, although only 1% of the *loci* are sex-limited, 56.6% of the *lethal alleles* detectable through inbreeding are at sex-limited loci. Values of z comparable to those obtained by Kerr can readily be obtained by inserting the appropriate values in eqn. (16). A similar increase occurs for sex-limitation at autosomal loci, but it is much smaller. Thus, for the conditions for eqn. (16), but for autosomal loci, $z = 0.19$ rather than 0.57.

The above calculations refer only to genetic load *estimates*, and not to the actual load. For both autosomal and male-haploid loci, the load for both sex-limited and non-sex-limited loci is equal to the mutation rate if calculated in terms of the loss of *genetic material*. Sex limitation at male-haploid loci halves the genetic load in terms of the loss of *individuals* (going from $1.5u$ to $0.75u$).

Even though male-haploidy exaggerates the proportion of detectable sex-limited loci, real sex-limited characteristics do of course occur in spider mites. Perhaps the most important sex-limited characteristic known for spider mites is diapause ability, which is restricted to females and for which there is extensive genetic variability (Helle, 1968 and Chapter 1.3.2).

Inbreeding effects

It is often thought that there should be little inbreeding depression in male-haploids because of their genetic system. But how true is this? A cursory examination raises doubts. Inbreeding effects result from the increased homozygosity of inbreeding raising the frequency of genotypes with lower fitness relative to the frequency in an outbred population. These effects can be partitioned into those due to completely recessive deleterious alleles, to deleterious alleles with some effect in heterozygotes, and to alleles in polymorphic equilibrium.

Let us start by comparing the mortality due to completely recessive deleterious genes in a population made completely homozygous with the mortality from the same cause in its outbred precursor (Crow and Kimura, 1970c). For an autosomal locus, and considering a completely recessive allele, the loss in the outbred population is, as noted above, equal to the mutation rate, u. In a completely inbred population, all individuals are homozygotes, and $(u/s)^{1/2}$ of them will be homozygotes for the deleterious allele. The loss is the product of s (the selective disadvantage to homozygotes carrying the deleterious allele) and the frequency of homozygotes. Hence, the increase in mortality due to inbreeding in autosomal systems for completely recessive genes is given as a ratio by

$$I_{A1} = s(u/s)^{1/2}(1/u) = (s/u)^{1/2} \tag{17}$$

Now, sex-limited autosomal loci show an increase of the frequency of deleterious recessives to $(2u/s)^{1/2}$ (Crozier, 1976), so that, in the case of limitation to females, the increase in mortality is given by

$$I_{A2} = s(2us)^{1/2}/(2u) = (s/2u)^{1/2} \tag{18}$$

By similar reasoning the equivalent formulae for the effects of inbreeding on females in male-haploid species is obtained

$$I_{M1} = s/(3u) \tag{19}$$

$$I_{M2} = [2s/(3u)]^{1/2} \tag{20}$$

For the sake of comparison, let $s = 0.01$ and $u = 10^{-4}$. The autosomal cases then yield $I_{A1} = 10$ and $I_{A2} = 7$, and the male-haploid ones yield $I_{M1} = 33$ and $I_{M2} = 8$. Sex-limitation thus has qualitatively similar effects on inbreeding depression in both systems, but affects male-haploids more.

In the case of incompletely recessive deleterious genes, let the fitness of the (female!) heterozygotes be $(1 - hs)$, with h thus being a measure of the dominance of a over A. For autosomal systems, the frequency of such deleterious alleles can be shown to be

$$q_A = u/(sh) \tag{21}$$

so that the ratio of the loss in inbred populations to that in outbred populations is approximately

$$I_{A3} = (su/sh)/(2u) = 1/(2h) \tag{22}$$

For male haploids, the loss in females of the deleterious allele will be greater than that for males because, especially for rare alleles, the frequency of female heterozygotes will greatly exceed that of the male hemizygotes. It can readily be shown that the approximate frequency in females of a deleterious allele with some effect in heterozygotes is

$$q_{MF} = 3u/[s(1 + h)] \tag{23}$$

and the ratio of inbred to outbred loss is

$$I_{M3} = [3u/(1 + h)]/[6uh/(1 + h)] = 1/(2h) \tag{24}$$

which is the same answer as given by eqn. (22)!

Although eqns. (17)—(24) impel the conclusion that inbreeding effects are to be expected in male haploid genetic systems, and to degrees at least equalling those of autosomal systems, this conclusion has to be seen as tentative. This unhappy uncertainty stems from the difficulty (to me, at least) of assessing the effects of balanced polymorphisms. The difficulty arises from the fact that, if differential selection occurs between the sexes, then the loss in inbred females may not be much greater, and may even be less, than in outbred females. 'Other things being equal' there would be no difference between the 2 systems, but if, as Curtsinger (1980) suggests, male-haploids may be more likely to have differing selection values in males compared to females, the inbreeding effects of male-haploids might well be less than in autosomal systems. We can still proceed a little further, however, and investigate the effects of overdominant loci, supposing these to be less numerous by a certain factor, say $(1 - k)$, in male-haploids than in autosomal systems. If we consider the load to be carried by overdominant loci in which the 2 homozygotes have equal fitnesses, differing from that of the heterozygote by a small amount s, then the fitness of individuals in an outbred autosomal population, letting the number of overdominant loci be n, is given by

$$V_{A1} = (1 - s)^{n/2} \tag{25}$$

and in the inbred equivalent

$$V_{A2} = (1 - s)^n \tag{26}$$

The fitness ratio of inbred to outbred is then

$$V_{A2}/V_{A1} = (1 - s)^{n/2} \tag{27}$$

Making similar calculations for male-haploids involves replacing n by kn in eqns. (26) and (27), so that the ratio of the inbreeding effects in autosomal systems to those in male-haploid systems becomes

$$R_{A/M} = (1 - s)^{n(1-k)/2} \tag{28}$$

It is thus not easy to be certain about the expected inbreeding effects in male-haploids compared with autosomal systems. While this uncertainty arises chiefly from the difficulties imposed by polymorphic loci, the assumptions of the models above are easily violated for deleterious genes. In particular, these models assume that the population is forced to move from an outbred to an inbred state; if it is already inbred, no inbreeding depression will occur. There are thus various explanations for failures to find inbreeding depression in male-haploids, as reported by Schulten (Chapter 2.1.2.4) for various mites and by Biemont and Bouletreau (1980) for the hymenopterous parasitoid *Leptopilina boulardi* (Barb., Caton & Keln.-Pill.). However, inbreeding effects have been found in both *Tetranychus urticae* Koch (Helle, 1965) and the honey bee *Apis mellifera* (Brueckner, 1978, 1980; Moritz, 1982). Helle and Moritz both found these effects to be manifested in a reduced ability to hatch haploid as well as fertilized eggs, indicating maternal effects. A possible basis for these effects would be via the RNA stores of the egg, which are maternally derived. These maternal effects, if general, would reduce the usefulness of sex-ratio data for studying inbreeding depression.

Male-haploidy may speed evolution

Hartl (1972) found that new favourable mutations should be fixed one-third more rapidly at male-haploid than at autosomal loci, and one-third more slowly than at completely haploid loci. Hartl's assumptions in arriving at this conclusion were weak selection, panmixia, additive gene action (no dominance or overdominance) and absolute correlation of fitnesses between the sexes. While these assumptions rob Hartl's conclusion of wide generality, it is probably qualitatively correct.

Comparative studies of evolutionary rates in male-haploids compared with male-diploid systems with similar biological attributes are lacking, but there is certainly no lack of reports of rapid evolution in spider mites (De Boer, 1979; Gould, 1979; Helle and Overmeer, 1973; Overmeer and Van Zon, 1976).

There are practical considerations arising from the conclusion that male-haploids evolve more rapidly than male-diploids. Other things being equal, we would therefore expect male haploids to evolve insecticide resistance, for example, more rapidly than totally autosomal organisms of otherwise similar biology. It would therefore be unwise to rely on male-diploid predators of spider mites to 'keep up' with them in dealing with insecticides aimed at the spider mites: it is more likely that the spider mites will 'outrun' their predators in such a race.

Theoretical analyses indicate a reduced scope for polymorphism under male-haploidy

The apparent reduced scope for overdominance because of the lack of male heterozygotes intuitively suggests that male-haploids are likely to have fewer polymorphisms maintained by selection than organisms with both sexes diploid. A number of authors have quantified this intuition analytically; references to earlier studies are given by Pamilo (1979), Curtsinger (1980), and Pamilo and Crozier (1981). Rather than present a detailed account, the basic approach to the problem will be outlined and readers are referred to the works cited; this account is based mostly on the approach of Pamilo and Crozier (1981).

An autosomal locus has 6 possible genotypes, which, following Mandel (1971), may be designated

	Females			Males		
Genotype	AA	Aa	aa	AA	Aa	aa
Fitness	a_1	h_1	b_1	a_2	h_2	b_2

There are thus 2 independent fitnesses for each sex. Ascertaining the presence or absence of stable equilibria is a complex task (Owen, 1953; Mandel, 1971); up to 2 may occur.

For male haploid loci, 5 genotypes occur, with fitnesses which may be designated (1) following Mandel (1959) or (2) following Curtsinger (1980)

	Females			Males	
Genotype	AA	Aa	aa	A	a
Fitnesses (1)	a	h	b	a^*	b^*
Fitnesses (2)	$1 + f$	$1 + h$	$1 - f$	$1 + rf$	$1 - rf$

There are now not 4 but 3 independent fitnesses, 2 in females and 1 in males, under either scheme of fitnesses. The conditions for selection to maintain a polymorphism under this scheme are well known to be

$$aa^* < [h(a^* + b^*)]/2 > bb^* \tag{29}$$

and likewise the equilibrium allele frequencies in the 2 sexes to be, for females

$$p_F = [h(a^* + b^*) - 2bb^*]/[2h(a^* + b^*) - 2aa^* - 2bb^*] \tag{30a}$$

and for males

Chapter 1.3.4. references, p. 219

$$p_\text{M} = a^* p_\text{F} / [b^* + p_\text{F} (a^* - b^*)] \tag{30b}$$

The problem is to determine the relative levels of polymorphism under the 2 kinds of inheritance. Pamilo (1979) and Curtsinger (1980) charted the conditions under which polymorphism will occur in either scheme, under defined changes of the fitnesses, and Pamilo and Crozier (1981) used random numbers to generate the fitnesses. These studies, as did earlier ones, all found a reduction in the level of polymorphism in male-haploid loci relative to autosomal ones, but with some complexities. Thus, it is worthwhile to ask about the effects of varying the strength of selection or the correlation between genotypes within sexes (e.g., of a with h, and h with b) or between sexes (e.g., a with a^*). One can also partition the resulting genetic variation into P, the probability of a locus being polymorphic, H_P, the mean heterozygosity at polymorphic loci, and H_A, the mean heterozygosity at all loci. One can then obtain *variation ratios* of each polymorphism measure between the 2 genetic systems, e.g., $P_\text{MALE-HAPLOID} / P_\text{AUTOSOMAL}$.

Pamilo and Crozier (1981) found that, for their simulations, P and H_A were markedly less for male haploid than for autosomal loci, and that the disparity increased with increasing strength of selection and with increasing correlation between the sexes or with increasing correlation of genotypes within sexes. H_P values for the 2 systems were, however, comparable.

To paraphrase Li (1967a), analyses such as those above tend to be 'naturally artificial despite the attempt to be artificially natural'. Pamilo and Crozier (1981) found that the distribution of allele frequencies predicted by their models departed markedly from that actually observed in hymenopteran populations. It is perhaps risky to try to guess at such factors as the real relationship between the fitnesses of the 2 sexes.

Overview of male-haploid electrophoretic variation

Although some studies of electrophoretic genetic variation in mites have appeared (Sula and Weyda, 1983; Ward et al., 1982), none have included enough loci for valid comparison with autosomal systems. Such comparisons are possible with hymenopterans, however, and hymenopterans are indeed less polymorphic than male-diploid insects for all 3 measures of polymorphism (Pamilo and Crozier, 1981). However, *Drosophila* species are unusually highly polymorphic, and removing the *Drosophila* species somewhat increased the variation ratios for P and H_A, and resulted in H_P for male-haploids being substantially greater than that for male diploid insects (Pamilo and Crozier, 1981; for a parallel discussion see also Ward, 1980). The differences between observed male haploid and autosomal loci are therefore not clear-cut, although they do tend towards the qualitative predictions made by Pamilo and Crozier (1981).

The selection models discussed on page 209 are fixed-fitness ones. Other factors may influence the levels of genetic variation. Frequency- and density-dependent selection, and any mode of selection which is dependent on microhabitat variation, is unlikely to differ on average in its relative effectiveness in male-haploid and autosomal systems, and hence is not indicated as a possible explanation for the observed differences. Other possible explanations, considered by Pamilo and Crozier (1981), involve most polymorphisms being a result of mutation—selection balance, smaller effective population size in male-haploids, greater 'effective' linkage in male-haploids, and the effects of more rapid fixation of favourable mutations.

Equation (11) predicts that the frequency of deleterious alleles in autosomal systems will be greater than those in male-haploid systems, with this difference increasing with (s/u). Pamilo and Crozier (1981) found the same qualitative trend using the charge-state model of Ohta and Kimura (1975), but noted that the difference becomes small with decreasing (s/u).

As shown above (page 204), male haploids will have smaller effective population sizes than comparable autosomal systems if the number of males and females are equal. For such populations which are small, it is therefore possible that the number of effectively neutral polymorphisms (polymorphisms in which the fitness differentials are so small that allele frequencies are determined primarily by random genetic drift) will be reduced by 25% in male-haploids compared with male-diploids. However, this difference lessens with increasingly female-biased sex ratios, common in male-haploids such as mites (Hamilton, 1967), so that the role of effective population size in reducing male haploid genetic variation is uncertain.

Lester and Selander (1979) argue that the reduction of genetic variation in male-haploids stems from an increased 'hitch-hiking' effect in male-haploids. 'Hitch-hiking' is the carriage of an allele to high frequencies because it is linked to an allele at a different locus that is increasing in frequency under selection. The effect of hitch-hiking is generally to reduce genetic variation as neutral alleles are carried either to fixation by their 'carriers', or close enough to it often to be lost through drift. Male-haploidy could increase this 'hitch-hiking' effect in 2 ways. One way is by its influence on effective population size but, as shown above, there are reservations about this. The other is through an effective reduction of recombination because one sex is haploid. However, Pamilo and Crozier (1981) point out that the rates of recombination are unknown in most male-haploids and are considerable in those that are known. Furthermore, there is no recombination between loci on the same chromosome of males of most *Drosophila* species, and yet these species are about the most genetically polymorphic of all bisexual animals!

There remains the possibility that a number of these factors combine together to produce a reduction of genetic variation in male-haploids. An important factor synergistic with any 'hitch-hiking' effect would be the greater speed of evolution suspected for male-haploid systems (page 208). Such favourable alleles would be expected to carry linked alleles to fixation at a greater rate than would similar alleles at autosomal loci, because there would be less time for recombination to break down the linkage disequilibrium between them.

Maximization of fitness

One of the most heuristic aids to evolutionary understanding has been Wright's adaptive topography concept (see, e.g., Wright, 1977), which posits that the average fitness of a population invariably increases under selection. This model was developed for an autosomal locus with fitnesses exactly the same in each sex. If we designate dp as the change in allele frequency, and a, h, and b as the fitnesses of the genotypes AA, Aa, and aa, then the average fitness of the population is definable as

$$W = p^2 a + 2pqh + q^2 b \tag{31}$$

and the change in allele frequency as

$$dp = [p(1-p)/2W](dW/dp) = [p(1-p)/2](d \ln W/dp) \tag{32}$$

It can readily be shown from eqn. (32) that equilibrium points for p correspond to maxima for W. The surface of possible W values over all p values constitutes the adaptive topography, with high points on the

Chapter 1.3.4. references, p. 219

topography constituting adaptive peaks. Populations under selection always move 'uphill' on the topography, eventually reaching the neighbourhood of a peak. The peak reached may not be the highest one available, but rather depends on the starting position of the population; history is thus important in evolution. Wright (see, e.g., 1977, 1982; see also Hartl, 1980) has emphasized the potential role of drift in shifting small populations off low peaks, allowing them the possibility of reaching higher ones.

I know of no attempt to derive the equivalent of eqn. (32) for a truly general model of an autosomal locus, one with the fitnesses free to vary between the sexes. Such efforts have, however, been made for male-haploid loci, with Li (1967b; see also Crozier, 1979) defining a function that is maximized under natural selection at a male-haploid locus

$$M = aa^* p_F^2 + h(a^* + b^*) p_F (1 - p_F) + bb^* (1 - p_F)^2 \qquad (33)$$

It can readily be shown that maxima of M do occur at the equilibrium points predicted by eqn. (29). Owen (1984) has remarked that the maxima of M are lower than those for W, i.e., that differential selection between sexes at an autosomal locus results in a lower average fitness than at an autosomal locus with the same fitnesses in the 2 sexes.

Although the adaptive topography concept is important as a qualitative aid to understanding evolution (see, e.g., Crozier et al., 1972), it does not hold exactly for many situations. Thus, inbreeding requires further formulation of the outcome of selection (see Li, 1967a), and for multiple loci it only holds approximately if selection and linkage are weak. Furthermore, selection conditions may change so rapidly as to make the attainment of any peak unlikely (Colgan and Cheney, 1980).

Internal versus relational balance

General concept of coadaptation

The shifting balance theory of Wright follows naturally from the realization that, to paraphrase John Donne, no locus is an island, complete unto itself, but rather is embedded in a genome together with several to many thousands of its fellows. The selective process acting on an allele is therefore mediated by 2 forces: *relational* selection, involving the allele's interactions with other alleles at the same locus, and *internal* or epistatic selection, referring to fitness interactions with alleles at other loci. (The terms 'relational' and 'internal' balance (Mather, 1973) originally referred to interactions between and within chromosomes.) When a set of loci has come to possess a particular set of alleles through selective interactions between loci, we call the system *coadapted*. Such a system can be thought of as occupying a multi-locus adaptive peak. It is a feature of adaptive topography dynamics that different populations under the same selection conditions can arrive at different adaptive peaks. The consequence of this is that crosses between them lead to populations with lower average fitness than the parents, because the offspring population has allele frequencies divergent from those yielding an adaptive peak. Wallace (1981a) provides an extensive review of studies of coadaptation, and notes that evidence for it has been found frequently, in the form of reduction in fitness in hybrid populations. Such reduction does not usually affect the F_1 generation, but becomes apparent in the subsequent F_2 or back-cross generation, indicating that selection of individual chromosomes as functioning units has been significant.

Coadaptation is not restricted to diploid organisms, but has also been found, as might be expected, in bacteria selected for chloramphenicol resistance (Cavalli and Maccacaro, 1952, cited in Wallace, 1981b).

Although many population geneticists, including myself, intuitively expect that coadaptation should be a universal phenomenon, a number of studies searching for it failed to find it, most notably in our own species (Chung et al., 1966). These failures may not indicate that coadaptation is absent, but rather that an inappropriate measure was used (Wallace, 1981c), or, as may be so in our own case, that the populations are all distant from *any* adaptive peak due to rapid environmental change.

Finally, it is worth noting that attempts to find evidence for coadaptation in the form of electrophoretically characterized loci in linkage disequilibrium have generally failed, except for loci included within inversions. Yet, as mentioned above, breeding experiments have frequently demonstrated the phenomenon, even in the same species for which electrophoretic evidence was vainly sought. This apparent paradox is resolvable if the number of loci involved in coadaptation do not, in fact, form a large part of the genome, so that they are unlikely to be detected by electrophoretic surveys. Population geneticists are coming increasingly to realize that the number of loci which are important in evolutionary changes may be quite small, of the order of scores rather than thousands of loci (Thoday and Thompson, 1976; Thompson, 1977; Templeton, 1981). (Another possible reason for failing to find linkage disequilibrium is that electrophoresis does not distinguish between all alleles present, because the products of different alleles can have identical electrophoretic mobilities, and thus much disequilibrium may go undetected (Zouros et al., 1977).)

Coadaptation in male-haploids?

Given that coadaptation occurs in many organisms with both sexes diploid, and in completely haploid bacteria (above), we expect it to occur in male-haploids such as spider mites as well. In fact, for loci expressed in both sexes, coadaptation seems likely to be relatively more important in male-haploid than in autosomal systems because of the reduction in scope for overdominance.

Coadaptation has not been reported from hymenopterans, perhaps because it has seldom been looked for, or looked for only in species that probably inbreed habitually (see, e.g., Biemont and Bouletreau, 1980). Coadaptation has been found frequently in spider mites, as reflected in a greater reduction of F_2 than F_1 fitness (see, e.g., Helle and Overmeer, 1973; Overmeer and Van Zon, 1976), but generally is not as clearly demonstrated as the usually accompanying incompatability between strains of different geographic origins: coadaptation is generally superseded by speciation (page 208). Various insecticide-resistance selection experiments have, however, been interpreted in terms of the build-up of linkage disequilibrium and coadaptation (McEnroe and Harrison, 1968; McEnroe and Naegele, 1968).

HIGHER-LEVEL SELECTION AND THE EFFECTS OF POPULATION STRUCTURE

Sex ratio

Fisher (1930) showed that the sex ratio can be explained as the product of selection rather than as an automatic consequence of the sex-determining mechanisms (50% production of X- and Y-bearing sperm). Thus, in a population with an excess of females, males enjoy greater success individually than do individual females, and so couples that produce more males will have more grandchildren than do couples producing fewer males. Only a sex ratio of 1:1 females:males will always do as well or better than any other. At a population sex ratio of 0.5 (proportion of females), all sex ratios are equally successful.

Fisher also argued, and this has been borne out by subsequent investigations (Charnov, 1982a), that selection adjusts the sex ratio until investment in females returns the same pay-off in ultimate reproductive success as investment in males. The result of such selection is for the same effort or expenditure to be made in each sex.

The above arguments are valid for organisms living in large random-mating populations. Spider mites, on the contrary, probably tend in nature to live in small sib-mated groups (McEnroe, 1969; Mitchell, 1973). It is also relevant (to the ease with which they can adjust the sex ratio) that they obey the classic rules they share with Hymenoptera (Hamilton, 1967; Crozier, 1977) under which females are potentially the sole determiners of the sex ratio, through the 'decision' whether or not to release stored sperm to fertilize eggs before oviposition. Males fight over those females in the last pre-adult stage (Potter et al., 1976), and females effectively mate only once because any subsequent matings are ineffective (Helle, 1967). The dispersers tend strongly to be mated females (Mitchell, 1973; McEnroe, 1969).

Charnov (1982a) discusses the evolution of sex ratios in spider mite populations, pointing out that local mate competition (LMC) is a powerful explanation for the observed trends of high sex ratios at low densities and lower ones at high densities. Under the LMC model, if the members of one sex among a sibship compete with each other for mates more than does the other, then the competing sex is devalued as regards the pay-offs for investment in it. The result is that selection favours reduced investment in it. The pattern is also clearly seen in organisms other than spider mites (Hamilton, 1967). Hartl (1971) points out that in inbreeding parasitoid Hymenoptera, the equilibrium number of males per brood of exclusively sib-mating species is 1, or, rather, that number which guarantees one surviving adult male, provided that he can mate with all his sisters. Colwell (1981) has pointed out that the sex ratio of inbreeding species, or species with small temporary breeding groups, is determined by selection at 2 levels: within and between groups. Traditional 'Fisherian' selection favours a sex ratio of 0.5 *within* groups, but selection *between* groups favours groups which produce higher proportions of females. In accordance with the LMC and group selection models, the sex ratios of spider mites and other organisms varies inversely with population density of the founding females (Charnov, 1982b). As required for such a selection model, Mitchell (1972) demonstrated genetic differences between strains for the sex ratio, although it is worth noting that in hymenopterous parasitoids such as *Nasonia vitripennis* (Walk.) such changes in sex ratio reflect the *individual* response to density (Charnov, 1982c).

The arguments from selection theory concerning the sex ratio do not conflict with studies showing an influence of mortality or amount of sperm store on sex ratio (see, e.g., Wrensch and Young, 1983), because these factors will certainly mediate the response to selection. But the trends in sex ratio under population density etc. are strongly in favour of the operation of selection of the LMC or group selection types.

Finally, Bull (1983, cited in Charnov, 1982d) points out that sex ratios in phytoseiid mites are likely to follow different selection dynamics to tetranychids, because phytoseiids apparently lay diploid eggs, with the paternal set of chromosomes being eliminated in male-destined eggs (Schulten, Chapter 2.1.2.4). If sex ratio is still under exclusive maternal control, the same dynamics apply as for tetranychids, but if the *zygote* has control, the result is a strong *intrinsic* shift to female-biased sex ratios.

Sociobiological aspects of spider mite biology

The only form of social behaviour reported for spider mites is that concerned directly with courtship and mating. Nevertheless, these organisms present a number of features of biology that make them potentially useful for tests of many evolutionary models, and it seems worthwhile to mention these approaches.

The chief concern of sociobiologists at present is the evolutionary origin and maintenance of altruistic and co-operative behaviour (see, e.g., Crozier, 1982). Co-operative behaviour leads to increased reproductive success for both participants, whereas altruistic behaviour involves an altruist acting in such a way as to reduce its own reproductive success in favour of that of a beneficiary. The web-spinning behaviour characteristic of spider mites is the only one that might be considered under either heading. If any individuals 'specialize' in web-spinning, then they would qualify as altruists in the above sense, but it seems *a priori* that all participate equally, in which case it could be classed as, at most, co-operative. However, it is doubtful whether anyone has considered spider mites from a sociobiological viewpoint, so that it is premature to rule out the existence of appropriate behaviours. A cautionary example here is that of aphids, which form clonal colonies for several generations. For a long time, sociobiologists 'explained away' the lack of altruistic behaviour on the grounds that there was not much one aphid could do to help another. Now it is known that pemphigid aphids frequently have soldier castes active against insects and mammals threatening their colonies (Aoki, 1982).

The reason why sociality was sought in clonal aphids is that in clonal organisms what one colony member does for another is done (genetically) as though for itself, because they are genetically identical. Therefore, as long as the ratio of the help given (h_B) to the beneficiary of the act to the cost (c_A) to the altruist exceeds unity, altruism in clonal organisms will be selected for. This form of selection, depending on the interactions between relatives, is termed *kin selection*, and is also explicable in terms of Hamilton's (1964) updating of the concept of Darwinian fitness. This update leads to the idea of *inclusive*, or *absolute*, fitness being defined as

$$J_1 = S_1 + \sum_{i \neq 1} G_{i1} \, dS_{1i} \tag{34}$$

where J_1 is the inclusive fitness of individual 1;

$\quad$ S_1 is the personal reproductive success of individual 1;

$\quad$ G_{i1} is the relatedness of individual i to individual 1;

$\quad$ dS_{1i} is the effect of individual 1 on the personal reproductive success of individual i;

and $\quad \sum_{i \neq 1} G_{i1} \, dS_{1i}$ is the *inclusive fitness effect*, or *extended fitness*, of individual 1.

Equation (34) leads to an understanding that selection can favour reduced personal reproductive success if the behaviour in question increases the extended fitness sufficiently. From eqn. (34) can be derived the fundamental theorem of kin selection (Crozier, 1982), usually called 'Hamilton's Rule', which specifies the threshold value of the h/c value for a type of behaviour to be selected for

$$h_B G_{BA} > c_A \tag{35}$$

Elsewhere (Crozier, 1982) I review the history of assessment of eqn. (35) as an explanation for hymenopteran sociality, noting that attention can be focused either on selection of the offspring to assist their mother

---

Chapter 1.3.4. references, p. 219

or on the mother to produce helping offspring. Theoretical endeavours show that this latter, *parental manipulation*, model has a lower threshold under formulae analogous to eqn. (35) and hence is a stronger force than offspring-level kin selection as an explanation for hymenopteran eusociality. However, for offspring to be susceptible to 'maternal coercion' implies that they are facultatively altruistic, requiring a standard precursor kin-selection stage anyway (Crozier, 1982; Craig, 1983).

Wade (1980) points out that kin selection is itself readily resolvable as a form of group selection: within kin groups, selection will always militate against altruistic behaviour, whereas selection between groups favours altruism. The outcome follows Hamilton's Rule. Group selection has been extensively dealt with in various publications (see, e.g., Wilson, 1980; Boorman and Levitt, 1980; Uyenoyama and Feldman, 1980), and spider mites seem to be at least as well suited for experiments to test the theory as those insects which have been extensively used for this purpose (see, e.g., Wade, 1979).

As should now be apparent, relatedness is a key concept to understanding of the evolution of social behaviour. Assessment of relatedness in natural populations can be carried out by pedigree tracing (Jacquard, 1974) or by regression analysis of the genetic resemblances between relatives (Pamilo and Crozier, 1982; Crozier et al., 1984).

The sociobiology of spider mites has received a significant boost from the recent discovery (Y. Saito, in prep.) that adults of *Schizotetranychus celarius* (which lives in a multi-chambered web nest; see 1.4.4) eject or kill invading predatory phytoseiid mites, thus protecting immature mites in the nest (which are not always their own offspring). If further experiments show that these adults are acting to protect the immatures rather than just themselves, then the case would clearly be one of parental care.

SPECIATION

Remarks on the general genetics of speciation

Descendant organisms may be of low fitness after genetic experiments for a variety of reasons, and it is worthwhile now to distinguish between some of these.

As we have seen, *inbreeding depression* occurs as a result of increased homozygosity for either alleles completely deleterious and maintained by mutation in the base population, or for alleles maintained by their higher fitness as heterozygotes. Inbreeding depression, of course, results from the inbreeding of members of a usually random-mating population.

Coadaptation, because of its essential nature as inter-population genetic differentiation built up by selection, is broken down by crosses between populations of the same species. Coadaptation does not involve major discordancies in the modes of action of alleles, but rather is a fine-tuning of existing variation to local conditions. It is generally not manifested as a depression of F_1 but rather of F_2 or back-cross fitness, indicating that chromosomes probably function as distinct molecular entities in the cell; the allelic content of neighbouring loci in the same chromosome affects the relative biochemical efficiency of the alleles at a locus.

Reproductive isolation carries coadaptation further, and involves basic incompatability between the genetic material of different populations, to the extent that F_1 fertility or viability, or both, are markedly depressed, usually to the point of preventing gene flow between the parental populations. This incompatability may not manifest itself as noticeable physiological sterility or inviability; behavioural disturbances may prevent apparently healthy hybrids from reproducing.

The bringing about of reproductive isolation is the process of speciation. Species are characterized by possessing separate gene pools and following separate evolutionary tendencies (Mayr, 1970: 12—13; Wiley, 1978; Grant, 1981: 46—63; Crozier, 1983). Clearly, there are operational problems in being sure of one's decision! Both genetic and ecological isolation are therefore criteria in deciding whether or not 2 entities are separate species. The phenomena (often dubiously termed 'mechanisms') leading to genetical isolation are many, but have been reviewed often (see, e.g., Dobzhansky, 1970; Mayr, 1970; Futuyma, 1979).

There is a wide diversity of speciation mechanisms that have been proposed and for which there is evidence. Templeton (1981) reviews the mechanisms of speciation, and the genetics of reproductive isolation. As noted above, evolutionary phenomena can involve few loci, and Templeton finds that, although many loci *may* be involved, speciation *can* be the result of changes in only a few loci (say 10 or fewer). The older view is that speciation *necessarily* involves a major reorganization of the genome (Carson, 1983) and requires that there be a hidden kind of genetic material within which the reorganization takes place (because there is little electrophoretic evidence for the postulated reorganization (see, e.g., Nei, 1980), which seems implausible (Crozier, 1983)). Templeton suggests that speciation mechanisms can be divided broadly into adaptive and non-adaptive modes. By 'adaptive' is meant that speciation is a result of divergent evolution under selection, the populations adapting to different environments (although if they adapted to the same habitat but in different places the adaptive topography idea would suggest that reproductive isolation could still arise between them). By 'non-adaptive' he meant the production of genetic differentiation by some process, particularly random genetic drift caused by a temporary drastic reduction in population size. More recently, Nei et al. (1983) have suggested that speciation may result from the fixation by drift in large populations of incompatible alleles in different populations, and Dover (1982; see also Rose and Doolittle, 1983) that intra-genomic selection for transposable elements and similar DNA types of variable numbers within chromosomes may result in chromosomes from different populations differing sufficiently to impair meiosis.

There is laboratory evidence from *Drosophila* for both selection (Kilias and Alahiotis, 1982; Ehrman, 1983) and bottleneck (temporary population size reduction) factors leading to incipient speciation (Powell, 1978). The bottleneck effect will not lead to speciation unless it involves a usually large population with plentiful genetic variation. The bottleneck, by drastically altering the genetic environment of the loci that do remain polymorphic, leads to a changed genetic constitution that may result in genetic isolation. Spider mite populations in agricultural and adjacent (McEnroe, 1970) settings can be large, and may remain so long enough to permit bottleneck events to result in speciation; the response of sex ratio to density variation (page 213) indicates adaptation to a range of density levels.

A factor involved in *Drosophila* speciation that is, curiously, seldom considered by Templeton and other speciation theorists is that of cytoplasmically-determined sterility. It now seems certain, following the extensive studies of Ehrman (Ehrman and Kernaghan, 1971), that the rapid development of F_1 hybrid male sterility between *Drosophila* populations

Chapter 1.3.4. references, p. 219

results from genetic change not in the *Drosophila* nuclear genome, but in the genome of cytoplasmic symbionts. Consistent with the symbiont hypothesis, rearing the flies on media containing antibiotics improves the fertility of the hybrid males (Kernaghan and Ehrman, 1970; Powell, 1982). Of course, cytoplasmically based sterility has been noted in insects other than *Drosophila*, notably in *Culex* mosquitoes (Barr, 1980, cited in Powell, 1982).

Apparent rapid speciation is a feature of spider mites

Tetranychus urticae strains are often found to be wholly or partially reproductively isolated. De Boer (Chapter 1.3.3) reviews the evidence for this phenomenon, so my discussion will be brief.

The reproductive isolation observed could result from a number of causes. Firstly, it may not indicate rapid evolution at all, but may arise from *T. urticae* being, not 1 species, but rather a cluster of sibling species. Thus, the form known as *T. cinnabarinus* could just represent a morphologically unusually well-delineated species in the complex. The situation suggested under this hypothesis is paralleled in another male haploid group, the hymenopterous parasitoid genus *Trichogramma* (Nagarkatti, 1977). The failure of one deliberate attempt to induce inter-strain sterility (De Boer, Chapter 1.3.3) may reflect the undoubtedly stochastic nature of the process. That the 'patchwork quilt' effect of incompatibilities between closely-adjacent greenhouse populations was not found in a non-agricultural area (De Boer, 1980) could indicate either that greenhouse conditions are more variable than natural ones and hence support a greater diversity of spider mite species or, more plausibly, that the greenhouse populations have evolved rapidly under the twin selection pressures of new host plant species and insecticide applications.

Among hypotheses implicating the recent evolutionary origin of strain inter-sterility is that of such isolation being the result of rapid adaptive divergence under selection to the changed conditions of the greenhouse. This idea is supported by the plethora of such forms from greenhouses, but this may reflect the relatively small amount of work done on 'natural' populations (but cf. De Boer, Chapter 1.3.3). Alternatively, speciation may be induced by cycles of population expansion and contraction (Powell, 1978), likely to occur often in greenhouses! This latter model presupposes that the populations are large long enough to build up and adapt to higher levels of genetic variation than presumably occurred in pre-agricultural times, although the finding (De Boer, 1979) of high frequencies of lethals in a laboratory population indicates that population sizes may be higher than generally thought. Both 'genetic' types of speciation imply that derivative populations will be different from each other.

Recent findings from work on *Drosophila* are generally relevant to studies on speciation. The male progeny resulting from crosses between females of *D. melanogaster* strains held for a long time in the laboratory and males of strains recently collected from the wild tend strongly to exhibit a syndrome of sterility and other phenomena called *hybrid dysgenesis*. Hybrid dysgenesis stems from the possession by the recently-collected strains of transposable genetic elements lacking in the older strains (Kidwell, 1983). While it is possible that the time of collection phenomenon reflects successive sampling during a time of rapid invasion of the field populations by the transposable elements, it is generally regarded as more probable that these elements were present all along, and that there is a tendency to lose them in laboratory culture.

The last mechanism to consider is that of change in cytoplasmic factors, such as symbionts. Lasting maternal effects have frequently been demonstrated in spider mites (Helle and Overmeer, 1973), implicating such factors as a cause of speciation. However, other experiments (De Boer, 1982) clearly implicated a major role for nuclear genes.

My impression from the work so far is that speciation probably is rapid in spider mites, and that it stems from a number of causes. Now that these are becoming clear, it will be easier to design experiments to distinguish between them. One obvious experiment would be to investigate the impact of antibiotics on sterility barriers.

Finally, it is appropriate to end with a note on a difference between the male-haploid and autosomal genetic systems with respect to speciation and hybridization between species. Because male-haploid males are uniparental, in distinction to the biparental males of 'normal' species, F_1 hybrid males do not exist. Consequently, all interspecific hybridizations can only progress by back-crosses: F_2 generations are strictly not possible. Modellers of speciation processes have yet to take this factor into account.

ACKNOWLEDGEMENTS

I thank Drs Pekka Pamilo and Wim Helle for cogent comments on the manuscript, and Dr Ian A. Boussy for letting me see a manuscript of his on spider mite genetic biology.

REFERENCES

Aoki, S., 1982. Soldiers and altruistic dispersal in aphids. In: M.D. Breed, C.D. Michener and H.E. Evans (Editors), The Biology of Social Insects. Westview Press, Boulder, CO, pp. 154—158.

Barr, A.R., 1980. Cytoplasmic incompatibility in natural populations of a mosquito, *Culex pipiens* L. Nature (London), 283: 71.

Biemont, C. and Bouletreau, M., 1980. Hybridization and inbreeding effects on genome coadaptation in a haplo-diploid hymenopteran: *Cothonaspis boulardi* (Eucoilidae). Experientia, 36: 45—46.

Boorman, S.A. and Levitt, P.R., 1980. The Genetics of Altruism. Academic Press, New York.

Brueckner, D., 1978. Why are there inbreeding effects in haplo-diploid systems? Evolution, 32: 456—458.

Brueckner, D., 1980. Hoarding behaviour and life span of inbred, non-inbred and hybrid honeybees. J. Apicultural Res., 19: 35—41.

Bull, J.J., 1983. The Evolution of Sex Chromosomes and Sex Determination. In preparation.

Carson, H.L., 1983. Speciation as a major reorganization of polygenic balances. In: C. Barigozzi (Editor), Mechanisms of Speciation. Liss, New York, pp. 411—433.

Cavalli, L.L. and Maccacaro, G.A., 1952. Polygenic inheritance of drug-resistance in the bacterium *Escherichia coli*. Heredity, 6: 311—331.

Charnov, E.L., 1982a. The Theory of Sex Allocation. Princeton University Press, Princeton, NJ, pp. 89—92.

Charnov, E.L., 1982b. The Theory of Sex Allocation. Princeton University Press, Princeton, NJ, p. 91.

Charnov, E.L., 1982c. The Theory of Sex Allocation. Princeton University Press, Princeton, NJ, pp. 79—87.

Charnov, E.L., 1982d. The Theory of Sex Allocation. Princeton University Press, Princeton, NJ, p. 88.

Chung, C.S., Morton, N.E. and Yasuda, N., 1966. Genetics of inter-racial crosses. Ann. N.Y. Acad. Sci., 134: 666—687.

Clegg, M.T. and Cavener, D.R., 1982. Dynamics of correlated genetic systems. VII. Demographic aspects of sex-linked transmission. Am. Nat., 120: 108—118.

Colgan, D.J. and Cheney, J., 1980. The inversion polymorphism of *Keyacris scurra* and the adaptive topography. Evolution, 34: 181—192.

Colwell, R.K., 1981. Group selection is implicated in the evolution of female-biased sex ratios. Nature (London), 290: 401—404.

Cornette, J.L., 1977. Sex-linked genes in age-structured populations. Heredity, 40: 291—297.

Craig, R., 1983. Subfertility and the evolution of eusociality by kin selection. J. Theor. Biol., 100: 379—397.

Crow, J.F. and Kimura, M., 1970a. An Introduction to Population Genetics Theory. Harper and Row, London, pp. 47—50.

Crow, J.F. and Kimura, M., 1970b. An Introduction to Population Genetics Theory. Harper and Row, London, pp. 109—111.

Crow, J.F. and Kimura, M., 1970c. An Introduction to Population Genetics Theory. Harper and Row, London, p. 303.

Crozier, R.H., 1975a. Animal Cytogenetics 3. Insecta. 7. Hymenoptera. Gebrüder Bornträger, Berlin, pp. 73—74.

Crozier, R.H., 1975b. Animal Cytogenetics 3. Insecta. 7. Hymenoptera. Gebrüder Bornträger, Berlin, p. 75.

Crozier, R.H., 1976. Why male-haploid and sex-linked genetic systems seem to have unusually sex-limited genetic loads. Evolution, 30: 623—624.

Crozier, R.H., 1977. Evolutionary genetics of the Hymenoptera. Annu. Rev. Entomol., 22: 263—288.

Crozier, R.H., 1979. Genetics of sociality. In: H.R. Hermann (Editor), Social Insects, Vol. 1. Academic Press, New York, NY, pp. 223—286.

Crozier, R.H., 1982. On insects and insects: twists and turns in our understanding of the evolution of sociality. In: M.D. Breed, C.D. Michener and H.E. Evans (Editors), The Biology of Social Insects. Westview Press, Boulder, CO, pp. 4—10.

Crozier, R.H., 1983. Genetics and insect systematics: retrospect and prospect. In: E. Highley and R.W. Taylor (Editors), Australian Systematic Entomology: A Bicentenary Perspective. CSIRO, Melbourne, pp. 80—92.

Crozier, R.H., Briese, L.A., Guerin, M.A., Harris, T.R., McMichael, J.L., Moore, C.H., Ramsey, P.R. and Wheeler, S.R., 1972. Population genetics of hemoglobins S, C, and A in Africa: Equilibrium or replacement? Am. J. Hum. Genet., 24: 156—167.

Crozier, R.H., Pamilo, P. and Crozier, Y.C., 1984. Relatedness and microgeographic genetic variation in *Rhytidoponera mayri* an Australian arid-zone ant. Behav. Ecol. Sociobiol., 15: 143—150.

Curtsinger, J.W., 1980. On the opportunity for polymorphism with sex-linkage or haplodiploidy. Genetics, 96: 995—1006.

De Boer, R., 1979. Investigations concerning the development of reproductive incompatibilities between populations of the spider mite, *Tetranychus urticae*. Z. Angew. Entomol., 87: 113—121.

De Boer, R., 1980. Genetic affinities between spider mite *Tetranychus urticae* populations in a non-agricultural area. Entomol. Exp. Appl., 28: 22—28.

De Boer, R., 1982. Laboratory hybridization between semi-incompatible races of the arrhenotokous spider mite *Tetranychus urticae* Koch (Acarina: Tetranychidae). Evolution, 36: 553—560.

Dobzhansky, T., 1970. Genetics of the Evolutionary Process. Columbia University Press, New York, p. 314

Dover, G., 1982. Molecular drive: a cohesive mode of species evolution. Nature (London), 299: 111—117.

Dyson, J.G., 1965. Natural Selection of the Two Mutant Genes Honey and Orange in Laboratory Populations of *Habrobracon juglandis*. Unpublished Ph. D. Thesis, North Carolina State University.

Ehrman, L., 1983. Fourth report on natural selection for the origin of reproductive isolation. Am. Nat., 121: 290—293.

Ehrman, L. and Kernaghan, R.P., 1971. Microorganismal basis of infectious hybrid male sterility in *Drosophila paulistorum*. J. Hered., 62: 67—71.

Fisher, R.A., 1930. The Genetical Theory of Natural Selection. Oxford University Press, Oxford.

Futuyma, D.J., 1979. Evolutionary Biology. Sinauer, Sunderland, MA, p. 192.

Gould, F., 1979. Rapid host range evolution in a population of the phytophagous mite *Tetranychus urticae*. Evolution, 33: 791—802.

Grant, V., 1981. Plant Speciation, 2nd edn. Columbia University Press, New York.

Hamilton, W.D., 1964. The genetical evolution of social behaviour. I and II. J. Theor. Biol., 7: 1—52.

Hamilton, W.D., 1967. Extraordinary sex ratios. Science, 156: 477—488.

Hartl, D.L., 1971. Some aspects of natural selection in arrhenotokous populations. Am. Zool., 11: 309—325.

Hartl, D.L., 1972. A fundamental theorem of natural selection for sex linkage or arrhenotoky. Am. Nat., 106: 516—524.

Hartl, D.L., 1980. Principles of Population Genetics. Sinauer, Sunderland, MA, pp. 327—339.

Helle, W., 1965. Inbreeding depression in an arrhenotokous mite (*Tetranychus urticae* Koch). Entomol. Exp. Appl., 8: 299—304.

Helle, W., 1967. Fertilization in the two-spotted spider mite (*Tetranychus urticae*: Acari). Entomol. Exp. Appl., 10: 103—110.

Helle, W., 1968. Genetic variability of photoperiodic response in an arrhenotokous mite (*Tetranychus urticae*). Entomol. Exp. Appl., 11: 101—113.

Helle, W. and Overmeer, W.P.J., 1973. Variability in tetranychid mites. Annu. Rev. Entomol., 18: 97—120.

Jacquard, A., 1974. The Genetic Structure of Populations. Springer Verlag, Berlin, pp. 102—140 (Translated by D. Charlesworth and B. Charlesworth).

Kernaghan, R.P. and Ehrman, L., 1970. Antimycoplasmal antibiotics and hybrid sterility in *Drosophila paulistorum*. Science, 169: 63—64.

Kerr, W.E., 1976. Population studies in Hymenoptera. 2. Sex-limited genes. Evolution, 30: 94—99.

Kidwell, M.G., 1983. Evolution of hybrid dysgenesis determinants in *Drosophila melanogaster*. Proc. Natl. Acad. Sci. USA, 80: 1655—1659.

Kilias, G. and Alahiotis, S.N., 1982. Genetic studies on sexual isolation and hybrid sterility in long-term cage populations of *Drosophila melanogaster*. Evolution, 36: 121—131.

Lester, L.J. and Selander, R.K., 1979. Population genetics of haplo-diploid insects. Genetics, 92: 1329—1345.

Li, C.C., 1967a. Genetic equilibrium under selection. Biometrics, 23: 397—484.

Li, C.C., 1967b. The maximization of average fitness by natural selection at a sex-linked locus. Proc. Natl. Acad. Sci. USA, 57: 1260—1261.

Mandel, S.H., 1959. Stable equilibrium at a sex-linked locus. Nature (London), 183: 1347—1348.

Mandel, S.H., 1971. Owen's model of a genetical system with differential viability between sexes. Heredity, 26: 49—63.

Mather, K., 1973. Genetical Structure of Populations. Chapman and Hall, London, pp. 132—134.

Mayr, E., 1970. Populations, Species and Evolution. Harvard University Press, Cambridge, pp. 12—13; 55—68.

McEnroe, W.D., 1969. Spreading and inbreeding in the spider mite. J. Hered., 60: 343—345.

McEnroe, W.D., 1970. A natural outcrossing swarm of *Tetranychus urticae*. J. Econ. Entomol., 63: 822—823.

McEnroe, W.D. and Harrison, R., 1968. The role of linkage in the response of an arrhenotokous spider mite, *Tetranychus urticae*, to directed selection. Ann. Entomol. Soc. Am., 61: 1239—1247.

McEnroe, W.D. and Naegele, J.A., 1968. The coadaptive process in an organophosphorus-resistant strain of the two-spotted spider mite, *Tetranychus urticae*. Ann. Entomol. Soc. Am., 61: 1055—1059.

Mitchell, R., 1972. The sex ratio of the spider mite *Tetranychus urticae*. Entomol. Exp. Appl., 15: 299—304.

Mitchell, R., 1973. Growth and population dynamics of a spider mite (*Tetranychus urticae* K., Acarina: Tetranychidae). Ecology, 54: 1349—1355.

Moritz, R.F.A., 1982. Maternale Effekte bei der Honigbiene (*Apis mellifera* L.). Z. Tierz. Zuectungsbiol., 99: 139—148.

Nagarkatti, S., 1977. Biosystematics of *Trichogramma* and *Trichogrammatoides* species. Annu. Rev. Entomol., 22: 157—176.

Nagylaki, T., 1981. The inbreeding effective population number and the expected homozygosity for an X-linked locus. Genetics. 97: 731—737.

Nei, M., 1980. Stochastic theory of population genetics and evolution. In: C. Barigozzi (Editor), Vito Volterra Symposium on Mathematical Models in Biology. Springer Verlag, Berlin, pp. 17—47.

Nei, M., Maruyama, T. and Wu, C.-I., 1983. Models of evolution of reproductive isolation. Genetics, 103: 557—579.

Ohta, T. and Kimura, M., 1975. Theoretical analyses of electrophoretically detectable polymorphisms: models of very slightly deleterious mutations. Am. Nat., 109: 137—145.

Overmeer, W.P.J. and Van Zon, A.Q., 1976. Partial reproductive incompatibility between populations of spider mites (Acarina: *Tetranychidae*). Entomol. Exp. Appl., 20: 225—236.

Owen, A.R.G., 1953. A genetical system admitting of two distinct stable equilibria under natural selection. Heredity, 7: 97—102.

Owen, R.E., 1984. Properties of the mean fitness at a sex-linked or haplo-diploid locus in a polymorphic equilibrium. Seen as manuscript.

Pamilo, P., 1979. Genic variation at sex-linked loci: quantification of regular selection models. Hereditas, 91: 129—133.

Pamilo, P. and Crozier, R.H., 1981. Genic variation in male-haploids under deterministic selection. Genetics, 89: 199—214.

Pamilo, P. and Crozier, R.H., 1982. Measuring relatedness in natural populations: methodology. Theor. Popul. Biol., 21: 171—193.

Potter, D.A., Wrensch, D.L. and Johnston, D.E., 1976. Aggression and mating success in male spider mites. Science, 193: 160—161.

Powell, J.R., 1978. The founder-flush speciation theory: an experimental approach. Evolution, 32: 465—474.

Powell, J.R., 1982. Genetic and Nongenetic Mechanisms of Speciation. In: C. Barigozzi (Editor), Mechanisms of Speciation. Liss, New York, pp. 67—74.

Rose, M.R. and Doolittle, W.F., 1983. Molecular biological mechanisms of speciation. Science, 220: 157—162.

Sula, J. and Weyda, F., 1983. Esterase polymorphism in several populations of the two-spotted spider mite, *Tetranychus urticae* Koch. Experientia, 39: 78—79.

Templeton, A.R., 1981. Mechanisms of speciation — a population genetic approach. Annu. Rev. Ecol. Syst., 12: 23—48.

Thoday, J.M. and Thompson, J.N., 1976. The number of segregating genes implied by continuous variation. Genetica, 46: 335—344.

Thompson, J.N., 1977. Analysis of gene numbers and development in polygenic systems. Stadler Genet. Symp., 9: 63—82.

Uyenoyama, M. and Feldman, M.W., 1980. Theories of kin and group selection: a population genetics perspective. Theor. Popul. Biol., 17: 380—414.

Wade, M.J., 1979. The primary characteristics of *Tribolium* populations group selected for increased and decreased population size. Evolution, 33: 749—764.

Wade, M.J., 1980. Kin selection: its components. Science, 210: 665—667.

Wallace, B., 1981a. Basic Population Genetics. Columbia University Press, New York, pp. 513—539.

Wallace, B., 1981b. Basic Population Genetics. Columbia University Press, New York, pp. 515—516.

Wallace, B., 1981c. Basic Population Genetics. Columbia University Press, New York, p. 537.

Ward, P.S., 1980. Genetic variation and population differentiation in the *Rhytidoponera impressa* group, a species complex of ponerine ants (Hymenoptera: Formicidae). Evolution, 34: 1060—1076.

Ward, P.S., Boussy, I.A. and Swincer, D.E., 1982. Electrophoretic detection of enzyme polymorphism and differentiation in three species of spider mites (*Tetranychus*) (Acari: Tetranychidae). Ann. Entomol. Soc. Am., 75: 595—598.

Wiley, E.O., 1978. The evolutionary species concept reconsidered. Syst. Zool., 27: 17—26.

Wilson, D.S., 1980. The Natural Selection of Populations and Communities. Benjamin Cummings, Menlo Park, CA.

Wrensch, D.L. and Young, S.S.Y., 1983. Relationship between primary and tertiary sex ratio in the two-spotted spider mite (Acarina: Tetranychidae). Ann. Entomol. Soc. Am., 76: 786—789.

Wright, S., 1969a. Evolution and the Genetics of Populations. Vol. 2. The Theory of Gene Frequencies. Chicago University Press, Chicago, IL, pp. 211—214.

Wright, S., 1969b. Evolution and the Genetics of Populations. Vol. 2. The Theory of Gene Frequencies. Chicago University Press, Chicago, IL, pp. 211—220.

Wright, S., 1977. Evolution and the Genetics of Populations. Vol. 3. Experimental Results and Evolutionary Deductions. Chicago, IL, pp. 449—452.

Wright, S., 1982. Character change, speciation, and the higher taxa. Evolution, 36: 427—443.

Zouros, E., Golding, G.B. and MacKay, T.C.F., 1977. The effect of combining alleles into electrophoretic classes on detecting linkage disequilibrium. Genetics, 85: 543—550.

Chapter 1.4 Ecology

1.4.1 Webbing

U. GERSON

INTRODUCTION

Webbing in the Arachnida appears to have evolved at least twice, involving different spinning organs. Spiders, weaving their silk from abdominal spinnerets, represent one evolutionary avenue. The pseudoscorpions and the mites, whose modified salivary glands produce their webs, represent the other. However, as only a few species in just a handful of families in merely a single order of mites — the Prostigmata — web, it could be argued that acarine weaving had evolved independently from that of the pseudoscorpions. The similarity in amino acid composition of spider mite and pseudoscorpion silks (Hazan et al., 1975b) could thus be seen as signifying phylogenetic relationships and/or convergence, resulting from similar components in the ancient salivary glands of both groups being used for the same purpose.

Once an animal has produced an external, non-living excretion which confers upon it some advantage, natural selection would favour the additional development of that product. And as the latter becomes of importance in the animal's life, selection would bring about appropriate morphological, physiological, ecological and behavioural changes in the excreting organism. This indeed appears to have happened with spider mites once they had begun to web, and 2 essays in this book are concerned with that topic. The present chapter, which is a partial continuation of a former review (Gerson, 1979) deals mainly with the webbing of individual mite species and how the product is used. The other essay (Saitô, Chapter 1.4.4) describes how webbing has affected, and has been affected by, spider mite life types. There is no clear separation between the 2 topics, making some overlap unavoidable.

Three subjects related to webbing will not be discussed. These are the silk-producing glands (Alberti and Crooker, Chapter 1.1.2), the chemical composition of the webs (Hazan et al., 1975b) and methods for quantitatively estimating or measuring their production (Gerson, 1979).

A perusal of the scant data on webbing by members of other Prostigmatic families suggests that this excretion initially served to protect eggs, for mate finding and for hunting. These functions may therefore be regarded as the more primitive ones and will be discussed first.

PROTECTION OF EGGS

Members of the families Bdellidae, Camerobiidae and Cheyletidae cover their eggs with webs (Wallace and Mahon, 1972; Bolland, 1983; Avidov et al., 1968, respectively). This is also the only known spinning activity of

Chapter 1.4.1. references, p. 230

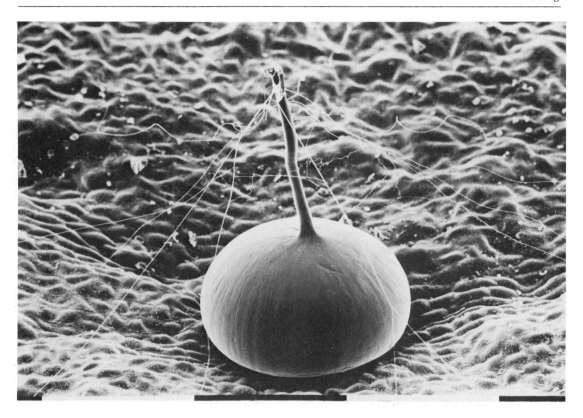

Fig. 1.4.1.1. Scanning electron microscope micrograph of an egg of *Panonychus citri* (McGregor) held by 'guy ropes'. Bar 0.1 mm.

Tenuipalpoides dorychaeta Pritchard and Baker (Singer, 1966), a member of the primitive tribe Tenuipalpoidini in the sub-family Tetranychinae. *Aponychus corpuzae* Rimando and *Eurytetranychus japonicus* Ehara are 2 other primitive Tetranychines whose only webbing activity revolves around the eggs (Saitô, 1983). The eggs of *Panonychus citri* (McGregor), a poor weaver, are webbed down by 'guy ropes' (Fig. 1.4.1.1). Eggs of heavy weavers are invariably covered by webbing (Saitô, 1983).

The main function of this webbing is probably to protect eggs from mite predators, many of which are voracious egg feeders.

Life history studies on *Tetranychus cinnabarinus* (Boisduval) (Hazan et al., 1974) revealed a very close correlation between estimated amounts of webbing and numbers of eggs deposited. Upon comparing hatch percentages between eggs left in the webbing and eggs removed therefrom, a slight but significant decline was found in the latter group at very low and very high relative humidities (Figs. 1—2 in Hazan et al., 1975a). One of the functions of the webs, in this case, would thus be to regulate humidity in the immediate vicinity of the eggs.

MATE FINDING

Courtship- or spermatophore-associated webbing by males of Trombidiidae was described by Tevis and Newell (1962) and Moss (1960), respectively. Maturing *Tetranychus urticae* Koch utilize their webs for various courtship-associated purposes. While the pre-moulting last female nymph spins the

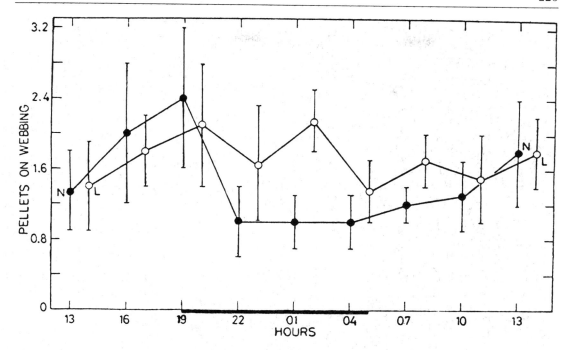

Fig. 1.4.1.2. Webbing of females of *Tetranychus cinnabarinus* (Boisduval) (estimated as pellets on webbing, see Hazan et al., 1974) as affected by continuous light (L, ○) and by a 8:16 dark:light regime (N, ●). Mites held at 24°C, 38% relative humidity throughout. Black bar on abcissa between 19 and 05 h denotes dark period.

silken cover under which it will transform, it also secretes a sex pheromone (Cone et al., 1971; Chapter 1.4.3). Male behaviour is modified by the web, its movements towards the moulting nymph becoming more linear (Penman and Cone, 1972). After locating its potential mate the male spins an extensive mat over it, remaining nearby. The web thus serves as a means of physical contact between the sexes, should the male become separated from its intended mate. The web may also serve as a pheromone substrate or carrier (Penman and Cone, 1974). The latter authors interpreted the silk spun by males around the moulting nymphs as a form of territoriality. Potter et al. (1976) observed intraspecific fights between waiting males and reported that they may at times apply strands of silk to the mouthparts and legs of their opponents, forcing them to withdraw. The male which spins more silk faster thus wins, a factor selecting for more and faster spinning. Such web-associated fighting appears to be the only aggressive behaviour in spider mites. It could be interpreted as a modification of the hunting behaviour of the Bdellidae and Cunaxidae, which use silk to capture their prey (Alberti and Ehrnsberger, 1977).

LOCOMOTION AND DISPERSAL

Saitô (1977a and unpublished data, 1976) demonstrated a very close relationship between the walking activity and the amounts of webs produced by *T. urticae*, *P. citri* and *Oligonychus ununguis* (Jacobi). He postulated that these mites spun silk threads whenever walking. Later (Saitô, 1979) he further postulated that the threads serve as 'lifelines' throughout the mites' lives on their host plants. These 'lifelines' are used by blown-off or dispersing mites, as the threads are taken up by winds and the animals drift along in the air streams (Fleschner et al., 1956). *Tetranychus* may also disperse within

Chapter 1.4.1. references, p. 230

thick (2—4 in.) units of silk, each enclosing many live mites (Wene, 1956). Spider mites in greenhouses may form silk ropes when leaving heavily infested plants (Hussey and Parr, 1963). Mites concentrate at the upper plant apices to form a silken ball, and when they start dropping off, they spin threads which thicken as more and more mites descend. Hussey and Parr (1963) counted 1350 protonymphs, 156 last-instar nymphs and 14 females of *Tetranychus* in 1 silken rope. They further reported that less roping took place under saturation conditions as compared to somewhat lower (70%) relative humidities.

Gerson and Aronowitz (1981a) and Penman and Chapman (1983) observed pesticide-induced spindown of *T. cinnabarinus* and *T. urticae*, respectively. This may well enhance mite dispersal (see below).

COLONIZATION AND HOST PLANT EFFECTS

As a dispersing *Tetranychus* female reaches a suitable plant and begins to feed, it also commences to web. The eggs, as noted, are deposited in or under the webs. The emergent larvae and nymphs likewise spin. Consequently a canopy of silk is formed on the substrate. This canopy, increasing as the mite family increases, serves as a nest, i.e. as a place to live in, for the colony. Once established, mites tend to remain within their web (Foott, 1962), which defines the limits of their colony. Many mite species spin throughout their lives (Hazan et al., 1974; Y. Saitô, unpublished work, 1976), and even diapausing females may hide in webbed shelters. The amounts of silk spun depend on various factors. These include air humidity and temperature (Hazan et al., 1974; Saitô, 1977b), substrate smoothness (more web being produced on rougher surfaces), and species of host plant. Gerson and Aronowitz (1981b) reported that 7 host plant species elicited 7 different spinning patterns from *T. cinnabarinus* under similar conditions. Although most silk was produced by mites feeding on the best host, and least on the worst, the amounts of webbing spun on the intermediate plants appeared to be affected by other unknown factors. Finally, spinning is not necessarily dependent on feeding, as starving females were observed to produce much silk (Hazan et al., 1974).

THE EFFECT OF LIGHT

More web was spun by *T. cinnabarinus* under diurnal light conditions as compared to continuous darkness (Hazan et al., 1974). Gerson and Aronowitz (unpublished work, 1979) compared diurnal webbing patterns of that mite to those produced by it under continuous light conditions. Mites placed in the dark immediately reduced their spinning (Fig. 1.4.1.2), and when returned to the light, the increase in webbing was rather low, but mites kept continuously in the light spun even amounts of silk throughout. The high correlation usually obtained between amounts of webbing (estimated as faecal pellets on webbing) and fecundity (Hazan et al., 1974; Fig. 1.4.1.3, N) during the normal diurnal cycle did not hold for mites kept in continuous light (Fig. 1.4.1.3, L). It is assumed that continuous light, by inducing continuous spinning, caused an imbalance between webs and eggs.

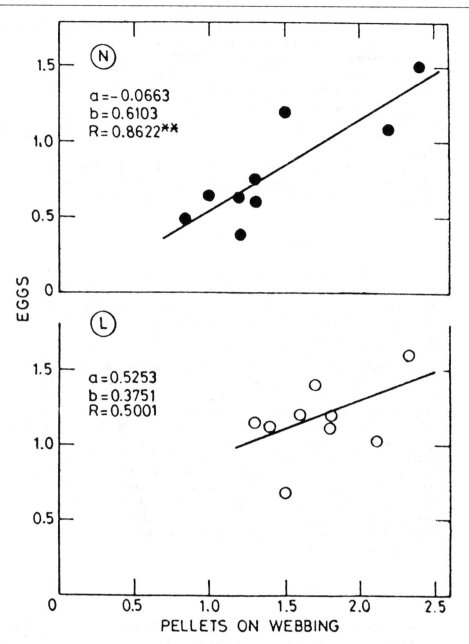

Fig. 1.4.1.3. Number of eggs deposited by females of *Tetranychus cinnabarinus* (Boisduval) as related to webbing (estimated as in Fig. 1.4.1.2), under conditions of continuous light (L) and a 8:16 dark:light regime (N). Mites held at 24°C, 38% relative humidity throughout.

THE WEBBING AS PROTECTANT

Climate

Webs protect mites and entire colonies from being blown off by winds and wetted by rains (Davis, 1952; Linke, 1953). The survival of webbed colonies on the plants under adverse climatic conditions contributes to their

Chapter 1.4.1. references, p. 230

subsequent rapid population increase (Davis, 1952). Cast moults and dirt particles, accumulating in and on webs, may increase the protectiveness of the silk cover (Reeves, 1963).

Interspecific relationships

Because strongly webbing species, like *Tetranychus* spp., colonize plant leaves, they may displace other plant-feeding mites or hinder their settlement there. Foott (1962) documented the displacement of *Panonychus ulmi* (Koch), a weak spinner, by *T. urticae*, and Hoyt (1969) noted that silk produced by *T. mcdanieli* (McGregor) trapped *P. ulmi*, as well as reducing its feeding and oviposition sites. *Bryobia*, a non-spinning spider mite, was displaced by *Tetranychus* (Georgala, 1955), as the former could not traverse leaf surfaces webbed by the latter.

Natural enemies

Webbing protects *Tetranychus* from various general feeders (Gerson, Chapter 2.2.2). Non-specific phytoseiid predators become entangled in the webs and tend to avoid them (McMurtry et al., 1970). On the other hand, specific predators appear to be attracted to and unhindered by the silk. Ladybirds of the genus *Stethorus* easily penetrate webs of *Tetranychus* and *Oligonychus* and feed there (Putman, 1955; Tanigoshi and McMurtry, 1977, respectively). Several Phytoseiidae are attracted to and thrive under spider mite webbing. Takafuji and Chant (1976) explored the reactions of 2 of these predators to webs of *T. pacificus* McGregor. One, the less specialized *Iphiseius degenerans* Berlese, consistently avoided leaf surfaces covered by silken threads, never placing any of its eggs onto such webs. The more specialized *Phytoseiulus persimilis* Athias-Henriot, on the other hand, invariably distributed itself on silk-covered leaves, where it placed its eggs. Further evidence of the importance of silk was presented by Schmidt (1976), who found that *Tetranychus* webs attracted *P. persimilis* more than prey eggs or exuviae. Another phytoseiid, *Metaseiulus occidentalis* (Nesbitt), changes its behaviour upon encountering silk and other 'cues' deposited by *Tetranychus* (Hoy and Smilanick, 1981). The predator slows its rate of walking, increases its rate of turning and enters a specialized 'search mode' until it has located the prey.

The relationship between webbing and successful predatory activity of four phytoseiids was analysed by Sabelis (1981). He found that *M. occidentalis* and *Amblyseius bibens* Blommers had about the same rate of success on webbed as on unwebbed leaves, that *A. potentillae* Garman was better on unwebbed leaves and that *P. persimilis* did best on webbed ones (Sabelis, 1981, Fig. 33). The latter species was the most efficient of all 4 on webbed leaves.

These sets of observations suggest that some predators have 'turned the tables' on webbing spider mites, using the latter's defence mechanism to locate them. In other words, the predators have co-evolved with the webbing. As such adaptation would be restricted to specialized natural enemies, the ability of predators of unknown feeding habits to survive within the webbing could be used as an indication of their specialization to that prey.

WEBBING AND ACARICIDES

Webs may act as a protective canopy against pesticide particles, holding spray droplets and dust grains away from the mites. Davis (1952) and Linke

(1953) urgently advocated the destruction of webs during chemical control procedures.

Some pesticides induced *T. urticae* to let itself down from sprayed leaves by a thread, behaviour termed 'spindown' by Gemrich et al. (1976). As mites often disperse by forming similar threads (see above), the possibility of pesticides bringing about mite dispersal became obvious.

Length of spindown threads produced by *T. cinnabarinus* when exposed to several experimental acaricides, Cyhexatin (Plictran) and Dienochlor (Pentac), was about the same (6—7 mm), but the percentage of mites spinning down was quite different (Gerson and Aronowitz, 1981a). Cyhexatin induced about half of the mites to spin down, whereas Dienochlor induced less than 20%. However, as thread length remained the same, an 'all-or-nothing' reaction was postulated for pesticide-invoked spindown.

Penman and Chapman (1983) identified the spindown of *T. urticae* with avoidance of and repellency by the synthetic pyrethroid Fenvalerate. Chemical residues induced spindown of many mites from treated leaves. Experiments with entire plants confirmed that spindown was the most significant component of Fenvalerate-induced dispersal. Mites spinning down from treated plants in the field would be in a better position to be disseminated by winds. This is probably one of the mechanisms contributing to pyrethroid-induced spider mite outbreaks.

DISCUSSION

This brief review of spider mite webbing serves to point out many areas of basic as well as applied research in which more data are needed.

Gutierrez et al. (1970) noted the connection between the evolutionary status of various Tetranychinae and their webbing. Saitô (Chapter 1.4.4) postulated that the lifestyle of spinning species has been affected and is affecting their webbing. More basic research is thus required on spider mite—web co-evolution. Saitô and Takahashi (1982) studied 2 sympatric populations of *Schizotetranychus celarius* (Banks), noting that they differed in the lengths of 2 pairs of setae and in web nest size. Observations on the females suggested that these setae were sensors for recognizing the nest roof, the population with the longer setae having the nest with the higher roof. There was also clear post-mating reproductive isolation between the 2 forms, leading Saitô and Takahashi to conclude that the 2 populations were sibling species. More data on the involvement of webbing in spider mite speciation would be of interest.

Turnbull (1973) wrote that the webs of sedentary spiders could be seen as extensions of their sensory system. The webs of spider mites appear to serve many of the functions that spiders' webs have. Research on the possible sensory roles of spider mite webbing is thus indicated.

The webs of *T. cinnabarinus* are the only ones which have so far been chemically analysed (Hazan et al., 1975b) and even these, only in regard to their amino acids. Further data on the proteinaceous and non-proteinaceous components of spider mite webs could lead to a better understanding of systematic relationships and to chemicals capable of dissolving the webs with no phytotoxic side effects.

The effect of environmental conditions on web production was studied in only a very few species (Hazan et al., 1974; Saitô, 1977b and unpublished data, 1976). Much more information is clearly needed here.

The interactions of predators with the webbing are of obvious economic importance. One immediate question is why certain phytoseiids become entangled in webs, whereas others traverse them with ease. Finally, the development of spray additives which would dissolve the webbing could lead to reductions in pesticide dosages applied while resulting in as good, or better, chemical control of spider mites. Furthermore, such additives could enhance the effect of non-specialized predators.

When the very first spider mite was described by Linnaeus, he called it 'telarius', meaning the weaver or spinner. Vernacular names in many languages attest to a recognition of the spider-like spinning of 'spider mites'; thus, 'tetranyques tisserands' in French, 'arañuelas rojas' in Spanish or 'spinmilben' in German (Merino-Rodriguez, 1966). Nevertheless, spider mite webbing has long been a neglected topic (with the exception of André, 1932). Only recently has it emerged as possibly the outstanding attribute of those Tetranychidae which spend their entire lives within their webs.

ACKNOWLEDGEMENT

I wish to thank Dr Y. Saitô, of the Institute of Applied Zoology, Faculty of Agriculture, Hokkaido University, for placing unpublished data as well as English translations of his papers at my disposal.

REFERENCES

Alberti, G. and Ehrnsberger, R., 1977. Rasterelektronenmikroskopische untersuchungen zum Spinnvermögen der Bdelliden und Cunaxiden (Acari, Prostigmata). Acarologia, 19: 55—61.

André, M., 1932. La sécrétion de la soie chez les Acariens. Societé Entomologique de France, Livre du Centenaire, pp. 457—472.

Avidov, Z., Blumberg, D. and Gerson, U., 1968. *Cheletogenes ornatus* (Acarina: Cheyletidae), a predator of the chaff scale on citrus in Israel. Isr. J. Entomol., 3: 77—94.

Bolland, H.R., 1983. A description of *Neophyllobius aesculi* n. sp. and its developmental stages (Acari: Camerobiidae). Entomol. Ber. (Amsterdam), 43: 42—47.

Cone, W.W., Predki, S. and Klostermeyer, E.C., 1971. Pheromone studies of the two-spotted spider mite. 2. Behavioural response of males to quiescent deutonymphs. J. Econ. Entomol., 64: 379—382.

Davis, D.W., 1952. Influence of population density on *Tetranychus multisetis*. J. Econ. Entomol., 45: 652—654.

Fleschner, C.A., Badgley, M.E., Ricker, D.W. and Hall, J.C., 1956. Air drift of spider mites. J. Econ. Entomol., 49: 624—627.

Foott, W.H., 1962. Competition between two species of mites. I. Experimental results. Can. Entomol., 94: 365—375.

Gemrich, E.G., II, Lamar Lee, B., Tripp, T.L. and Van de Streek, E., 1976. Relationship between formamidine structure and insecticidal, miticidal and ovicidal activity. J. Econ. Entomol., 69: 301—306.

Georgala, M.B., 1955. The biology of orchard mites in the Western Cape Province. Union S. Afr., Dept. Agric. Tech. Serv., Sci. Bull., 360.

Gerson, U., 1979. Silk production in *Tetranychus* (Acari: Tetranychidae). In: J.G. Rodriguez (Editor), Recent Advances in Acarology, Vol. 1. Academic Press, New York, NY, pp. 177—188.

Gerson, U. and Aronowitz, A., 1981a. Spider mite webbing. Part IV: The effect of acaricides on spinning by the carmine spider mite *Tetranychus cinnabarinus* (Boisduval). Pestic. Sci., 12: 211—214.

Gerson, U. and Aronowitz, A., 1981b. Spider mite webbing. V. The effect of various host plants. Acarologia, 22: 277—281.

Gutierrez, J., Helle, W. and Bolland, H.R., 1970. Étude cytogénétique et réflexions phylogénétiques sur la famille des Tetranychidae Donnadieu. Acarologia, 12: 732—751.

Hazan, A., Gerson, U. and Tahori, A.S., 1974. Spider mite webbing. I. The production of webbing under various environmental conditions. Acarologia, 16: 68—84.

Hazan, A., Gerson, U. and Tahori, A.S., 1975a. Spider mite webbing. II. The effect of webbing on egg hatchability. Acarologia, 17: 270—273.

Hazan, A., Gertler, A., Tahori, A.S. and Gerson, U., 1975b. Spider mite webbing. III. Solubilization and amino acid composition of the silk protein. Comp. Biochem. Physiol. B, 51: 457—462.

Hoy, M.A. and Smilanick, J.M., 1981. Non-random prey location by the phytoseiid predator *Metaseiulus occidentalis*: Differential responses to several spider mite species. Entomol. Exp. Appl., 29: 241—253.

Hoyt, S.C., 1969. Population studies of five mite species on apple in Washington. In: G.O. Evans (Editor) Proc. 2nd Int. Congr. Acarol., Akademiai Kiado, Budapest, pp. 117—133.

Hussey, N.W. and Parr, W.J., 1963. Dispersal of the glasshouse red spider mite *Tetranychus urticae* Koch (Acarina, Tetranychidae). Entomol. Exp. Appl., 6: 207—214.

Linke, W., 1953. Investigation of the biology and epidemiology of the common spider mite, *Tetranychus althaeae* v. Hanst. with particular consideration of the hop as the host. Hoefchen-Briefe (Engl. Ed.), 6: 181—232.

McMurtry, J.A., Huffaker, C.B. and van de Vrie, M., 1970. Ecology of tetranychid mites and their natural enemies: A review. I. Tetranychid enemies: Their biological characters and the impact of spray practices. Hilgardia, 40: 331—390.

Merino-Rodriguez, M., 1966. Lexicon of Plant Pests and Diseases. Elsevier, Amsterdam, 351 pp.

Moss, W.W., 1960. Description and mating behaviour of *Allothrombium lerouxi*, new species (Acarina: Trombidiidae), a predator of small arthropods in Quebec apple orchards. Can. Entomol., 92: 898—905.

Penman, D.R. and Chapman, R.B., 1983. Fenvalerate-induced distributional imbalances of two-spotted spider mite on bean plants. Entomol. Exp. Appl., 33: 71—78.

Penman, D.R. and Cone, W.W., 1972. Behaviour of two-spotted spider mites in response to quiescent female deutonymph and to web. Ann. Entomol. Soc. Am., 65: 1289—1293.

Penman, D.R. and Cone, W.W., 1974. Role of web, tactile stimuli and female sex pheromone in attraction of male two-spotted spider mites to quiescent female deutonymphs. Ann. Entomol. Soc. Am., 67: 179—182.

Potter, D.A., Wrensch, D.L. and Johnston, D.E., 1976. Guarding, aggressive behaviour and mating success in male two-spotted spider mites. Ann. Entomol. Soc. Am., 69: 707—711.

Putman, W.L., 1955. Bionomics of *Stethorus punctillum* Weise (Coleoptera: Coccinellidae) in Ontario. Can. Entomol., 87: 9—33.

Reeves, R.M., 1963. Tetranychidae infesting woody plants in New York State, and a life history study of the elm spider mite, *Eotetranychus matthyssei* n. sp. Mem. Cornell Univ. Agric. Exp. Stn., Ithaca, NY, 380.

Sabelis, M.W., 1981. Biological control of two-spotted spider mites using phytoseiid predators. Part 1. Modelling the predator—prey interaction at the individual level. Pudoc, Wageningen.

Saitô, Y., 1977a. Study on the spinning behaviour of the spider mite (Acarina: Tetranychidae). I. Method for the quantitative evaluation of the mite webbing, and the relationship between webbing and walking. Jpn. J. Appl. Entomol. Zool., 21: 27—34 (in Japanese, with English summary).

Saitô, Y., 1977b. Study on the spinning behaviour of the spider mite (Acarina: Tetranychidae). II. The process of secretion and the webbing on the various stages of *Tetranychus urticae* Koch under different environmental conditions. Jpn. J. Appl. Entomol. Zool., 21: 150—157 (in Japanese, with English summary).

Saitô, Y., 1979. Study on spinning behaviour of spider mites. III. Responses of mites to webbing residues and their preferences for particular physical conditions of leaf surfaces (Acarina: Tetranychidae). Jpn. J. Appl. Entomol. Zool., 23: 82—91 (in Japanese, with English summary).

Saitô, Y., 1983. The concept of 'life types' in Tetranychinae. An attempt to classify the spinning behaviour of Tetranychinae. Acarologia, 24: 377—391.

Saitô, Y. and Takahashi, K., 1982. Study on variation of *Schizotetranychus celarius* (Banks). II. Comparison of mode of life between two sympatric forms (Acarina: Tetranychidae). Jpn. J. Ecol., 32: 69—78 (in Japanese, with English summary).

Schmidt, G., 1976. Der Einfluss der von den Beutetieren hinterlassenen Spuren auf Suchverhalten und Sucherfolg von *Phytoseiulus persimilis* A.H. (Acarina: Phytoseiidae). Z. Angew. Entomol., 82: 216—218.

Singer, G., 1966. The bionomics of *Tenuipalpoides dorychaeta* Pritchard and Baker (1955) (Acarina, Trombidiformes, Tetranychidae). Univ. Kans. Sci. Bull., 46: 625—645.

Takafuji, A. and Chant, D.A., 1976. Comparative studies of two species of predacious phytoseiid mites (Acarina: Phytoseiidae), with special reference to their responses to the density of their prey. Res. Popul. Ecol., 17: 255—310.

Tanigoshi, L.K. and McMurtry, J.A., 1977. The dynamics of predation of *Stethorus picipes* (Coleoptera: Coccinellidae) and *Typhlodromus floridanus* on the prey *Oligonychus punicae* (Acarina: Phytoseiidae, Tetranychidae). Part I. Comparative life history and life table studies. Hilgardia, 45: 237—261.

Tevis, L., Jr. and Newell, I.M., 1962. Studies on the biology and seasonal cycle of the giant red velvet mite, *Dinothrombium pandorae* (Acari, Trombidiidae). Ecology, 43: 497—505.

Turnbull, A.L., 1973. Ecology of the true spiders (Araneomorphae). Annu. Rev. Entomol., 18: 305—348.

Wallace, M.M.H. and Mahon, J.A., 1972. The taxonomy and biology of Australian Bdellidae (Acari). I. Subfamilies Bdellinae, Spinibdellinae and Cytinae. Acarologia, 14: 544—580.

Wene, G.P., 1956. *Tetranychus marianae* McG., a new pest of tomatoes. J. Econ. Entomol., 49: 712.

1.4.2 Dispersal

G.G. KENNEDY and D.R. SMITLEY

INTRODUCTION

In natural ecosystems, populations of tetranychid mites generally consist of widely scattered individuals or small groups of individuals which are in dynamic equilibrium with their natural enemies and rarely cause severe damage to their host plant. In many modern agroecosystems, populations of certain tetranychid species are characterized by very high rates of increase and frequently reach extremely high densities. If left uncontrolled they are capable of completely destroying large populations of their host plants. There are a number of factors which contribute to population differences between natural and agricultural ecosystems, including the suppression of natural enemies of tetranychid mites by modern pesticides, the stimulative effects of certain pesticides on mite population growth, and the use of cultural procedures which improve plant growth and tetranychid nutrition (Huffaker et al., 1969; McMurtry et al., 1970; van de Vrie et al., 1972; Dittrich et al., 1974). In addition, in modern agroecosystems the patterns of spatial and temporal availability as well as the available biomass of host plants (crops) are vastly different than in natural systems. The mechanisms evolved by tetranychid mites to cope with the high degree of spatial and temporal variation in host suitability and availability characteristic of natural ecosystems are highly efficient for exploiting the spatially and temporally predictable super-abundance of suitable host plants characteristic of modern agroecosystems.

Tetranychid mites have well-developed dispersal mechanisms which enable their populations to spread throughout and fully exploit individual host plants as well as to spread over large areas and colonize widely separated plants. In addition to its role in efficient habitat utilization, dispersal is an important mechanism of escape from natural enemies and has been clearly shown to contribute significantly to the persistence and stability of predator/prey systems (Huffaker, 1958; Huffaker et al., 1963; Pimentel et al., 1963; Takafuji, 1977).

Dispersal in the context of this chapter shall be considered as movement of individuals away from the mite colony in which they developed. Thus, under the broad heading of dispersal we will consider both intra-plant movements leading to colonization of previously uninfested plant parts and inter-plant movements leading to colonization of previously uninfested plants.

ADAPTATIONS FOR DISPERSAL

When dispersal is under genetic control, in general, selection for dispersal will occur if reproductive success is higher among dispersing individuals than

---

Chapter 1.4.2. references, p. 240

among those that remain with the colony (Baker, 1978). However, in highly unstable habitats, random or non-selective dispersal may confer adaptive advantage even if there is a high mortality associated with dispersal (Kuno, 1981).

Tetranychid mite populations are characterized by cycles of initial colonization by a mated female followed by rapid population growth and localized (i.e. individual leaf, leaf cluster or plant) host exploitation with subsequent dispersal or migration to a new resource. Thus, the ultimate success of the colony requires the production of a large number of potential colonizers. Tetranychids possess a number of adaptations which appear to increase the production of potential colonizers.

Intra-plant dispersal, at least in some species, results from a tendency for a portion of the pre-reproductive females to emigrate from the leaf on which they developed, regardless of population density on that leaf (Hussey and Parr, 1963; Coates, 1974). At least for *Tetranychus urticae* Koch, ovipositing females show less tendency to emigrate from a leaf than do pre-reproductive females. Presumably, dispersal by pre-reproductive females has the effects of prolonging the supply of food available to the established colony, more fully exploiting the available colonization sites on the host and 'spreading the risk' of colony extinction resulting from locally adverse biotic or abiotic factors.

Some intra-plant and a significant portion of inter-plant dispersal by tetranychid mites involves dispersal as aerial plankton. Although these mites possess a number of behavioral adaptations which increase the probability of being carried aloft on air currents, once airborne the dispersal appears to be entirely passive. The probability of one or more dispersers from a colony landing on a suitable host plant and successfully founding a new colony is related to both the abundance and distribution of suitable hosts within the dispersal range of the mites, and the number of dispersers. Mitchell (1970) and Wrensch and Young (1978, 1983) have identified a number of adaptations possessed by tetranychid mites which appear to compensate for the low probability of reaching a new host. These include: (1) a wide host range, which increases the probability of a disperser arriving at a suitable host; (2) mating prior to dispersal; (3) an increase in the number of female eggs produced and an increase in the mortality rate of males during development as the host leaf deteriorates; (4) males are smaller and less abundant than females on a deteriorating host so that most of the standing crop biomass is females.

It is not known whether dispersers possess any physiological differences from non-dispersers which might enable them to withstand prolonged periods away from a suitable host and greater extremes of abiotic conditions. Information of this type would be extremely valuable to our understanding of the biology and ecology of the Tetranychidae.

DISPERSAL BEHAVIORS

The initiation of dispersal by tetranychid mites in most instances is not a passive phenomenon, but involves identifiable behaviors in response to specific stimuli. While the consequences of those behaviors and the population distributions that result may differ between species and produce different patterns of host utilization (e.g. Wanibuchi and Saito, 1983), many of the behaviors appear to be common among the Tetranychidae.

Crawling

Crawling is a common means of dispersal between the various portions of a host plant. In dense aggregations of host plants, crawling is probably also important in inter-plant movement within a host patch or aggregation. Fields of single crops characteristic of modern agriculture probably act as very large host plant aggregations. To the extent that the canopies of individual plants intertwine, they may serve as a single very large plant to a mite population. A large proportion of the spread of mites from the focus of initial coloniz-ation may result from mites crawling from plant to plant through intertwined foliage and over the ground. *T. urticae* populations invade fields of maize from weeds growing around the field margins and spread throughout the field largely by crawling between plants (Brandenburg and Kennedy, 1982a; Margolies and Kennedy, 1985). Some of this crawling between plants occurs over the soil surface since banding the base of the maize plants with a sticky material greatly reduces the number of mites colonizing the plants. No specific behavior leading to mites crawling down the plant to the soil surface for subsequent dispersal has, as yet, been identified. Those mites crawling over the soil surface may have dropped from the foliage actively or passively. Once on the soil surface, the rate at which the mites can crawl varies with soil type, ranging from $5\,\text{cm}\,\text{h}^{-1}$ to about $6\,\text{m}\,\text{h}^{-1}$ (McGregor, 1913; Parker, 1913). The direction of crawling appears to involve movement of dispersing mites towards illumination. In contrast, diapause forms move away from light (Hussey and Parr, 1963).

Under conditions of low density, populations of *T. urticae* and *Panony-chus citri* McGregor summer forms spread throughout unoccupied plant growth largely by crawling (McEnroe and Dronka, 1971; Wanibuchi and Saito, 1983). Dispersal by crawling to other plants or plant parts has also been shown to occur in response to both population density and the presence of predators (Bernstein, 1984).

Under conditions of high density and plant damage, individuals undergo a change in behavior pattern which leads to dispersal from the plant. In the case of *T. urticae*, mites in the dispersal phase manifest a positive photo-tactic response which is not shown by mites in the sedentary or non-dispersal phase. The initiation of the dispersal phase appears to be a response to food shortage and dessication interacting with some unknown factors (Suski and Naegele, 1966). The response is intensified under conditions of low relative humidity in the plant microclimate such as that which results from extensive mite feeding on the foliage. This phototactic response results in dispersal phase mites moving up the plant and concentrating around the periphery of their host (Suski and Naegele, 1966) where, presumably, they are more exposed to winds which might lead to their aerial dispersal. McEnroe and Dronka (1971) have suggested that the shift from sedentary to dispersal phases is a response to dessication of the mite and ambient relative humidity, which they hypothesized are perceived by a stretch receptor and a humidity receptor in the mite.

In their study of inter-plant dispersal of *T. urticae* in glasshouses, Hussey and Parr (1963) reported that the mites moved up the plant and began to abandon the host when all the apical foliage became damaged. They observed masses of dispersing mites forming at the apices of the foliage and dropping from the plant on webs. This 'roping' or spinning down behavior occurred only in still air. The mites leaving the plant in this way presumably crawled in search of other plants once they reached the ground. Hussey and Parr concluded that, since most *T. urticae* dispersed from their host by 'roping' in still air, dispersal on air currents could be ignored in the glasshouse.

Chapter 1.4.2. references, p. 240

Aerial dispersal

The aerial transport of mites has been well documented, with mites being captured at altitudes as high as 10 000 feet (Coad, 1931, cited in Fleschner et al., 1956). It is common among the Tetranychidae and appears to involve at least 2 different behavioral mechanisms.

A number of species, including *P. citri*, *Oligonychus punicae* (Hirst), *O. ununguis* (Jacobi), and *Eotetranychus sexmaculatus* (Riley), have been observed spinning down from the foliage on silk threads during periods of gentle breeze which breaks the silk threads and carries the mites aloft (Fleschner et al., 1956; Wanibuchi and Saito, 1983). This behavior, although referred to as 'ballooning' by some workers, is different than the 'ballooning' response of true spiders (Arachnida). Ballooning in spiders involves the spider raising its abdomen and exuding a web which is caught and drawn out by an air current until the tension becomes so great that it carries the spider aloft (Main, 1976). In contrast, the tetranychids, which employ silk threads in their aerial dispersal, affix the thread to a substrate, hang from the thread in the air, and are carried off by the wind. To avoid confusion, the term ballooning should not be used to describe this type of aerial dispersal behavior in spider mites.

Although Hussey and Parr (1963) described 'roping' or spinning down behavior in *T. urticae*, which frequently involves masses or aggregations of mites and occurs only in still air, they did not associate that behavior with aerial dispersal. While such 'roping' aggregations of *T. urticae* may occassionally be carried aloft by a sudden gust of wind, there is no evidence that *T. urticae* aerially disperses primarily by first spinning down on silk threads (Fleschner et al., 1956; Boyle, 1957). Indeed, *T. urticae* exhibits much more complex behavior leading to aerial dispersal (Smitley and Kennedy, 1985). It is similar in many ways to that exhibited by the predaceous mite, *Amblyseius fallacis* (Garman) and certain scale insects (Johnson and Croft, 1976; Washburn and Washburn, 1984).

In *T. urticae* this behavior involves raising the forelegs upright to assume a dispersal posture (Fig. 1.4.2.1). This posture clearly facilitates aerial dispersal as 50% of the adult female mites adopting this posture and facing into the wind were carried aloft by a wind of 1.5 m s^{-1}, whereas none of the mites in the normal standing posture were carried aloft.

Adult females are the stage most commonly observed to be aerially dispersing; although the nymphal stages are also observed, albeit less frequently. Aerial dispersal of males is rarely observed (Brandenburg and Kennedy, 1982a). Dispersal posturing is manifest by all active stages except adult males. The response is most frequently observed among adult females and there is a positive relationship between wind velocity and the proportion of mites assuming the dispersal posture. This response to wind velocity by adult females levels off at approximately 1.5 m s^{-1}. Nymphal stages of *T. urticae* also manifest this dispersal posturing, but do so less frequently than adult females. Larvae only rarely display this behavior. The reduced response of the immature stages may be due, in part, to their smaller size and the boundary layer wind velocity gradient which exists around a leaf surface. Because of the boundary layer, the smaller stages would experience much less air movement than adult females and in the case of larvae the air speed at the leaf surface may rarely be sufficient to stimulate dispersal posturing. In addition to these boundary layer effects, larvae and protonymphs may be less responsive to a given wind velocity.

While it is not certain that the mites exhibiting this dispersal posturing are equivalent to the 'dispersal phase' mites of Suski and Naegele (1966) and

Fig. 1.4.2.1. *T. urticae* in dispersal posture with forelegs raised. This posture is only manifest in the presence of wind and light. The directional orientation is independent of wind direction, but involves a negative phototactic response such that the mite orients to face away from the light source. In the field this results in the mite facing downward on the plant stems and leaf edges with forelegs uplifted.

Green + phase mites of McEnroe and Dronka (1971), the same stimuli appear to be involved in conditioning the mites to manifest this response. Dessication, host plant condition and population density are involved in conditioning the response, with dessication playing perhaps the greatest role. Smitley and Kennedy (1985) found that when adult females were denied access to foliage and held for 4 h at 95% relative humidity, only 1.7% of them assumed the dispersal posture when exposed to wind and light, whereas 14% of those similarly held at 25% relative humidity responded. In separate experiments involving an array of different population densities on bean plants, they found, using regression analysis, that egg density and leaf chlorophyll content (an indicator of feeding injury) accounted for 72% of the variation in the proportion of adult females posturing ($R^2 = 0.72$). It is not known if the relationship to density is attributable to greater deterioration of foliage at higher mite densities or if it involves more direct effects of high density, such as increased webbing and increased mite to mite contact.

Dispersal posturing by *T. urticae* only occurs in the presence of both wind and light and typically occurs after the mites have concentrated at the upper portions of the plant and leaf apices as a result of the positive phototactic response described by Suski and Naegele (1966) and McEnroe and Dronka (1971). The directional orientation involves a negative response to light which results in the mites facing away from the light source, standing still and raising their forelegs. There is no apparent directional response to wind, but wind is necessary to condition the negative response. Under field conditions this results in a vertical orientation such that the mites are facing downwards toward the ground. Presumably, this increases the probability of being carried aloft on updrafts. Dispersal on updrafts would generally lead to greater horizontal displacement of dispersers than would dispersal

Chapter 1.4.2. references, p. 240

on down drafts or horizontal air currents where the likelihood of landing on adjacent plants similar in condition to the original host would seemingly be great.

In laboratory studies involving individual maize plants, Silberman (1983) captured aerially dispersing *T. urticae* throughout the period of population growth on the plants, but observed that greatest aerial dispersal occurred after the mite populations had spread throughout all available foliage. In the field, Margolies and Kennedy (1985) rarely captured *T. urticae* aerially dispersing from infested maize fields until the populations reached the upper leaves of the maize plant. This is readily explained; until the mite population expands to the upper leaves, it is well protected from wind by surrounding plants and a very high proportion of mites which actually did aerially disperse would be intercepted by other leaves and plants making up the crop canopy.

In the field, dense aggregations or masses of mites of the type described by Hussey and Parr (1963) are generally not observed until after the peak in aerial dispersal. Although mites in these aggregations are readily carried aloft when exposed to a breeze (Brandenburg and Kennedy, 1982a; Smitley and Kennedy, unpublished) the formation of these aggregations is not a prerequisite to aerial dispersal. Indeed, they appear to play a minor role in aerial dispersal and their actual function is not known.

Although aerial dispersal has the potential of carrying spider mites great distances (see Johnson, 1969), and long-distance dispersal certainly occurs, most aerially dispersing mites probably fall from the air stream fairly soon after they are carried aloft.

Taylor (1960) demonstrated that the density of airborne insects drops off rapidly with height above a crop and follows a linear log/log relationship. The limited data available for aerially dispersing *T. urticae* suggest that this linear log/log relationship between aerial density and height is also true for spider mites (Boykin and Campbell, 1984). The distribution of *T. urticae* in peanut fields, which results from aerial dispersal from adjacent maize fields, lends support to the idea that most aerially dispersing individuals are not carried far. Although mite colonies are distributed rapidly throughout entire peanut fields, the greatest populations and damage to the crop are generally found closest to the source of aerially dispersing mites (Brandenburg and Kennedy, 1982a; Margolies and Kennedy, unpublished).

Phoretic dispersal

Although phoretic dispersal is known to occur among some groups of mites (Binns, 1982), there is no evidence that phoresy is important in the dispersal of tetranychid mites. The lack of evidence for phoresy among the Tetranychidae does not necessarily mean that phoretic dispersal does not occur or is not important. The entire subject has received so little attention that, at present, no conclusions are possible.

It seems likely that some degree of phoretic dispersal does occur, although it may involve more serendipity than evolved behavior patterns. One would be surprised if birds or large insects alighting on foliage heavily infested with spider mites and covered with webbing did not carry off some webbing and mites which might later be left behind when the animals visited other plants. Similarly, animals brushing against heavily infested plants are very likely to pick up mites which may subsequently be transferred to another plant. The dense aggregations of *T. urticae* which occur around the periphery of heavily damaged plants would seemingly provide an ideal mechanism for phoretic dispersal, although there is no evidence that this is their function.

Experience has shown that spider mites can be spread both between and within fields on farm machinery. We also have seen instances where careless pest management scouts have gone from maize fields heavily infested with *T. urticae* into previously uninfested cotton and left a trial of mite-infested plants through the cotton field.

Upon more detailed investigation, we are likely to find that phoretic dispersal commonly takes place. Its importance in the life histories of the various tetranychids and the specific behavioral responses, if any, that result in phoretic dispersal are worthwhile areas of inquiry.

AGRICULTURAL IMPLICATIONS OF DISPERSAL

As an integral component of the life system of tetranychid mites, dispersal cannot be ignored. The role of dispersal in the dynamics and pest status of *T. urticae* in a maize/peanut agroecosystem has been intensively studied and illustrates the importance of dispersal in the population dynamics and pest status of this agriculturally important species (Brandenburg and Kennedy, 1981, 1982a, b; Margolies and Kennedy, 1985).

In Chowan County, North Carolina, *T. urticae* from populations overwintering on non-cultivated vegetation around field borders disperse in the spring, primarily by crawling, into adjacent cultivated fields, regardless of the crop growing in the field. For a variety of reasons, the populations become established only on maize where they spread through the field and increase on individual plants until the infestations reach the top of the crop canopy. It is at this time that large numbers of mites aerially disperse from maize and rather suddenly appear on a number of other plants, including the peanut crop where they frequently become a serious problem and must be controlled. Later in the season, the mites disperse from peanut to overwintering sites in non-cultivated vegetation around peanut field margins. Since fields planted in peanut one year are commonly planted in corn the next, mites dispersing from the overwintering sites in the spring have a high likelihood of encountering maize.

The severity of mite problems experienced in peanut are related to the intensity of the mite infestation in nearby corn. In those years in which a large mite population fails to develop in maize or when large mite populations develop in maize but are decimated by an epizootic of the pathogenic fungus *Neozygites floridana* Weiser and Muma (Brandenburg and Kennedy, 1982b; Smitley, unpublished) before large-scale aerial dispersal occurs, the frequency of occurrence and severity of mite problems in peanuts is reduced. It is clear that the sizes of populations which develop on any particular host are related not only to the events which take place on that particular host, but also to events which previously took place on other spatially distant hosts earlier in the season.

An understanding of the patterns of large-scale movement of mites between hosts may suggest ways of modifying the habitat or agroecosystem to reduce mite populations on specific crops. In the maize/peanut system described above, it may prove possible to manage the vegetation around field margins to reduce overwintering populations or to find alternative crop rotation schemes which are less conducive to the development of very large mite populations early in the season. Similarly, a knowledge of the patterns and importance of dispersal is necessary for estimating the risk of a particular field experiencing a mite problem.

Wide area and large-scale dispersal of tetranychid mites has important implications for the occurrence and management of acaricide resistance.

Chapter 1.4.2. references, p. 240

Several studies have shown that immigration of susceptible individuals into a deme under selective pressure for resistance can substantially reduce the development of resistance (Comins, 1977; Georghiou and Taylor, 1977a,b; Taylor and Georghiou, 1979; Tabashnik and Croft, 1982). Thus, one might expect fewer problems with acaricide resistance in those agroecosystems in which the target mite species show extensive dispersal between hosts which receive no acaricide treatments and acaricide-treated crops.

Certain insecticides and acaricides have been shown to stimulate dispersal of spider mites from treated foliage (Gerson and Aronowitz, 1981; Penman et al., 1981; Penman and Chapman, 1983). Iftner and Hall (1983) hypothesized that in addition to destroying predaceous arthropods, the pyrethroid insecticides stimulate mite population growth by causing the dispersal of dense colonies into a large number of isolated colonies each with a low density and a higher reproductive potential.

Thus, dispersal among the Tetranychidae is a significant factor in their importance as agricultural pests. A knowledge of dispersal mechanisms and the importance of dispersal in the life history and population dynamics of individual species is essential to their efficient long-term management as agricultural pests.

CONCLUDING REMARKS

Although we have learned a great deal about the dispersal of tetranychid mites, most of the detailed information relates to a single species, *T. urticae*. More information is needed on the specific dispersal responses of other tetranychid species and on the role of dispersal in their population ecology and their importance as agricultural pests. We should expect to find differences among tetranychid species and those differences should be related to the different habitat requirements and life history strategies of the various species. A comparative genetic analysis among tetranychid species of the dispersal response and associated life history characters could provide considerable insight into the ecology, evolution and pest status of tetranychid mites. An understanding of the role of dispersal in the population ecology of individual species of agricultural importance will help to make the population management of these pest species both more efficient and more effective over the long term.

REFERENCES

Baker, R.R., 1978. The Evolutionary Ecology of Animal Migration. Hodder and Stoughton, London, 1012 pp.

Bernstein, C., 1984. Prey and predator emigration responses in the acarine system *Tetranychus urticae — Phytoseiulus persimilis*. Oecologia (Berlin), 61: 134—142.

Binns, E.S., 1982. Phoresy as migration — some functional aspects of phoresy in mites. Biol. Rev., 57: 571—620.

Boykin, L.S. and Campbell, W.V., 1984. Wind dispersal of the two-spotted spider mite (Acari: Tetranychidae) in North Carolina peanut fields. Environ. Entomol., 13: 221—227.

Boyle, W.W., 1957. On the mode of dissemination of the two-spotted spider mite, *Tetranychus telarius*. Proc. Hawaii. Entomol. Soc., 16: 261—268.

Brandenburg, R.L. and Kennedy, G.G., 1981. Overwintering of the pathogen *Entomophthora floridana* Weiser and Muma and its host *Tetranychus urticae* Koch. J. Econ. Entomol., 74: 428—431.

Brandenburg, R.L. and Kennedy, G.G., 1982a. Intercrop relationships and spider mite dispersal in a corn/peanut agroecosystem. Entomol. Exp. Appl., 32: 269—276.

Brandenburg, R.L. and Kennedy, G.G., 1982b. Relationship of *Neozygites floridana*

(Entomophthorales: Entomophthoraceae) to two-spotted spider mite populations in field corn. J. Econ. Entomol., 75: 691—694.

Coates, T.J.D., 1974. The influence of some natural enemies and pesticides on various populations of *Tetranychus cinnabarinus* (Boisduval), *T. lombardinii* Baker and Pritchard and *T. ludeni* Zacher (Acari: Tetranychidae) with aspects of their biologies. Entomology Memoir No. 42, Department of Agricultural Technical Services, Republic of South Africa. 40 pp.

Comins, H.N., 1977. The development of insecticide resistance in the presence of immigration. J. Theor. Biol., 64; 177—197.

Dittrich, V., Streibert, P. and Bathe, P.A., 1974. An old case reopened; mite stimulation by insecticide residues. Environ. Entomol., 3: 534—540.

Fleschner, C.A., Badgley, M.E., Richer, D.W. and Hall, J.C., 1956. Air drift of spider mites. J. Econ. Entomol., 49: 624—627.

Georghiou, G.P. and Taylor, C.E., 1977a. Genetic and biological influences in the evolution of insecticide resistance. Environ. Entomol., 70: 319—323.

Georghiou, G.P. and Taylor, C.E., 1977b. Operational influences in the evolution of insecticide resistance. Environ. Entomol., 70: 653—658.

Huffaker, C.B., 1958. Experimental studies on predation: dispersion factors and predator—prey oscillations. Hilgardia, 27: 343—383.

Huffaker, C.B., Shea, K.P. and Herman, S.G., 1963. Experimental studies on predation. Complex dispersion and levels of food in an acarine predator—prey interaction. Hilgardia, 34: 305—329.

Huffaker, C.B., M. van de Vrie, and McMurtry, J.A., 1969. The ecology of tetranychid mites and their natural control. Annu. Rev. Entomol., 14: 125—174.

Hussey, N.W. and Parr, W.J., 1963. Dispersal of the glasshouse red spider mite *Tetranychus urticae* Koch (Acarina, Tetranychidae). Entomol. Exp. Appl., 6: 207—214.

Johnson, C.G., 1969. Migration and Dispersal of Insects by Flight. Methuen, London, 763 pp.

Johnson, D.T. and Croft, B.A., 1976. Laboratory study of the dispersal behavior of *Amblyseius fallacis* (Acarina: Phytoseiidae). Ann. Entomol. Soc. Am., 69: 1019—1023.

Kennedy, G.G. and Margolies, D.C., 1985. Considerations in the management of mobile arthropod pests in diversified agroecosystems. Bull. Entomol. Soc. Am. In press.

Kuno, E., 1981. Dispersal and the persistence of populations in unstable habitats: a theoretical note. Oecologia (Berlin), 49: 123—126.

Main, B.Y., 1976. Spiders. William Collins, Sydney, 296 pp.

Margolies, D.C. and Kennedy, G.G., 1985. Movement of the two-spotted spider mite *Tetranychus urticae* Koch (Acari: Tetranychidae), among hosts in a corn—peanut agroecosystem. Entomol. Exp. Appl. 37: 55—61.

McEnroe, W.D. and Dronka, K., 1971. Photobehavioral classes of the spider mite *Tetranychus urticae* (Acarina: Tetranychidae). Entomol. Exp. Appl., 14: 420—424.

McGregor, E.A., 1913. The red spider on cotton. U.S. Dep. Agric., Bur. Entomol. Circ. 172, 22 pp.

Mitchell, R., 1970. An analysis of dispersal in mites. Am. Nat., 104: 425—431.

Parker, W.B., 1913. The red spider mite on hops in the Sacramento Valley of California. U.S. Dep. Agric., Bur. Entomol. Bull., 117: 1—41.

Pimentel, D., Nagel, W.P. and Madden, J.L., 1963. Space—time structure of the environment and survival of parasite—host systems. Am. Nat., 97: 141—167.

Smitley, D.R. and Kennedy, G.G., 1985. A photo-oriented aerial dispersal behaviour of *Tetranychus urticae* (Acari: Tetranychidae) enhances escape from the leaf surface. Ann. Entomol. Soc. Amer., 78: (in press).

Silberman, L., 1983. Aerial dispersal of the two-spotted spider mite, *Tetranychus urticae* (Koch), from its host plant. Master of Science Thesis, North Carolina State University, Raleigh, North Carolina, 68 pp.

Suski, Z.W. and Naegele, J.A., 1966. Light response in the two-spotted spider mite. II. Behavior of the sedentary and dispersal phases. In: Recent Advances in Acarology, Cornell Press, Ithaca, New York, 480 pp.

Tabashnik, B.E. and Croft, B.A., 1982. Managing pesticide resistance in crop—arthropod complexes: interactions between biological and operational factors. Environ. Entomol., 11: 1137—1144.

Takafuji, A., 1977. The effect of successful dispersal of a phytoseiid mite *Phytoseiulus persimilis*. Anthias—Henriot (Acarina: Phytoseiidae) on the persistance of the interactive system between the predator and its prey. Res. Popul. Ecol., 18: 210—222.

Taylor, L.R., 1960. The distribution of insects at low levels in the air. J. Anim. Ecol., 29: 45—63.

Taylor, C.E. and Georghiou, G.P., 1979. Suppression of insecticide resistance by alteration of dominance and migration. J. Econ. Entomol., 72: 105—109.

Van de Vrie, M., McMurtry, J.A. and Huffaker, C.B., 1972. Ecology of tetranychid mites and their natural enemies: a review. III. Biology, ecology, and pest status and host relations of tetranychids. Hilgardia, 41: 343—432.

Wanibuchi, K. and Saito, Y., 1983. The process of population increase and patterns of resource utilization of two spider mites, *Oligonychus ununguis* (Jacobi) and *Panonychus citri* (McGregor), under experimental conditions (Acari: Tetranychidae). Res. Popul. Ecol., 25: 116—129.

Washburn, J.O. and Washburn, L., 1984. Active aerial dispersal of minute wingless arthropods: exploitation of boundary-layer velocity gradients. Science, 223: 1088—1089.

Wrensch, D.L. and Young, S.S.Y., 1978. Effects of density and host quality on rate of development, survivorship and sex ratio in the carmine spider mite. Environ. Entomol., 7: 499—501.

Wrensch, D.L. and Young, S.S.Y., 1983. Relationship between primary and tertiary sex ratio in the two-spotted spider mite (Acarina: Tetranychidae). Ann. Entomol. Soc. Am., 76: 786—789.

1.4.3 Mating and Chemical Communication

W.W. CONE

MATING

Male precopulatory behaviour (aggression, attraction, guarding, male webbing, territoriality)

In *Tetranychus*, female deutonymphs (penultimate instar) spin strands of silk into a pad on a leaf surface, settle and become quiescent during their final moult. Wandering males, perhaps guided by silk webbing (Penman and Cone, 1972) and a sex attractant (Cone et al., 1971a) discover the quiescent females. These females become increasingly attractive over time (Cone et al., 1971b; Potter et al., 1976a, b) and as ecdysis approaches, almost all are attended by males. Ewing (1914) described this behaviour and postulated that this premature courting instinct no doubt is of value to the species (obviously, to individuals within a population) in detaining the male until the female transforms, after which copulation soon takes place. Ewing (1914) confirmed the findings of Morgan (1897) that unfertilized females produce haploid (male) eggs. This result is well known for most tetranychid mites. Obviously the detention of males for mating purposes influences the number of female eggs produced and has significant consequences for subsequent population development.

Perkins (1897) stated that several males may mate with a single female, each taking his turn without aggressive interactions. He later described males in combat over a female. Potter et al. (1976b) found that aggressive interactions between competing males of *Tetranychus urticae* Koch are frequent and determine the ultimate success or failure in mating. They characterized agonistic encounters in order of increasing intensity and decreasing frequency as follows:

(1) intruder retreating without aggression;

(2) one-sided aggressive response or threat, usually by the guarding male and ending with retreat of the intruder;

(3) moderate to intense interaction involving both males and ending with retreat of one;

(4) intense fighting resulting in injury to or death of 1 combatant. They concluded that

> male aggression does not generate any specific dominance order but gives rise to a series of random confrontations. Guarding pharate females represents a form of ephemeral territoriality; the procurement and retention of such territory being dependent on fighting ability.

Chapter 1.4.3. references, p. 250

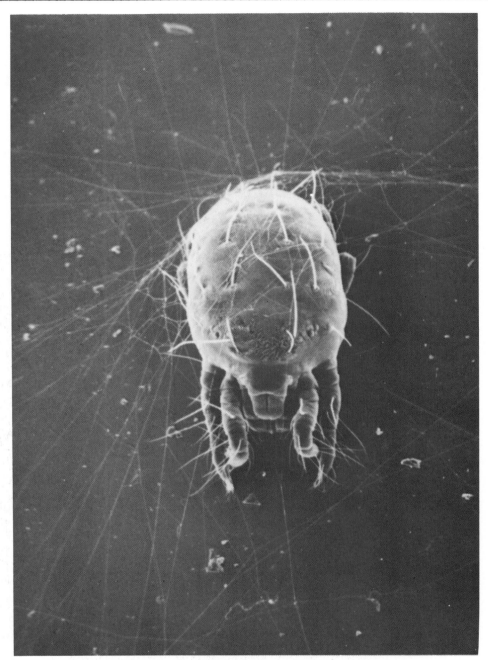

Fig. 1.4.3.1. Quiescent female deutonymph, *T. urticae*, with strands of web spun over her. Photo by D.R. Penman.

Penman and Cone (1974b) showed that males spin their own web over the quiescent deutonymph (Fig. 1.4.3.1). This helped to explain several aspects of male behaviour with regard to the resting female. The webbing consisted of strands of silk spun over the female and radiating outward 2—3 mm where they were attached to the leaf. The male in its 'guarding' position (Ewing, 1914; Laing, 1969; Cone et al., 1971a; Saba, 1973) is then able to detect movements or disturbances within the area. Such movements may be the arrival of a second male attracted to the resting pharate female, or may be the onset of ecdysis. The latter event stimulates the male to stay very close to the emerging female. Laing (1969) observed that just prior to the moult of the deutonymph, when it had taken on a silvery colour apparently because

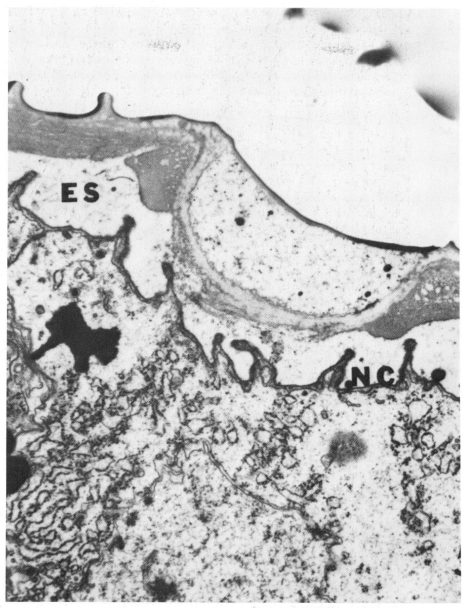

Fig. 1.4.3.2. A section through a cuticular lyrifissure of a quiescent female deutonymph of *T. urticae* showing the ecdysial space (ES) and the new cuticle (NC). Photo by D.R. Penman.

of air between the new and old cuticle (Fig. 1.4.3.2), the male remained on or was at least touching the female. As the deutonymphal exoskeleton split, the male often aided the female in freeing her from the exuvium. Mating took place as soon as the posterior part of the exuvium had been shed and sometimes before the female was entirely free of the anterior portion of the old exoskeleton. The ecdysal suture is dorsal, transverse and about midway along the longitudinal axis of the body.

Adult males reared from collections of eggs obtained from single females emerge 0.5—1.0 days before the females (Laing, 1969; Saba, 1973). Since copulation occurs immediately after ecdysis, this ensures that nearly all females are mated and become potential colonizers.

Saba (1973) working with *Tetranychus tumidus* Banks found essentially the same biology. Males emerged first and were observed close to or on top of the female teleiochrysalis (pharate female or quiescent deutonymph). Mating took place immediately after emergence and lasted approximately 1 min. The preoviposition period was 1 day at both 24° and 29°C and was not influenced by mating (i.e. both mated and unmated females laid eggs; the unfertilized females produced male eggs). Shih et al. (1976) working with *T. urticae* reported similar results and stated:

> this behaviour increases the probability of mating and might enhance the reproductive potential under natural conditions.

Channabasavanna and Channabasavanna (1980), working with *Tetranychus ludeni* Zacher, reported similar results. Mating lasted 1.5—2.5 min. Usually the female mated once. Attempts by males to mate with older females were observed but were generally futile.

Saba (1962) found males mated with 10-day old virgin females, and the females later produced fertilized eggs (female progeny).

Female precopulatory behaviour

Little information exists regarding selection of a resting site by the female deutonymphs prior to spinning their pad. However, Penman and Cone (1972) showed that males were able to find females more quickly if the females had deposited a web on the substrate prior to becoming quiescent, as opposed to placing previously quiescent females on a clean substrate. This suggested that the female deutonymph may aid the male in finding her by leaving web as a cue. It also suggested that a predator might also use the web to find its prey.

The act of mating in tetranychid mites

The male and female external genital structures involved in the act of mating have been described in detail in Chapters 1.1.1 and 1.1.2.

General description

The mating process is somewhat different for tetranychids than for many of the Acari. The male slips under the female from behind and clasps her legs (III or IV) with his front pair (Fig. 1.4.3.3). The male opisthosoma is then strongly reflexed upward to bring the extruded aedeagus in contact with the female genital opening. Coupling is accomplished and sperm transferred in a relatively short period of time (0.5—2.5 min) (Cagle, 1949; Laing, 1969; Banerjee, 1979). This mating posture may make it difficult for intruding males to dislodge the male. The shape, form and size of the aedeagus may have some benefit in keeping the sexes coupled during copulation.

Frequency, multiple matings, interrupted matings

Potter and Wrensch (1978) identified 3 factors that intensify male competition for mating:

(1) males are sexually long-lived and individual females are available only briefly;

(2) males accumulate and are in excess during most of the colonizing episode; and

(3) male response is concentrated on those quiescent female deutonymphs closest to emergence. These authors correlated the duration of first mating prior to interruption with numbers of genetically marked female offspring (the result of a second mating) and found second matings to be ~ 75% effective after 30 s, ~ 60% effective after 60 s, 47% effective after 120 s and 31% effective after 150 s.

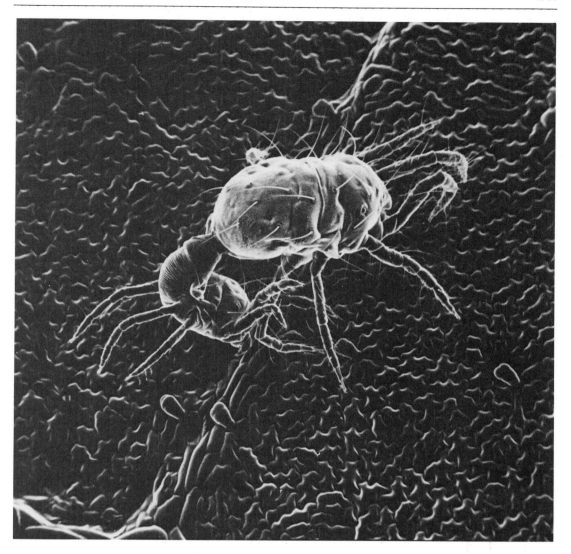

Fig. 1.4.3.3. The mating position of male and female *T. urticae*. (Courtesy A.M. Feldman, ITAL, The Netherlands).

Second matings following a 'complete' first mating ($\bar{x} = 286\,\text{s}$) were totally ineffective, similar to results obtained by Boudreaux (1963) and Helle (1967). Under crowded conditions, 14% of the females are doubly inseminated at ecdysis.

The mechanism by which complete first matings preclude fertilization by later insemination remains obscure. Boudreaux (1963) proposed a 'mating barrier'. Helle (1967) hypothesized that the sperm supply from the first mating determines the success or failure of the second mating. Overmeer (1972) suggested that the female only accepts a certain amount of sperm.

CHEMICAL COMMUNICATION BETWEEN THE SEXES

Pheromones

The term pheromone has been used (Cone et al., 1971a, b; Cone and Pruszynski, 1972; Cone, 1979) in conjunction with mite behaviour and

Chapter 1.4.3. references, p. 250

seems consistent with the terminology proposed by Nordlund and Lewis (1976).

The attraction of male *T. urticae* to pharate females resting inside the deutonymph exoskeleton has been documented in the preceding section (mating). This strong attraction appeared to be the result of a chemical with sex-attracting capabilities (Cone et al., 1971a). Since 1971, a series of tests have been conducted to determine the identity of the sex attractant. Most of the earlier investigations concerned with identification of sex attractants for Lepidoptera and Coleoptera depended heavily upon the development of a bioassay. Therefore, attempts were made to develop bioassay procedures for male *T. urticae* (Cone et al., 1971b; Cone and Pruszynski, 1972). Cone (1979) presented a summary of methods for the preparation of extracts and bioassay procedures.

The normal male response to quiescent deutonymphs was described by Cone et al. (1971b). Later, Penman and Cone (1972) quantified male response to female deutonymphs using front-surfaced mirrors and long-exposure infrared photography. The data obtained showed that

(1) males alone in the test area moved in convoluted patterns with no trend toward linear motion;

(2) males released in arenas with pharate females moved across the surface in more linear travel. The coefficient for angular change was $r = 0.928$; the coefficient of a straight line was $r = 0.902$;

(3) deutonymphal web immediately induced more linear travel compared to the motion of males in the presence of pharate females alone;

(4) a two-component system (web plus female sex pheromone) stimulated male searching behaviour and increased the linearity of male motion during the test period. A later study (Penman and Cone, 1974b) concluded that

(1) deutonymphal web aids the male in finding the pharate female;

(2) the female sex pheromone acts as an attractant until initial male contact, thereafter acting as an arrestant;

(3) the effective holding or arrestant range is about 2 body lengths of the female (1.8—2.3 mm).

Pheromone production—reception studies

This study was initiated to determine the site of production (female) and the receptors (male). Observations by many workers described the male over the female with male legs I and II wrapped around the female idiosoma (hovering behaviour).

Scanning electron microscopy revealed the presence of 3 pairs of cuticular slits located dorsolaterally on the idiosoma of the pharate female. These were hypothesized to be the source of pheromone production. Subsequent observation established the presence of 3 pairs of slits (lyrifissures) in all stages of male and female *T. urticae* which somewhat weakened the hypothesis. Penman and Cone (1974a) succeeded in producing transverse sections through the cuticular slits and found them to be pit-like with a thin epicuticular covering. A fibrous basal area was present but there was no indication of glandular or secretory tissue. The lyrifissures probably function as mechanoreceptors. However, the porous nature of the basal portion may indicate a secondary role in pharate females as a passive release site for the attractant. Since duplex setae (originally hypothesized in this study to be pheromone receptors) are found on both male and female mites, the general production—reception picture was substantially weakened.

Observations by Ewing (1914) and studies by the present author show that the male is very attracted to the mid-dorsal area where the transverse suture will develop at eclosion. There is much rostral and pedipalpal probing of the area accompanied by twitching of legs I—III. Also, since the male frequently attempts to mate with fresh exuviae, it seems likely that the moulting fluid or the fluid between the old and new cuticles may have an active role in male sex attraction.

Regev and Cone (1975a) submitted evidence for farnesol as 1 component of the male sex attractant system in *T. urticae.* Later (1976b) they correlated farnesol content with the total egg production by individual female *T. urticae.*

Webbing and kairomones

Schmidt (1976) showed that female *Phytoseiulus persimilis* Athias-Henriot spend much time in and around the webbing of their prey *T. urticae* but whether the reaction was caused by tactile perception or chemoreception was not determined. Recent studies (Hislop and Prokopy, 1981; Hoy and Smilanick, 1981; Sabelis and Van de Baan, 1983) showed that kairomones are involved in predator arrestment and predator attraction to prey. Sabelis et al. (1984) demonstrated that the predators responded to volatile sub-stance(s) in an air stream and that attractive substances were extractable in solvents. Whether or not male tetranychid mites respond to similar cues to find new colonies and mate with new females (out-crossing) has not been tested.

Host plant influence

The relationship of the host plant to sex attraction and fecundity of insects and mites has scarcely been investigated. As future plant breeding programs develop, an understanding of specific resistance/susceptibility mechanisms, particularly with regard to the reproductive biology of pest species, will be important. Ultimately, and hopefully, this knowledge might progress to a point where plant breeders could be advised to maximize or minimize the content of certain chemicals in the crosses they make and thereby incorporate factors that would naturally regulate pest species at a density below the economic injury level.

Numerous studies (Maxwell, 1972; Tulisalo, 1972; Patterson et al., 1975; Gould, 1978; De Ponti and Garretsen, 1980) have shown varietal differences to tetranychid infestation for several crop plants. Few have explored specific mechanisms. Regev and Cone (1975b) related chemical differences in hop varieties (based on farnesol content) to susceptibility to two-spotted spider mite development. Regev (1978) indicated a difference in farnesol content of strawberry varieties and susceptibility to infestation by *Tetranychus cinnabarinus* (Boisduval). Gunson and Hutchins (1982) found essentially the reverse for both hops and strawberries. These different findings are unresolved at this time but S. Regev (personal communication, 1982) has indicated a possible explanation concerning concentrations of the tested chemicals. Regev and Cone (1976a) reported that male *T. urticae* were attracted to 10 ppm synthetic nerolidol but were not attracted to 100 ppm nerolidol or the tested rates of geraniol. Citronellol at 10 ppm was highly attractive to male *T. urticae* (Regev and Cone, 1980).

As genotypic differences are observed in varieties or selections of plant species, as interpreted by differential mite population development, it seems logical to search for a chemical basis for those differences. The success of such an approach will depend heavily on the development of successful bioassay techniques. That in turn will depend on greater knowledge of tetranychid biology.

Chapter 1.4.3. references, p. 250

REFERENCES

Banerjee, B., 1979. The mating speed in *Oligonychus gossypii* (Zacher) (Tetranychidae: Acarina). Int. J. Invertebr. Reprod., 1: 201—204.

Boudreaux, H.B., 1963. Biological aspects of some phytophagous mites. Annu. Rev. Entomol., 8: 137—154.

Cagle, L.R., 1949. Life history of the two-spotted spider mite. Va. Agric. Exp. Stn., Tech. Bull., 113: 1—31.

Channabasavanna, P. and Channabasavanna, G.P., 1980. Life history of *Tetranychus ludeni* (Acari: Tetranychidae) under field conditions. Indian J. Acarol., 4: 41—48.

Cone, W.W., 1979. Pheromones of Tetranychidae. In: J.G. Rodriguez (Editor), Recent Advances in Acarology, Vol. 2. Academic Press, New York, NY, pp. 309—317.

Cone, W.W., McDonough, L.M., Maitlen, J.C. and Burdajewicz, S., 1971a. Pheromone studies of the two-spotted spider mite, *Tetranychus urticae* Koch. I. Evidence of a sex pheromone. J. Econ. Entomol., 64: 355—358.

Cone, W.W., Predki, S. and Klostermeyer, E.C., 1971b. Pheromone studies of the two-spotted spider mite, *Tetranychus urticae* Koch. II. Behavioral response of male *T. urticae* to quiescent deutonymphs. J. Econ. Entomol., 64: 379—382.

Cone, W.W. and Pruszynski, S., 1972. Pheromone studies of the two-spotted spider mite, *Tetranychus urticae* Koch. III. Response of males to different host tissues, age, searching area, sex-ratios and solvents in bioassay trials. J. Econ. Entomol., 65: 74—77.

De Ponti, O.M.B. and Garretsen, F., 1980. Resistance in *Cucumis sativus* L. to *Tetranychus urticae* Koch. 7. The inheritance of resistance and bitterness and the relation between these characters. Euphytica, 29: 513—523.

Ewing, H.E., 1914. The common red spider or spider mite. Oreg. Agric. Exp. Stn. Bull., 121: 1—95.

Gould, F., 1978. Predicting the future of resistance of crop varieties to pest populations: a case study of mites and cucumbers. Environ. Entomol., 7: 622—626.

Gunson, F.A. and Hutchins, R.F.N., 1982. Absence of farnesol in strawberry and hop foliage. J. Chem. Ecol., 8: 785—796.

Helle, W., 1967. Fertilization in the two-spotted spider mite (*Tetranychus urticae*: Acari). Entomol. Exp. Appl., 10: 103—110.

Hislop, R.G. and Prokopy, R.J., 1981. Mite predator responses to prey and predator-emitted stimuli. J. Chem. Ecol., 7: 895—904.

Hoy, M.A. and Smilanick, J.M., 1981. Non-random prey location by the phytoseiid predator *Metaseiulus occidentalis*. Differential responses to several spider mite species. Entomol. Exp. Appl., 29: 241—253.

Laing, J.E., 1969. Life history and life table of *Tetranychus urticae* Koch. Acarologia, Facsimile Hors Ser., 11: 32—42.

Maxwell, F.G., 1972. Host plant resistance to insects — nutritional and pest management relationships. In: J.G. Rodriguez (Editor), Insect and Mite Nutrition. North Holland, Amsterdam, pp. 599—609.

Morgan, H.A., 1897. Observations on the cotton mite. La. Agric. Exp. Stn., Bull, 48: 130—135.

Nordlund, D.R. and Lewis, W.J., 1976. Terminology of chemical-releasing stimuli in intra-specific and interspecific interactions. J. Chem. Ecol., 2: 211—220.

Overmeer, W.P.J., 1972. Notes on the mating behaviour and sex ratio control of *Tetranychus urticae* Koch (Acarius: Tetranychidae). Entomol. Ber. (Amsterdam), 32: 240—244.

Patterson, C.G., Knavel, D.E., Kemp, T.R. and Rodriguez, J.G., 1975. Chemical basis for resistance to *Tetranychus urticae* Koch in tomatoes. Environ. Entomol., 4: 670—674.

Penman, D.R. and Cone, W.W., 1972. Behaviour of male two-spotted spider mites in response to quiescent female deutonymphs and to web. Ann. Entomol. Soc. Am., 65: 1289—1293.

Penman, D.R. and Cone, W.W., 1974a. Structure of cuticular lyrifissures in *Tetranychus urticae*. Ann. Entomol. Soc. Am., 67: 1—4.

Penman, D.R. and Cone, W.W., 1974b. Role of web, tactile stimuli, and female sex pheromone in attraction of male two-spotted spider mites to quiescent female deuto-nymphs. Ann. Entomol. Soc. Am., 67: 179—182.

Perkins, C.H., 1897. The red spider. In: Report of the Entomologist, 10th Annu. Rep. Vermont Agric. Exp. Stn., pp. 75—86.

Potter, D.A. and Wrensch, D.L., 1978. Interrupted matings and the effectiveness of second inseminations in the two-spotted spider mite. Ann. Entomol. Soc. Am., 71: 882—885.

Potter, D.A., Wrensch, D.L. and Johnston, D.E., 1976a. Guarding, aggressive behaviour, and mating success in male two-spotted spider mites. Ann. Entomol. Soc. Am., 69: 707—711.

Potter D.A., Wrensch, D.L. and Johnston, D.E., 1976b. Aggression and mating success in male spider mites. Science, 193: 160—161.

Regev, S., 1978. Differences in farnesol content in strawberry varieties and their susceptibility to the carmine spider mite *Tetranychus cinnabarinus* (Boisd.) (Acari: Tetranychidae). Entomol. Exp. Appl., 24: 22—26.

Regev, S. and Cone, W.W., 1975a. Evidence of farnesol as a male sex attractant of the two-spotted spider mite, *Tetranychus urticae* Koch (Acarina: Tetranychidae). Environ. Entomol., 4: 307—311.

Regev, S. and Cone, W.W., 1975b. Chemical differences in hop varieties vs. susceptibility to the two-spotted spider mite. Environ. Entomol., 4: 697—700.

Regev, S. and Cone, W.W., 1976a. Analyses of pharate female two-spotted spider mites for nerolidol and geraniol: evaluation for sex attraction of males. Environ. Entomol., 5: 133—138.

Regev, S. and Cone, W.W., 1976b. Evidence of gonadotropic effect of farnesol in the two-spotted spider mite *Tetranychus urticae*. Environ. Entomol., 5: 517—519.

Regev, S. and Cone, W.W., 1980. The Monoterpene Citronellol, as male sex attractant of the two-spotted spider mite, *Tetranychus urticae* (Acarina: Tetranychidae). Environ. Entomol., 9: 50—52.

Saba, F., 1962. Kopulations frequenz, Giftempfindlichkeit und Heterogenität von Weibchen und Männchen bei *Tetranychus urticae* Koch. Anz. Schaedlingskd., 35: 141—142.

Saba, F., 1973. Life history and population dynamics of *Tetranychus tumidus* in Florida (Acarina: Tetranychidae). Fla. Entomol., 57: 46—63.

Sabelis, M.W., Afman, B. and Slim, P.J., 1984. Location of distant spider mite colonies by *Phytoseiulus persimilis* Athias-Henriot (Acarina: Phytoseiidae): Localization and extraction of a Kairomone. Proc. 6th Int. Congr. Acarol., Edinburgh, Vol. I: 431—440.

Sabelis, M.W. and Van de Baan, H.E., 1983. Location of distant spider mite colonies by phytoseiid predators: demonstration of specific kairomones emitted by *Tetranychus urticae* and *Panonychus ulmi*. Entomol. Exp. Appl, 33: 303—314.

Schmidt, G., 1976. Der Einfluss der von den Beutetieren hinterlassen Spuren auf Suchverhalten und Sucherfolg von *Phytoseiulus persimilis* Athias-Henriot (Acarina: Phytoseiidae). Z. Angew. Entomol., 82: 216—218.

Shih, C.T., Poe, S.L. and Cromroy, H.L., 1976. Biology, life table and intrinsic rate of increase of *Tetranychus urticae*. Ann. Entomol. Soc. Am., 69: 362—364.

Tulisalo, U., 1972. Resistance to the two-spotted spider mite, *Tetranychus urticae* Koch (Acarina, Tetranychidae) in the genera *Cucumis* and *Citrullus* (Cucurbitaceae). Ann. Entomol. Fenn. (Suom. Hyonteistiet. Aikak.), 38: 60—64.

1.4.4 Life Types of Spider Mites

Y. SAITÔ

INTRODUCTION

Most Tetranychine species produce silk threads and their spinning behaviour characterizes their lives. Gerson (Chapter 1.4.1) contributed to this topic and his earlier review (Gerson, 1979) is also recommended to the reader. In this chapter this topic will be approached from a different point of view. The present author will first show that tetranychines have evolved various behavioural patterns in response to the webs. An attempt to classify the spider mite mode of life into 'life types', mostly through web characters, will then be presented. Although most data are based on previous work (Saitô, 1983), some additional information concerning mite life types will be included.

MITE RESPONSE TO ITS WEB

It has been reported previously (Saitô, 1979b) that there is much diversity of mite behaviour in response to web development on newly invaded leaves. Figs. 1.4.4.1—1.4.4.4 show the behaviour of 4 species, *Oligonychus ununguis*

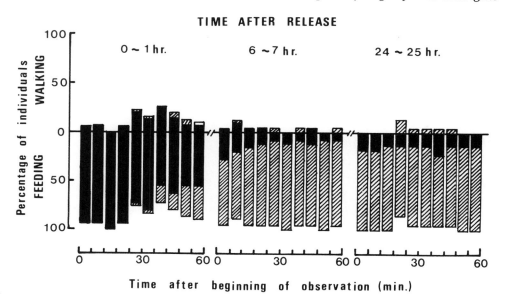

Fig. 1.4.4.1. Relationship between walking and feeding behaviour of females of *Oligonychus ununguis* and webs produced on chestnut leaf surfaces. (■) No relation with webs, (▨) under the webs, (▣) within the webs, (□) on the webs. (From Saitô, 1979b).

Chapter 1.4.4. references, p. 264

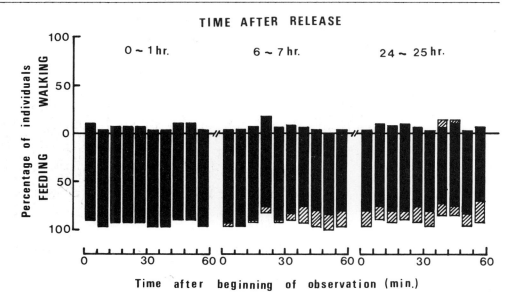

Fig. 1.4.4.2. Relationship between walking and feeding behaviour of females of *Panonychus citri* and webs produced on citrus leaf surfaces. Legend as Fig. 1.4.4.1. (From Saitô, 1979b).

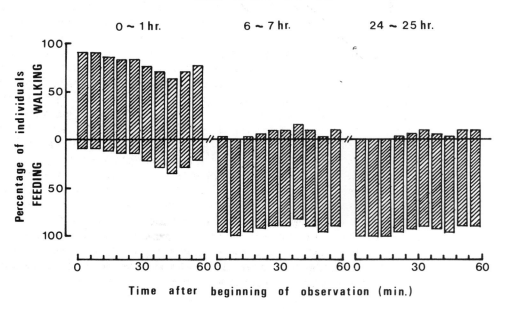

Fig. 1.4.4.3. Relationship between walking and feeding behaviour of females of *Schizotetranychus celarius* and webs produced on sasa leaf surfaces. Legend as Fig. 1.4.4.1. (From Saitô, 1979b).

(Jacobi), *Panonychus citri* (McGregor), *Schizotetranychus celarius* (Banks) and *Tetranychus urticae* Koch, towards webbing residues. These figures illustrate that each species responds to its own web in its own way. The females of *O. ununguis* spin threads while walking (Saitô, unpublished work, 1978). After a while these threads, produced over leaf depressions, accumulate to form a sort of woven 'roof'. When this web has been con-structed, the mites enter this rough nest-like web and start feeding and

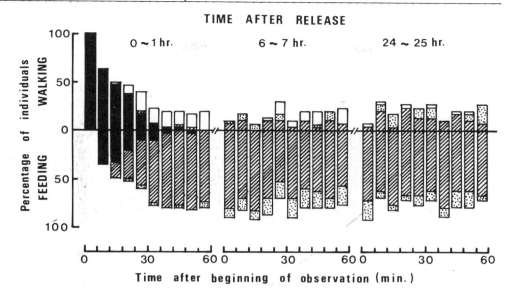

Fig. 1.4.4.4. Relationship between walking and feeding behaviour of females of *Tetranychus urticae* and webs produced on red clover leaf surfaces. Legend as Fig. 1.4.4.1. (From Saitô, 1979b).

resting (Fig. 1.4.4.1). In *P. citri*, however, the pattern of walking and feeding of females on newly infested leaf surfaces does not change with time (Fig. 1.4.4.2). The females of this species also produce silk threads during walking (Saitô, 1977a). These threads accumulate on the leaves and sometimes attain a web-like shape over the leaf depression, as in *O. ununguis*, but *P. citri* avoids its own threads and seldom lives under them.

A close relationship between mite behaviour and its webs was observed in *S. celarius* (Fig. 1.4.4.3). The females search for depressed portions for nesting as soon as they invade a new leaf (this searching requires only a few seconds, so is not shown in Fig. 1.4.4.3). After finding the depression, they cover it with closely woven nest webs. Once the web is constructed to a certain extent (after 6—7 h in Fig. 1.4.4.3), the females turn to feeding under the web and rarely leave it. The walking behaviour during the first hour (Fig. 1.4.4.3) reflects the web-constructing behaviour of this species.

The behavioural patterns of *T. urticae* shown in Fig. 1.4.4.4 are intermediate between those of *O. ununguis* and *S. celarius*. The females also walk actively and spin threads (Saitô, 1977a) when they invade a new host leaf (Fig. 1.4.4.4). After a while, the threads, which are produced over the leaf surface and among the leaf hairs, accumulate to become a complicated web. With the accumulation of threads and the formation of the web, the females become inactive and tend to feed under these webs (after 6—7 h in Fig. 1.4.4.4). A characteristic habit of this species is that it uses its threads in different ways, unlike the other 3 species. *T. urticae* freely walks on, in and under its complicated webs, and feeds under them.

From these observations, it is apparent that the spider mite threads serve in *O. ununguis*, *T. urticae* and *S. celarius* as the material for building structures, useful for life in microhabitats. This does not apply to *P. citri*. For *P. citri*, the threads only serve as lifelines for walking on the lower surface of the leaf.

Figure 1.4.4.5 shows the various patterns of webs and threads produced by different species of spider mite, as obtained from the literature and from Figs. 1.4.4.1—1.4.4.4. The spinning behaviour of Tetranychinae is divided into 2 basic categories, dragging and weaving, as shown in Fig. 1.4.4.5.

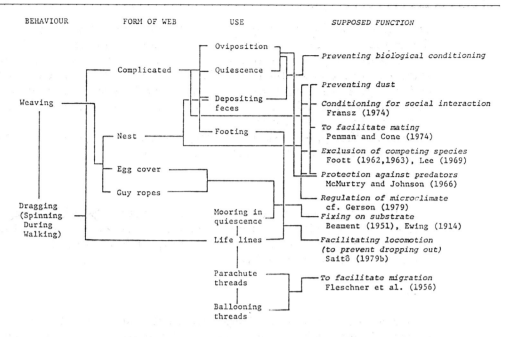

Fig. 1.4.4.5. Spinning behaviour and web structures occurring in the Tetranychinae, and their supposed functions. (After Saitô, 1983).

Weaving generally results in 3 web structures. Dragging is basically different from weaving, because the threads of dragging serve mainly as lifelines or ballooning threads, while the threads of weaving are the material used in constructing web structures. Weaving links up certain functions when the web structure is completed (although dragging may also result in webs). The relationships between spinning behaviour, web structure, use and supposed function are indicated schematically in Fig. 1.4.4.5.

Four forms can be distinguished in the web structures: nest web, complicated web, egg cover and guy ropes. Nest web and complicated web threads occasionally are used for deposition of a faeces. Complicated web threads are used in oviposition (Hazan et al., 1974) as well as during quiescence. Threads are often useful for mooring of quiescent stages (Ewing, 1914), and also serve as lifelines when the mites walk on the leaf lower surface (Saitô, 1977b). When overpopulation and subsequent leaf deterioration cause the mites to migrate, the lifeline threads are used by adult females of some species as ballooning threads (Fleschner et al., 1956; Wanibuchi and Saitô, 1983).

The functions of the various threads and webs, as postulated by the respective authors, are listed in the last column of Fig. 1.4.4.5. It seems easy to postulate many 'plausible' functions concerning the respective threads and web patterns, but solid evidence that these functions are actually related to the evolution of individual species is rather scarce. To obtain such evidence, we first need to develop a method to describe in a uniform way the web pattern of each mite species in relation to its life pattern.

THE CONCEPT OF LIFE TYPES

The concept of 'life types', defined earlier (Saitô, 1983), must first be explained. In Saitô (1983), 8 items were selected in order to characterize the spider mite life with respect to the patterns of webbing: (1) structure of web; (2) density of web; (3) site for oviposition; (4) egg cover produced by females; (5) site where mites enter quiescent stages; (6) preferred site for feeding and walking; (7) spinning behaviour during walking; (8) site for defaecation. By using these items, it became possible to divide the life patterns of many tetranychines into 3 main groups, called 'life types' and symbolically indicated by LW (little web type), CW (complicated web type) and WN (web nest type). Each life type includes several subtypes. Hereafter, the characteristics of each life type and subtype will be shown, one by one, with specific representatives.

The LW life type consists of 4 subtypes. This life type is the simplest, and may be the basic life type of the Tetranychinae.

LW-f

Mites of this subtype spin no threads or webs. Eggs are deposited under dense plant hairs along the veins (Fig. 1.4.4.6). A single species, *Aponychus firmianae* Ma et Yuan, shows this subtype (Table 1.4.4.1), which is the simplest type known from Japan.

LW-j

Species of this subtype have only webbed egg covers, as they spin no threads during walking (Fig. 1.4.4.6). This subtype is obviously more complex than LW-f. *Aponychus corpuzae* Rimando and *Eurytetranychus japonicus* Ehara represent this subtype (Table 1.4.4.1).

LW-c

Mites of this subtype spin silk threads while walking on the leaf and also weave egg guy ropes, but make no other complex web structures (Fig. 1.4.4.6). LW-c is considered to be more advanced than LW-f and LW-j, as spinning continues during the entire life-cycle. This type is represented by the 2 common pest species *Panonychus citri* and *Panonychus ulmi* (Koch) (Table 1.4.4.1). The function of the guy ropes is to maintain the egg in an

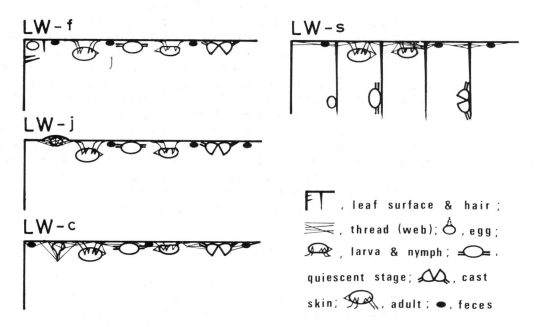

Fig. 1.4.4.6. Schematic expression of LW (little web) life type including 4 subtypes. (After Saitó, 1983).

Chapter 1.4.4. references, p. 264

TABLE 1.4.4.1

Life type of tetranychid mite species and its host plant species

Mite	Life type	Host plant	Host type[a]
Aponychus			
firmianae Ma et Yuan	LW-f	*Firmiana simplex*	A
corpuzae Rimando	LW-j	*Sasa senanensis*	B
Eotetranychus			
celtis Ehara	CW-r	*Celtis sinensis*	A
querci Reeves	WN-u	*Tilia japonica*	A
shii Ehara	WN-s	*Castanopsis sieboldii*	C
suginamensis (Yokoyama)	WN-s	*Morus bombysis*	A
tiliarium (Hermann)	WN-t	*Alnus japonica*	A
tiliarium	CW-r	*Alnus hirsuta*	A
uncatus Garman	WN-u	*Betula platyphylla*	A
sp. 1	CW-p	*Quercus dentata*	A
sp. 2	CW-u	*Tilia maximowicziana*	A
sp. 3	CW-u	*Ulmus laciniata*	A
Eurytetranychus			
japonychus Ehara	LW-j	*Quercus glauca*	C
Oligonychus			
biharensis (Hirst)	WN-u	Unidentified	C
rubicundus Ehara	CW-r	*Miscanthus sinensis*	D
ununguis (Jacobi)	WN-u	*Castanea crenata*	A
uruma Ehara	CW-r	Unidentified	B
Panonychus			
citri (McGregor)	LW-c	*Citrus* sp.	C
ulmi (Koch)	LW-c	*Malus pumila*	A
(Sasanychus)			
akitanus Ehara	CW-a	*Sasa senanensis*	B
Schizotetranychus			
bambusae Reck	CW-b	*Phyllostachys nigra*	B
celarius (Banks)	WN-c	*Sasa senanensis*	B
cercidiphylli Ehara	WN-u	*Cercidiphyllum japonicum*	A
leguminosus Ehara	WN-t	*Wisteria sinensis*	A
recki Ehara	WN-r	*Sasa senanensis*	B
schizopus (Zacher)	WN-t	*Salix subfragilis*	A
sp. 1	WN-s	*Quercus glauca*	C
Tetranychus			
cinnabarinus (Boisduval)	CW-u	*Rubus* sp.	E
desertorum Banks	CW-u	*Solidago altissima*	E
kanzawai Kishida	CW-u	*Althaea rosea*	E
piercei McGregor	CW-u	*Pandanus tectorius*	C
urticae Koch	CW-u	*Sambucus sieboldiana*	A
viennensis Zacher	CW-u	*Quercus mongolica*	A
Yezonychus			
sapporensis Ehara	LW-s	*Sasa senanensis*	B

[a] A, broad-leaved deciduous tree; B, bamboo plant; C, broad-leaved evergreen tree; D, perennial herbacious plant; E, annual plant.

upright position for a brief period, in order to waterproof it (Beament, 1951). The spinning threads produced while walking serve as 'lifelines' (Saitô, 1979b), i.e. facilitate mite locomotion over leaf surfaces and prevent the mite from falling off its host plant. When mite population density increases and the plant deteriorates, the lifelines may serve as ballooning threads for the migration of adult females, as described by Fleschner et al. (1956) and Wanibuchi and Saitô (1983).

LW-s

Yezonychus sapporensis Ehara represents this subtype (Table 1.4.4.1). The species spins lifelines during walking. Eggs and moulting stages are attached to the tips of leaf hairs (Fig. 1.4.4.6). Because of these special habits, LW-s is considered the most complex subtype among LW and it differs to some extent from the other 3 subtypes. The oviposition and moulting sites may help to protect the immobile mite stages from predators.

The second life type, CW, is characterized by complicated webs which are quite different from LW (Fig. 1.4.4.7). The fundamental feature of this life type lies in the web which is a three-dimensional and irregularly complicated structure. Five subtypes are recognizable.

CW-p

This type is considered to be the simplest one; it occurs in *Eotetranychus* sp. 1 (Table 1.4.4.1). This species produces silk threads during walking; the threads accumulate on the leaf surface. The mites tend to walk on the web, and deposit their eggs and pass their quiescence on the leaf surface under the web. In this subtype, the web seems to serve as footing when mites traverse the leaf surface. The web and threads produced during walking facilitate mite locomotion over the leaf.

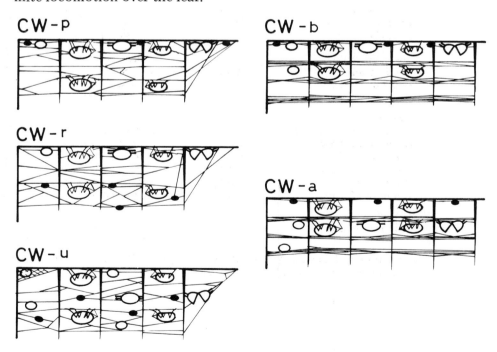

Fig. 1.4.4.7. Schematic expression of CW (complicated web) life type including 5 sub types. (After Saitô, 1983).

Chapter 1.4.4. references, p. 264

CW-r

This subtype may be directly derived from CW-p. Whereas the complicated web in CW-p serves only as footing for mites, that in CW-r is used as sites on which they deposit their faeces (Fig. 1.4.4.7). Four species represent this subtype, as given in Table 1.4.4.1. The placement of faeces on the threads may serve to prevent the inhabited surface from deterioration through the accumulation of faeces. The other functions of the web and threads are the same as in CW-p. CW-r seems to be more complicated than CW-p, because of the additional function of the web.

CW-u

In this subtype the mites construct a highly complicated and irregular web on the leaf surface. In addition to this, a few species, such as *Tetranychus viennensis* Zacher, sometimes weave an egg cover. When the population density becomes high, numerous faecal pellets, eggs and cast skins accumulate on the complicated web threads. These habits, especially that of depositing faeces on the threads, have attracted some workers' attention (Hazan et al., 1974). This subtype is represented by the two-spotted spider mite, which is known to be a destructive pest of many crops, and is common in the genus *Tetranychus* (Table 1.4.4.1). Two *Eotetranychus* species also belong to this subtype. It is obvious that here the web serves to prevent the leaves from becoming soiled, as well as having several functions which were noted in regard to CW-r.

CW-b

The subtype CW-b (as well as CW-a, described below) is believed to be somewhat different from CW-p, -r and -u. For the time being it is categorized within the CW life type. The mites of CW-b construct a dense web which appears stratiform rather than complicated. *Schizotetranychus bambusae* Reck represents this subtype (Table 1.4.4.1). The eggs are usually oviposited on the leaf surface under the web (Fig. 1.4.4.7), although some are laid on the web when the population increases. The threads are extremely thin, and the mites usually move between the web layers. The web of CW-b seems to serve as footing for moving around and as a defence mechanism against predators.

CW-a

Panonychus akitanus Ehara represents the CW-a subtype (Table 1.4.4.1). The web structure of CW-a is approximately the same as that of CW-b. The mites deposit all their eggs on the web, and quiescent stages are found attached there (Fig. 1.4.4.7). These habits are reminiscent of LW-s, in which eggs and quiescent stages are found on the tip of leaf hairs (Fig. 1.4.4.6). These habits were always observed, regardless of population density. Hence, it is thought that they may serve to protect against predators. The mites of this subtype can easily walk between web layers; the web seems to facilitate locomotion and serves as footing. CW-a is considered to be more advanced than CW-b.

The complicated web probably developed as a consequence of thread accumulation produced by walking mites. Therefore, it is reasonable to consider the CW life type as more complex than LW. In the CW, the weaving behaviour which results in a particular web structure was rarely observed.

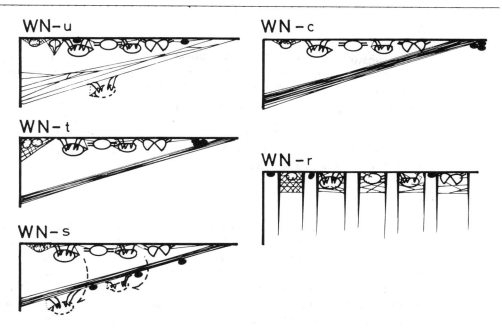

Fig. 1.4.4.8. Schematic expression of WN (web nest) life type including 5 subtypes. (After Saitô, 1983).

Weaving behaviour predominates in the WN life type. Although many of the species of this type retain the habit of spinning while walking (dragging behaviour), in highly specialized species it is completely suppressed.

WN-u

This is the simplest subtype (Fig. 1.4.4.8). Superficially it resembles subtype CW-p, but it differs from the latter in that the mites of subtype WN-u always feed and usually walk under the webs. The process of web construction appears to be similar to the CW life type, but differs from the latter because mite behaviour in WN-u on the webless leaf and under the webs is quite different, as is seen in *O. ununguis* (see Fig. 1.4.4.1 and text). This suggests that the WN-u subtype may represent the process of changing from CW to WN. Egg guy ropes sometimes appear in *O. ununguis* (which inhabits the upper surface of leaves), while egg cover webs appear in the other 4 species (Table 1.4.4.1) (these species inhabit the lower surface of leaves). The web of WN-u serves to shelter the mites from adverse climatic factors and from predators. The functions of threads spun during walking are the same as in the CW life type. In *O. ununguis*, these threads change their function when the population density becomes high and resources deteriorate, at which time they become ballooning threads for adult females (Wanibuchi and Saitô, 1983).

WN-t

Mites representing the WN-t subtype are characterized by their habit of depositing faeces near the margin of their web nest (Fig. 1.4.4.8). *Schizotetranychus schizopus* (Zacher) and 2 other species belong to this subtype (Table 1.4.4.1). The mites seldom move out of the nest web, and feed and oviposit only under the web of the nest. Woven egg covers were commonly observed in these species. The web of WN-t is comparatively dense and probably serves as shelter against adverse weather conditions, predation

Chapter 1.4.4. references, p. 264

and dust. The role of the egg cover is not known. The habit of depositing faeces near the margin of the nest web may be an adaptation to prevent fouling of the infestation site.

WN-s

The subtype WN-s is characterized by the mites' habit of depositing faeces on their nest web threads (Fig. 1.4.4.8). The mites usually live under a very densely woven roof, but leave the nest to defaecate on the roof. Three species, namely *Eotetranychus shii* Ehara, *Schizotetranychus* sp. and *Eotetranychus suginamensis* (Yokoyama) represent this subtype. The web has the same functions as in WN-t, but in addition, the threads serve as the surface on which faeces are deposited. This habit probably evolved in order to prevent the rather limited resource becoming soiled, and it also serves to repel predators. Furthermore, there is the possibility that faecal pellets may function as 'anchors' in stabilizing webbing (Gerson, 1979). The WN-s mites have somewhat flattened bodies as an apparent adaptation to their life type.

WN-c

The most specialized subtype of the Tetranychinae can be seen in WN-c (Fig. 1.4.4.8). The mites construct an extremely densely woven roof (nest) over depressions on the lower leaf surface, as mentioned above (see Fig. 1.4.4.3 and text). Spinning during walking is completely suppressed in this subtype and weaving is restricted to the construction of nest and egg covers. One or 2 sites (along the 'gutter' over which the woven roof is constructed) outside and near the woven nest are used for defaecation. All individuals living together in the same nest always deposit their faeces in these fixed sites. Only a single species, *S. celarius*, represents this subtype. The woven roof of the nest may well protect the mites against their predators (U. Gerson, personal communication, 1980; Saitô and Takahashi, 1982) and preserves the resource from faecal foulness. WN-c is undoubtedly more complex than WN-s, -t and -u.

WN-r

This subtype is characterized by cells constructed for individual mites. A larva, a nymph or an adult constructs a small nest web around 3 or 4 leaf hairs, where it feeds or rests. The egg is invariably covered with a highly dense web like a curtain, woven by its mother. The webs of the other stages are more loosely constructed. These web patterns are qualitatively different from those spun by mites representing the other subtypes in the WN life type. For the present, this subtype is categorized within the WN life type, but its position is not quite certain. WN-r mites were not able to infest hairless leaf surfaces because of their special webbing habits, and they spun threads during walking. Their individual webs may serve to protect them against predators, as well as against adverse climatic conditions. The spinning threads function as lifelines when the mite walks over the leaf. *Schizotetranychus recki* Ehara represents this subtype (Table 1.4.4.1).

Life type WN appears to be qualitatively different from CW, but it is uncertain whether it is more derivative than CW. Both the WN and CW life types appear to be derived from LW, and seem to have developed in a parallel way within the Tetranychinae.

CONCLUSION

Although many observations concerning the life patterns of spider mites in relation to their webbing have been reported, they were usually rather fragmentarily included in many reports on spider mite biology. The reason for the present attempt was the lack of, and therefore the necessity of obtaining a description of the life patterns of various species in a 'uniform way' in order to classify the life types mentioned here. Once we have conformity in expressing spider mite life patterns, it will be possible to compare them interspecifically. Through such comparisons, we may be able to advance in understanding spider mite adaptations and evolution. In addition, the life type is useful as a criterion for identifying spider mite species, as mentioned by Reeves (1963).

Helle et al. (1970) stated that the production of silk was an important evolutionary step within the Tetranychidae. If that is true, it may be very important to see the relationship between the phylogenetic status of the mite species and their life types. In the more ancestral forms (according to Gutierrez et al., 1971), such as *Eurytetranychus*, *Aponychus* and *Panonychus* (except for the sub-genus *Sasanychus*), only 1 or 2 simple life subtypes were recognized. On the other hand, only a single subtype (CW-u) is recognized in *Tetranychus*, which appears to be the most advanced genus. Much variation in subtypes was seen in *Eotetranychus* and *Schizotetranychus*, and not in *Oligonychus*. The trend that the more ancestral genera have simpler types, while the more advanced ones represent more complex types seems to agree with the proposition of Gutierrez et al. (1971). On the other hand, many diversities seen in the lives of *Eotetranychus* and *Schizotetranychus* are considered to be related to the specialization (or adaptation) of these species to their specific host plants. The fact that the most advanced genus, *Tetranychus*, does not show any diversities in its life type seems to be related to its feeding habit and host plant peculiarities. The host plants of *Tetranychus* spp. appear mainly to be herbaceous (Ehara and Shinkaji, 1975; Saitô, 1979a) in Japan (except *T. viennensis* and *Tetranychus ezoensis* Ehara). They feed on an extremely wide range of host plants. This feeding habit does not induce the species to specialize on a certain host plant. Their CW-u life type is probably convenient for the generalistic mode of life of these species through the effective use of time-limited food resources (such as annual plants).

No theory is yet available for understanding the process of adaptation in the Tetranychidae. In this chapter the author has tried to summarize observations about the life types of some Tetranychinae, but it goes without saying that the species discussed are only a fraction of all Tetranychidae. The data are thus incomplete, and further studies concerning life types are necessary.

Recently (see Chapter 1.4.1), many reports have been published regarding the relationships between phytoseiid efficiency and prey webbing patterns. From the above it seems that most spider mite life types co-evolved with specific predators. This implies close interactions between spider mite life characteristics (i.e. life types) and the behaviour of specific phytoseiid predators. Studies along these lines might be fruitful from the basic as well as the applied points of view.

ACKNOWLEDGEMENTS

I wish to express my sincere gratitude to Professor H. Mori, Institute of Applied Zoology, Faculty of Agriculture, Hokkaido University, and to

Chapter 1.4.4. references, p. 264

Professor U. Gerson, Entomology Department, Faculty of Agriculture, Hebrew University of Jerusalem, for their criticisms, valuable suggestions and correction of the manuscript.

REFERENCES

Beament, J.W.L., 1951. The structure and formation of the egg of the fruit tree red spider mite *Metatetranychus ulmi* Koch. Ann. Appl. Biol., 38: 137—154.

Ehara, S. and Shinkaji, N., 1975. An Introduction to Agricultural Acarology. Zenkoku Noson Kyoiku Kyokai, Tokyo, 328 pp (in Japanese).

Ewing, H.E., 1914. The common red spider or spider mite. Oreg., Agric. Exp. Stn., Bull., 121: 1—95.

Fleschner, C.A., Badgley, M.E., Ricker, D.W. and Hall, J.C., 1956. Air drift of spider mites. J. Econ. Entomol., 49: 624—627.

Foott, W.H., 1962. Competition between two species of mites. I. Experimental results. Can. Entomol., 94: 365—375.

Foott, W.H., 1963. Competition between two species of mites. II. Factors influencing intensity. Can. Entomol., 95: 45—57.

Fransz, H.G., 1974. The Functional Response to Prey Density in an Acarine System. Pudoc, Wageningen, 143 pp.

Gerson, U., 1979. Silk production in *Tetranychus* (Acari; Tetranychidae). In: J.G. Rodriguez (Editor), Recent Advances in Acarology, vol. 1. Academic Press, New York, NY, pp. 177—188.

Gerson, U., 1985. Webbing. In: W. Helle and M. Sabelis (Editors), Spider Mites and Their Control. Elsevier, Amsterdam, Ch. 1.4.1.

Gutierrez, J., Helle, W. and Bolland, H.R., 1971. Étude cytogénétique et réflexions phylogénétiques sur la famille des Tetranychidae Donnadieu. Acarologia, 12: 732—751.

Hazan, A., Gerson, U. and Tahori, A.S., 1974. Spider mite webbing. I. The production of webbing under various environmental conditions. Acarologia, 16: 68—84.

Helle, W., Gutierrez, J. and Bolland, H.R., 1970. A study on sex determination and karyotypic evolution in Tetranychidae. Genetica, 41: 21—32.

Lee, B., 1969. Cannibalism and predation by adult males of the two-spotted spider mite *Tetranychus urticae* Koch (Acarina; Tetranychidae). J. Aust. Entomol. Soc., 8: 210.

McMurtry, J.A. and Johnson, H.G., 1966. An ecological study of the spider mite *Oligonychus punicae* (Hirst) and its natural enemies. Hilgardia, 37: 363—402.

Penman, D.R. and Cone, W.W., 1974. Role of web, tactile stimuli, and female sex pheromone in attraction of male two-spotted spider mites to quiescent female deutonymphs. Ann. Entomol. Soc. Am., 67: 179—182.

Reeves, R.M., 1963. Tetranychidae infesting woody plants in New York State, and a life history study of the elm spider mite, *Eotetranychus matthyssei* n. sp. Mem. Cornell Univ. Agric. Exp. Stn., Ithaca, NY, 380: 1—99.

Saitô, Y., 1977a. Study on spinning behaviour of spider mites (Acarina: Tetranychidae). I. Method for quantitative evaluation of the mite webbing, and the relationship between webbing and walking. Jpn. J. Appl. Entomol. Zool., 21: 27—34 (in Japanese, with English summary).

Saitô, Y., 1977b. Study on spinning behaviour of spider mites (Acarina: Tetranychidae). II. The process of silk secretion and the webbing on the various stages of *Tetranychus urticae* Koch under different environmental conditions. Jpn. J. Appl. Entomol. Zool., 21: 150—157 (in Japanese, with English summary).

Saitô, Y., 1979a. Comparative studies on life histories of three species of spider mites (Acarina: Tetranychidae). Appl. Entomol. Zool., 14: 83—94.

Saitô, Y., 1979b. Study on spinning behaviour of spider mites. III. Responses of mites to webbing residues and their preferences for particular physical conditions of leaf surfaces (Acarina: Tetranychidae). Jpn. J. Appl. Entomol. Zool., 23: 82—91 (in Japanese, with English summary).

Saitô, Y., 1983. The concept of 'life types' in Tetranychinae — An attempt to classify the spinning behaviour of Tetranychinae. Acarologia, 24: 377—391.

Saitô, Y. and Takahashi, K., 1982. Study on variation of *Schizotetranychus celarius* (Banks) II. Comparison of mode of life between two sympatric forms (Acarina: Tetranychidae). Jpn. J. Ecol., 32: 69—78 (in Japanese, with English summary).

Wanibuchi, K. and Saitô, Y., 1983. The process of population increase and patterns of resource utilization of two spider mites, *Oligonychus ununguis* (Jacobi) and *Panonychus citri* (McGregor), under experimental conditions (Acarina; Tetranychidae). Res. Popul. Ecol., 25: 185—198.

1.4.5 Reproductive Strategies

M.W. SABELIS

INTRODUCTION

Several species of the Tetranychidae inhabit ephemeral and patchily distributed resources, such as weeds. Examples are *Tetranychus evansi* Baker and Pritchard feeding on nightshade in California and Brazil (Qureshi et al., 1969), *Tetranychus neocaledonicus* André feeding on *Ipomoea* spec. in Madagascar (Gutierrez, 1976) and *Oligonychus pratensis* (Banks) feeding on Blue Grama grass in steppe systems in Colorado (Congdon and Logan, 1983). Once arrived on their host plant these spider mites give rise to a rapidly increasing population; but, sooner or later, this trend ends abruptly either because: (1) the life span of the host plant is short; or (2) the host plant becomes overexploited by the spider mites; or (3) the spider mites are outcompeted by other herbivores; or (4) natural enemies and fungal diseases eliminate the spider mite population; or because of (5) harsh weather conditions (wind and rain). Hence, local persistence is essentially impossible and selection for genotypes that have a greater chance of reaching new resources will be intense. To find new host plants spider mites should cover larger distances than possible by locomotion only. Being tiny and wingless creatures, transport by air currents or by larger organisms (with better dispersal capacities than the mites) is the only solution. As phoretic transport has been reported only occasionally and the phoretics were not organisms specifically searching for food sources that are also favourable to the spider mites, it can be inferred that dispersal is random with respect to the position of the host plants and colonization chances will probably be low. Therefore, selection will favour genotypes coding for a higher capacity of population increase (and thus an increased number of dispersers carrying these genes) and genotypes coding for a greater ability to find a host plant after passive long-range dispersal. There is an extensive literature dealing with the intrinsic rate of increase of tetranychid species. Most of these reports deal with species of the colonizing type, because these are usually the ones causing damage of economic importance. To return to the above examples, *T. evansi* is an important pest of tomatoes, *T. neocaledonicus* is a pest in cotton and cassava and *O. pratensis* is a serious pest in maize, sorghum and wheat. Their capacity of increase has been studied extensively in the context of pest control. However, almost nothing is known of their ability to select and find host plants. In this chapter current knowledge of host plant selection and capacity of increase will be discussed and placed in an evolutionary perspective. The main goal is to trace the adaptations that have resulted from natural selection under the regime of rapidly succeeding phases of colonization, population growth, dispersal and local extinction.

Chapter 1.4.5. references, p. 276

THE NON-EQUILIBRIUM STATUS OF LOCAL SPIDER-MITE POPULATIONS

As emphasized by Caswell (1982), the overall selective pressure on a given trait depends on how the selective pressures vary over the evolutionary history. The colonizing species in the examples presented above are almost permanently in the increase phase. The selective pressures are thus immediately recognized and meet the classic notion of r_m selection. However, the converse argument that non-equilibrium populations must be r_m selected, does not hold (Caswell, 1982). There may well be non-equilibrium species, whose populations spend much of their evolutionary history in the decline phase. If the decline phase is longer than the increase phase, then non-equilibrium populations are subject to selective pressures that meet the classic notion of K selection. This postulate has been elaborated mathematically by Caswell (1982), who also showed that the patterns of selective pressure in equilibrium populations and those in declining non-equilibrium populations are not only similar, but they are even intensified in the latter case: a non-equilibrium species may be more K selected than an equilibrium species. Therefore, to detect the adaptiveness of a given trait it is certainly worthwhile to compare non-equilibrium species that have different local dynamics.

Unfortunately, there is little published information on local population dynamics of tetranychid species that experience less rapid successions of colonization, growth, dispersal and local extinction. Clearly, this is because these species are of less economic importance. I presume that the tetranychids living on dwarf bamboo in Japan may be the best example currently available. Saito and co-workers (Saito and Ueno, 1979; Gotoh, 1983) studied the spider mites inhabiting bamboo leaves, one of which is *Schizotetranychus celarius* (Banks). Dwarf bamboo plants are long-lived and rather abundant in Japan, so that the chance of finding new resources after local extinction is high. Under these circumstances one may expect less intensive selection for increased r_m than under circumstances of ephemeral, scarce and patchily distributed food sources. Consequently, the host plant will be overexploited much less frequently. Females of *S. celarius* deposit their eggs underneath a shelter-like web, which is generally thought to be helpful in resisting adverse weather conditions, competing herbivores and predators. *Typhlodromus bambusae* Ehara is a phytoseiid predator that has overcome this barrier. Upon invasion of the predator into the nest web, the female and male spider mites leave their nest, but return later and the males kill the predator's offspring after they emerge from the eggs (Saito, 1984). Undoubtedly, this set of traits will contribute to the local persistence of *S. celarius* but presumably not to an equilibrium situation due to leaf deterioration, harsh weather, etc. Therefore, decline of local populations may well be an important period in the evolutionary history of *S. celarius*. It is not possible to classify all tetranychid species according to the type of local dynamics and the level of colonization because firm data on their ecology are lacking. The work of Saito, Ueno and Gotoh on *S. celarius* will therefore be an important benchmark study.

INTRINSIC RATE OF INCREASE

Table 1.4.5.1 summarizes most of the literature on the intrinsic rate of increase, r_m, of tetranychid species. The value of r_m characterizes the potential or innate capacity of increase for the case of idealized populations in idealized environments, i.e. constant climate, unlimited food, no interference and a stable age distribution. Taylor (1979) argued that the time needed for convergence to the stable age distribution is probably too long. He estimated that the time required to approach the stable age distribution within 5% was ca. 90 days for *Tetranychus desertorum* Banks and ca. 140 days for *Tetranychus mcdanieli* McGregor. As these rank among the most prolific tetranychid species, stable age distributions may generally not be found in spider mite populations. Carey (1982) tested this hypothesis. He calculated the stable age distributions of 6 *Tetranychus* and 1 *Panonychus* species (i.e. 66% eggs, 26% immatures and 8% adults) and compared these with reported stage distributions of growing mite populations. He convincingly demonstrated that natural mite populations are often quite near to the stable age distribution. This result is very important in that it supports the hypothesis that populations of *Tetranychus* and *Panonychus* are usually in the increase phase under natural conditions. Hence, r_m selection is expected to be the prevailing selective force.

Such a test is currently not available for *S. celarius*, but, as argued above, there are good reasons to assume that selection leading to increased r_m will be less intense than in *Tetranychus* species. Indeed, Table 1.4.5.1 shows that not only *S. celarius*, but also other *Schizotetranychus* species living under comparable circumstances have lower r_m values than *Tetranychus* species. *Panonychus* species, however, have r_m values that are close to those of *Schizotetranychus* species. Saito (1979) argued that *Panonychus* species live on woody plants, which are long-lived, and have sufficient food reserves to initiate regenerative growth after defoliation. Hence, because woody plants are a more persistent food source than weeds, colonization chances will be higher and r_m selection will be less intense than in the case of *Tetranychus* species feeding on weeds. Saito's argument may be correct, but I need to be a little bit more convinced that local populations of *Panonychus* species are relatively more persistent than local populations of *Tetranychus* species. As pointed out before, there are many factors that may violate local persistence (predation, competition, harsh weather conditions, etc.). Long life span of the host plant is certainly not a guarantee for local persistence. Nevertheless, Saito's view does have intuitive appeal.

There is more evidence in support of Saito's hypothesis. Firstly, *Oligonychus pratensis* (Banks) feeding on grasses has a higher r_m than *Oligonychus ununguis* (Jacobi) feeding on chestnut leaves. Secondly, *Aponychus corpuzae* Rimando feeding on bamboo leaves has only a slightly higher r_m than *S. celarius* feeding on the same host plant. Thirdly, Gotoh (1983) presented an interesting explanation for the higher r_m of *Schizotetranychus schizopus* (Zacher) feeding on *Salix*, relative to 2 other *Schizotetranychus* species feeding on *Wisteria* and *Cercidiphyllum*. He argued that *S. schizopus* may have to disperse more frequently than the other 2 species because *Salix* gradually defoliates starting from early June, whereas *Wisteria* and *Cercidiphyllum* defoliate almost all at the same moment in autumn.

Several important differences cannot be explained, but this is merely because adequate information on environmental conditions is lacking. For example, the r_m values of *Tetranychus* species range from 0.201 to 0.29 day^{-1}, for which I do not see a clear reason on the basis of the information currently available. More rigorous tests of Saito's theory are required. Agreement between theory and facts does not prove anything, but it should be a challenge to test it again in other situations or to formulate and test alternative hypotheses. In the latter respect differences in the food quality or in the defense of the host plant against herbivores should be considered as a potential alternative to Saito's theory.

Chapter 1.4.5. references, p. 276

TABLE 1.4.5.1

Net reproduction (R_0), intrinsic rate of increase (r_m) and finite rate of increase (λ) of 18 tetranychid species at $25°C \pm 1°C$[a]. The genera are ranked according to the mean value of r_m, just as the species within genera. The life table parameters are defined by Birch (1948). Generation time is not presented because the commonly used calculation by $\ln (R_0)/r_m$ is not very informative (see May, 1976; Pielou, 1977)

Genera and species	Host plant	R_0	r_m (day^{-1})	λ (= e^{r_m})	Reference
Tetranychus					
desertorum Banks	Cotton seedlings	111.1	0.290	1.336	Nickel, 1960[c]
neocaledonicus André	Cotton	57.2	0.260	1.297	Gutierrez, 1976
urticae Koch[b]	Cucumber (susceptible line)		0.282	1.326	Rauwerdink et al., 1985
	Red Clover	65.0	0.259	1.295	Saito, 1979
	Cotton		0.241	1.272	Guttierez, 1976
	Cotton cotyledons	74.8	0.219	1.245	Carey and Bradley, 1982
	Cucumber (resistant line)		0.218	1.243	Rauwerdink et al., 1985
pacificus McGregor	Bush Lima bean	108.3	0.290	1.340	Takafuji and Chant, 1976
	Cotton cotyledons	44.6	0.207	1.229	Carey and Bradley, 1982
cinnabarinus (Boisduval)	Bean	109.6	0.270	1.310	Gerson and Aronowitz, 1980
	Bean	27.4	0.197	1.210	Hazan et al., 1973[d]
turkestani (Ugarov and Nikolski)	Cotton cotyledons	46.8	0.203	1.225	Carey and Bradley, 1982
mcdanieli McGregor	Bean	75.1	0.201	1.223	Tanigoshi et al., 1975
viennensis Zacher	Cherry plum	36.0	0.136	1.144	Skorupska and Boczek, 1984
Oligonychus					
pratensis (Banks)	Corn	71.2	0.290	1.340	Perring et al., 1984
	Blue Grama grass	20.1	0.203	1.260	Congdon and Logan, 1983[c,d]
punicae (Hirst)	Avocado	43.1	0.220	1.245	Tanigoshi and McMurtry, 1977
ununguis (Jacobi)	Chestnut	24.7	0.178	1.195	Saito, 1979
Aponychus					
corpuzae Rimando	Dwarf Bamboo	53.8	0.181	1.198	Saito and Ueno, 1979
Panonychus					
ulmi (Koch)	Apple		0.185	1.203	Herbert, 1981b[e]
	Apple		0.180	1.197	Rabbinge, 1976[f]
citri (McGregor)	Citrus	28.3	0.171	1.186	Yasuda, 1982[g]
	Citrus	24.4	0.162	1.176	Saito, 1979
Schizotetranychus					
schizopus (Zacher)	*Salix* sp.	27.5	0.206	1.229	Gotoh, 1983
leguminosus Ehara	*Wisteria* sp.	27.1	0.173	1.189	Gotoh, 1983
celarius (Banks)	Dwarf bamboo	67.6	0.162	1.176	Saito and Ueno, 1979
cercidiphylli Ehara	*Cercidiphyllum* sp.	26.0	0.160	1.173	Gotoh, 1983

LIFE-HISTORY EVOLUTION

The intrinsic rate of increase, r_m, can be determined from the life-history components of the organism (Birch, 1948), such as the developmental rate, the ovipositional rate, the survival rate and the proportion of females in the offspring (sex ratio). Standard proportional increases in each of these components will increase r_m, but their effects on r_m can be very different. An important example is the effect of the developmental rate relative to the ovipositional rate. The higher r_m, the more sensitive it is to changes in developmental rate relative to proportionally equivalent changes in the rate of reproduction. This demographic rule has been demonstrated by several authors (e.g. Lewontin, 1965) and analytically proven by Caswell and Hastings (1980). In the context of evolution it can thus be predicted that the higher r_m, the more it will pay to increase r_m further by increasing the developmental rate. In other words the pressure of natural selection to decrease developmental time will intensify with r_m. Similarly, we may expect the selection pressure to intensify with increasing rates of reproduction. The latter hypothesis can be tested in a more rigorous way than the former, because data on r_m are scarcer than data on developmental and ovipositional rates. Because the ovipositional rate of tetranychids varies drastically with the age of the ovipositing female, a choice of either the life-time average or the mean maximum value of the ovipositional rate should be made. The latter is the better choice, because generally egg production peaks early in the ovipositional period and then it is more important in determining r_m than the life-time average of the ovipositional rate. Hence, as shown in Fig. 1.4.5.1, the relation between the developmental rate and the maximum rate of oviposition is ascertained by using data from 26 species at $25 \pm 1°C$. It shows that the developmental rate initially increases with the peak ovipositional rate, but above peak productions of 5 to 6 eggs per day the developmental rate seems to level off. This phenomenon is not unexpected, because it is conceivable that there are physiological constraints to speeding up development: 'Developmental time will certainly not be reduced to zero'. Apparently, the maximum ovipositional rate can be driven up far beyond 6 eggs per day at 25°C. Of course, when developmental time is at its minimum and survival is near 90% even after peak oviposition, then selection can only act to increase the ovipositional rate further until physiological constraints impose an upper limit. It is noteworthy that the relation in Fig. 1.4.5.1 does not reveal trade-offs between the maximum ovipositional rate and the developmental rate. The highest values of the ovipositional rate are not associated with lower developmental rates.

The above inference on the constraints to maximization of the developmental rate has an important flaw: the variation in developmental rate in the supposed maximum region is rather large, the egg-to-egg period varying from 9 to 12.5 days. If natural selection on the developmental rate has

Table 1.4.5.1 (continued)

[a] If r_m was determined at several humidity levels, only the highest value is presented in the table.
[b] Important additional references with data at other temperatures are: Laing, 1969; Bengston, 1970; Shih et al., 1976; Jesiotr and Suski, 1976; Jesiotr, 1979; Herbert, 1981a; Feldmann, 1981.
[c] Sex ratio assumed to be 1.0.
[d] Sex ratio assumed to be 0.5.
[e] Extrapolated values using the regression line presented by Herbert (1981b).
[f] Calculated from Rabbinge's data.
[g] Sex ratio assumed to be 0.7.

Chapter 1.4.5. references, p. 276

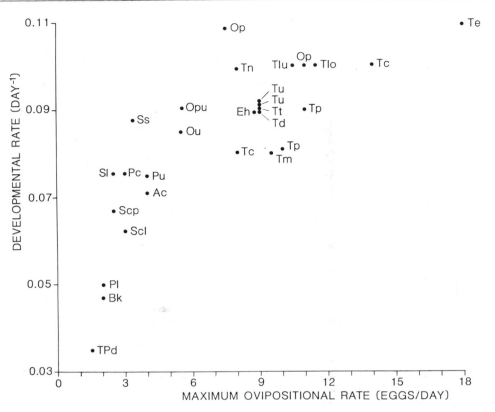

Fig. 1.4.5.1. The relationship between the mean (egg-to-egg) developmental rate and the mean maximum of the ovipositional rate in the course of the oviposition period. The data are obtained from experiments at ca. 25°C with 24 tetranychid species: (Ac) *Apony-chus corpuzae* Rimando (Saito and Ueno, 1979); (Bk) *Bryobia kissophila* Van Eyndhoven (Sabelis, unpublished data); (Eh) *Eotetranychus hicoriae* (McGregor) (Micinski et al., 1979); (Op) *Oligonychus pratensis* (Banks) (Congdon and Logan, 1983; Perring et al., 1984); (Opu) *Oligonychus punicae* (Hirst) (Tanigoshi and McMurtry, 1977); (Ou) *Oli-gonychus ununguis* (Jacobi) (Saito, 1979); (Pl) *Petrobia latens* (Müll.) (Cox and Lieber-man, 1960); (Pc) *Panonychus citri* (McGregor) (Saito, 1979); (Pu) *Panonychus ulmi* (Koch) (Rabbinge, 1976); (Scl) *Schizotetranychus celarius* (Banks) (Saito and Ueno, 1979); (Scp) *Schizotetranychus cercidiphylli* Ehara (Gotoh, 1983); (Sl) *Schizotetranychus leguminosus* Ehara (Gotoh, 1983); (Ss) *Schizotetranychus schizopus* (Zacher) (Gotoh, 1983); (Tc) *Tetranychus cinnabarinus* (Boisduval) (Hazan et al., 1973; Coates, 1974); (Td) *Tetranychus desertorum* Banks (Nickel, 1960); (Te) *Tetranychus evansi* Baker and Pritchard (Qureshi et al., 1969); (Tlo) *Tetranychus lombardinii* Baker and Pritchard (Coates, 1974); (Tlu) *Tetranychus ludeni* Zacher (Coates, 1974); (Tm) *Tetranychus mcdanieli* McGregor (Tanigoshi et al., 1975); (Tn) *Tetranychus neocaledonicus* André (Gutierrez, 1976); (Tp) *Tetranychus pacificus* McGregor (Takafuji and Chant, 1976; Carey and Bradley, 1982); (Tt) *Tetranychus turkestani* Uvarov and Nikolski (Carey and Bradley, 1982); (Tu) *Tetranychus urticae* Koch (Saito, 1979; Sabelis, 1981; Carey and Bradley, 1982); (TPd) *Tenuipalpoides dorychaeta* Pritchard and Baker (Singer, 1966).

been intense so as to drive it to its physiological maximum, then why is this variation so large? Possible causes may be: (1) differences in ambient tem-perature during the experiments; (2) differences in experimental methods; and (3) differences in the geographical regions and the associated environ-mental conditions that prevailed during the evolutionary history of the different tetranychid species. These causes can be eliminated by considering the data originating from comparative studies that have been carried out under the same experimental conditions, and that have focussed on at least 2 tetranychid species inhabiting the same geographical area. There are 3 comparative studies on *Tetranychus* spp.: the work of Carey and

TABLE 1.4.5.2

Life history components of three tetranychid species inhabiting cotton fields in the San Joaquin Valley in California (Carey and Bradley, 1982). Experimental conditions: 24°C; 50—65% RH; cotton cotyledons as host plants

Species	Mean egg-to-adult period of the female (days)	Preoviposition period (days)	Maximum oviposition per day per female	Total number of eggs per female
Tetranychus turkestani Uvarov and Nikolski	10.7	1.17	9.5	84.6
Tetranychus urticae Koch	10.5	1.08	10.5	103.3
Tetranychus pacificus McGregor	10.5	1.17	11.0	78.9

TABLE 1.4.5.3

Life history components of two tetranychid species inhabiting Madagascar (Gutierrez, 1976). Experimental conditions: 25°C; 50% RH; cotton seedlings as hosts

Species	Mean egg-to-adult Period of the female (days)	Total number of eggs produced per female
Tetranychus urticae Koch	9.2	65.5
Tetranychus neocaledonicus André	9.2	80.3

TABLE 1.4.5.4

Life history components of three tetranychid species inhabiting South Africa (Coates, 1974). Experimental conditions: 25°C; 50% RH; bean leaves as hosts

Species	Mean egg-to-adult period of the female (h)	Total number of eggs produced per female
Tetranychus cinnabarinus (Boisduval)	235	106
Tetranychus lombardinii Baker and Pritchard	235	86
Tetranychus ludeni Zacher	233	59

Bradley (1982) on 3 *Tetranychus* spp. inhabiting cotton fields in California's San Joaquin Valley (Table 1.4.5.2); the work of Gutierrez (1976) on 2 *Tetranychus* spp. feeding on weeds, cotton and cassava in Madagascar (Table 1.4.5.3); and the work of Coates (1974) on 3 *Tetranychus* spp. inhabiting cotton fields in South Africa (Table 1.4.5.4). Each of these studies shows that the interspecific differences in developmental time are no more than 1 or 2 h (!!!), whereas the differences in total number of eggs produced are very large (up to a factor of 2 !!!). These data convincingly corroborate the hypothesis that colonizing species such as *Tetranychus* spp. have been subject to such severe selection that developmental rates have been driven to their physiological maximum. Relative to the developmental rate the total

Chapter 1.4.5. references, p. 276

number of eggs produced is expected to be subject to much less severe selection (Lewontin, 1965). The large variation in this trait is therefore in agreement with the theory.

Using Caswell and Hasting's model it can be calculated that for r_m values above ca. 0.1 day^{-1} an increase in developmental rate is more important in increasing r_m than an equivalent increase in the ovipositional rate. However, for r_m values below 0.1 day^{-1} the reverse is true. Oviposition will then be subject to the most severe selection (relative to the developmental rate). Therefore, it may be hypothesized that tetranychid species with r_m values close to the turnover point will have a small variation in the ovipositional rate and a large variation in developmental rate. However, there are no reports on tetranychid species with such low r_m values (at 25°C). According to Fig. 1.4.5.1, *Bryobia* spp., *Petrobia* spp. and *Tenuipalpoides* spp. would be good candidates for r_m values of about 0.1 day^{-1} or even lower. Unfortunately, there are no comparative life-history studies of these species available. Among the tetranychid species studied by Saito (1979), Saito and Ueno (1979) and Gotoh (1983) there are 6 species with r_m values of ca. 0.17 day^{-1}. Their mean developmental rates range from 0.07 to 0.1 day^{-1}, whereas their mean ovipositional rates range from 1.9 to 4.3 eggs per day. This variation in developmental rate clearly contrasts with the extremely small variation found in *Tetranychus* spp. (as shown in Tables 1.4.5.2—1.4.5.4), whereas the variation in ovipositional rates is of similar magnitude. These facts are in agreement with Caswell and Hasting's predictions.

If r_m selection prevails, it would be profitable to produce as many daughters as possible, provided that their fertilization is ensured. However, under random mating this would in turn give rise to an evolutionary gambit, because genes coding for producing more sons would spread in the population, in case the females are in excess. Fisher (1930) therefore proposed that sex ratio evolves as a product of selection maximizing the contribution of the individual's genes to future generations. When the sex ratio in the offspring is 0.5, a mother cannot increase her relative contribution of genes by altering the sex ratio of her offspring away from 0.5. Therefore, the fitness of the mites may be increased even at the expense of their capacity of increase. However, if tetranychid females tend to isolate their brood so as to increase the chances of brother—sister mating, then they would increase their fitness by producing many daughters and just enough sons to fertilize them. Hamilton (1967) predicted that the sex ratio will evolve to

TABLE 1.4.5.5

Changes of the sex ratio in the course of a colonizing episode (Wrensch and Young, 1978; Wrensch, 1979)

Treatment	Stage of colonization				
	Upon founding	Early	Middle	Late	Dispersing
Leaf quality during parental development	Poor	Good	Good	Good	Poor
Leaf quality during oviposition and development of offspring	Good	Good	Good	Poor	Poor
Density	Low	Low	High	High	High
Proportion of females in the colony	0.795	0.727	0.545	0.595	0.660

$(n-1)/2n$ depending on the number of colony foundresses (n). Charnov (1982) tested this theory for the case of *Tetranychus* mites using data from Wrensch and Young (1978) and Zaher et al. (1979). These data show that sex ratio is female biased, but tends to 0.5 at high spider mite density (Table 1.4.5.5). At high density (causing food scarcity) the frequency of matings between relatives will decrease and therefore a shift to the Fisherian sex ratio of 0.5 is to be expected. Because the sex ratio shifts among the adults indeed reflect shifts in the sex ratio in the egg phase (Wrensch and Young, 1983), Charnov (1982) was right in concluding that Hamilton's theory is in agreement with the data on sex ratio shifts in *Tetranychus* spp. (Caution should be exercised, however, as Wrensch and Young (1983) did not clarify how their 1983 experiment relates to their 1978 experiment; for example, they did not specify female density).

The interpretations of Charnov have certainly been a breakthrough in our understanding of the sex ratio control by *T. urticae*, but some questions remain unanswered. (1) Why is the sex ratio of *T. urticae* biased to 0.75 females and not further away from 0.5? Wrensch and Flechtmann (1978) reported a sex ratio as extreme as 0.90 for a strain of the carmine spider mite, whereas usually the sex ratio equals 0.7—0.8 (Wrensch and Young, 1978). (2) Why are the interspecific differences in sex ratio so large? See for example sex ratios reported by Carey and Bradley (1982) and Saito and Ueno (1979). The latter question may be answered by using Hamilton's theory. For example, Saito and Ueno (1979) report that *S. celarius* 'families' live in woven nests separate from each other, whereas *A. corpuzae* females deposit their eggs at relatively long distances from each other. This suggests that sibmating is the rule in *S. celarius* and that sibmating will be less frequent in *A. corpuzae*. The sex ratios produced by females of these species are in close agreement with the above speculations on mating structure of the populations. The sex ratio of *S. celarius* appears to be heavily skewed towards females (overall sex ratio in the offspring = 0.84), whereas the sex ratio of *A. corpuzae* has only a slight female bias (overall sex ratio = 0.63). These results indicate that Hamilton's theory may explain sex ratio evolution in the Tetranychidae, but before accepting the theory, more work should be done to elucidate the mating structure of spider mite populations.

HOST PLANT SELECTION, HOST PLANT CHANGE AND REPRODUCTIVE SUCCESS

The chances of colonizing suitable host plants are likely to be in favour of genotypes coding for a high reproductive capacity, as well as for the ability to select suitable host plants and to overcome resistance barriers of potentially suitable host plants. Because spider mites disperse passively via wind currents causing randomized mite landings, Byrne et al. (1982) concluded that the process of host plant selection should be viewed in terms of host plant acceptance and not in terms of host finding. Thus, when the mites land on a host plant, they probe and accept it or wait for the wind to carry them off again. Admittedly, spider mites are small creatures and cannot travel very far by locomotion, but if the ability to locate host plants from a distance leads to increased reproductive success, then natural selection will favour any improvement in the ability to find suitable host plants. Sabelis (unpublished data) found that young females of *T. urticae* can easily cover several metres within a day and preliminary experiments in Y-tube olfactometers (see Sabelis and Van de Baan, 1983) indicated that these females were able to walk towards the arm containing air blown over Lima bean

Chapter 1.4.5. references, p. 276

leaves, when the alternative arm contained humid air only. Whether this behavioural response to olfactory perception has a function in host plant selection and location from a distance is still to be elucidated, but this ability has been shown for other mites, such as phytoseiids (see Chapter 2.1.3.1). Hence, there are good reasons for pursuing this topic.

Once landed or arrived on their host plant spider mites start piercing the leaf and subsequently they continue feeding or attempt to move away (Dabrowski and Marczak, 1972). Dabrowski and co-workers observed that on dandelion leaves hungry females of *T. urticae* began feeding after a much longer initial period than on bean leaves, whereas on petunia leaves movements were disorderly and feeding sites were changed frequently. These behavioural observations correlated perfectly with measurements of survival and oviposition on these host plants. It was not until the investigations of Byrne et al. (1982) that host plant preference was assessed in two-choice situations. They found that *Mononychellus tanajoa* (Bondar) preferred susceptible varieties of cassava to resistant varieties. These experiments suggest that spider mites acquire an estimate of host plant quality during initial piercing (before oviposition starts) and select the most profitable host plant in terms of reproductive success.

Natural selection will not only favour genotypes coding for improved ability of finding suitable host plants, but also genotypes with a broader range of suitable host plant species. Therefore, it is of interest to consider how spider mite populations respond when forced to feed on one particular host plant species. Such experiments have usually been done by transferring spider mites reared on bean to another host plant species and then recording life-history parameters (Dabrowski and Marczak, 1972; Gerson and Aronowitz, 1980). As the response is tested within the first generation, natural selection has been left out of the scope of these studies. The importance of selection, however, is suggested by the existence of intraspecific variability in host plant range. Perhaps the results of Van de Bund and Helle (1960) can be explained in this way. If *T. urticae* and *T. cinnabarinus* are conspecific (Dupont, 1979), then the experimental results of Van de Bund and Helle were the first that shed some light on the matter of host range evolution. They found that *T. urticae* populations sampled from rose, bean, *Chelidonium*, *Urtica*, *Sambucus* and *Althaea* thrived on each other's host plants, but not on carnations, whereas *T. cinnabarinus* populations sampled from carnations thrived on all plant species that were hosts of *T. urticae*. This difference indicates that selection may have led to a broader host plant range of the population living on carnations. Very similar differences have been found with respect to *T. urticae* populations originating from a laboratory culture on Lima bean and from greenhouse tomatoes. The former strain did not survive on tomato, whereas the latter strain did thrive both on bean and tomato (Sabelis, unpublished data).

That spider mite strains can become adapted to new host plant species through selection was first postulated by Jesiotr and Suski (1976) and Jesiotr (1979). Two-spotted spider mites that had been reared for 2 years on Baccara roses were transferred to leaves of Golden Saxa beans and then reared for 10 generations, after which they were transferred back to rose again. The r_m value measured upon return to rose appeared to be lower than that of the original strain that had fed permanently on Baccara roses. After 2 to 3 generations the r_m value of the transferred strain had increased to the same level as the original strain. These results led Jesiotr to suggest that selection modified the genetic make-up of the spider mite population. If this is true then Jesiotr's experiments also indicate that there is some trade-off between traits that are favourable on bean, and those on rose.

However, without knowledge of the underlying causes, this remains conjectural. An excellent study on the micro-evolutionary processes involved in host-range evolution was subsequently published by Gould (1979). He divided a population of *T. urticae* in two and reared one half on Lima bean, whereas the other half was placed in a plant 'community' consisting of Lima bean and a somewhat toxic host, i.e. mite-resistant cucumber. The fitness of the 2 populations was monitored over 21 months. The line maintained in the cucumber—bean community had higher fitness on cucumber and lower fitness on bean than the line maintained on bean only. Additionally, the line maintained in the cucumber—bean community had higher survivorship on 2 or 3 other marginal hosts tested (tobacco and potato, but not plantain). This indicates that as a herbivore population's plant community becomes more diverse, interaction effects may become important in structuring host range through cross-adaptations to sets of plants.

COMPROMISES UNDERLYING REPRODUCTIVE SUCCESS

If spider mites were allowed to channel all their energy into reproduction, the intrinsic rate of increase would be higher than that attained in the experiments discussed above. Presumably the value of r_m is likely to be a compromise between the energy spent in development and reproduction on the one hand, and, on the other, the energy spent in: (1) probing new host plant species; (2) avoiding toxic host plants; (3) detoxification of secondary compounds of the host plant; (4) defense against other phytophagous arthropods competing for food; (5) defense against predators and diseases; and (6) defense against harsh weather conditions. Dabrowski and Marczak (1972) found that hungry females of *T. urticae*, after release on a leaf of a tobacco line with high nicotine and alkaloid content, became very 'irritated' and then paralyzed after a few initial piercings. To my knowledge no one has ever attempted to study avoidance responses in relation to toxic host plants. It would be interesting to study the mite's take-off and locomotory behaviour on these plants. Moreover, to understand the decisions made by the spider mites it is important to estimate the costs involved in tolerating or detoxifying certain plant compounds.

The production of a web is believed to be functional in defense against competing herbivores, predators and adverse weather conditions (Chapter 1.4.1). That the energy allocated to web production influences the ovipositional rate, is suggested by Gerson and Aronowitz (1981). They argue that for both products the mites require quantities of nitrogenous substances. It would be interesting to investigate how spider mites allocate their energy to the production of a web relative to the production of eggs under various circumstances. Gerson and Aronowitz found that on more suitable hosts relatively more resources are allocated by the mite to egg production than to webbing and vice versa on less suitable plants. The possible reasons for the differential allocation remain to be elucidated. The same applies to the large interspecific differences in silk production. For example, *Panonychus* spp. produce very little silk compared with *Tetranychus* spp. If an effective barrier against predators is formed by investing much silk in the construction of a complex web, then why has it not evolved in *Panonychus* spp.? Perhaps there are alternative strategies that require less energy. For example, spider mites, such as *P. ulmi*, may use the silken thread to lower themselves when a predator attacks, or they may render themselves unprofitable to their predators by spacing out their progeny over the leaf surface, or by sequestering secondary plant compounds that render them

Chapter 1.4.5. references, p. 276

toxic to their predators. By considering conceivable alternative strategies of defense we may be able to understand differential allocation of nutrients to either silk production or reproduction.

Perhaps the most clear example of a compromise is the induction and termination of diapause in spider mites inhabiting the temperate regions. By entering into diapause the mites delay their residual reproductive effort and, if winter eggs are produced, additional costs are involved relative to the production of summer eggs, e.g., because of the 'armoured' shell structure of the winter egg. Therefore, natural selection will favour genotypes coding for postponed diapause, as long as this trait promotes their contribution to future generations. Similar considerations apply to the termination of diapause. How spider mites determine when to enter and when to terminate diapause is discussed by Veerman in Chapter 1.4.6. However, the optimality of the mite's decisions in terms of reproductive success has not yet been investigated.

The above examples prompt the thought that there are important trade-offs underlying energy allocation by spider mites. Whether the allocation employed by spider mites really leads to a maximization of reproductive success, cannot be answered. Our lack of knowledge on the energy needed to achieve certain ecologically relevant goals hampers further progress in understanding the reproductive strategies of spider mites.

ACKNOWLEDGEMENT

I thank Martin Brittijn for drawing the figure.

REFERENCES

Bengston, M., 1970. Effect of different varieties of the apple host on the development of *Tetranychus urticae* (Koch). Queensl. J. Agric. Anim. Sci., 27: 95—114.

Birch, L.C., 1948. The intrinsic rate of natural increase of an insect population. J. Anim. Ecol., 17: 15—26.

Byrne, D.H., Guerrero, J.M., Bellotti, A.C. and Gracen, V.E., 1982. Behavior and development of *Mononychellus tanajoa* (Acari: Tetranychidae) on resistant and susceptible cultivars of cassava. J. Econ. Entomol., 75(5): 924—927.

Carey, J.R., 1982. Demography of the two-spotted spider mite, *Tetranychus urticae* Koch Oecologia (Berlin), 52: 389—395.

Carey, J.R. and Bradley, J.W., 1982. Developmental rates, vital schedules, sex ratios and life tables for *Tetranychus urticae*, *T. turkestani* and *T. pacificus* (Acarina: Tetranychidae) on cotton. Acarologia, 23: 333—345.

Caswell, H., 1982. Life history theory and the equilibrium status of populations. Am. Nat., 120: 317—339.

Caswell, H. and Hastings, A., 1980. Fecundity, developmental time, and population growth rate: an analytical solution. Theor. Popul. Biol., 17: 71—79.

Charnov, E.L., 1982. The Theory of Sex Allocation. Monographs in Population Biology 18, Princeton University Press, Princeton, New Jersey, 355 pp.

Coates, T.J.D., 1974. The influence of some natural enemies and pesticides on various populations of *Tetranychus cinnabarinus* (Boisduval), *T. lombardinii* Baker & Pritchard and *T. ludeni* Zacher (Acari: Tetranychidae) with aspects of their biologies. Entomology Memoir No. 42, Department of Agricultural Technical Services, Republic of South Africa, 40 pp.

Congdon, B.D. and Logan, J.A., 1983. Temperature effects on development and fecundity of *Oligonychus pratensis* (Acari: Tetranychidae). Environ. Entomol., 12: 359—362.

Cox, H.C. and Lieberman, F.V., 1960. Biology of the brown wheat mite. J. Econ. Entomol., 53(5): 704—708.

Dabrowski, Z.T. and Marczak, Z., 1972. Studies on the relationship of *Tetranychus urticae* Koch and host plants. I. Effect of plant species. Pol. Pismo Entomol., XLII (4): 821—855.

Dupont, L.M., 1979. On gene flow between *Tetranychus urticae* Koch, 1836 and *Tetranychus cinnabarinus* (Boisduval) Boudreaux, 1956 (Acari: Tetranychidae): Synonomy between the two species. Entomol. Exp. Appl., 25: 297—303.

Feldmann, A.M., 1981. Life table and male mating competitiveness of a wild type and a chromosome mutation strain of *Tetranychus urticae* in relation to genetic pest control. Entomol. Exp. Appl., 29: 125—137.

Fisher, R.A., 1930. The Genetical theory of Natural Selection. Oxford University Press, Oxford, 291 pp.

Gerson, U. and Aronowitz, A., 1980. Feeding of the carmine spider mite on seven host plant species. Entomol. Exp. Appl., 28: 109—115.

Gerson, U. and Aronowitz, A., 1981. Spider mite webbing V. The effects of various host plants. Acarologia, 22(3): 277—281.

Gotoh, T., 1983. Life-history parameters of three species of *Schizotetranychus* on deciduous trees (Acarina: Tetranychidae). Appl. Entomol. Zool., 18: 122—128.

Gould, F., 1979. Rapid host range evolution in a population of the phytophagous mite *Tetranychus urticae* Koch. Evolution, 33(3): 791—802.

Gutierrez, J., 1976. Etude biologique et écologique de *Tetranychus neocaledonicus* André (Acariens, Tetranychidae). Travaux et Documents de l'ORSTOM 57, ORSTOM, Paris, 173 pp.

Hamilton, W.D., 1967. Extraordinary sex ratios. Science, 156: 477—488.

Hazan, A., Gerson, U. and Tahori, A.S., 1973. Life history and life tables of the carmine spider mite. Acarologia, 15(3): 414—440.

Herbert, H.J., 1981a. Biology, life tables and innate capacity for increase of the two-spotted spider mite, *Tetranychus urticae* (Acarina: Tetranychidae). Can. Entomol., 113: 371—378.

Herbert, H.J., 1981b. Biology, life tables, and intrinsic rate of increase of the European red mite, *Panonychus ulmi* (Acarina: Tetranychidae). Can. Entomol., 113: 65—71.

Jesiotr, L.J., 1979. The influence of the host plants on the reproduction potential of the two-spotted spider mite, *Tetranychus urticae* Koch (Acarina: Tetranychidae). II. Responses of the field population feeding on roses and beans. Ekol. Pol., 27(2): 351—355.

Jesiotr, L.J. and Suski, Z.W., 1976. The influence of the host plants on the reproduction potential of the two-spotted spider mite *Tetranychus urticae* Koch (Acarina: Tetranychidae). Ekol. Pol., 24: 407—411.

Laing, J.E., 1969. Life history and life table of *Tetranychus urticae* Koch. Acarologia, 9: 32—42.

Lewontin, R.C., 1965. Selection for colonizing ability In: H.G. Baker and G.L. Stebbins (Editors), The Genetics of Colonizing Species. Academic Press, New York, pp. 79—94.

May, R.M., 1976. Estimating *r*: a pedagogical note. Am. Nat., 110: 496—499.

Micinski, S., Boethel, D.J. and Boudreaux, H.B., 1979. Influence of temperature and photoperiod on development and oviposition of the pecan leaf scorch mite, *Eotetranychus hicoriae*. Ann. Entomol. Soc. Am., 72: 649—654.

Nickel, J.L., 1960. Temperature and humidity relationships of *Tetranychus desertorum* Banks with special reference to distribution. Hilgardia, 30(2): 41—100.

Perring, T.M., Holtzer, T.O., Kalisch, J.A. and Norman, J.M., 1984. Temperature and humidity effects on ovipositional rates, fecundity and longevity of adult female Banks grass mites (Acari: Tetranychidae). Ann. Entomol. Soc. Am., 77: 581—586.

Pielou, E.C., 1977. Mathematical Ecology. John Wiley & Sons, New York, 385 pp.

Qureshi, A., Oatman, R. and Fleschner, C.A., 1969. Biology of the spider mite, *Tetranychus evansi*. Ann. Entomol. Soc. Am., 62(4): 898—903.

Rabbinge, R., 1976. Biological control of fruit-tree red spider mite. Simulation Monographs, Pudoc, Wageningen, The Netherlands, 228 pp.

Rauwerdink, J.B., Sabelis, M.W. and De Ponti, O.M.B., 1985. Life history studies *Tetranychus urticae* Koch on a two-spotted spider mite susceptible and resistant cucumber line. J. Exp. Appl. Acarol. In press.

Sabelis, M.W., 1981. Biological control of two-spotted spider mites using phytoseiid predators. Part I: Modelling the predator—prey interaction at the individual level. Agricultural Research Reports 910. Pudoc, Wageningen, The Netherlands, 242 pp.

Sabelis, M.W. and Van de Baan, H.E., 1983. Location of distant spider-mite colonies by phytoseiid predators: Demonstration of specific kairomones emitted by *Tetranychus urticae* and *Panonychus ulmi* (Acarina: Tetranychidae). Entomol. Exp. Appl., 33: 303—314.

Saito, Y., 1979. Comparative studies on life histories of three species of spider mites (Acarina: Tetranychidae). Appl. Entomol. Zool., 14(1): 83—94.

Saito, Y., 1984. Prey kills predator! — counterattack of tetranychid mite against phytoseiid predator. XVII International congress of Entomology, Hamburg, West Germany.

Saito, Y. and Ueno, J., 1979. Life history studies on *Schizotetranychus celarius* (Banks) and *Aponychus corpuzae* Rimando as compared with other tetranychid mite species (Acarina: Tetranychidae). Appl. Entomol. Zool., 14(4): 445—452.

Shih, C.T., Poe, S.L. and Cromroy, H.L., 1976. Biology, life table and intrinsic rate of increase of *Tetranychus urticae*. Ann. Entomol. Soc. Am., 69(2): 362—364.

Singer, G., 1966. The bionomics of *Tenuipalpoides dorychaeta* Pritchard and Baker (1955) (Acarina, Trombidiformes, Tetranychidae). Univ. Kans. Sci. Bull., 46(17): 625—645.

Skorupska, A. and Boczek, J., 1984. Biology, ecology and demographic parameters of the hawthorn spider mite on various host plants. Prace Naukowe Instytutu Ochrony Roslin, 26: 119—143.

Takafuji, A. and Chant, D.A., 1976. Comparative studies of two species of predaceous phytoseiid mites (Acarina: Phytoseiidae), with special reference to their responses to the density of their prey. Res. Popul. Ecol., (Kyoto), 17: 255—310.

Tanigoshi, L.K. and McMurtry, J.A., 1977. The dynamics of predation of *Stethorus picipes* (Coleoptera: Coccinellidae) and *Typhlodromus floridanus* on the prey *Oligonychus punicae*(Acarina: Phytoseiidae, Tetranychidae). Part I. Comparative life history and life table studies. Hilgardia, 45(8): 237—261.

Tanigoshi, L.K., Hoyt, S.C., Browne, R.W. and Logan, J.A., 1975. Influence of temperature on population increase of *Tetranychus mcdanieli* (Acarina: Tetranychidae). Ann. Entomol. Soc. Am., 68(6): 972—986.

Taylor, F., 1979. Convergence to the stable age distribution in populations of insects. Am. Nat., 113(4): 511—530.

Van de Bund, C.F. and Helle, W., 1960. Investigations on the *Tetranychus urticae* complex in North-West Europe (Acari; Tetranychidae). Entomol. Exp. Appl., 3: 142—156.

Wrensch, D.L., 1979. Components of reproductive success in spider mites In: J.G. Rodriguez (Editor), Recent Advances in Acarology, Vol. 1. Academic Press, New York, pp. 155—164.

Wrensch, D.L. and Flechtmann, C.H.W., 1978. Extraordinary sex ratio in a carmine spider mite. Cienc. Cult. (Maracaibo), 31: 1039—1040.

Wrensch, D.L. and Young, S.S.Y., 1978. Effects of density and host quality on the rate of development, survivorship, and sex ratio in the carmine spider mite. Environ. Entomol., 7: 499—501.

Wrensch, D.L. and Young, S.S.Y., 1983. Relationship between primary and tertiary sex ratio in the two-spotted spider mite (Acarina: Tetranychidae). Ann. Entomol. Soc. Am., 76: 786—789.

Yasuda, M., 1982. Influence of temperature on some of the life cycle parameters of the citrus red mite, *Panonychus citri* (McGregor) (Acarina: Tetranychidae). Jpn. J. Appl. Entomol. Zool., 26: 52—57.

Zaher, M.A., Shehata, K.K. and El-Khatib, H., 1979. Population density effects on biology of *Tetranychus arabicus*, the common spider mite in Egypt. In: J.G. Rodriguez (Editor), Recent Advances in Acarology, Vol. 1, Academic Press, New York, pp. 507—509.

1.4.6 Diapause

A. VEERMAN

DIAPAUSE IN TETRANYCHID MITES: CHARACTERISTICS AND OCCURRENCE

Mites, like insects, show various seasonal adaptations, which enable them to survive periods of unfavourable environmental conditions such as low winter temperatures and periods of heat or drought in the summer. The life cycles of many species, particularly in temperate and northern latitudes, are characterized by an alternation of periods of active development with periods of rest. In many instances this rest indicates a physiological condition known as diapause. Diapause has been defined as a genetically determined state of suppressed development, the expression of which may be controlled by environmental factors (Beck, 1980). Diapause differs from simple quiescence in that quiescence is a direct response to deleterious physical conditions and is terminated as soon as the environmental conditions are favourable again. Diapause, on the other hand, typically begins long before the onset of unfavourable conditions and may not be terminated until long after the disappearance of such conditions (Beck, 1980). The importance of diapause lies not only in ensuring survival through unfavourable seasons, but also in the regulation of seasonal phenologies by the synchronization of life cycles and the determination of patterns of voltinism (Danilevskii, 1965; Tauber and Tauber, 1978). The terms hibernation an aestivation are generally used to denote a state of dormancy (diapause or quiescence) which occurs during winter and summer months, respectively (Saunders, 1982).

For a large number of tetranychid mites dormant stages have been described, many of which have been demonstrated or may reasonably be inferred to represent a true diapause state. Both the hibernal and aestival type of diapause have been found among tetranychid mites, and in some cases hibernation and aestivation have been demonstrated in the same species (Table 1.4.6.1). In each species of spider mite, diapause occurs only at a specific growth stage, either the egg stage or the adult female. Embryonic or egg diapause has been found in members of the genera *Bryobia*, *Petrobia*, *Schizonobia*, *Aplonobia*, *Eurytetranychus*, *Panonychus*, *Schizotetranychus* and *Oligonychus*; reproductive or adult diapause has been found in members of the genera *Eotetranychus*, *Platytetranychus*, *Neotetranychus* and *Tetranychus*, and in 1 member of the genus *Oligonychus*.

Table 1.4.6.1 lists the species of spider mites for which a dormant stage has been described. No attempt has been made to cover the older literature on spider mites dealing with the description of overwintering stages and observations on overwintering behaviour; in particular, for the long-known species of great economic importance a selection had to be made from the literature available. Various aspects of diapause in spider mites have been

Chapter 1.4.6. references, p. 310

TABLE 1.4.6.1

Occurrence of diapause in tetranychid mites. Selected references are given for the various species for which a dormant stage has been demonstrated or may reasonably be inferred to represent a true diapause state (A: aestival diapause, H: hibernal diapause)

Species	Type	Diapause stage	Ref.
Bryobia praetiosa Koch	A	Egg	Anderson and Morgan (1958); Jeppson et al. (1975)
	H	Egg	Venables (1943); Mathys (1957)
Bryobia rubrioculus (Scheuten)	H	Egg	Zacher (1920); Summers (1950); Böhm (1954); Mathys (1957); Anderson and Morgan (1958); Herbert (1965); Jeppson et al. (1975)
Bryobia ribis Thomas	H	Egg	Mathys (1957); Jeppson et al. (1975)
Petrobia harti (Ewing)	A	Egg	Garcia Mari and del Rivero (1982)
Petrobia apicalis (Banks)	A	Egg	Zein-Eldin (1956); Brooking (1957, cited in Boudreaux, 1963); Jeppson et al. (1975)
Petrobia latens (Müller)	A	Egg	Baker and Pritchard (1953)
	H	Egg	Fenton (1951); Baker and Pritchard (1953); Lees (1961)
Schizonobia sycophanta Womersley	H	Egg	A.Q. Van Zon (personal communication, 1973) and author's own observations
Aplonobia juliflorae Tuttle and Baker	A	Egg	W. Helle (personal communication, 1983)
Eurytetranychus buxi (Garman)	H	Egg	Ries (1935); Jeppson et al. (1975)
Panonychus ulmi (Koch)	H	Egg	Zacher (1920); Geijskes (1939); Cagle (1946); Kuenen (1949); Lees (1950)
Panonychus citri (McGregor)	H	Egg	Shinkaji (1979); Takafuji and Morimoto (1983); Morimoto and Takafuji (1983); Takafuji and Kamezaki (1984)
Eotetranychus tiliarium (Hermann)	H	Adult ♀	Zacher (1920, 1921); Geijskes (1939)
Eotetranychus hicoriae (McGregor)	H	Adult ♀	Micinski et al. (1979); Jackson et al (1983)
Eotetranychus uncatus Garman	H	Adult ♀	Pritchard and Baker (1952); Ubertalli (1955); Jeppson et al. (1975)
Eotetranychus querci Reeves	H	Adult ♀	Reeves (1963)
Eotetranychus carpini (Oudemans)	H	Adult ♀	Pritchard and Baker (1952); Jeppson et al. (1975); Schruft (Chapter 3.2.9)
Eotetranychus hirsti Pritchard and Baker	H	Adult ♀	Jeppson et al. (1975)
Eotetranychus matthyssei Reeves	H	Adult ♀	Reeves (1963); Jeppson et al. (1975)
Eotetranychus populi (Koch)	H	Adult ♀	Zacher (1920); Jeppson et al. (1975)
Eotetranychus pruni (Oudemans)	H	Adult ♀	Jeppson et al. (1975)
Eotetranychus willametti (McGregor)	H	Adult ♀	Jeppson et al. (1975)
Eotetranychus yumensis (McGregor)	A	Adult ♀	Jeppson et al. (1975)
Platytetranychus multidigituli (Ewing)	H	Adult ♀	English and Snetsinger (1957); Jeppson et al. (1975)
Neotetranychus rubi Trägårdh	H	Adult ♀	Gutierrez et al. (1970)
Schizotetranychus schizopus Zacher	H	Egg	Zacher (1920); Geispitz (1968)
Oligonychus ilicis (McGregor)	H	Egg	Mague and Streu (1980)

TABLE 1.4.6.1. (Continued)

Species*	Type	Diapause stage	Ref.
Oligonychus ununguis (Jacobi)	H	Egg	Zacher (1921); Geijskes (1939); Löyttyniemi (1970); Shinkaji (1975a)
Oligonychus newcomeri (McGregor)	H	Egg	Reeves (1963); Jeppson et al. (1975)
Oligonychus yothersi (McGregor)	H	Egg	Jeppson et al. (1975)
Oligonychus pratensis (Banks)	H	Adult ♀	André (1942); Malcolm (1955); Jeppson et al. (1975)
Tetranychus urticae Koch	H	Adult ♀	Von Hanstein (1902); Weldon (1910); Geijskes (1939); André (1942); Cagle (1949); Gasser (1951)
	A	Adult ♀	Srivastrava and Mathur (1962); Geispitz and Orlovskaja (1971)
Tetranychus cinnabarinus (Boisduval)	H	Adult ♀	Boudreaux (1956); Dosse (1964, 1967); Helle and Overmeer (1973); Dupont (1979)
Tetranychus desertorum Banks	H	Adult ♀	Jeppson et al. (1975)
Tetranychus kanzawai Kishida	H	Adult ♀	Osakabe (1962); Jeppson et al. (1975); Uchida (1980)
Tetranychus turkestani (Ugarov and Nikolski)	H	Adult ♀	Pritchard and Baker (1952); Mellott and Connell (1965); Jeppson et al. (1975)
Tetranychus mcdanieli McGregor	H	Adult ♀	Jeppson et al. (1975)
Tetranychus pacificus McGregor	H	Adult ♀	Lamiman (1935); Pritchard and Baker (1952); Jeppson et al. (1975)
Tetranychus viennensis Zacher	H	Adult ♀	Zacher (1921); Razumova (1967); Jeppson et al. (1975)
Tetranychus canadensis (McGregor)	H	Adult ♀	Jeppson et al. (1975)
Tetranychus schoenei McGregor	H	Adult ♀	Baker and Pritchard (1953); Jeppson et al. (1975)

*Note added in proof. A table listing as yet unpublished data on the diapause of many additional (mainly Japanese) species of Tetranychinae is to be found in the Ph.D Thesis of Gotoh (1984).

reviewed by Lees (1959), Boudreaux (1963), Parr and Hussey (1966), Huffaker et al. (1969), Helle and Overmeer (1973), Jeppson et al. (1975) and Hussey and Huffaker (1976). In this chapter, diapause in spider mites will be discussed with particular emphasis on the physiological aspects of the induction and termination of diapause.

The occurrence of diapause in spider mites may differ greatly, even in closely related species in the same geographical area, as was shown by Mathys (1957) in his study of the biology of the various species of *Bryobia* occurring in Switzerland. *Bryobia kissophila* van Eyndhoven, the species living on ivy, was found to have no diapause at all; the succession of generations appears to proceed without interruption throughout the year, the duration of the life cycle merely being lengthened during the winter. According to Mathys (1957) development of *B. kissophila* may continue even at temperatures close to 0°C. *Bryobia cristata* Dugès, a species living on grasses and other herbaceous plants, also seems to develop continuously throughout the year; during the winter Mathys (1957) found all developmental stages of *B. cristata* tightly packed together in sheltered places. However, some of the eggs laid by females of this species in May and June did not develop into a summer generation, but apparently were diapause eggs which remained dormant until the following spring. Little is known of the biology of *Bryobia praetiosa* Koch in Europe, a species also living on grasses and herbaceous plants, but according to Mathys (1957) it may have 2 generations per year, separated by a hibernal and an aestival diapause, both passed in the egg stage. *Bryobia rubrioculus* (Scheuten) and *Bryobia ribis* Thomas, which infest fruit trees and bush fruits, respectively, both overwinter as diapause eggs, the difference between these species being that the former is multivoltine and the latter univoltine. The phenology of several *Bryobia* species in the U.S.A. has been studied by Anderson and Morgan (1958) and by Snetsinger (1964). Undoubtedly, *Bryobia* would present an interesting case for reflection on the evolution of diapause in spider mites. For information on the occurrence of diapause in other genera of the Tetranychidae, the reader is referred to the references listed in Table 1.4.6.1.

Diapause is not just an arrest of development or of reproductive activity; the diapause condition has often been characterized as a syndrome comprising various aspects of biology and physiology. In spider mites, differences between diapause and non-diapause forms may affect colouration, morphology, physiology and behaviour. One of the most conspicuous differences between 'summer' and 'winter' forms of many spider mites is the more intense yellow, orange or red pigmentation of diapause forms compared with the often much weaker or even quite different colouration of non-diapause forms (Colour plate II, see preliminary pages). The colour change observed in diapausing mites has often been attributed to changes in the metabolism of the leaves of the host plant (von Hanstein, 1902; Ewing, 1914; Reiff, 1949; Gasser, 1951; Linke, 1953; Fritzsche, 1959; Jeppson et al., 1975), but analyses of the pigments of *Tetranychus urticae* Koch have shown the orange and red pigments to be hydroxyketo-carotenoids which are not found in plant leaves (Veerman, 1974). These carotenoids are present in greater quantities in overwintering females than in summer females, possibly because of dissolution of carotenoids in lipids or the lipid moiety of lipoproteins, which are present in increased amounts in diapausing mites. Definite proof of the independence of the winter colouration of the mites from the pigment metabolism of the host plant was obtained when it proved possible to induce diapause in spider mites which were reared under short-day conditions on an artificial diet from the egg stage onward; the diapausing females which developed on the membrane showed the orange—red colouration characteristic of the winter form of *T. urticae* (L.P.S. Van der Geest, personal communication, 1983, and author's own observations).

Morphological differences between diapause and non-diapause forms may be found in the structure of the egg shell in the case of embryonic diapause and in the structure of the cuticular ridges in the case of adult diapause. Whereas in some genera (e.g. *Bryobia* (Mathys, 1957) and *Panonychus* (Beament, 1951)) the summer and winter eggs are rather alike in general appearance, this is by no means true of *Petrobia* (Fenton, 1951; Lees, 1961; Garcia Mari and del Rivero, 1982). The summer eggs of *Petrobia latens* (Müller) are cherry-red, subspherical, lightly ribbed and surmounted by a tapering 'whisker', whereas the glistening white winter eggs consist of a basal rounded portion and an expanded cap (Fig. 1.4.6.1). The distinctive appearance of the diapause egg is caused by the great quantity of wax in the outer layer of the egg covering, in comparison with the non-diapause egg, and by its complex structure. The waterproofing wax forms an impermeable cap overlying a porous 'aeroscopic sponge' which surrounds the ovum and which communicates with the exterior only via a restricted opening in the middle of the cap. No doubt the complete wax envelope serves to limit

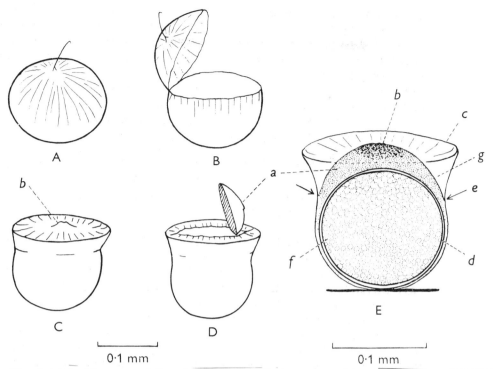

Fig. 1.4.6.1. The eggs of *Petrobia latens* (Müller). A, B, unhatched and hatched non-diapause eggs. C, D, unhatched and hatched diapause eggs. E, diapause egg viewed in optical section showing respiratory air spaces. Abbreviations: a, porous wax dome; b, 'boss' in centre of cap; c, rim of cap; d, 'shell' layer; e, point of fracture of dome; f, egg contents; g, air reservoir between cap and dome. (From Lees, 1961.)

evaporation while at the same time permitting the free uptake of oxygen (Lees, 1961).

In the fruit tree red spider mite *Panonychus ulmi* (Koch) the shells of the summer and winter egg are similar in construction and both are waterproofed by a wax layer which is secreted by the developing embryo into the inside of the shell; a difference is that winter eggs are held up in the female until a later stage of embryonic development, so that they have already received the inner wax layer at the moment of oviposition, whereas in summer eggs the waterproofing wax is secreted only 6 h after laying (Beament, 1951).

A morphological difference between diapausing and non-diapausing females which, according to Pritchard and Baker (1952) appears in all *Tetranychus* and *Eotetranychus* species living in temperate and northern climates, is found in the structure of the integumentary striae on the dorsal body surface. In the summer forms these cuticular ridges appear broken, showing semicircular or triangular lobes; in the winter forms these ridges are not incised and therefore are devoid of lobes (Plate 1.4.6.2). According to Boudreaux (1958) the lobes would increase the rate of evaporation of body moisture through the increased evaporative surface provided. The lack of lobes on the cuticle of diapausing mites might then be important for the conservation of water and could minimize the chance of desiccation during hibernation. Anatomical differences in the integument, midgut and ovary of active and diapausing females of *T. urticae* are illustrated in Plates 1.4.6.2 and 1.4.6.3.

Physiological differences between diapause and non-diapause forms have been found in the amount of fat or sugar present, as well as in the rate of respiration. According to McEnroe (cited in Boudreaux, 1963) fat is deposited

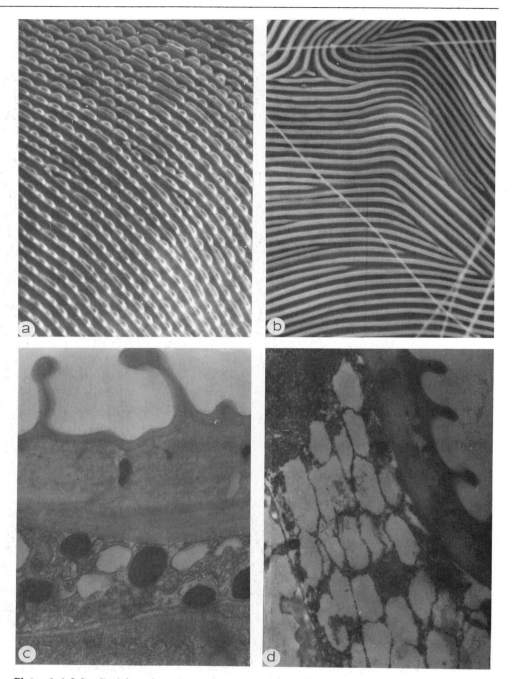

Plate 1.4.6.2. Cuticle of active (a, surface, × 2.700; c, cross-section, × 36.000) and diapausing (b, surface, × 2.700; d, cross-section, × 19.000) females of *Tetranychus urticae*. (By courtesy of F. Weyda (c, d) and F. Weyda and P. Berkovský (a, b).)

in the body tissues of diapausing *T. urticae*; water sufficient for maintaining life is probably obtained through oxidation of diapause fat in overwintering females. Oxygen consumption in diapausing *T. urticae* was shown to be very low compared to that of active females (McEnroe, 1961; Geispitz and Orlovskaja, 1971). An accumulation of sorbitol was found by Sømme (1965) in overwintering eggs of *P. ulmi*; the increased sorbitol content probably causes a depression of the supercooling points of the eggs.

Cold hardiness has been observed to be increased in diapausing mites in comparison with the respective non-diapausing forms. Diapause females of

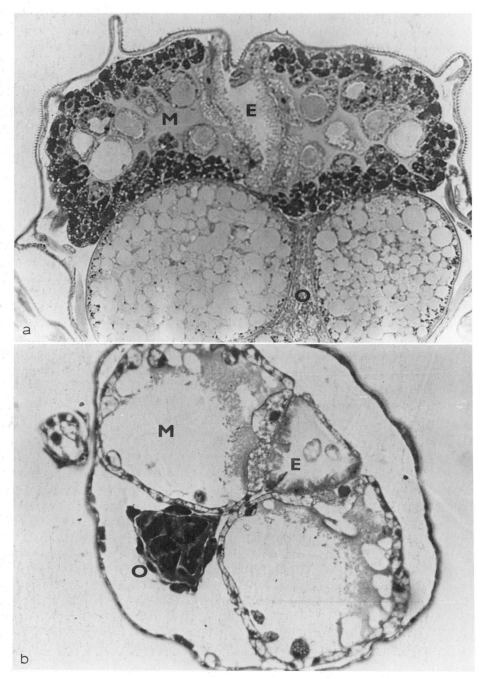

Plate 1.4.6.3. Whole-body cross-section of active (a, × 900) and diapausing (b, × 800) female of *Tetranychus urticae*. M, midgut; E, excretory organ; O, ovary with developing (a) or arrested (b) oocytes. (By courtesy of F. Weyda.)

T. urticae overwintering outdoors have been reported to withstand minimum temperatures of − 13°C (Lloyd, 1922) and − 27°C (Bondarenko, 1958); in laboratory tests van de Bund and Helle (1960) and Geispitz and Orlovskaja (1971) showed that diapause females of *T. urticae* could survive temperatures of − 22°C and − 24°C, respectively. When exposed to − 32°C, however, no mites survived (Bondarenko, 1958). According to Helle (1962) diapausing females could be preserved for at least 8 months at − 2°C, on condition that they were protected from drying. The importance of high humidity to survive winter diapause was demonstrated for *T. urticae* by Parr and Hussey (1966); at 6°C many mites held at relative humidities of 75% and 93% survived for more than 8 months, nearly twice as long as those kept in drier air with a relative humidity of only 40%.

Stenseth (1965) showed that survival at a constant temperature of −15°C was much higher for diapausing females than for active females of *T. urticae*. While freezing is fatal to the mites, they are able to survive temperatures below 0°C in a supercooled state. The supercooling point was shown to be −18.7°C for diapausing females and −22.4°C for active females; no correlation was found between the supercooling points of the different forms and their survival at constant low temperatures (Stenseth, 1965). Different results were reported by Geispitz et al. (1971), who found higher supercooling points in active females (in the range −12.6 to −16.9°C) than in diapausing females (−18.3 to −24.4°C) of a Russian strain of *T. urticae*.

The effect of low temperatures on the mortality of winter eggs has been investigated for the fruit tree red spider mite *P. ulmi* (Lienk and Chapman, 1958; MacPhee, 1961; Sømme, 1966) and the brown mite *B. rubrioculus* (MacPhee, 1963, 1964). Low temperature was shown to be an important direct mortality factor for both *P. ulmi* and *B. rubrioculus*. Winter eggs of both species have a considerable ability to supercool, but are killed by freezing. The mean supercooling points of 2 strains of *P. ulmi* studied by MacPhee (1961) in Canada were found to be −31°C and −37°C, respectively; the lowest supercooling points recorded for 2 Norwegian localities by Sømme (1965) were −31.4°C and −33.2°C. For *B. rubrioculus* a mean supercooling point of −32°C was found (MacPhee, 1963). The supercooling point is usually defined as the temperature at which freezing occurs at a rapid cooling rate (Sømme, 1966); as nucleation in a supercooled animal depends on both temperature and time (Salt, 1961; Sømme, 1982), freezing may occur after prolonged exposures to temperatures above the supercooling point. This may explain the finding that mortality is also strongly affected by the duration of exposure to low temperatures (MacPhee, 1961, 1964).

Differences in 'wetting' capacity and tolerance to pesticides between active and diapausing spider mites are probably caused by differences in cuticular properties of the 2 forms. Tibilova (1932, cited in Bondarenko, 1958) showed that diapausing females of *T. urticae* were still alive after 100 h of submersion in water, whereas summer females did not survive a stay under water of 10 h. A higher tolerance to a number of pesticides in diapausing females of *T. urticae*, in comparison with non-diapausing females, has been reported by several authors (Bondarenko, 1958; Fritzsche, 1959; Parr and Hussey, 1966; den Houter, 1976; Hürková and Weyda, 1982).

Mites which are induced to diapause, or, in species with an embryonic diapause, to lay diapause eggs, show distinct differences in behaviour in comparison with the summer forms. For instance, summer females of *P. ulmi* are relatively inactive and may deposit all their eggs on 1 leaf. Females laying winter eggs, on the other hand, become highly active at the maturation of each egg and walk from the leaf on which they have been feeding. After many searching movements they deposit the egg on the bark of the stem, often under the petiole or in crevices near dormant buds. Because of the constant alternation between leaf and bark, females are often seen feeding on a different leaf each day (Lees, 1953a).

In *T. urticae*, females which have been induced to diapause feed only very little before they leave the host plant in search of hibernation sites; mating takes place on the plant after the last ecdysis, but no eggs are laid (Linke, 1953; Nuber, 1961). Once the colour has changed and the hindgut has been emptied the mites become positively geotactic (Foott, 1965) and negatively phototactic (Bondarenko, 1958; Hussey and Parr, 1963; Parr and Hussey, 1966) and start to migrate from the plants. According to McEnroe (1971) the negative photoresponse of the diapausing females is a response to red and green light as well as to the ultraviolet region of the spectrum.

In greenhouses the mites have been found overwintering in cracks and crevices in the house structure, in supporting stakes, in hollow stems and in straws on the beds, in irrigation equipment, door locks and pipe fittings (Hussey, 1972; French and Ludlam, 1973). Outside, overwintering females of *T. urticae* have been found, often in great numbers, in clods of earth in apple orchards (Weldon, 1910), hop gardens (Massee, 1942; Linke, 1953), blackcurrant plantations (Collingwood, 1955), and in clay soils (Helle, 1962); however, it appeared that in most cases mortality of mites overwintering in the soil was high. Overwintering females have also been found in the cracks and under the bark of poles (Massee, 1942; Nuber, 1961; Helle, 1962), in dried leaves (Massee, 1942; Nuber, 1961), in straw (Collingwood, 1955), in the stalks of hop plants remaining in the field (Nuber, 1961), and in hollow withered flower stems (Helle, 1962). On woody host plants, females of *T. urticae* have been observed to hibernate concealed under the bark (Helle, 1962; Uchida, 1980).

A remarkable phenomenon reported by several authors (see, e.g., von Hanstein, 1902; Zacher, 1920, 1932; Carter, 1956) is the mass migration in the autumn of millions of winter females of *Eotetranychus tiliarium* (Hermann) on heavily infested lime trees down the trunk of the trees into their winter quarters on the basal part of the trees and in the soil. Zacher (1922) noted that the mites did not yet possess the characteristic winter colouration during the downward migration, whereas the red pigmentation had already disappeared when the mites migrated upwards again in the spring.

Further data on overwintering behaviour and hibernation sites of other species of tetranychid mites may be found in the references listed in Table 1.4.6.1.

ENVIRONMENTAL FACTORS GOVERNING THE ONSET OF DIAPAUSE

Until the early 1950s it was generally believed that the condition of the host plant was the major factor determining diapause in spider mites (see, e.g., Reiff, 1949; Gasser, 1951; Pritchard and Baker, 1952; Fritzsche, 1959). The role of photoperiod in the induction of diapause was first demonstrated by Lees (1950) and Miller (1950) for *P. ulmi* and by Bondarenko (1950) for *T. urticae*. The importance of daylength as a controlling factor for the induction of diapause has now been shown for a number of species, viz. *Petrobia apicalis* (Banks) (Glancey, 1958, cited in Boudreaux, 1963), *Eotetranychus hicoriae* (McGregor) (Micinski et al., 1979), *Schizotetranychus schizopus* Zacher (Geispitz, 1968), *Oligonychus ununguis* (Jacobi) (Shinkaji, 1975a), *Tetranychus cinnabarinus* (Boisduval) (Boudreaux, 1956), *Tetranychus kanzawai* Kishida (Uchida, 1980) and *Tetranychus viennensis* Zacher (Razumova, 1967). For some of the above species photoperiodic response curves have been determined, 3 of which are shown in Fig. 1.4.6.2. The photoperiodic reactions are of the so-called long-day type, short days resulting in the incidence of diapause and long days promoting diapause-free development. The populations of *P. ulmi*, *T. urticae* and *T. viennensis* represented in Fig. 1.4.6.2 all show a sharply defined critical daylength of about 14 h. The main difference among the 3 species is found in their response to very short daylengths, where the reaction to the photoperiod is more or less weakened in *P. ulmi* and *T. viennensis*, but not in *T. urticae*. On

Chapter 1.4.6. references, p. 310

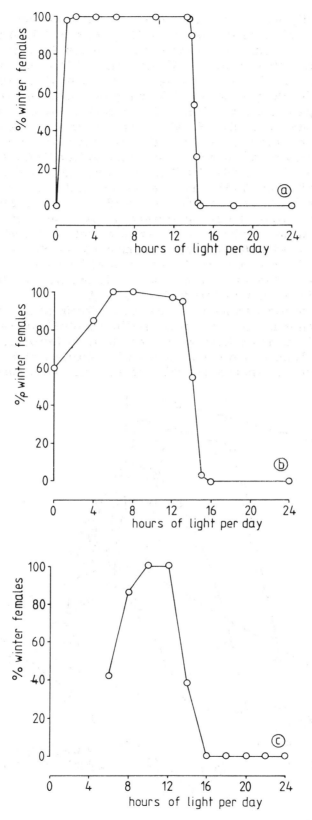

Fig. 1.4.6.2. Photoperiodic response curves for the induction of diapause in: (a) *Tetranychus urticae* Koch (from Veerman, 1977a); (b) *Panonychus ulmi* (Koch) (from Lees, 1952); (c) *Tetranychus viennensis* Zacher (from Razumova, 1978).

the other hand, diapause is absent in *T. urticae* in continuous darkness, whereas about 60% diapause still occurs in *P. ulmi* under this regime. The photoperiodic response curve of *O. ununguis* as determined by Shinkaji (1975a) strongly resembles in its overall shape that of *T. urticae* shown in Fig. 1.4.6.2a, the only differences being found in the critical daylength, which was 12.5 h for the population of *O. ununguis* tested, and in the fact that the latter species shows full diapause induction in continuous darkness. Although the extremes of the photoperiodic response curves have no significance for the natural responses of the mites, the results may be of interest for interpretation of the mechanism of photoperiodic time measurement. As shown by Lees (1953a) for *P. ulmi* and by L.P.S. Van der Geest (personal communication, 1983) and the present author for *T. urticae*, photoperiod has a direct effect on the mites and does not act through the medium of the host plant. It was also shown that photoperiodic induction depends on the absolute daylength experienced by the mites and not on a gradual change in daylength. This was demonstrated for *P. ulmi* by Lees (1953a) and for *T. viennensis* by Volkovich and Razumova (1982), in experiments where the mites were exposed to increasing and decreasing daylengths, either below or above the critical daylength.

Although photoperiod was shown to be the predominant factor for the regulation of diapause in the above species and will probably prove to be of equal importance in many of those species for which the factors controlling the evocation of diapause have not yet been investigated, other environmental variables such as temperature and food also exert an influence. The

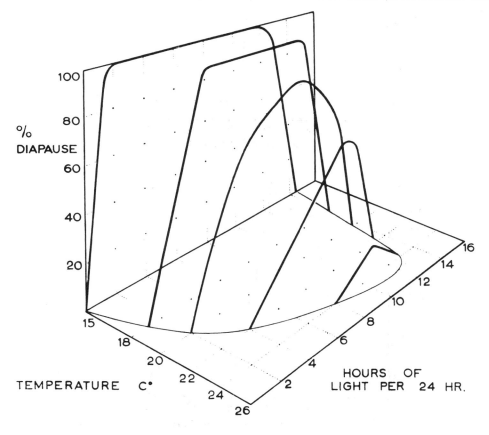

Fig. 1.4.6.3. The effect of temperature on the photoperiodic response curve in a population of *Tetranychus urticae* Koch from Holland. (From Helle, 1962.)

Chapter 1.4.6. references, p. 310

effect of various constant temperatures on diapause induction was determined by Lees (1953a) for *P. ulmi*, by Geispitz (1968) for *S. schizopus* and by Bondarenko and Kuan Khay-yuan (1958) and by Helle (1962) for *T. urticae*. As shown in Fig. 1.4.6.3 for *T. urticae*, higher temperatures tend to suppress the incidence of diapause over the complete range of photoperiods; if the temperature is high enough (25°C in the case of the Dutch strain of *T. urticae*) no diapause is found at any photoperiod. In *P. ulmi*, the response was greatly reduced at 25°C, although about 20% diapause was still found at photoperiods with photophases of 8 and 12 h (Lees, 1953a). Low temperatures have the reverse effect: at 10°C, 45% of the females of *P. ulmi* produced diapause eggs even in a light/dark regime with a 16 h photophase (*LD 16*:8) (Lees, 1953a). High temperatures and long-day photoperiods appear to act in concert to avert diapause, whereas the combination of low temperatures and short-day photoperiods tends to promote diapause.

The action of temperature in relation to the light and dark phases of the cycle of illumination has been studied by Lees (1953a) for *P. ulmi*, by Shinkaji (1975a) for *O. ununguis* and by Veerman (1977a) for *T. urticae*. In the case of *O. ununguis* the occurrence of diapause appeared to depend on the daily mean temperature and not on whether the phases of high and low temperature coincided with either the light or the dark phase. On the other hand, in *P. ulmi* and, even more pronounced, in *T. urticae*, the dark phase proved to be temperature sensitive, whereas high temperatures during the light phase were entirely without influence (Table 1.4.6.2).

TABLE 1.4.6.2

Effect on diapause induction in *T. urticae* of a thermoperiod, in combination with a photoperiod or in continuous darkness

Experiment No.	Regime	Diapause (%)	Number of mites tested
1	*LD 12*(25°): 12(25°)	0	548
2	*LD 12*(15°): 12(15°)	100	199
3	*LD 12*(15°): 12(25°)	1	529
4	*LD 12*(25°): 12(15°)	98	524
5	*LD 12*(20°): 12(20°)	95	316
6	*LD 0*: 24(20°)	0	425
7	*LD 0*: 24; 12(25°): 12(15°)	2	302

From Veerman (1977a)

No thermoperiodic effect on diapause induction was found in *T. urticae*; when mites were reared in a thermoperiod consisting of twelve-hourly phases of 25°C and 15°C in continuous darkness, the incidence of diapause was negligible; however, almost complete diapause was found when the same thermoperiod was combined with a *LD 12*:12 light/dark cycle, in which the low-temperature phase coincided with the dark phase, or when the mean temperature of 20°C was combined with the same light/dark regime (Table 1.4.6.2). This shows that at least in *T. urticae* a thermoperiod cannot replace photoperiod as an environmental signal for the regulation of diapause; thermoperiodic induction of diapause has been shown to occur in some insect species (Saunders, 1982).

Apart from the possible direct effect of temperature on the induction mechanism, especially during the hours of darkness, temperature may also indirectly influence the incidence of diapause by its effect on the rate of development of the mites, which determines the duration of the period during which the mites are sensitive to photoperiodic induction. If this sensitive period becomes very short, it is conceivable that the number of light/dark cycles experienced by the mites will be insufficient for attaining full diapause.

The developmental stages sensitive to photoperiodic induction have been determined for 4 species of spider mite, mainly by transferring mites in different stages of development from diapause-inducing to diapause-suppressing conditions and vice versa. In *P. ulmi* eggs, larvae and protonymphs were found to be insensitive to photoperiod. The deutonymphal instar showed the greatest sensitivity, but egg-laying females, if exposed to antagonistic conditions, could still be induced to 'switch-over' to the alternative egg type (Lees, 1953a). In *O. ununguis* the larva as well as the nymphal stages were sensitive to the photoperiod, but 2 consecutive developmental stages had to be exposed to short-day photoperiods for induction to take place (Shinkaji, 1975a). For *T. urticae*, more or less conflicting results have been obtained, possibly because of differences in the experimental set-up. Bondarenko (1950) produced winter forms by exposing deutonymphs and adult females to a short-day regime; Boudreaux (1956), on the other hand, obtained winter-type females by exposure of larvae and protonymphs to a short-day photoperiod. Veerman (1977a) found the greatest sensitivity in the protonymph and a sharp decline in sensitivity in the deutonymph. Parr and Hussey (1966) and Veerman (1977a) showed that a rather high sensitivity to photoperiod is already present in the larva. No sensitivity was ever reported to be present in the egg stage, except by Gotoh and Shinkaji (1981) in their work with a Japanese strain of *T. urticae*. Reversal of determination by an antagonistic regime, as mentioned above for *P. ulmi*, was also demonstrated to occur in *T. urticae*, although here the reversal occurred prior to the expression of the photoperiodic response in the young adult female (Veerman, 1977a). For *T. viennensis*, Volkovich and Razumova (1982) concluded that sensitivity to daylength appears towards the end of the larval stage, reaches a maximum in the protonymph and declines rapidly after the moult to the deutonymphal stage.

A third factor that may influence the onset of diapause is the quality of the food plant. Diapausing females of *T. urticae* have been observed at an earlier date or in greater numbers on older or senescent leaves than on the green leaves of various host plants (see, e.g., Reiff, 1949; Linke, 1953; Collingwood, 1955; Bondarenko, 1958; Kingham, 1960, cited in Parr and Hussey, 1966; Gould and Kingham, 1965). An influence of the host plant on the incidence of diapause could be inferred also from the work of Bengston (1965), who observed appreciable numbers of winter females of *T. urticae* on deciduous hosts in Queensland in Australia at the relatively low latitude of 28°S, whereas on evergreen hosts the species persisted mainly as the active two-spotted form. Similar dual overwintering behaviour has been reported by Dosse (1967) for *T. cinnabarinus* (the red form of the *T. urticae* complex) in the Lebanon; in this case, however, both active and diapause females were found on the same plants, the winter females remaining on the plants (*Viola* sp.) during the whole winter.

Experiments in which *T. urticae* was reared under contrasting nutritional conditions (young leaves versus senescing leaves) mostly showed that food has only a minor influence (Nuber, 1961; Parr and Hussey, 1966; author's own observations, 1983); only when photoperiod and temperature are becoming critical does the inadequacy of food have a noticeable impact on diapause induction. A more profound effect was found by Bengston (1965), who compared the incidence of diapause in *T. urticae* reared under various photoperiods on either young mature bean leaves or senescing apple leaves; with the marginal nutrition, diapause was increased at all photo-

Chapter 1.4.6. references, p. 310

periods tested, although a clear photoperiodic response was still present under both nutritional regimes.

As remarked by Parr and Hussey (1966), the red forms of *T. urticae* observed by several authors (Reiff, 1949; Gasser, 1951; Bondarenko, 1958; Lees, 1953a) on leaves senescing either through old age or mite-feeding damage, even in the summer or under long-day photoperiods, could have been starving females which are brick-red in colour and so resemble the hibernating form. This might also explain the observation by Gasser (1951) that summer females could be transformed into winter females in 5—7 days at any time of the year by transferring the mites to withering leaves.

For *P. ulmi* Lees (1953a) showed that with an adequate food supply the incidence of diapause is determined solely by photoperiod and temperature. If the food supplies are restricted, however, females laying diapause eggs appear even when photoperiod and temperature are such as to prevent diapause. The same was observed by Shinkaji (1975a) for *O. ununguis*; in this species, winter-egg laying females appeared either under crowded conditions or on 'bronzed' leaves previously infested with a large mite population, even when both photoperiod and temperature tended to prevent diapause. In these cases nutrition appears to operate as a limiting factor, diapause being induced regardless of photoperiod and temperature if food intake falls below a certain threshold.

Under natural conditions the incidence of diapause will be the result of the concerted action of photoperiod, temperature and nutrition affecting the mites during the developmental stages sensitive to diapause induction. Undoubtedly, photoperiod is the predominant factor in most cases. However, the relative importance of the 3 environmental factors mentioned apparently varies from species to species; e.g., whereas in some species nutrition exerts only a modifying influence, tending to direct development towards diapause if photoperiod and temperature are critical, in others food inadequacy may overrule the influence of photoperiod completely (Kuenen, 1949; Lees, 1953a).

Aestivation has been reported for only a few species of spider mite, both in the egg stage (*B. praetiosa, Petrobia harti* (Ewing), *P. apicalis, P. latens, Aplonobia juliflorae* Tuttle and Baker) and as adult females (*Eotetranychus yumensis* (McGregor), *T. urticae*) (Table 1.4.6.1). Little is known of the factors controlling the initiation of an aestival diapause in these mites. For *P. apicalis*, Zein-Eldin (1956) concluded that the initiation of diapause egg laying was independent of photoperiod, temperature and food availability; the number of generations without diapause would be controlled endogenously. However, Glancey (1958, cited by Boudreaux, 1963) showed that diapause in this species is under photoperiodic control; he was able to force females hatched from diapause eggs into diapause-egg production by exposing them during their development to daily photoperiods with a photophase of 14 h. According to Brooking (1957, cited by Boudreaux, 1963), it is the lengthening of the days during the spring which induced females of *P. apicalis* to produce diapause eggs. Temperature seemed to be unimportant, but weakened food plants would induce diapause independently of the photoperiod.

The diapause egg of *P. latens*, which is quite characteristic in appearance and completely different from the non-diapause egg (Fig. 1.4.6.1), is reported to serve for overwintering in Oklahoma but for aestivating in California (Baker and Pritchard, 1953). Lees (1961) showed that the production of diapausing winter eggs of *P. latens* in England is not under photoperiodic control; the controlling factors remain uncertain. The same can be said for *Schizonobia sycophanta* Womersley, which lays 2 types of egg in Holland, the production of which could not be shifted towards either type by rearing the mites under various photoperiodic regimes (A.Q. Van Zon, personal communication, 1973 and author's own observations).

Aestivating females of *T. urticae* which behave like true diapausing females have been reported from India by Srivastrava and Mathur (1962). However, no morphological examination of the striae has been made and no mention is made of the colour of the aestivating mites. In the U.S.S.R. Geispitz and Orlovskaja (1971) produced orange—red females of *T. urticae* by rearing the mites at high temperatures (28—30°C) and under constant illumination. Comparison of these 'aestivating' females with both active females, reared under a 20 h photophase at 18°C, and orange—red diapausing females reared under a 12 h photophase at the same temperature, showed that the 'aestivating' females were much more resistant to desiccation than both the 'summer' and 'winter' females; their oxygen consumption was much reduced compared to the summer females, but their resistance to temperatures below zero was intermediate between that of active females and 'short-day' diapausing females. The authors concluded, in view of these physiological characteristics, that the 'long-day' diapausing females of *T. urticae* should be considered to be in a state of true diapause. No mention was made, however, of the occurrence of these 'long-day' diapausing females either outdoors or under greenhouse conditions. As stated by Hussey (1972), aestivating females of *T. urticae* have not been encountered under glass; starving mites resembling diapause females have been found in greenhouses in the summer, but these have only a limited capacity for survival.

ENVIRONMENTAL FACTORS GOVERNING THE TERMINATION OF DIAPAUSE

Successful control of spider mites hatching from winter eggs or emerging from their winter quarters may form an important means of limiting the development of infestations later during the year. It is not surprising therefore that the first appearance in the spring, both under field and greenhouse conditions, as well as factors influencing the phenology of spider mites, have been studied for several species of economic importance. For example, in the case of *P. ulmi*, attempts have been made to predict the time of hatching of winter eggs by using the sums of effective temperatures accumulated by the mites (Tsugawa et al., 1961, 1966; Light et al., 1968; Cranham, 1972, 1973; Trottier and Herne, 1979; Herbert and McRae, 1982). Here some observations will be summarized for the species investigated, made either in the field under various local climatic conditions, or using material brought into the laboratory at regular intervals from populations overwintering outside. In addition, some results will be discussed of experiments with spider mites kept under constant conditions in the laboratory from the start of the experiments onwards.

The time of hatching of winter eggs of *B. rubrioculus* has been studied by Anderson and Morgan (1958) and Herbert (1965) in orchards in Canada. In British Columbia the eggs of *B. rubrioculus* were found to hatch 2—3 weeks earlier than those of *P. ulmi* on the same trees (Anderson and Morgan, 1958). Winter eggs of the clover mite, *B. praetiosa*, from the same area, began to hatch very early in the spring; usually they hatched a month or more before those of the brown mite, *B. rubrioculus*. Aestivated eggs of *B. praetiosa* began to hatch in September, the peak emergence of the larvae following the first period of cool weather in the autumn (Anderson and Morgan, 1958). In Europe Mathys (1957) succeeded in breaking the diapause of winter eggs of

Chapter 1.4.6. references, p. 310

the univoltine *B. ribis* from January onwards by transferring the eggs to a regime with temperatures of 15°C and above.

For *P. apicalis*, Zein-Eldin (1956) showed that aestivating eggs require storage at high temperatures and a relatively dry atmosphere for a period of 2—3 months after laying, before low temperature and high humidity can break diapause.

Termination of diapause in the conifer spider mite, *O. ununguis*, has been studied by Löyttyniemi (1970) in the south of Finland and by Shinkaji (1975b) in Japan. In both cases it was found that diapause is already ended in the latter half of the winter; e.g., for the Japanese population of *O. ununguis* it was shown that most winter eggs fail to hatch before January, when brought into the laboratory and incubated at 25°C, but hatchability increased during the following months and reached a maximum between late February and early March. In the field, 50% hatch of the winter eggs was not observed until the end of April and the beginning of May in 2 consecutive years (Shinkaji, 1975b).

In *P. ulmi* the influence of winter temperatures on the mortality of winter eggs has been studied by Lienk and Chapman (1958) in the United States, by MacPhee (1961) in Canada and by Kozár (1974) in Hungary. The time of hatching of winter eggs of *P. ulmi* in the field and the influence of the temperature of hatching has been investigated by Lees (1950, 1952, 1953a), Becker (1952), Tsugawa et al. (1961, 1966), Light et al. (1968), Cranham (1971, 1972, 1973) and Trottier and Herne (1979). Lees (1953a) found that a very lengthy cold treatment was required to break diapause in *P. ulmi*, the maximum percentage hatch occurring only after the eggs had been chilled for 150—200 days.

An artificial means of breaking the diapause in winter eggs of *P. ulmi* was reported by Dierick (1950), who showed that winter eggs collected at the end of November could be forced to hatch at 20°C after dipping the eggs for 20 min in xylol. Lees (1953a) supposed that penetration of the solvent through the shell in subtoxic quantities might serve as a stimulant to embryonic development. However, J.E. Cranham (personal communication, 1984) notes that he was unable to repeat these results.

Hueck (1951) found that in *P. ulmi* light may provide a stimulus for the larva to break the shell of the winter egg. He incubated eggs at 25°C and 80% relative humidity both in daylight and in the dark; the mean percentage hatch obtained was 75% in daylight compared to 52% in darkness. Becker (1952) also reported a somewhat reduced hatch in darkness. In contrast, J.E. Cranham (personal communication, 1984) found that exposure to natural photoperiods (daylight) during chilling, or to daily photoperiods with photophases of 16 h (strip lighting) during incubation at 21°C, when compared to continuous darkness, did not increase the percentage hatch or rate of hatch.

J.E. Cranham (personal communication, 1984) studied the effect of relative humidity during chilling on the hatching of winter eggs of *P. ulmi*. At relative humidities of 35, 60 and 97% during chilling at − 5, 0 and 5°C and with subsequent incubation at 21°C and 60% relative humidity, he found significant differences associated with the temperature and duration of chilling, but not with the 3 humidities. He concludes that survival of ~ 6 months storage at humidities as low as 35% is evidence of the waterproof nature of diapause eggs. From the work of Kuenen (1946), Andersen (1947) and J.E. Cranham (personal communication, 1984) it appears that relative humidity during incubation does have an effect on the percentage hatch, although the effect is not large over the range of humidities from about 60—90%; however, at low (below 40%) or high (above 90%) humidities, hatch was greatly reduced.

An extensive study of the effect of temperature on hatching of the winter eggs of *P. ulmi* from English apple orchards has been made by Cranham (1971, 1972, 1973). He showed that, as the chilling period was increased from 60 to over 200 days, the total percentage hatch on subsequent incubation at 21°C increased, and the mean incubation time and its variance decreased. The same has been reported by Tsugawa et al. (1966) for *P. ulmi* in Japan. Cranham (1972) found that maxima in the total percentage hatch were reached after 200—230 days of chilling. After 260—300 days, eggs kept at 5°C had become non-viable; the hatching of eggs stored at 0°C began to decline after 230 days and was negligible after 400 days. Temperatures of 0 and 5°C appeared to lie within the optimum range for the chilling requirement to terminate diapause and were more effective than − 5°C and 9°C, however long the period of chilling. The upper and lower limits for the reaction were found to be close to 15 and − 10°C (Fig. 1.4.6.4). With chilling temperatures of − 5, 0 and 5°C, the mean incubation time declined progressively to minima of 17, 9 and 6.5 days, respectively. The threshold temperature for complete development to eclosion was found to be 7°C. The minimum value of 9 days found after chilling at 0°C was the same as that required for full morphogenesis in summer eggs at 21°C. This indicates that diapause was fully completed but that no morphogenesis could occur at 0°C. At 5°C, on the other hand, the earlier stages of morphogenesis could occur, to the extent of reducing the time for 50% hatch to 6.5 days. At − 5°C the last stages of diapause development apparently could not take place, judging from the fact that the minimal value for the mean incubation time was 8 days more than the 9 days required for morphogenesis at the incubation temperature of 21°C. It is evident, however, that the final stages of diapause development did occur at the higher temperature in those eggs that hatched after being transferred from − 5 to 21°C.

In Fig. 1.4.6.4, data are shown also for the percentage hatch of winter eggs over the range of incubation temperatures after chilling at 5°C for 169 days, and the percentage hatch of the non-diapause summer eggs over the effective range of temperature. These data therefore illustrate and compare the ranges of temperature which are effective for (a) chilling, (b) post-diapause development in the winter eggs, and (c) morphogenesis in the non-diapause summer eggs. Diapause development clearly takes place only within a certain temperature range with a well-defined temperature optimum, which lies well below the optimum temperature for morphogenesis. Only around 9°C do the ranges of both processes overlap, and Cranham (1972) showed that some eggs stored continuously at 9°C hatched after 200 days, but the total percentage hatch was smaller and occurred more slowly than when eggs were chilled at 0 or 5°C and then incubated at 9°C. A temperature of 27°C appeared to be close to the upper limit for post-diapause development, which is notable because more than 90% of the summer eggs hatched at 27°C. According to Cranham (1972) the final processes in the termination of diapause and those of early morphogenesis occur concurrently in *P. ulmi*. He concluded that diapause development in *P. ulmi* involves reactions which progress steadily to completion and which at no point release the egg abruptly from the diapause state. Thus, late stages of diapause development, which can be slowly completed at temperatures below the threshold for morphogenesis, can also be completed more rapidly over a range of temperatures above the threshold. If the chilling requirement provides the 'coarse adjustment' of the mechanism controlling the time of hatch, the 'fine adjustment' is provided by the gradually declining heat sum required to complete diapause development. The effect of this is to put a 'brake' on the rate of development, a brake which is gradually released by rising field temperatures and only more

Chapter 1.4.6. references, p. 310

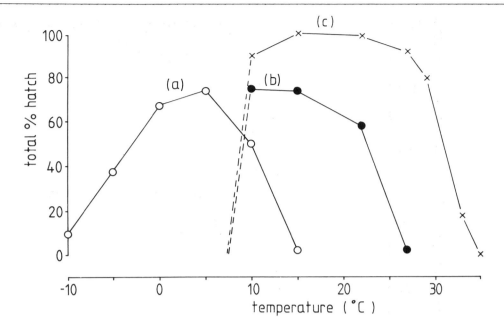

Fig. 1.4.6.4. The ranges of temperature effective in *Panonychus ulmi* (Koch) for: (a) the termination of diapause by chilling (percentage hatch at 21°C after chilling), (b) post-diapause development of winter eggs, and (c) development of the summer eggs. (From Cranham, 1972.)

slowly at low temperatures. This acts to prevent premature hatch if the weather is unseasonably warm in March or April; alternatively, if cold weather is unusually prolonged, the hatch will occur faster when it does turn warm (Cranham, 1972).

In *T. urticae*, observations on the termination of diapause under natural and laboratory conditions have been reported by Lees (1953a), Linke (1953), Bondarenko (1958), Nuber (1961), Saba (1961), Helle (1962), Dubynina (1965), Gould and Kingham (1965), Parr and Hussey (1966), Dosse (1967), Geispitz et al. (1971), Glinyanaya (1972), Veerman (1977b) and Uchida (1980). A great variation was found in the period of chilling required to break diapause in *T. urticae* from different localities. Lees (1953a) showed that diapause was ended in an English strain of *T. urticae* after exposure to temperatures below 10°C for 100 days. Bondarenko (1958), on the other hand, found that in Leningrad mites it required only 55 days at 3—6°C to break diapause, and in England Parr and Hussey (1966) showed that after a cold rest at 7°C of only 14 days, 80% of the females of *T. urticae* terminated diapause when transferred onto fresh foliage at 25°C and a daylength of 16 h. Both Helle (1962) and Parr and Hussey (1966) observed that a certain percentage of newly diapausing females reverted to active females without previous cold-treatment when placed at 25°C. Parr and Hussey (1966) concluded that there are probably wide differences in diapause intensity, even between strains from similar climates. However, one point which makes comparisons of the results of different authors rather difficult is the fact that in most cases the daylength at which the mites were kept during attempts to reactivate them has not been specified. As Glinyanaya (1972) showed for *T. viennensis* and *T. urticae* from the U.S.S.R. and Veerman (1977b) for *T. urticae* from Holland, whether or not diapause can be terminated at incubation temperatures of from 15 to 25°C depends very strongly on the photoperiod, at least during the first 3 or 4 months of diapause (Fig. 1.4.6.5). When diapausing mites were transferred

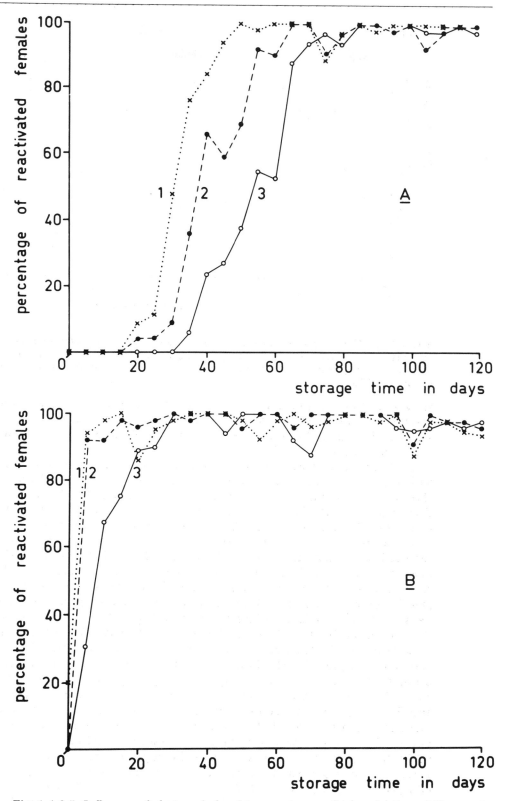

Fig. 1.4.6.5. Influence of photoperiod and temperature on the termination of diapause in *Tetranychus urticae* Koch. Reactivation of diapause females at 3 different temperatures (1, 23°C; 2, 20°C; 3, 17°C) in relation to the cold rest during which the mites were stored at 4°C. A, cold rest and reactivation in short days (*LD 10*:14); B, cold rest and reactivation in long days (*LD 16*:8). (From Veerman, 1977b.)

Chapter 1.4.6. references, p. 310

at regular intervals from cold storage at 4°C to fresh leaves and incubated at 3 different temperatures (17, 20 and 23°C) under either a short-day (*LD 10*:14) or a long-day (*LD 16*:8) regime, all mites terminated diapause within a few days under the long-day regime after a previous cold-treatment of only a few weeks, whereas a cold rest of between 1 and 3 months, dependent on the subsequent incubation temperature, was required before the mites terminated diapause under the short-day regime (Fig. 1.4.6.5). Glinyanaya (1972) showed that the critical daylength for the photoperiodic termination of diapause is the same as that for the induction of diapause. In view of the fact that sensitivity to the photoperiod gradually disappears a few months after the initiation of diapause, both Glinyanaya (1972) and Veerman (1977b) concluded that the ecological significance of this photoperiodic sensitivity should not be sought in the termination of diapause, but rather in its maintenance. Sensitivity to photoperiod during the first few months of diapause may well serve to retain the diapause state during the autumn until the arrival of low winter temperatures. These experiments suggest that diapause development is ended already at the beginning of the winter, the mites remaining dormant till spring under the influence of prevailing low temperatures. These findings are in agreement with the observations of several authors that diapause is ended by midwinter, as judged by the rapid return to feeding and egg-laying and loss of the winter colouration in mites brought into the laboratory and put onto fresh foliage at room temperature (see, e.g., Nuber, 1961; Dosse, 1967; Uchida, 1980, for both *T. urticae* and *T. kanzawai*). For species with an egg diapause it was found that hatchability of winter eggs, brought into the laboratory and incubated at higher temperatures, already reached a maximum in the middle of the winter (see, e.g., Mathys (1957) for *B. ribis*, Tsugawa et al. (1966) for *P. ulmi*, and Löyttyniemi (1970) and Shinkaji (1975b) for *O. ununguis*).

Working with a Russian strain of *T. urticae*, Dubynina (1965) found from 1—3 rather irregular maxima and minima in the percentage of mites which returned to the active form on fresh bean leaves, in experiments in which the capability to terminate diapause was tested for diapausing females which had spent a cold period of variable duration at temperatures between — 5°C and 10°C. The number of 'cycles', as well as the duration of each cycle, which could be 25—45 days, appeared to be dependent on the storage temperature. The ecological significance of this 'secondary rest', as it was called by Dubynina (1965), is not clear. It should be mentioned that no specification was given of the temperature and the daylength at which the reactivation experiments were performed. As shown in the work on *T. urticae* by Glinyanaya (1972) in the U.S.S.R. and Veerman (1977b) in the Netherlands, daylength may have a profound effect on the termination of diapause during the first 100 days or so; no trace of cyclicity was found in either study.

Many points of interest concerning diapause development and the termination of diapause in spider mites deserve further investigation. One of these is the question of whether or not there is any connection between the inductive regime and the intensity or duration of diapause. One result worth mentioning in this context is the finding of Cranham (1972) that in *P. ulmi* the date of oviposition may have a small influence on the date of hatching, but the extended hatching period of winter eggs as found in England is not determined by the length of the oviposition period.

INTRASPECIFIC VARIABILITY IN DIAPAUSE RESPONSE

The variability of the photoperiodic reaction in spider mites has been studied by Russian acarologists in particular; they showed that the diapause response is not a constant characteristic of a given species, but may show wide variations, e.g. between populations of different geographic origin, or even within 1 population over a number of successive generations. In this section the most interesting results will be discussed briefly, with some emphasis on those points which still need clarification or where conflicting results have been obtained.

Bondarenko and Kuan Khay-yuan (1958) were the first to show that the critical daylength for the photoperiodic induction of diapause differed in populations of *T. urticae* obtained from different localities in the U.S.S.R. Mites from low latitudes responded to much shorter critical daylengths than did those from higher latitudes, the critical daylength decreasing by 1 h for each 3 degrees fall in latitude. This change in critical daylength is clearly of selective advantage, mites living at higher latitudes compensating for the longer summer daylengths and the earlier onset of winter with a higher diapause threshold. Further work by Geispitz (1960, 1968) showed that populations from the same latitude may also show considerable differences in photoperiodic response (Fig. 1.4.6.6). Undoubtedly, local climatic conditions are the most important factor determining the date of entry into diapause, and this is reflected by the differences in the photoperiodic response curves of populations which, although living at the same latitude, are exposed to widely different climates (Fig. 1.4.6.6, populations 3 and 4). The importance of adaptation to local conditions and the overriding influence of photoperiod was also demonstrated for *T. viennensis* by Danilevskii and Kuznetsova (1968); they showed that mites transferred from Leningrad to the Black Sea coast all diapaused in midsummer, notwithstanding the favourable conditions of temperature and food which allowed the local population of *T. viennensis* to develop completely without diapause. Comparable results were obtained by Gotoh and Shinkaji (1981) with different populations of *T. urticae* in Japan.

In regions with a mild winter climate, *T. urticae* does not diapause but survives the winter on various crops or weeds in the active form; this has

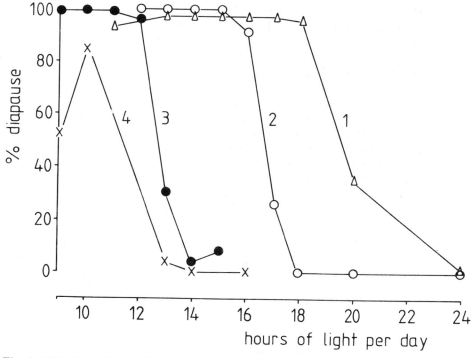

Fig. 1.4.6.6. The effect of latitude on the photoperiodic response of *Tetranychus urticae* Koch. Photoperiodic response curves determined at 18°C for 4 populations of mites: 1, from Chibinsk (67°N); 2, from Leningrad (60°N); 3, from Kirghisia (43°N); 4, from Abchasia (43°N). (From Geispitz, 1968.)

Chapter 1.4.6. references, p. 310

been reported by McGregor and McDonough (1917) for the Southeast U.S.A., later confirmed by Brandenburg and Kennedy (1981) for North Carolina, by Fenner (1962) for *T. urticae* from the Upper Murray in Australia and by Gotoh and Shinkaji (1981) for different populations from Japan. In laboratory experiments it was found that a photoperiodic response was absent in a population of *T. urticae* originating from Egypt (Author's own observations, 1983).

Some controversy exists over whether or not diapause occurs in *T. cinnabarinus*, the widely distributed red form of the *T. urticae* complex, which occurs mainly in regions with mild winter climates and in the subtropics. In northern regions it is found in greenhouses only, where it has never been observed to enter diapause; attempts to induce diapause in these greenhouse forms have been unsuccessful (Hussey and Parr, 1958; van de Bund and Helle, 1960). Diapause does occur in *T. cinnabarinus*, however, as was shown in photoperiodic induction experiments by Boudreaux (1956) with a strain from the United States (the exact origin of which is not mentioned), by Helle and Overmeer (1973) and Dupont (1979) with a strain from Ferrara in Italy, and by M. Vaz Nunes (personal communication, 1984) with a strain from Thessaloniki in Greece. Dosse (1964, 1967) was the first to describe diapausing females of *T. cinnabarinus* from the field in the Lebanon. At an altitude of 1500 m, all females living on apple trees were found to hibernate as diapausing forms hidden under the bark of the trees; at a height of 1000 m both active and diapausing females were seen on *Viola* spp. during the winter, and in the coastal plain of the Lebanon no diapause was found to occur at all. Hazan et al. (1971), working with a population of *T. cinnabarinus* from Israel, reported that attempts to induce diapause in these mites were totally unsuccessful. It would seem that the potential for diapause is easily lost in *T. cinnabarinus* under conditions favouring homodynamic development.

Considerable intra-populational changes in the photoperiodic reaction were observed for the first time by Geispitz (1960) in 2 strains of *T. urticae*, 1 from Leningrad and 1 from Tashkent, after rearing the mites for several months in the laboratory under continuous light at 25°C and 15°C, respectively. The changes consisted of a shift in the critical daylength as determined at 18°C; in the Leningrad strain the threshold decreased from 17 h to 14.5 h within 5 months of laboratory rearing, but in the Tashkent strain it increased from about 14.5 h to almost 17 h within 7 months. According to Geispitz (1960) the direction of the change in critical daylength might depend on the temperature at which the cultures have been kept. This has not been verified, however, by rearing the Leningrad strain at the lower temperature and the Tashkent strain at the higher temperature.

Dubynina (1965) found that in both a natural and a greenhouse strain of *T. urticae* from Moscow, and in a greenhouse strain from Tashkent, no diapause could be induced in the first generation of mites produced by females which had been in diapause; the capacity to diapause under short-day conditions (*LD* 10:14 and 18°C) returned only gradually in the following 3—5 generations. The ecological significance of this lack of tendency to diapause in the spring generations would be the avoidance of a re-entry into diapause of the first post-diapause generations in the spring, when the temperature is favourable for development but the daylength is still below the critical threshold (Dubynina, 1965). These results have not been confirmed for any other population of *T. urticae*, however. There are a number of reports, both for field and greenhouse populations, in which it is stated that the photoperiodic reaction does not change in successive generations, regardless of whether the previous generations have been in diapause or not. On cucumbers in England, Lloyd (1922), Speyer (1924) and Parr and Hussey (1966) observed that females of the first post-diapause generation often re-entered diapause, particularly in early heated cucumber houses, where the primary emergence of the mites took place in January and the daylengths during the following weeks were sufficiently short to induce diapause. Nuber (1961) was able to induce diapause in the first post-diapause generation of *T. urticae* from a hop plantation in West Germany, and Hussey (1972), in experiments with an English strain of *T. urticae*, found 72%, 92% and 73% diapause in the first, second and third generation respectively, following reactivation from diapause. Tjallingii (cited in Helle and Overmeer, 1973), using the same experimental procedures as Dubynina (1965), succeeded in inducing diapause in the post-diapause generation of 2 populations of *T. urticae*, 1 from Holland and 1 from Celje in Yugoslavia. Also, Veerman (1977a) found that the Dutch strain 'Sambucus', which had been kept in the laboratory for over 15 years at the time of the experiments, showed a constant photoperiodic reaction, regardless of whether the preceding generations had diapaused or not. The observations of Dubynina (1965) are not in agreement with those of Geispitz et al. (1971) either; they were able to induce diapause at any time of the year in samples from a natural population of *T. urticae* in the vicinity of Leningrad.

In the latter experiment Geispitz et al. (1971) studied the annual dynamics of the photoperiodic reaction of a natural population of *T. urticae*; they determined a series of photoperiodic response curves at 25°C, one for almost every month of the year. Mites taken directly from an unspecified natural habitat served as parents for each experimental generation. The curves show considerable variation, in terms of both threshold and response level. The photoperiodic responses appeared to be weakened from August to December; response curves determined during this period showed an unusual biapical shape. The seasonal variation found is surprising, since it appears to have no ecological significance. The highest tendency to diapause is found in February and the lowest in October, which is quite the opposite of what Dubynina (1965) established for the Moscow strains of *T. urticae*. The results of Geispitz et al. (1971) with the Leningrad population of *T. urticae* differ from those of other authors in still another respect; Bondarenko and Kuan Khay-yuan (1958), also working with a Leningrad strain of *T. urticae*, and Helle (1962), using a Dutch strain of the same species, found 2.8% and 0% diapause, respectively, at 25°C whereas Geispitz et al. (1971) obtained high percentages of diapause at this temperature under various short daylengths and at all times of the year.

In subsequent papers Geispitz et al. (1972a, b) developed the idea that the biapicality found in the photoperiodic response curve of the Leningrad population of *T. urticae* at certain times of the year could be a fundamental property of all photoperiodic response curves. Still later experiments showed 3 peaks of high diapause incidence in a photoperiodic response curve determined in August for the Leningrad population; an almost identical response curve was obtained for a population of *T. viennensis* from Voronezh, also determined at 25°C in the first half of June (Geispitz et al., 1976). The authors suggest that all types of photoperiodic response curves may be explained on the basis of a variable expression of these 3 maxima in the tendency to diapause, as found in *T. urticae* and *T. viennensis*, the first maximum lying at photoperiods with photophases of 6—11 h, the second of 11—17 h, and the third at photoperiods with unnaturally long photophases of 22—24 h. According to this model, all types of photoperiodic response in insects and mites (long-day, short-day, intermediate) may have a common basis. The value of the model seems to be limited, however; it may indeed

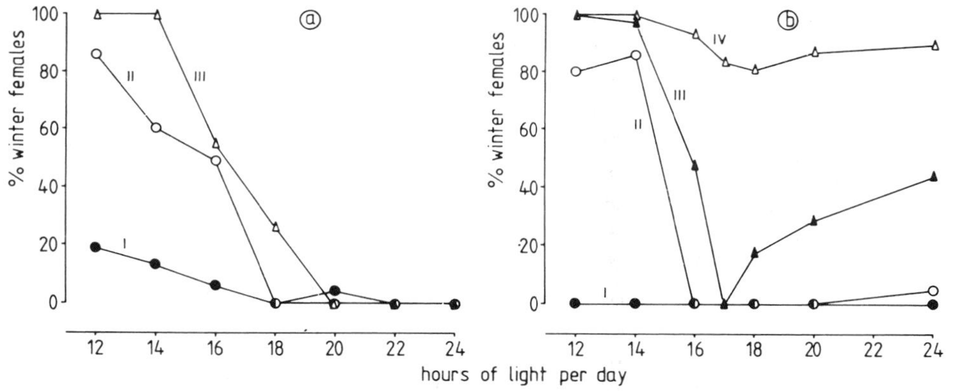

Fig. 1.4.6.7. Changes in the photoperiodic responses of successive generations of: (a) *Panonychus ulmi* (Koch) (from Razumova, 1967 and Geispitz et al., 1978) and (b) *Schizotetranychus schizopus* Zacher (from Geispitz, 1968). Both were natural populations from Leningrad. Photoperiodic response curves determined at 25°C. Generation number indicated by Roman numerals.

generate all conceivable response curves, but this obviously results from the fact that the 3 maxima in the tendency to diapause are not strictly defined and may be shifted at will. The model does not explain the variable expression of the 3 maxima at different times of the year, and no attempt has been made to explain photoperiodic responses in non-diel cycles, i.e. cycles with a period length other than 24 h.

Considerable changes in sensitivity to photoperiod in a number of successive generations were also found by Razumova (1967) in *P. ulmi* and by Geispitz (1968) in *S. schizopus* (Fig. 1.4.6.7). In both cases, experimental animals of each generation were taken directly from natural populations living in Leningrad on apple and willow respectively. In both species a photoperiodic response was practically absent in the first (post-diapause) generation at 25°C. The tendency to diapause increased in the following generations and in *S. schizopus* diapause was almost obligatory in the 4th generation for all daylengths tested (Fig. 1.4.6.7b). Also, at lower temperatures (18 and 15°C) considerable differences were still found in the responses of the successive generations; the experiments at 15°C showed, however, that sensitivity to photoperiod was not absent in the first generation, since up to 100% diapause was found in both species under a short-day regime (*LD 12*:12) at this temperature. The results of Razumova (1967) are in contradiction with those of Lees (1950, 1953a), who remarked that in *P. ulmi* from England the probability that diapause will occur in any given generation is neither enhanced nor diminished by the diapause history of the previous generations. Rekk (1959) noted for both *P. ulmi* and *T. viennensis* that in some parts of the U.S.S.R. a part of the population may already re-enter diapause in the first post-diapause generation.

In another experiment, Geispitz (1968) showed that in *S. schizopus* the tendency to diapause also depends strongly on the temperature experienced by the preceding generation. In an experiment performed under continuous illumination, the incidence of diapause at 25°C appeared to be proportionately higher, the higher the temperature at which the preceding (post-diapause) generation had been reared. For *P. ulmi*, Razumova (1969) reported the opposite effect, namely that diapause was higher over a range of short-day as well as long-day photoperiods at 18°C in mites whose parental generation had been reared at 15°C, in comparison with mites whose parents had been kept at 25°C, both under continuous illumination.

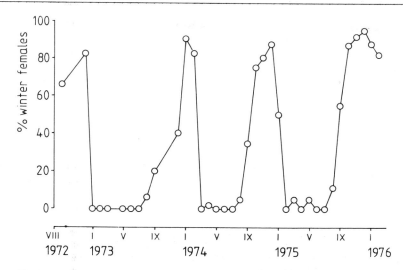

Fig. 1.4.6.8. Annual rhythm in photoperiodic sensitivity in a population of *Tetranychus viennensis* Zacher from Yalta. Photoperiodic responses determined at *LD 12*:12 and 25°C; mite culture kept at 20°C under constant illumination for over 3 years. (From Razumova, 1978.)

Annual changes in photoperiodic sensitivity were observed by Razumova (1978) in a population of *T. viennensis* from Yalta. A mite culture kept at 20°C under continuous illumination for over 3 years was sampled at regular intervals to obtain the photoperiodic responses with *LD 12*:12 at 25°C. The results presented in Fig. 1.4.6.8 show a rhythmic change in the tendency to diapause, high incidences of diapause being found from about October— January; no diapause was found from about February—August under the test conditions chosen. The same experiment was performed with mites cultured for 2 years at 15, 18 and 25°C, the tests again being done at *LD 12*:12 and 25°C. The results obtained over the 2 year period showed a great similarity for the 4 temperatures chosen, which could mean that the rhythm is either extremely well temperature compensated or even temperature independent. The data have not been analysed further but, notwithstanding certain variations in the widths of the peaks, it would seem that the period of the rhythm is very close to 12 months. According to Razumova (1978) the rhythm is endogenous and circannual; changes in the tendency to diapause in the course of the seasons apparently occur independently of the number of generations per year.

Circannual rhythms have been demonstrated in some long-lived insect species (Saunders, 1982), but a rhythm encompassing several generations within each cycle of the rhythm, as in *T. viennensis*, has not been found in any animal before. Short-lived insects such as aphids can also distinguish spring from autumn, but this is accomplished by means of an apparently non-rhythmic 'interval-timer', which prevents fundatrices and their descendents in young clones from producing sexuales in response to short-day photoperiods that prevail in spring. Interestingly, the interval-timer response is independent of generation number; the timer always starts its operation in the first post-diapause generation, i.e. in the fundatrices (Lees, 1960, 1979). In view of the uniqueness of the rhythm found in *T. viennensis*, arguments for its endogeneity should be carefully considered. Since the mite culture was maintained under constant conditions for over 3 years it may be expected to have been in free-run during the test period. Nevertheless, there are no indications of any alterations in the period of the rhythm, and different constant temperatures do not seem to have even a slight effect on the period length. No phase-shifts have been observed in the rhythm at any of the experimental temperatures, and no attempts to induce phase-shifts have been made; in all cases the rhythm remained exactly synchronized with the external seasons. In view of the above considerations there still remains some doubt as to the endogenous character of the annual rhythm in photoperiodic sensitivity in *T. viennensis*, since the possibility that some as yet unknown 'Zeitgeber' is involved has not been ruled out completely.

With respect to the influence of the temperature experienced by previous generations on the tendency to diapause, observed in *S. schizopus* and *P. ulmi*, as discussed earlier, it is interesting to note that this influence apparently does not exist in *T. viennensis*; in her experiments Razumova always found a maximum in the tendency to diapause during the winter months, irrespective of the temperature at which the preceding generations of mites had been reared (Razumova, 1978; Geispitz et al., 1978).

In similar experiments, Geispitz et al. (1974, 1976) found 2 peaks in the tendency to diapause in a population of *T. viennensis* from Sochi, 1 broad maximum from October to February, and 1 (smaller) in July and August. In a later paper, Geispitz et al. (1978) compared the seasonal variability of 4 populations of *T. viennensis* of different geographic origin. Apart from the population from Yalta, studied by Razumova (1978), which showed a maximum in its tendency to diapause during the winter months, and the population from Sochi, which showed a summer and a winter peak, they studied populations from Leningrad and from Voronezh. It appeared that under the test regime of *LD 12*:12 and 25°C, mites of the Leningrad population all diapaused at any time of the year, whereas the photoperiodic reaction of the Voronezh population was only slightly weakened in spring. The seasonal variation in sensitivity to photoperiod seems most pronounced in southern populations, and may be completely absent in mites from the northern part of their range.

The various changes in photoperiodic sensitivity during the seasons, as found in different species of spider mites in the U.S.S.R., are explained by Geispitz et al. (1978) on the basis of an endogenous annual rhythm in the tendency to diapause. As discussed above, the endogenous character of the annual rhythm needs confirmation from further experiments; it would seem possible, using some of the mite populations mentioned above (e.g., the Moscow strain of *T. urticae* studied by Dubynina (1965)), to discriminate between the possibilities that the changes in photoperiodic responses in a number of successive generations are based on an endogenous circannual rhythm, or on some kind of 'interval-timer', like the one described by Lees (1960, 1979) for the aphid *Megoura viciae* Buckton. Further research will be needed to solve the evident controversies and to find an answer to the intriguing questions posed above about the variability of the photoperiodic reaction in spider mites.

Genetic variability with respect to factors controlling diapause has been studied in *T. urticae* by Helle (1962, 1968), Geispitz (1968) and Geispitz et al. (1972a). Selection against diapause under short-day photoperiods resulted in a strong reduction in the number of diapause forms within a few generations (Helle, 1962). In other experiments Helle (1968) showed that from 19 inbred lines of *T. urticae*, derived from isolated females, the photoperiodic response curves showed remarkable differences after 7 generations of sib-mating. In 6 lines the photoperiodic response appeared to be weakened over the range of photoperiods tested (*LD 12*:12—*LD 14*:10), whereas 2 lines showed a decrease in the critical daylength of about 20 min; no shifts of the threshold in the opposite direction were found. His data illustrate the maintenance of genetic variability for factors controlling the photoperiodic reaction in a population of limited size over a long period of environmental constancy (200—300 generations reared in the laboratory under continuous light and high temperature). Reciprocal mass crosses between lines with a saturated photoperiodic response (HR) and a line with a low-level response (LR) showed that HR was dominant over LR.

Also using sib-selection, Geispitz (1968) found a decrease in critical day-length of about 1 h even after 2 selection rounds in a Leningrad strain of *T. urticae*; after 5 generations of selection the photoperiodic response had decreased to such an extent that about 20% diapause was found only at *LD 12*:12. Using a strain from Kirghizia, Geispitz (1968) found an increase in the incidence of diapause from 10% to 47% after 1 generation of selection. However, in view of the variability of the diapause response in successive generations of some Russian strains of mites, it is questionable whether the increase observed may be described to the applied selection procedure.

Mass selections performed with the strain from Kirghizia at a photoperiod of *LD 12*:12, close to the critical daylength which was 12.5 h for this strain, resulted in practically diapause-free development of the mites after only 3 generations over the whole range of photoperiods from *LD 5*:19 to *LD 24*:0. Different results were obtained by Geispitz et al. (1972a) in mass selection experiments with a Leningrad population of *T. urticae*. Selection at *LD 4*:20 for 6 generations resulted in a photoperiodic response curve, as determined in the 7th generation, showing a strong decline in diapause incidence from about 90% to almost complete absence of diapause at photoperiods with photophases of 4—8 h, but only a very slight influence was observed on the critical daylength. After a further selection for 2 generations, but now performed at *LD 14*:10, the form of the photoperiodic response curve again changed, but this time selection appeared to have the opposite effect: the short-day part of the curve had not changed, but the critical daylength had shifted from 16 h to about 13 h. Although the differences in the results obtained with different strains of *T. urticae* are difficult to explain, the studies on the genetic variability of the photoperiodic reaction do reveal the wealth of genetic variability which is present in this respect in various populations of this species of spider mite.

Evidence for a genetic basis of differences in diapause intensity and cold-hardiness in *P. ulmi* was presented by Cranham (1973) and MacPhee (1961), respectively. Cranham (1973) found differences in the chilling requirement of early- and late-hatching stocks from different orchards in England; his results suggest that differences in the intensity of diapause may be based on inherited variation. MacPhee (1961) found a strikingly greater cold-hardiness in *P. ulmi* from New Brunswick than in mites from Nova Scotia. Selection of the susceptible population by exposure to cold during 4 successive winters resulted in an increase in resistance to cold; the group which survived at $-34.4°C$ increased from 0.6% in the initial population to 91% of the eggs of the 4th selected population. Reciprocal crosses between the susceptible form from Nova Scotia and the resistant form from New Brunswick indicated that the factor for resistance to cold in *P. ulmi* is recessive.

THE PHYSIOLOGICAL MECHANISM OF THE INDUCTION OF DIAPAUSE

Little is known of the physiological mechanism of photoperiodic induction, and even its putative localization in the central nervous system awaits confirmation in mites. Just as in insects, the mechanism is thought to consist of a receptor part and an effector part. The receptor part conceivably comprises (1) a photoreceptor, which in all probability is intimately linked

Chapter 1.4.6. references, p. 310

with, or is part of the photoperiodic 'clock', (2) the clock itself, discriminating between long and short days (or nights), and (3) a photoperiodic 'counter', a mechanism which integrates the photoperiodic information contained in a number of successive photoperiodic cycles (Veerman and Vaz Nunes, 1984a, b). Mainly because of their small size, the effector part has not been studied in spider mites; consequently nothing is known of the humoral effectors, the neurohormones and hormones which may be concerned with the control of diapause in these mites. Techniques used in the study of the photoperiodic mechanism consist mainly in the application of widely differing photoperiodic regimes, based on diel (24 h) as well as nondiel cycles, both with and without short light interruptions in the dark phase. The results obtained in response to these regimes have been interpreted according to various theoretical models for the photoperiodic clock (see, e.g., Vaz Nunes and Veerman, 1979a, 1982a), and have also served for the development of a new model explaining photoperiodic induction in spider mites (Vaz Nunes and Veerman, 1982b; Vaz Nunes, 1983; Veerman and Vaz Nunes, 1984a, b).

Lees (1953a, b), working with the fruit tree red spider mite *P. ulmi*, was the first to analyse photoperiodic induction of diapause in spider mites. Using 'natural' fluorescent tubes as a light source, Lees (1953a) showed that the photoperiodic response is independent of light intensity, provided the illumination exceeds a threshold of 1—2 f.c. He determined the sensitivity to different wavelengths in the visible spectrum with the use of glass filters. The mites proved to be completely insensitive to high-intensity radiation in the orange, red and infrared regions of the spectrum, but diapause was influenced by radiation in the green, blue and near ultraviolet. As may be seen from the action spectrum presented in Fig. 1.4.6.9, the greatest sensitivity is shown to wavelengths in the blue region of the spectrum (Lees, 1953a).

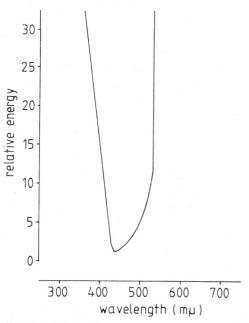

Fig. 1.4.6.9. Action spectrum curve showing the approximate relative energy required at different wavelengths to suppress diapause induction in *Panonychus ulmi* (Koch). The experimental regime consisted of 8 h standard illumination from a fluorescent tube, 8 h irradiation from a second source, provided with various glass filters, and 8 h darkness. (From Lees, 1953a.)

In experiments in which he varied the duration of the light and dark phases independently, Lees (1953b) showed that the response to any specific cycle depends upon the absolute lengths of both the light and dark component. *P. ulmi* appeared to be highly insensitive to light interruptions of up to 4 h given in the middle of a dark phase of 16 h. On the basis of the observation that with diapause-egg laying females 14 cycles of long days and short nights may have no effect, whereas 15 cycles cause the mites to switch over to summer eggs, Lees (1953b) supposed that synthesis and removal of some active substance during the light and dark phases of the photoperiodic cycle are cumulative processes with some residuum of active substance being carried over from 1 cycle to the next. Lees (1953b) noted, however, that a simple hypothesis of this kind is insufficient to explain all the results obtained in photoperiodic experiments with *P. ulmi*.

By studying the induction of diapause in several pigment mutants of *T. urticae*, which show various defects in their carotenoid metabolism (Plate 1.4.6.1), Veerman and Helle (1978) obtained the first indications that β-carotene might be necessary for photoperiodic induction. Diapause incidence in albino mutants was significantly lower than in the wild-type, when the mites were reared under diapause-promoting short days. Various crosses between albino and wild-type mites showed that a high incidence of diapause in albino mites was realized only when the albino daughters came from hybrid, phenotypically wild-type mothers. Apparently, a maternal effect was responsible for complete induction. The results could be explained by assuming that small amounts of carotenoids, deposited by the hybrid females in their eggs, are sufficient to allow photoperiodic induction to proceed normally in their albino offspring, which are not able to obtain β-carotene from their food themselves (Veerman, 1980a, b). In feeding experiments with the predaceous mite *Amblyseius potentillae* Garman, using carotenoid-deficient diets supplemented with various pure carotenoids or carotenoid derivatives, it was shown that only vitamin A and carotenoids with pro-vitamin A function are capable of restoring the photoperiodic response in the offspring of mites which had lost all sensitivity to photoperiod because of carotenoid deprivation. Apparently, vitamin A is essential for the photoperiodic reaction, and possibly functions as the photoreceptor pigment in the form of a rhodopsin-like protein complex in these mites (Van Zon et al., 1981; Veerman et al., 1983). The action spectrum, determined by Lees (1953a) for the suppression of the photoperiodic induction of diapause in *P. ulmi* (Fig. 1.4.6.9), does not seem to be in conflict with the view that, in at least some mite species, a rhodopsin-like pigment functions as the photoreceptor for the photoperiodic clock.

Working with a strain of *T. urticae* that had been reared in the laboratory for over 15 years, Veerman (1977a) obtained very sharp responses in light-break experiments, in answer to 1 h light interruptions 'scanning' the night of a LD *10*:14 light/dark cycle (so-called asymmetrical 'skeleton' photoperiods). A sharp bimodal response curve was obtained with 2 discrete points of sensitivity to the light breaks at 10 h after 'dusk' and 10 h before 'dawn', thus showing a remarkable similarity to results obtained in light-break experiments with some species of insects. In a series of night-interruption experiments in which 1 h light pulses were used to scan through nights of variable lengths (ranging from 11 to 18 h), the complete light/dark cycles all being based on a 24 h period, Vaz Nunes and Veerman (1979b) found 2 peaks of long-day effect in response to the light breaks in regimes with nightlengths of 11—15 h, but only 1 peak of long-day effect was observed with nightlengths longer than 15 h. In 3 series of experiments with symmetrical 'skeleton' photoperiods (regimes consisting of 2 light pulses of equal duration, the relative positions of which are systematically changed in the 24 h cycle in successive induction experiments), completely symmetrical unimodal response curves were obtained. The width of the peak in these response curves appeared to be a function of the duration of the light pulse used (1, 3 and 5 h, respectively, in the 3 series of experiments) (Vaz Nunes and Veerman, 1979b).

An important question is whether or not circadian rhythms are involved in the photoperiodic mechanism. The above experiments with asymmetrical and symmetrical skeleton photoperiods, although illustrative with respect to the systematic variation of the responses to short pulses of light, are not suited to reveal the possible involvement of the circadian system in photoperiodic induction, because all the experiments were based on regimes with cycle lengths of 24 h. Most of the evidence for the involvement of the circadian system in both plant and animal photoperiodism comes from experiments with non-diel photoperiods. The experimental design most frequently used is the so-called resonance technique, in which the light component of a light/dark cycle is held constant and the dark component is varied over a wide range in successive experiments, to provide cycles with period lengths up to 72 h or more. Rhythmic variations in the photoperiodic response with peaks about 24 h apart are interpreted as evidence for the involvement of circadian rhythmicity in the photoperiodic process studied. The results of such an experiment performed with *T. urticae*, in which the constant light phase has a duration of 8 h, are shown in Fig. 1.4.6.10. The sharply expressed 'resonance' peaks clearly show that circadian rhythmicity is somehow involved in the photoperiodic determination of diapause in *T. urticae*, notwithstanding the fact that the period of the oscillation concerned appears to be rather short, namely only about 20 h, as judged by the distance between the peaks (Veerman and Vaz Nunes, 1980).

Another approach which has sometimes been used to demonstrate a circadian effect on the photoperiodic response is to probe for light sensitivity with short pulses of light in extremely long nights; peaks of light sensitivity about 24 h apart are seen as evidence for the involvement of the circadian system. The results of a night-interruption experiment with a 1 h light pulse scanning through the night of a *LD 12:52* photoperiod, performed with the same strain of *T. urticae*, are presented in Fig. 1.4.6.11. The response curve

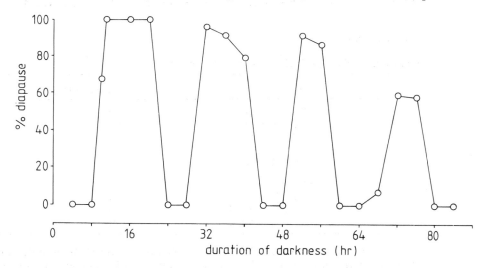

Fig. 1.4.6.10. Induction of diapause in *Tetranychus urticae* Koch in a resonance experiment. Photoperiodic regime consisting of a constant photophase of 8 h and scotophases which were varied over 4—84 h. (From Veerman and Vaz Nunes, 1980.)

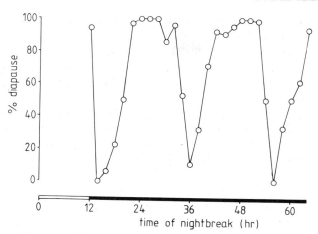

Fig. 1.4.6.11. Night-interruption experiment using *Tetranychus urticae* Koch with a 1 h supplementary light pulse systematically scanning the extended night of a *LD 12:52* light/dark cycle. (From Vaz Nunes and Veerman, 1982b.)

shows 3 peaks of sensitivity to the light-breaks in the extended night; the peak intervals of about 20 h agree well with the distance found between successive resonance peaks, as shown in Fig. 1.4.6.10. The results of this experiment therefore corroborate the evidence that the circadian system is involved in photoperiodic induction in *T. urticae* (Vaz Nunes and Veerman, 1982b).

The presence of some kind of photoperiodic counter mechanism in *T. urticae* was inferred from experimental data, which show that the percentage incidence of diapause in the population increases with an increasing number of long nights experienced by the developmental stages that are sensitive to the photoperiod (Table 1.4.6.3). Other experiments, in which various sequences of short-day (long-night) cycles were followed by long-day (short-night) cycles or by continuous darkness, demonstrated that long-day photoperiods exert an effect on induction as well, of about equal strength to that of short-day photoperiods, but working in the opposite direction: the inductive effect of a sequence of short-day cycles can be diminished or cancelled completely by a subsequent sequence of long-day cycles (Veerman, 1977a). One such series of experiments is shown in Table 1.4.6.3.

TABLE 1.4.6.3

Operation of photoperiodic counter in *T. urticae*

Regime	Diapause (%)
A[a] 1 LN + DD	0
2 LN + DD	13
3 LN + DD	35
5 LN + DD	68
7 LN + DD	95
B[b] 7 LN + 2 SN	58
7 LN + 4 SN	49
7 LN + 6 SN	4
7 LN + 8 SN	0

[a] A: Increasing number of long-night cycles (LN) followed by continuous darkness (DD).
[b] B: Seven long-night cycles (LN) followed by an increasing number of short-night cycles (SN).
Regimes A (at 18°C) and B (at 15°C) were applied during the developmental stages most sensitive to photoperiodic induction. Each experiment was based on ~ 50 mites.

Chapter 1.4.6. references, p. 310

The results of photoperiodic experiments with *T. urticae* for the greater part could not be described on the basis of the 'dual-system theory of the biological clock', a model proposed by Beck (1980) to explain photoperiodic time measurement in insects (Vaz Nunes and Veerman, 1979a). Also, the 'external coincidence model', developed by Pittendrigh (1972) and Saunders (1982) for the photoperiodic clock in insects could not explain all the results obtained with *T. urticae* (Vaz Nunes and Veerman, 1982a). Both models are based on the concept that photoperiodic time measurement is a function of the circadian system. As discussed above, the circadian system clearly is involved in photoperiodic induction in *T. urticae*; however, in view of the fact that circadian-based models for the photoperiodic clock were found to be inadequate to explain the results obtained with spider mites (Veerman and Vaz Nunes, 1984a), Vaz Nunes and Veerman (1982b) felt obliged to develop a new model, the 'hourglass timer oscillator counter model', based on another explanation of the role of the circadian system in photoperiodic induction. They elaborated the idea, first expressed by Pittendrigh (1972), that the circadian system might exert an influence on the expression of a photoperiodic response, without being involved in time measurement itself. The model is based on the assumption that, depending on how the circadian system is attuned to the external light/dark cycle, it may or may not interfere with the further processing of the clock's output, the accumulation of the daily bits of photoperiodic 'information'. As an alternative to circadian-based time measurement, the model was designed to investigate the possibility that photoperiodic time measurement is executed by some 'hourglass' clock, the kinetics of which are based on the hourglass timer developed by Lees (1973) for photoperiodic time measurement in the aphid *M. viciae*, an insect for which hourglass time measurement has been firmly established. The hourglass timer oscillator counter model is thus a theoretical model devised to test:

(1) whether or not photoperiodic induction in *T. urticae* can be explained as the combined result of hourglass time measurement and the accumulation, by a simple counter mechanism, of the output of this hourglass clock, and

(2) whether or not the influence of the circadian system on photoperiodic induction may be understood as a 'resonance' effect, expressed as an inhibition of the summation process that is performed by the photoperiodic counter. The kinetics of the model are described by Vaz Nunes and Veerman (1982b) and by Vaz Nunes (1983). A schematic representation of the operation of the model has been given by Veerman and Vaz Nunes (1984a). The hourglass timer oscillator counter model was shown to give an accurate description of all photoperiodic experiments performed with *T. urticae*, based on diel as well as non-diel photoperiods, both with and without light interruptions of the dark (Vaz Nunes and Veerman, 1982b, 1984; Vaz Nunes, 1983; Veerman and Vaz Nunes, 1984a). Preliminary tests showed that the model may even have a wider application, as the results of photoperiodic experiments with a number of insect species could also be explained on the basis of the spider mite model (Vaz Nunes, 1983). Nevertheless, extensive research will still be needed before the theoretical model may be translated into a concrete physiological mechanism for the photoperiodic induction of diapause in spider mites.

REFERENCES

Andersen, V.S., 1947. Untersuchungen über die Biologie und Bekämpfung der Obstbaum-spinnmilbe *Paratetranychus pilosus* Can. et Fanz. Thesis, University of Bonn.
Anderson, N.H. and Morgan, C.V.G., 1958. Life histories and habits of the clover mite,

Bryobia praetiosa Koch, and the brown mite, *B. arborea* M. & A. in British Columbia. Can. Entomol., 90: 23—42.

André, M., 1942. L'hivernation chez les Tétranyques et ses rapports avec la lutte contre ces Acariens phytophages. Bull. Mus. Natl. Hist. Nat., 14: 57—62.

Baker, E.W. and Pritchard, A.E., 1953. A guide to the spider mites of cotton. Hilgardia, 22: 203—234.

Beament, J.W.L., 1951. The structure and formation of the egg of the fruit tree red spider mite, *Metatetranychus ulmi* Koch. Ann. Appl. Biol., 38: 1—24.

Beck, S.D., 1980. Insect Photoperiodism, second edition. Academic Press, New York, NY.

Becker, H., 1952. Über den Einfluss konstanter Temperaturen, relativer Luftfeuchtigkeiten und Licht auf die Frühjahrsentwicklung der Wintereier der Obstbaumspinnmilbe *Paratetranychus pilosus* Can. et Franz. Anz. Schaedlingskd., 25: 116—118.

Bengston, M., 1965. Overwintering behaviour of *Tetranychus telarius* (L.) in the Stanthorpe district, Queensland. Queensl. J. Agric. Anim. Sci., 22: 169—176.

Böhm, H., 1954. Untersuchungen über die Biologie und Bekämpfung der roten Stachelbeermilbe (*Bryobia praetiosa* Koch). Pflanzenschutzberichte, 13: 161—176.

Bondarenko, N.V., 1950. The influence of shortened day on the annual cycle of development of the common spider mite. Dokl. Akad Nauk SSSR, 70: 1077—1080 (in Russian).

Bondarenko, N.V., 1958. Diapause peculiarities in *Tetranychus urticae* Koch. Zool. Zh., 37: 1012—1023 (in Russian).

Bondarenko, N.V. and Kuan Khay-yuan, 1958. Peculiarities in the origin of diapause in different geographical populations of the spider mite. Dokl. Akad. Nauk SSSR, 119: 1247—1250 (in Russian).

Boudreaux, H.B., 1956. Revision of the two-spotted spider mite (Acarina, Tetranychidae) complex, *Tetranychus telarius* (Linnaeus). Ann. Entomol. Soc. Am., 49: 43—48.

Boudreaux, H.B., 1958. The effect of relative humidity on egg-laying, hatching, and survival in various spider mites. J. Insect Physiol., 2: 65—72.

Boudreaux, H.B., 1963. Biological aspects of some phytophagous mites. Annu. Rev. Entomol., 8: 137—154.

Brandenburg, R.L. and Kennedy, G.G., 1981. Overwintering of the pathogen *Entomophthora floridana* and its host, the two-spotted spider mite. J. Econ. Entomol., 74: 428—431.

Cagle, L.R., 1946. Life history of the European red mite. Va. Agric. Exp. Stn. Tech. Bull., 98: 1—19.

Cagle, L.R., 1949. Life history of the two-spotted spider mite. Va. Agric. Exp. Stn. Tech. Bull., 113: 1—31.

Carter, C.I., 1956. Massing of *Tetranychus tiliarum* (Hermann) (Acarina, Tetranychidae) on the trunks of lime trees. Entomol. Mon. Mag., 92: 73—74.

Collingwood, C.A., 1955. The glasshouse red spider mite on blackcurrants. Ann. Appl. Biol., 43: 144—147.

Cranham, J.E., 1971. The effect of temperature on hatching of the winter eggs of fruit tree red spider mite, *Panonychus ulmi* (Koch). In: M. Daniel and B. Rosický (Editors), Proc. 3rd Int. Congr. Acarol., Prague, Dr. W. Junk B.V., The Hague, pp. 205—209.

Cranham, J.E., 1972. Influence of temperature on hatching of winter eggs of fruit-tree red spider mite, *Panonychus ulmi* (Koch). Ann. Appl. Biol., 70: 119—137.

Cranham, J.E., 1973. Variation in the intensity of diapause in winter eggs of fruit tree red spider mite, *Panonychus ulmi*. Ann. Appl. Biol., 75: 173—182.

Danilevskii, A.S., 1965. Photoperiodism and Seasonal Development of Insects. Oliver and Boyd, Edinburgh.

Danilevskii, A.S. and Kuznetsova, I.A., 1968. The intraspecific adaptations of insects to the climatic zonation. In: A.S. Danilevskii (Editor), Photoperiodic Adaptations in Insects and Acari. Leningrad State University, Leningrad, pp. 5—51 (in Russian).

Den Houter, J.S., 1976. The effect of acaricides on active and diapausing females of *Tetranychus urticae* Koch, the two-spotted spider mite. Z. Angew. Entomol., 81: 248—252.

Dierick, G.F.E.M., 1950. Breaking of diapause in the winter egg of the European red spider. Nature, 165: 900.

Dosse, G., 1964. Beobachtungen über Entwicklungstendenzen im *Tetranychus cinnabarinus*-Komplex (Acari, Tetranychidae). Pflanzenschutzberichte, 31: 113—128.

Dosse, G., 1967. Beiträge zum Diapause-Problem von *Tetranychus urticae* Koch und *Tetranychus cinnabarinus* Boisduval-Komplex im Libanon (Acarina, Tetranychidae). Pflanzenschutzberichte, 34: 129—138.

Dubynina, T.S., 1965. Entering the diapause and reactivation of the red spider mite *Tetranychus urticae* Koch (Acarina, Tetranychidae). Entomol. Obozr., 44: 287—292 (in Russian).

Dupont, L.M., 1979. On gene flow between *Tetranychus urticae* Koch, 1836 and *Tetranychus cinnabarinus* (Boisduval) Boudreaux, 1956 (Acari: Tetranychidae): synonymy between the two species. Entomol. Exp. Appl., 25: 297—303.

English, L.L. and Snetsinger, R., 1957. The biology and control of *Eotetranychus multidigituli* (Ewing) a spider mite of honey locust. J. Econ. Entomol., 50: 784—788.

Ewing, H.E., 1914. The common red spider mite. Oreg. Agric. Coll. Exp. Stn. Bull., 121: 1—95.

Fenner, T.L., 1962. Two-spotted mites. J. Dep. Agric. S. Austr. 66: 116—120.

Fenton, F.A., 1951. The brown wheat mite *Petrobia latens*. J. Econ. Entomol., 44: 996.

Foott, W.H., 1965. Geotactic response of the two-spotted spider mite, *Tetranychus urticae* Koch (Acarina, Tetranychidae). Proc. Entomol. Soc. Ont., 95: 106—108.

French, N. and Ludlam, F.A.B., 1973. Observations on winter survival and diapausing behaviour of red spider mite (*Tetranychus urticae*) on glasshouse roses. Plant Pathol., 22: 16—21.

Fritzsche, R., 1959. Morphologische, biologische und physiologische Variabilität und ihre Bedeutung für die Epidemiologie und Bekämpfung von *Tetranychus urticae* Koch. Biol. Zentralbl., 5: 521—576.

Garcia Mari, F. and del Rivero, J.M., 1982. El ácaro *Tetranychina harti* (Ewing) en España. Bol. Serv. Plag., 8: 55—62.

Gasser, R., 1951. Zur Kenntnis der gemeinen Spinnmilbe *Tetranychus urticae* Koch. I. Mitteilung: Morphologie, Anatomie, Biologie und Oekologie. Mitt. Schweiz. Entomol. Ges., 24: 217—262.

Geijskes, D.C., 1939. Beiträge zur Kenntnis der europäischen Spinnmilben (Acari, Tetranychidae), mit besonderer Berücksichtigung der niederländischen Arten. Meded. Landbouwhoogesch. Wageningen, 42: 1—68.

Geispitz, K.F., 1960. The effect of the conditions in which preceding generations were reared on the photoperiodic reaction of geographical forms of the cotton spider mite (*Tetranychus urticae* Koch). Tr. Petergof. Biol. Inst. Leningrad. Gosudarst. Univ., 18: 169—177 (in Russian).

Geispitz, K.F., 1968. Genetic aspects of variation of photoperiodic adaptations. In: A.S. Danilevskii (Editor), Photoperiodic Adaptations in Insects and Acari. Leningrad State University, Leningrad, pp. 52—79 (in Russian).

Geispitz, K.F., Glinyanaya, E.I., Dubynina, T.S., Kvitko, N.V., Pidzhakova, T.V., Razumova, A.P., Sapozhnikova, F.D., Simonenko, N.P. and Taranets, M.N., 1978. Annual endogenous rhythm of changes of photoperiodic reaction in Arthropoda and its connection with exogenous factors. Entomol. Obozr., 57: 731—745 (in Russian).

Geispitz, K.F., Glinyanaya, E.I. and Sapozhnikova, F.D., 1972a. The general principle of the interrelation between the various manifestations of the photoperiodic reaction in Arthropoda. Dokl. Akad. Nauk SSSR, 202: 1229—1232 (in Russian).

Geispitz, K.F., Glinyanaya, E.I. and Sapozhnikova, F.D., 1972b. Interrelation of different types of photoperiodic response in arthropods. In: N.I. Goryshin (Editor), Problems of Photoperiodism and Diapause in Insects. Leningrad State University, Leningrad, pp. 74—88 (in Russian).

Geispitz, K.F., Glinyanaya, E.I., Sapozhnikova, F.D. and Simonenko, N.P., 1974. Correlation between endogenous and exogenous factors in the regulation of seasonal changes in the photoperiodic reaction of arthropods. Entomol. Obozr., 53: 523—534 (in Russian).

Geispitz, K.F. and Orlovskaya, E.E., 1971. Physiological characteristics of two types of diapause in the common spider mite *Tetranychus urticae* Koch (Acarina, Tetranychidae). In: Cold-hardiness in Insects and Mites, Abstr. Symp., Tartu (Estonian SSR), April 19—21, 1971, pp. 16—20 (in Russian).

Geispitz, K.F., Sapozhnikova, F.D. and Simonenko, N.P., 1976. Ecological principles of the regulation of the active and diapausing stages in seasonal turnover in arthropods. In: V.A. Zaslavskii (Editor), Photoperiodism in Animals and Plants. Zoological Institute of the Academy of Sciences of the USSR, Leningrad, pp. 26—46 (in Russian).

Geispitz, K.F., Sapozhnikova, F.D. and Taranets, M.N., 1971. Seasonal changes in photoperiodic reaction and physiological state of *Tetranychus urticae* Koch (Acarina, Tetranychidae). Entomol. Obozr., 50: 275—285 (in Russian).

Glinyanaya, E.I., 1972. The role of photoperiod in the reactivation of arthropods with summer and winter diapause. In: N.I. Goryshin (Editor), Problems of Photoperiodism and Diapause in Insects. Leningrad State University, Leningrad, pp. 88—102 (in Russian).

Gotoh, T., 1984. Ecological studies on diapause in Tetranchinae, with special reference to diapause attributes in relation to host plant phenology. Ph.D. Thesis, Hokkaido University (in Japanese).

Gotoh, T. and Shinkaji, N., 1981. Critical photoperiod and geographical variation of diapause induction in the two-spotted spider mite. *Tetranychus uricae* Koch (Arcarina: Tetranchidae). Jpn. J. App. Entomol. Zool., 25: 113—118 (in Japanese).

Gould, H.J. and Kingham, H.G., 1965. Observations on diapause in *Tetranychus urticae* infesting cucumbers. Plant Pathol., 14: 174—178.

Gutierrez J., Helle, W. and Bolland, H.R., 1970. Étude cytogénétique et réflexions phylogénétiques sur la famille des Tetranychidae Donnadieu. Acarologia, 12: 732—751.

Hazan, A., Tahori, A.S. and Gerson, U., 1971. Failures to induce diapause in an acaricide-susceptible strain of the carmine spider mite. Isr. J. Entomol., 6: 97—105.

Helle, W., 1962. Genetics of resistance to organophosphorus compounds and its relation to diapause in *Tetranychus urticae* Koch (Acari). Tijdschr. Plantenziekten, 68: 155—195.

Helle, W., 1968. Genetic variability of photoperiodic response in an arrhenotokous mite (*Tetranychus urticae*). Entomol. Exp. Appl., 11: 101—113.

Helle, W. and Overmeer, W.P.J., 1973. Variability in tetranychid mites. Annu. Rev. Entomol., 18: 97—120.

Herbert, H.J., 1965. The brown mite, *Bryobia arborea* Morgan and Anderson (Acarina: Tetranychidae), on apple in Nova Scotia. I. Influence of intratree distribution on the selection of sampling units. II. Differences in habits and seasonal trends in orchards with bivoltine and trivoltine populations. III. Distribution of diapause eggs. IV. Hatching of diapause eggs. Can. Entomol., 97: 1303—1318.

Herbert, H.J. and McRae, K.B., 1982. Predicting eclosion of overwintering eggs of the European red mite, *Panonychus ulmi* (Acarina: Tetranychidae), in Nova Scotia. Can. Entomol., 114: 703—712.

Hueck, H.J., 1951. Influence of light upon the hatching of winter-eggs of the fruit tree red spider. Nature, 167: 993—994.

Huffaker, C.B., van de Vrie, M. and McMurtry, J.A., 1969. The ecology of tetranychid mites and their natural control. Annu. Rev. Entomol., 14: 125—174.

Hürková, J. and Weyda, F., 1982. Response to pesticides of diapausing females of *Tetranychus urticae* (Acarina, Tetranychidae). Věstn. Česk. Spol. Zool., 46: 92—99.

Hussey, N.W., 1972. Diapause in *Tetranychus urticae* Koch and its implications in glasshouse culture. Acarologia, 13: 344—350.

Hussey, N.W. and Huffaker, C.B., 1976. Spider mites. In: V.L. Delucchi (Editor), Studies in Biological Control. Cambridge University Press, Cambridge, pp. 179—228.

Hussey, N.W. and Parr, W.J., 1958. A genetic study of the colour forms found in populations of the greenhouse red spider mite *Tetranychus urticae* Koch. Ann. Appl. Biol., 46: 216—220.

Hussey, N.W. and Parr, W.J., 1963. Dispersal of the glasshouse red spider mite *Tetranychus urticae* Koch (Acarina: Tetranychidae). Entomol. Exp. Appl., 6: 207—214.

Jackson, P.R., Hunter, P.E. and Payne, J.A., 1983. Biology of the pecan leaf scorch mite (Acari: Tetranychidae). Environ. Entomol., 12: 55—59.

Jeppson, L.R., Keifer, H.H. and Baker, E.W., 1975. Mites Injurious to Economic Plants. University of California Press, Berkeley, CA.

Kozár, F., 1974. The role of extreme temperature fluctuations in the population dynamics of overwintering eggs of *Panonychus ulmi* Koch. Acta Phytopathol. Acad. Sci. Hung., 9: 363—367.

Kuenen, D.J., 1946. Het winterei van het fruitspint (*Metatetranychus ulmi* Koch) en zijn bestrijding. Tijdschr. Plantenziekten, 52: 69—82.

Kuenen, D.J., 1949. The fruit tree red spider mite (*Metatetranychus ulmi* Koch) and its relation to its host plant. Tijdschr. Entomol., 91: 83—102.

Lamiman, J.F., 1935. The pacific mite, *Tetranychus pacificus* McG., in California. J. Econ. Entomol., 28: 900—903.

Lees, A.D., 1950. Diapause and photoperiodism in the fruit tree red spider mite (*Metatetranychus ulmi* Koch). Nature, 166: 874—875.

Lees, A.D., 1952. The physiology of diapause in the fruit tree red spider mite. Trans. 9th Int. Congr. Entomol., Amsterdam, Vol. 1, pp. 351—354.

Lees, A.D., 1953a. Environmental factors controlling the evocation and termination of diapause in the fruit tree red spider mite *Metatetranychus ulmi* Koch (Acarina: Tetranychidae). Ann. Appl. Biol., 40: 449—486.

Lees, A.D., 1953b. The significance of the light and dark phases in the photoperiodic control of diapause in *Metatetranychus ulmi* Koch. Ann. Appl. Biol., 40: 487—497.

Lees, A.D., 1959. Photoperiodism in insects and mites. In: R.B. Withrow (Editor), Photoperiodism and Related Phenomena in Plants and Animals. American Association for the Advancement of Science, Washington, DC, pp. 585—600.

Lees, A.D., 1960. The role of photoperiod and temperature in the determination of parthenogenetic and sexual forms in the aphid *Megoura viciae* Buckton — II. The operation of the 'interval timer' in young clones. J. Insect Physiol., 4: 154—175.

Lees, A.D., 1961. On the structure of the egg shell in the mite *Petrobia latens* Müller (Acarina: Tetranychidae). J. Insect Physiol., 6: 146—151.

Lees, A.D., 1973. Photoperiodic time measurement in the aphid *Megoura viciae*. J. Insect Physiol., 19: 2279—2316.

Lees, A.D., 1979. The maternal environment and the control of morphogenesis in insects. In: D.R. Newth and M. Balls (Editors), Maternal Effects in Development, 4th Symp. Br. Soc. Developmental Biology, Cambridge University Press, Cambridge, pp. 221—239.

Lienk, S.E. and Chapman, P.J., 1958. Effect of the winter temperatures of 1956—57 on the survival rate of European red mite eggs. J. Econ. Entomol., 51: 263.

Light, W.I.St.G., John, M.E., Gould, H.J. and Coghill, K.J., 1968. Hatching of the winter eggs of the fruit tree red spider mite (*Panonychus ulmi* Koch). Ann. Appl. Biol., 62: 227—239.

Linke, W., 1953. Untersuchungen über Biologie und Epidemiologie der Gemeinen Spinnmilbe, *Tetranychus althaeae* v.Hanst., unter besonderer Berücksichtigung des Hopfens als Wirtspflanze. Höfchenbriefe, 6: 185—238.

Lloyd, L., 1922. Red Spiders on Cucumbers and Tomatoes. Experimental and Research Station, Cheshunt, 7th Annu. Rep., 1921, The Cheshunt Press, Cheshunt, pp. 41—46.

Löyttyniemi, K., 1970. Zur Biologie der Nadelholzspinnmilbe (*Oligonychus ununguis* (Jacobi), Acarina, Tetranychidae) in Finnland. Acta Entomol. Fenn., 27: 1—64.

MacPhee, A.W., 1961. Mortality of winter eggs of the European red mite *Panonychus ulmi* (Koch) at low temperature, and its ecological significance. Can. J. Zool., 39: 229—243.

MacPhee, A.W., 1963. The effect of low temperatures on some predacious phytoseiid mites, and on the brown mite *Bryobia arborea* M. & A. Can. Entomol., 95: 41—44.

MacPhee, A.W., 1964. Cold-hardiness, habitat and winter survival of some orchard arthropods in Nova Scotia. Can. Entomol., 96: 617—625.

Mague, D.L. and Streu, H.T., 1980. Life history and seasonal population growth of *Oligonychus ilicis* infesting Japanese holly in New Jersey. Environ. Entomol., 9: 420—424.

Malcolm, D.R., 1955. Biology and control of the timothy mite, *Paratetranychus pratensis* (Banks). Wash. Agric. Exp. Stn., Tech. Bull., 17: 1—35.

Massee, A.M., 1942. Some important pests of the hop. Ann. Appl. Biol., 29: 324—326.

Mathys, G., 1957. Contribution à la Connaissance de la Systématique et de la Biologie du Genre *Bryobia* en Suisse Romande. Thèse, École Polytechnique Fédérale, Zurich.

McEnroe, W.D., 1961. The control of water loss by the two-spotted spider mite (*Tetranychus telarius*). Ann. Entomol. Soc. Am., 54: 883—887.

McEnroe, W.D., 1971. The red photoresponse of the spider mite *Tetranychus urticae* (Acarina, Tetranychidae). Acarologia, 13: 113—118.

McGregor, E.A. and McDonough, F.L., 1917. The red spider on cotton. U.S. Dep. Agric., Bull., 416: 1—72.

Mellott, J.L. and Connell, W.A., 1965. Notes on the life history of *Tetranychus atlanticus* (Acarina: Tetranychidae). Ann. Entomol. Soc. Am., 58: 379—382.

Micinski, S., Boethel, D.J. and Boudreaux, H.B., 1979. Influence of temperature and photoperiod on development and oviposition of the pecan leaf scorch mite, *Eotetranychus hicoriae*. Ann. Entomol. Soc. Am., 72: 649—654.

Miller, L.W., 1950. Factors influencing diapause in the European red mite. Nature, 166: 875.

Morimoto, N. and Takafuji, A., 1983. Comparison of diapause attributes and host preference among three populations of the citrus red mite, *Panonychus citri* (McGregor) occurring in the southern part of Okayama prefecture, Japan. Jpn. J. Appl. Entomol. Zool., 27: 224—228 (in Japanese).

Nuber, K., 1961. Zur Frage der Uberwinterung der gemeinen Spinnmilbe *Tetranychus urticae* Koch im Hopfenbau (Acari, Tetranychidae). Höfchenbriefe, 14: 6—15.

Osakabe, M., 1962. On the diapause of the tea red spider mite, *Tetranychus kanzawai* Kishida. Study of Tea in Japan, 26: 12—25 (in Japanese).

Parr, W.J. and Hussey, N.W., 1966. Diapause in the glasshouse red spider mite (*Tetranychus urticae* Koch): a synthesis of present knowledge. Hortic. Res., 6: 1—21.

Pittendrigh, C.S., 1972. Circadian surfaces and the diversity of possible roles of circadian organization in photoperiodic induction. Proc. Natl. Acad. Sci. USA, 69: 2734—2737.

Pritchard, A.E. and Baker, E.W., 1952. A guide to the spider mites of deciduous fruit trees. Hilgardia, 21: 253—287.

Razumova, A.P., 1967. Variability of photoperiodic reaction in a number of successive generations of spider mites (Acarina). Entomol. Obozr., 46: 268—272 (in Russian).

Razumova, A.P., 1969. Experimental analysis of the phenology of the red apple mite *Panonychus ulmi* Koch. Zool. Zh., 48: 212—217 (in Russian).

Razumova, A.P., 1978. Endogenous annual rhythm of variation of the photoperiodic reaction in *Tetranychus crataegi* (Acarina, Tetranychidae). Zool. Zh., 57: 530–539 (in Russian).

Reeves, R.M., 1963. Tetranychidae infesting woody plants in New York state, and a life history study of the elm spider mite *Eotetranychus matthyssei* n.sp. Mem. Cornell Univ. Agric. Exp. Stn., Ithaca, NY, 380: 1–100.

Reiff, M., 1949. Physiologische Merkmale bei Spinnmilben nach Veränderung des Blattstoffwechsels. Verh. Schweiz. Naturforsch. Ges., 129: 165–166.

Rekk, G.F., 1959. Identification of tetranychid mites. Fauna Trans-Caucasia. Akad. Nauk Gruz. SSR, Tbilisi, 1: 1–151 (in Russian).

Ries, D.T., 1935. A new mite (*Neotetranychus buxi* Garman) on boxwood. J. Econ. Entomol., 28: 55–62.

Saba, F., 1961. Über die Bildung der Diapauseform bei *Tetranychus urticae* Koch in Abhängigkeit von Giftresistenz. Entomol. Exp. Appl., 4: 264–272.

Salt, R.W., 1961. Principles of insect cold-hardiness. Annu. Rev. Entomol., 6: 55–74.

Saunders, D.S., 1982. Insect Clocks, second edition. Pergamon Press, Oxford.

Shinkaji, N., 1975a. Seasonal occurrence of the winter eggs and environmental factors controlling the evocation of diapause in the common conifer spider mite, *Oligonychus ununguis* (Jacobi), on chestnut (Acarina, Tetranychidae). Jpn. J. Appl. Entomol. Zool., 19: 105–111 (in Japanese).

Shinkaji, N., 1975b. Hatching time of the winter eggs and termination of diapause in the common conifer spider mite *Oligonychus ununguis* (Jacobi), on chestnut in relation to temperature (Acarina: Tetranychidae). Jpn. J. Appl. Entomol. Zool., 19: 144–148 (in Japanese).

Shinkaji, N., 1979. Geographical distribution of the citrus red mite, *Panonychus citri* and European red mite, *P. ulmi* in Japan. In: J.G. Rodriguez (Editor), Recent Advances in Acarology, Vol. 1. Academic Press, New York, NY, pp. 81–87.

Snetsinger, R., 1964. Variation in mites belonging to the genus *Bryobia*. Ann. Entomol. Soc. Am., 57: 220–226.

Sømme, L., 1965. Changes in sorbitol content and supercooling points in overwintering eggs of the european red mite (*Panonychus ulmi* (Koch)). Can. J. Zool., 43: 881–884.

Sømme, L., 1966. Mortality of winter eggs of *Metatetranychus ulmi* (Koch) during the winter of 1965/66. Norsk Entomol. Tidsskr., 13: 420–423.

Sømme, L., 1982. Supercooling and winter survival in terrestrial arthropods. Comp. Biochem. Physiol., 73A: 519–543.

Speyer, E.R., 1924. Entomological investigations. 1. Red spider (*Tetranychus telarius* L.). Experimental and Research Station, Cheshunt, 9th Annu. Rep., 1923. The Cheshunt Press, Cheshunt, pp. 69–78.

Srivastrava, B.K. and Mathur, L.M.L., 1962. Bionomics and control of castor mite. Indian J. Entomol., 24: 229–235.

Stenseth, C., 1965. Cold hardiness in the two-spotted spider mite (*Tetranychus urticae* Koch). Entomol. Exp. Appl., 8: 33–38.

Summers, F.M., 1950. Brown almond mites — overwintering eggs appear in June with three life cycles a year offering an advantage in control program. Calif. Agric., 4: 6.

Takafuji, A. and Kamezaki, H., 1984. Diapause incidence in eggs of the citrus red mite, *Panonychus citri* (M.) on pear twigs. Appl. Entomol. Zool., 19: 270–271.

Takafuji, A. and Morimoto, N., 1983. Diapause attributes and seasonal occurrences of two populations of the citrus red mite, *Panonychus citri* (McGregor) on pear (Acarina: Tetranychidae). Appl. Entomol. Zool., 18: 525–532.

Tauber, M.J. and Tauber, C.A., 1978. Evolution of phenological strategies in insects: a comparative approach with eco-physiological and genetic considerations. In: H. Dingle (Editor), Evolution of Insect Migration and Diapause. Springer-Verlag, New York, pp. 53–71.

Trottier, R. and Herne, D.H.C., 1979. Temperature relationships to forecast hatching of overwintered eggs of the European red mite, *Panonychus ulmi* (Acarina: Tetranychidae). Proc. Entomol. Soc. Ont., 110: 53–60.

Tsugawa, C., Yamada, M. and Shirasaki, S., 1961. Forecasting the outbreak of destructive insects in apple orchards. III. Forecasting the initial date of hatch in respect of the overwintering eggs of the European red mite, *Panonychus ulmi* (Koch), in Aomori prefecture. Jpn. J. Appl. Entomol. Zool., 5: 167–173 (in Japanese).

Tsugawa, C., Yamada, M., Shirasaki, S. and Oyama, N., 1966. Forecasting the outbreak of destructive insects in apple. VII. Termination of diapause in hibernating eggs of *Panonychus ulmi* (Koch) in relation to temperature. Jpn. J. Appl. Entomol. Zool., 10: 174–180 (in Japanese).

Ubertalli, J.A., 1955. Life history of *Eutetranychus uncatus* Garman. J. Econ. Entomol., 48: 47–49.

Uchida, M., 1980. Appearance time of diapausing females and termination of diapause in the two-spotted spider mite, *Tetranychus urticae* Koch and the Kanzawa spider mite, *Tetranychus kanzawai* Kishida on pear tree in Tottori district (Acarina: Tetranychidae). Jpn. J. Appl. Entomol. Zool., 24: 175—183 (in Japanese).

Van de Bund C.F. and Helle, W., 1960. Investigations of the *Tetranychus urticae* complex in north west Europe (Acari: Tetranychidae). Entomol. Exp. Appl., 3: 142—156.

Van Zon, A.Q., Overmeer, W.P.J. and Veerman, A., 1981. Carotenoids function in photoperiodic induction of diapause in a predacious mite. Science, 213: 1131—1133.

Vaz Nunes, M., 1983. Photoperiodic Time Measurement in the Spider Mite *Tetranychus urticae* Koch. PhD. Thesis, University of Amsterdam.

Vaz Nunes, M. and Veerman, A., 1979a. Photoperiodic time measurement in spider mites. I. Development of a two interval timers model. J. Comp. Physiol., 134: 203—217.

Vaz Nunes, M. and Veerman, A., 1979b. Photoperiodic time measurement in spider mites. II. Effects of skeleton photoperiods. J. Comp. Physiol., 134: 219—226.

Vaz Nunes, M. and Veerman, A., 1982a. External coincidence and photoperiodic time measurement in the spider mite *Tetranychus urticae*. J. Insect Physiol., 28: 143—154.

Vaz Nunes, M. and Veerman, A., 1982b. Photoperiodic time measurement in the spider mite *Tetranychus urticae*: a novel concept. J. Insect Physiol., 28: 1041—1053.

Vaz Nunes, M. and Veerman, A., 1984. Light-break experiments and photoperiodic time measurement in the spider mite *Tetranychus urticae*. J. Insect Physiol. 30: 891—897.

Veerman, A., 1974. Carotenoid metabolism in *Tetranychus urticae* Koch (Acari: Tetranychidae). Comp. Biochem. Physiol., 47B: 101—116.

Veerman, A., 1977a. Aspects of the induction of diapause in a laboratory strain of the mite *Tetranychus urticae*. J. Insect Physiol., 23: 703—711.

Veerman, A., 1977b. Photoperiodic termination of diapause in spider mites. Nature, 266: 526—527.

Veerman, A., 1980a. Induction and termination of diapause in the spider mite *Tetranychus urticae*. In: A.K. Minks and P. Gruys (Editors), Integrated Control of Insect Pests in the Netherlands. Pudoc, Wageningen, pp. 129—132.

Veerman, A., 1980b. Functional involvement of carotenoids in photoperiodic induction of diapause in the spider mite, *Tetranychus urticae*. Physiol. Entomol., 5: 291—300.

Veerman, A. and Helle, W., 1978. Evidence for the functional involvement of carotenoids in the photoperiodic reaction of spider mites. Nature, 275: 234.

Veerman, A., Overmeer, W.P.J., van Zon, A.Q., de Boer, J.M., de Waard, E.R. and Huisman, H.O., 1983. Vitamin A is essential for photoperiodic induction of diapause in an eyeless mite. Nature, 302: 248—249.

Veerman, A. and Vaz Nunes, M., 1980. Circadian rhythmicity participates in the photoperiodic determination of diapause in spider mites. Nature, 287: 140—141.

Veerman, A. and Vaz Nunes, M., 1984a. Photoperiod reception in spider mites: photoreceptor, clock and counter. In: R. Porter and G.M. Collins (Editors), Photoperiodic Regulation of Insect and Molluscan Hormones, Ciba Foundation Symp. No. 104. Pitman, London, pp. 48—64.

Veerman, A. and Vaz Nunes, M., 1984b. Photoperiodic time measurement in spider mites. In: D.A. Griffiths and C.E. Bowman (Editors), Acarology VI, Vol. 1. Ellis Horwood Ltd., Chichester, U.K., pp. 337—342.

Venables, E.P., 1943. Observations on the clover or brown mite *Bryobia praetiosa* Koch. Can. Entomol., 75: 41—42.

Volkovich, T.A. and Razumova, A.P., 1982. Photoperiodic reaction in *Tetranychus crataegi* Hirst. (Acarina, Tetranychidae) at constant and gradually changing day length. Entomol. Obozr., 61: 8—16 (in Russian).

Von Hanstein, R., 1902. Zur Biologie der Spinnmilben (*Tetranychus* Duf.). Z. Pflanzenkr., 12: 1—7.

Weldon, G.P., 1910. Life history notes and control of the common orchard mites *Tetranychus bimaculatus* and *Bryobia pratensis*. J. Econ. Entomol., 3: 430—434.

Zacher, F., 1920. Untersuchungen über Spinnmilben. Mitt. Kaiser. Biol. Anst. Land. Forstwirtsch., Heft, 18: 121—130.

Zacher, F., 1921. Untersuchungen über Spinnmilben. Mitt. Biol. Reichsanst. Land. Forstwirtsch. Heft, 21: 91—100.

Zacher, F., 1922. Biologie, wirtschaftliche Bedeutung und Bekämpfung der Spinnmilben. Verh. Deutsch. Ges. Angew. Entomol. E.V., 3: 5—10.

Zacher, F., 1932. Beiträge zur Kenntnis phytophager Milben. Zool. Anz., 97: 177—185.

Zein-Eldin, E.A., 1956. Studies on the legume mite, *Petrobia apicalis*. J. Econ. Entomol., 49: 291—296.

1.4.7 Effects on the Host Plant

A. TOMCZYK and D. KROPCZYŃSKA

INTRODUCTION

Injury to plants caused by spider mites and the specific reactions of the plants have attracted the attention of acarologists and plant physiologists during the last few decades. Data on various aspects of plant reactions to mite feeding have been published. These data include reaction of the entire plant organism, changes in the individual plant organs, tissues and cells and changes in chemical and biochemical processes. It should be realized that the reactions occurring in the plant after mite feeding, such as leaf colour changes, and reduction in growth rate, flower formation and yield, are only external symptoms which result from mechanical damage and biochemical alterations. These reactions to mite feeding can be understood only on the basis of a knowledge of the mutual interactions between mites and their host plants. Although much information on the changes in the host plant after mite feeding has now been accumulated, the mechanism of the mutual inter-action between pest and host plant is not sufficiently well known.

FEEDING SYMPTOMS

Spider mites are among the most common plant pests, mainly feeding on leaves, and sometimes causing damage to specific plant parts, e.g. cotyledons, fruits, flowers, fruit spurs or tips of shoots. *Tetranychus turkestani* Ugarov & Nikolski and *Tetranychus urticae* Koch, can develop dense populations on the cotyledons of cotton. *Mononychellus tanajoa* (Bondar) and *Mononychellus mcgregori* Flechtmann & Baker occur on the tips of cassava. *Eotetranychus yumensis* (McGregor) damages leaves, fruits and spurs of young lemon trees, while *E. lewisi* (McGregor) develops on the fruit of lemon trees during the ripening period. Symptoms of spider mite attack are variable and may depend, among other factors, on the mite species, the characteristics of the leaves, i.e. their anatomical structure or chemical composition, the weather during or shortly after attack, and the specific reactions in the plant to spider mite attack. Hot, dry weather during exposure to mite feeding seems to intensify the symptons of damage.

Most common is feeding on the underside of the leaves. Some species, however, feed on both sides of the leaves, while occasionally the upper side is preferred; for example *Oligonychus punicae* (Hirst) on avocado and *Bryobia rubrioculus* (Scheuten) on almond.

Typical symptoms of spider mite feeding are small, light coloured punctures which, on prolonged exposure, develop into irregularly shaped,

Chapter 1.4.7. references, p. 327

white or greyish-coloured spots. Changes in colour from yellowing to bronzing are often characteristic of mite attack. In some host plant species, leaf curling can be a result of mite feeding. Necrosis may also occur in young leaves, stems and even growing tips. Sometimes leaf burning and defoliation of the host plant can be observed. This type of damage was described for feeding by species of the genus *Eotetranychus* (Lal and Mukharji, 1979). The phenomenon of defoliation, however, also results from feeding by other spider mite species. *T. turkestani* and *T. urticae* damage the cotyledons and leaves of cotton and cause their abcission. The presence of an abcission layer in almond leaves attacked by *Tetranychus pacificus* McG. was demonstrated by Andrews and LaPré (1979). Leaves injured with a high density of *T. pacificus* were found to be 'brittle to the touch' (Andrews and LaPré, 1979).

Different host plants may show differences in the severity of symptoms after feeding by this species of spider mite; for instance, pear leaves damaged by *T. pacificus* turn black and appear to be burned. On other host plants, e.g. apple, grape and bean, the symptoms after feeding by the same species are typical for mite attack. Some other species, e.g. *Oligonychus punicae* on avocado leaves, apparently do not cause noticeable damage at relatively high population density levels.

DAMAGE

Morphological, histological and cytological aspects of damage

During feeding the mites penetrate the plant with their stylets and suck out the cell contents. The depth reached by the stylets is approximately 70—120 μm (Avery and Briggs, 1968a; Summers and Stocking, 1972; Sances et al., 1979a). The depth at which injury occurs is related to the length of the stylets, the feeding time and the population density; it also depends on the host plant characteristics (Sances et al., 1979a; Mothes and Seitz, 1982).

Low population densities of *T. urticae* Koch on strawberry leaves mainly damage the spongy mesophyll tissue; however, slight injury also may be caused to the lowest parenchyma cell layer. Higher densities of spider mite populations on the same plant increased the sphere of damage and more severe injury to the palisade parenchyma was then observed (Sances et al., 1979a; Kielkiewicz, 1981). Apple and pear leaves injured by *Panonychus ulmi* (Koch) feeding showed the greatest changes in the spongy parenchyma (Avery and Briggs, 1968b). In bean leaves inhabited by *T. urticae*, symptoms of injury occurred simultaneously in both spongy and palisade parenchyma layers. In addition to the damaged cells, some authors observed empty open spaces between the mesophyll layers after feeding by different species of mites (Tanigoshi and Davis, 1978; Mothes and Seitz, 1982).

A different picture of injury was observed when the mites fed on the upperside of the leaves. Almond leaves injured by *Bryobia rubrioculus* showed symptoms of damage in the first 2 layers of the palisade parenchyma; cells of the spongy parenchyma were unaffected (Summers and Stocking, 1972). Most authors did not report damage to the bundle sheaths; however, these showed changes in the surrounding cells (Summers and Stocking, 1972; Tanigoshi and Davis, 1978).

It is not completely clear how the stylets penetrate through the epidermis and the mesophyll layers. Avery and Briggs (1968b), in investigating damage in apple and plum leaves caused by *P. ulmi*, did not notice punctures in the epidermis and suggest that the stylets pass between the cells. *Tetranychus mcdanieli* McGregor, feeding on apple leaves, inserts the stylets through the walls of the epidermal cells (Tanigoshi and Davis, 1978). Similarly, *B. rubrioculus* feeding on almond leaves punctures the epidermal cells and 1 cell may have 2 or 3 punctures (Summers and Stocking, 1972). Epidermal cells of injured apple and strawberry leaves sometimes were flattened or deformed (Sances et al., 1979a). Deformation in strawberry leaves may be due to mechanical damage by puncturing, or can be an osmotic effect arising from damaged neighbouring cells, as suggested by Sances et al. (1979a).

A very important change which occurs in the epidermal cell layers is the disturbing effect on the function of the stomatal apparatus. Studies on strawberry plants damaged by *T. urticae* showed that destruction of the stomatal apparatus was not caused by mechanical damage to the epidermal layer, but results from injury to the spongy parenchyma. Dehydrated cells of this tissue cause a lack of turgor in the guard cells. This failure of the guard cells results in closing of the stomata. More than 60% of the stomata were closed in damaged strawberry leaves (Sances et al., 1979a). This phenomenon was also found in investigations by the present authors on chrysanthemum leaves injured by *T. urticae* feeding; the occurrence of closed stomata was more widespread on the younger foliage.

The thickness of injured leaves may be strongly reduced; a reduction in thickness of injured bean leaves of approximately 50% was reported by Mothes and Seitz (1982). Damaged tissue has a lower number of cells; in apple leaves damaged by *T. mcdanieli* this reduction can amount to 35—55% (Tanigoshi and Davis, 1978).

Cytological studies using light- and electron microscopy showed conspicuous changes in the damaged and adjacent cells. In apple leaves damaged by *T. mcdanieli* most of the injured cells were deformed, protoplasts were lacking and their walls were shrunken. In some cases only starch grains were present in the injured cells. Sometimes breaking of the walls of empty cells was observed and spaces filled with air were formed. In other damaged cells, amber-coloured protoplasts could be observed. Very characteristic for damaged tissue was the presence of electron-dense material concentrated along the cell walls. This electron-dense material may also fill the intercellular space and may be combined with coagulated material from other injured cells (Tanigoshi and Davis, 1978). The basis of these processes is not known. Cells with coagulated protoplasm also showed a degeneration of organelles, but these changes were not so great that the boundaries between the organelles could not be recognized. One of the reasons for changes in damaged cells may be probable breakdown of the tonoplasts; this would permit a direct contact between the vacuolar sap and the cytoplasm (Tanigoshi and Davis, 1978).

As well as the cells which are directly damaged by feeding activity of the mites, and which can be recognized by their lack of protoplasts or coagulated protoplasts, in some cases neighbouring cells were more or less changed (Tanigoshi and Davis, 1978). The protoplasts in these cells were not changed and had well-differentiated organelles. The chloroplasts were swollen and sometimes their thylakoids became cup-shaped, with stroma filling the cups. It is not clear whether the changes in the chloroplasts are brought about under the influence of disturbances in the osmotic pressure or directly, under the influence of saliva.

The destruction of chloroplasts in cells adjacent to damaged cells was also observed in strawberry leaves artificially inoculated with *T. urticae* (Kielkiewicz, 1981). This author also observed the presence of numerous small vacuoles along the tonoplasts. Also, in cells of bean leaves adjacent to damaged cells changes in the structure of organelles were noted. These changes were as follows: coagulation of the cytoplasm, degeneration of the

Chapter 1.4.7. references, p. 327

nucleus, alterations in the structure of the cell walls and even some degeneration processes in the chloroplasts. Degeneration processes in organelles increased in accordance with the number of days of mite feeding and were a function of time (Mothes and Seitz, 1982). Degeneration of chloroplasts and other cell structures may disturb some important physiological processes, consequently influencing the productivity of plants.

It is a matter of discussion whether the observed cytological changes are connected only with mechanical injury to the cells during the process of feeding or result from the toxic action of saliva injected by the mites prior to or during the process of feeding. The type and amount of damage may also be related to the mite species and the host plant variety. In some instances, the damage can be related exclusively to exhaustion of the cell contents. In almond leaves damaged by *B. rubrioculus*, Summers and Stocking (1972) observed that only cells penetrated by the mite's stylets were damaged.

It is possible also to explain some destructive changes in the cells entirely by a mechanical mechanism of damage; e.g. mechanical damage of a tonoplast may cause the release of some substance from vacuola to cytoplasm which can have a toxic effect on the organelles (Tanigoshi and Davis, 1978). However, it is very difficult, if not impossible, to differentiate this indirect effect of mechanical damage from the direct influence of toxic substances originating from saliva.

The occurrence of changes in the adjacent cells, which are not directly damaged, suggests that these changes are not of mechanical origin, and they can be assumed to be connected with the influence of mite saliva. It is also possible that cells injured by mite feeding influence adjacent cells through changes in the water balance of the mesophyll. Injured cells in damaged leaves interrupt contact of uninjured cells with each other, thus resulting in empty spaces between them. Such isolated cells can also be disturbed in their function. Another indirect influence of mite feeding can be related to damage of lisosome membranes, thus causing autolysis.

Mothes and Seitz (1982) give 3 different interpretations of spider mite damage:

(1) during feeding a vacuum is created which forces cells to burst;

(2) cells adjacent to injured cells may be influenced by mite saliva;

(3) degeneration processes in non-punctured cells are the result of reaction by damaged plant tissue.

In discussing the last concept we may suppose that mechanical and chemical damage of some of the cells can have a stressful influence either on the whole organ or on only those cells adjacent to the damaged cells.

Toxic effects of saliva

Histological studies of the digestive system of spider mites show that they have well-developed salivary glands and that these glands take part in the feeding act (Mothes and Seitz, 1981). Salivary secretions may be composed of mucous material, glycoproteins and liquid material. Andrews and LaPré (1979), studying the influence of *T. pacificus* on almond, conclude that the occurrence of defoliation as a reaction to mite feeding can be connected with a toxic effect of saliva on the plant. Similar suggestions were given by other authors studying damage to cotton plants by *T. urticae* and *T. turkestani*. Vehement reactions were observed after feeding by *T. turkestani*, as compared to *T. urticae*. It is possible that there are differences in the content or quality of the saliva injected by these 2 species (Simons, 1964). From 2 species of spider mite, *Eutetranychus orientalis* (Klein) and *Eotetranychus uncatus* Garman, attacking *Bauhinia variegata* L., only

E. uncatus caused characteristic burning of the leaves as a toxigenic effect on the plant. This mite may inject a toxic material into the plant (Lal and Mukharji, 1979).

Avery and Briggs (1968b) using $^{14}CO_2$ as a marker, observed that the process of feeding by *P. ulmi* is connected with the injection of certain substances which were translocated to other regions of the plant. These substances were transported to the roots and young shoots of plum trees. Storms (1971) studied feeding of *T. urticae* on bean leaves. He used ^{32}P-labelled mites and showed that some substance was injected into the leaves, and that this substance was translocated mainly to the growing parts of the plants. Storms suggested that the saliva has an important role in the degradation of the plant cell contents and that 1 or more proteolytic enzymes may be involved. He also does not exclude from consideration the presence of plant hormone substances in the mite saliva. It is quite possible that an increase in plant hormone content is a specific reaction to mite saliva; higher plant hormone content may stimulate physiological processes to compensate for losses due to mite damage. However, Sances et al. (1979a) on the contrary suggest that damage caused by *T. urticae* to strawberry was essentially mechanical, and excluded the presence of some toxic material during the feeding act. The chemical composition and the nature of the secretions are almost unknown; this is because it is extremely difficult to obtain a sufficient amount of saliva for chemical analysis. Only indirect investigations on damaged plant parts and the reaction of the plant to mite feeding permit a characterization of the nature of this substance and its influence on the host plant.

Chemical aspects of damage

Important changes in the chemical composition of the host plant after mite feeding have been reported, concerning both organic and inorganic components. The content of such basic mineral elements as nitrogen and phosphorus was reduced after a long period of mite feeding on apple (Herbert and Butler, 1973; Golik, 1975). Herbert and Butler (1973) also found a small decrease in potassium in damaged leaves; Golik (1975), however, did not observe any changes in potassium, calcium or magnesium content. In investigations by the present authors on chrysanthemum damaged by *T. urticae*, a decrease in potassium, nitrogen and phosphorus content was observed. Nitrogen content also was decreased slightly in *Tilia* leaves damaged by *Eotetranychus tiliarium* (Hermann). This may suggest that in injured leaves proteins can be degenerated, or that their synthesis is prohibited. Injured plants may also be restricted in their uptake and translocation of nitrogen; however, data to support this hypothesis are not available. Zukova (1963) showed that in leaves on which *T. urticae* had been feeding, the level of non-protein nitrogen was decreased.

It is not known whether the decrease in the mineral content of damaged leaves is due to a reduced uptake by the roots from the soil, or is caused by exhausting cells through feeding by spider mites. A decrease in nitrogen content may also be connected with an ageing process; this ageing process may be stimulated by mite feeding. Similarly, the often observed symptoms of early drying of damaged leaves can be attributed to changes in potassium content. Data on the relation between mite feeding and changes in the mineral composition of leaf tissue are scarce; more studies are needed to shed light on this aspect.

The concentration of organic compounds in the leaf tissue, e.g. proteins, carbohydrates, phenolic compounds, plant hormones and chlorophyll, can also be influenced by mite feeding. Changes in the concentration of these constituents may exert an influence on the physiology of the plant. Zukova (1963) observed a reduction of activity in amylase, proteolytic enzymes and a reduction of total protein content. She also observed accumulation of nitrogenous substances in the shoots and reduction of them in the roots. In leaves of linden trees damaged by *E. tiliarium*, a strong decrease in soluble protein content was found; however, the total N content was only slightly reduced. Information on the production of some organic compounds in damaged plants was obtained during studies by the present authors on carbon metabolism in photosynthesis. By using the ^{14}C method, it was found that in chrysanthemum leaves damaged by *T. urticae*, the total amount of labelled amino acids and soluble sugars increased and the uptake of ^{14}C in starch was decreased (Kolodziej et al., 1975). On extending these studies to 3 cultivars of chrysanthemum with different levels of resistance against *T. urticae*, an increase in reduced sugar content in injured leaves was found, simultaneously with a decrease in starch content (Tomczyk and Van de Vrie, 1982). Borichenko and Manolov (1982) demonstrated that in apple leaves injured by *P. ulmi* an increase of labelled amino acids occurred, while at the same time a decrease in the transport of carbohydrate sorbitol was found. The consequences of these changes will be discussed later.

In plants injured by spider mites, some substances are frequently found which have characteristics similar to plant hormones (Avery and Briggs, 1968b; Storms, 1971). It is possible that these substances are injected with the saliva, in order to stimulate certain physiological processes in such a way that the products can be used by the mites. A comparison of data available in the literature which give descriptions of growth and development processes of damaged plants, e.g. stimulation of growth after feeding by low mite populations, the phenomenon of earlier bud development in fruit trees in the year following the year of infestation, and also premature dropping of flowers, fruitlets and abcission of leaves (Van de Vrie et al., 1972), give evidence for this opinion. Avery and Lacey (1968) showed that in leaves injured by *P. ulmi*, contents of gibberellin-like substances were increased, but auxin-like substances were decreased with the exception of indole-3-acetic acid. The content of indole-3-acetic acid (IAA) was even higher in the damaged leaves; however, in shoot tips and in the internodia, the level was lower.

It is possible that the decrease in the hormone content is accomplished because of a destructive influence of IAA-oxidase which is catalyzed by monophenolic substances. "Thus it is possible that phenolic compounds are liberated by the feeding damage and subsequent enzymatic reactions. Some of them may be released into adjacent tissues and be translocated within the plant. Conditions favouring the oxidation of auxin may therefore be produced. Other compounds may possibly stimulate the production of gibberellin which appears to be higher in the shoot tips of infested plants" (Avery and Lacey, 1968).

Increased levels of phenolic compounds in strawberry leaves injured by *T. urticae* were found by Kielkiewicz (1981). This author also observed changes in the localization of these phenolic substances in the leaf tissues of the variety Macherauch's Frühernte. In the palisade parenchyma a decrease of phenols was observed but an increase was found in the upper and lower epidermis after mite feeding. At the same time an increase in the level of these phenolic compounds was observed in the veins. According to these findings, it seems possible that these kinds of change play an important role in the host plant—spider mite relationships. In many studies it is reported that in leaves damaged by spider mites the concentration of chlorophyll is changed. The first well-documented records (Van de Vrie et al., 1972) give data suggesting that the chlorophyll content is reduced in apple leaves after they are damaged by *P. ulmi*. In more recent studies, evidence has been presented which suggests that the decrease in chlorophyll content is correlated with the duration of the feeding period and the mite density (Atanasov, 1973; Zwick et al., 1976; Andrews and LaPré, 1979; Tomczyk and Van de Vrie, 1982; De Angelis et al., 1983a). Some authors, however, report only minor changes in chlorophyll content, even in severely damaged leaves (Poskuta et al., 1975; Sances et al., 1979b, 1982a; Kolodziej et al., 1979).

The main reason for reduction in chlorophyll content probably is the mechanical damage of the chloroplasts during feeding. The amount of loss in chlorophyll content depends on the spider mite species and the host plant variety. The greatest reduction in chlorophyll content can be observed when spider mites feed primarily on the upperside of the leaves because in this case the palisade parenchyma is damaged (Summers and Stocking, 1972; Sances et al., 1982b). When mites feed on the underside of the leaf, the spongy mesophyll, which contains less chloroplasts, is damaged. In such a case an identical feeding intensity results in a lower reduction in chlorophyll content (Poskuta et al., 1979; Sances et al., 1979b). Plourde et al. (1983) found that in leaves damaged by *P. ulmi* the total chlorophyll content was decreased and that there was no recovery of the chlorophyll content after the mites were removed. They demonstrated differences in reaction by susceptible and resistant apple selections and attributed these differences to differences in the intensity of mite feeding. De Angelis et al. (1983a), investigating the reasons for losses in extractable chlorophyll, are of the opinion that reduction in chlorophyll content results from cell disturbance and removal of chloroplasts, rather than the enhanced metabolism degradation of chloroplasts. There are also data from which we can expect that degradation of chlorophyll can be attributed to other causes than mechanical damage. It was also observed that in bean leaves damaged by *T. urticae*, chlorophyll changed into pharophytine (Van de Vrie et al., 1972).

A decrease in chlorophyll content of the leaves can also be attributed to changes in the chloroplast content of cells adjacent to damaged cells. Water stress induced by spider mite feeding can have an influence on chlorophyll metabolism in injured cells.

It has been shown that in plants damaged by spider mites, changes in secondary plant components may occur. In *Mentha piperita* L., De Angelis et al., (1983c) found a decrease in the amount of pulegone and an increase in menthol and neomenthol content after feeding by *T. urticae*. These authors suggest that the disturbance in the metabolism of these compounds is a consequence of water stress in mite-damaged plants

Physiological aspects of damage

Simultaneously with cytological and chemical changes in damaged plants, remarkable modifications in basic physiological processes such as transpiration and photosynthesis can be observed. Plants damaged by spider mite feeding lose large amounts of water by transpiration. It has been noticed that the intensity of this process increases with increasing mite damage (Atanasov, 1973; Golik, 1975). Boulanger (1958) observed that when *P. ulmi* fed on apple leaves, the increase in intensity of transpiration only occurred until a certain level was reached. The intensity of transpiration in severely damaged leaves was lower. Similarly, Sances et al. (1979b, 1981, 1982a, 1982b), studying strawberry plants damaged by *T. urticae* and avocado leaves injured by *O. punicae*, found an initial increase and a later decrease in the intensity of transpiration. The reasons for changes in the transpiration intensity can be damage to the

Chapter 1.4.7. references, p. 327

protective surface of the leaves and disturbance to the functioning of the stomatal openings (Sances et al., 1979a). The most recent data provided by De Angelis et al. (1982) show a difference in the influence of spider mite damage on daytime (stomatal) and night-time (cuticular) transpiration. *Mentha piperita* leaves damaged by *T. urticae* transpired less during the day than undamaged leaves. During darkness however, transpiration was 3 times higher in the damaged leaves as compared to undamaged leaves. As a consequence, water stress was induced in these plants.

The amount of water loss in infested plants depends mainly on the structure of the leaf surface which is damaged during the feeding. To some extent this also depends on the possibility of water uptake by the plants. A decrease in water uptake as a result of an increase in root resistance may be an important factor in developing water stress during the daytime. It is not clear whether closing of the stomatal openings results directly from mite injury (Sances et al., 1979a), or if it is to be considered as a reaction of the plant to water stress which occurs because of the intensity of night transpiration (De Angelis et al., 1983b). As a consequence of water stress, major metabolic changes in plants damaged by spider mites may result. It is well known that water stress stimulates hydrolytic processes and causes degradation of starch to simple sugars and proteins to amino acids. De Angelis et al. (1983b) demonstrated that in *Mentha piperita* leaves damaged by *T. urticae* an increase in the amount of soluble sugars may occur. These authors suggest that the reason for accumulation of these components can be water stress in damaged plants. However, the origin of these sugars is not explained, because the starch content is still the same and degradation was not observed. Tanigoshi and Davis (1978) showed that starch grains in the cells of apple leaves were unchanged after mite feeding. De Angelis et al. (1983b) do not reject the possibility that decreased transport of carbohydrates was responsible for maintaining starch at uninjured levels. We are of the opinion that the higher sugar content in damaged leaves may also be related to photosynthetic processes; this will be discussed later.

The decrease in assimilation of CO_2 is proportional to mite population density and duration of feeding and is also correlated to the chlorophyll content of the leaves. Sometimes a conspicuous decrease was observed when plants were severely damaged. Boulanger (1958) found a strong reduction in photosynthesis in apple leaves only when severe bronzing occurred after feeding by *P. ulmi*. The decrease in chlorophyll content was proportional to the decrease in photosynthesis. Poskuta et al. (1975), however, observed minor reductions in chlorophyll content accompanied by a strong reduction in CO_2 uptake in strawberry leaves damaged by *T. urticae*. Similarly, Sances et al. (1979b) and Hall and Ferrel (1975), in experiments with the same host plant species, did not observe any correlation between the decrease in chlorophyll content and the intensity of photosynthesis. These results prove that the decrease in chlorophyll content is not the only reason for the decrease in CO_2 assimilation. It is possible that in some plant species, other factors play a role in restricting CO_2 assimilation e.g. activity, induced by mite feeding, may also be involved. It is possible that these changes in enzyme activity can be caused by injected saliva or they may result from water stress.

Studies on carbon metabolism in photosynthesis by using ^{14}C showed important quantitative changes in the synthesis of the primary assimilation products such as soluble sugars, amino acids, organic acids and starch (Kolodziej et al., 1975; Borichenko and Manolov, 1982). Studies by Kolodziej et al. (1975) on detached chrysanthemum leaves damaged by *T. urticae* found differences in ^{14}C pathways during the beginning and the advanced stage of the feeding period. After a short feeding period they found a decrease in the incorporation of ^{14}C into soluble sugars and amino acids, and an increase in the incorporation of ^{14}C into starch. The total incorporation of ^{14}C in the assimilation process of CO_2 was increased. In the advanced period of mite feeding the incorporation of ^{14}C into starch was lower but a remarkable increase of labelled soluble sugars and amino acids was demonstrated. De Angelis et al. (1983b), working with populations of mites feeding for short periods, found a similar decrease in sugar content.

These results bring us to the suggestion that increased content of soluble sugars in damaged tissues can be a result of decrease in starch synthesis. In connection with the results of De Angelis et al. (1983b), where no decrease in the starch level was found in peppermint plants, it seems possible that in different host plants and at different mite densities, changes in the sugar content may be attributed to other mechanisms. From all the results of studies on photosynthesis, it is clear that changes in CO_2 uptake occur after spider mite feeding, accompanied by changes in the pathway of carbon in photosynthesis.

INFLUENCE ON GROWTH, FLOWERING AND CROPPING

As a consequence of damage to plant tissue and disturbance of plant physiological processes, changes in growth intensity, flowering and yield may be observed. The most common change is a retardation of the growth of all organs of damaged plants.

Shoots can be shorter and accumulate less dry matter. In fruit trees a slower rate of increase of the diameter of the trunk was reported. Sometimes a delayed effect in the rate of growth of the stems was observed in the year following injury (Van de Vrie et al., 1972; Golik, 1975; Barnes and Andrews, 1978; Jesiotr, 1978b, Papaioannou-Souliotis, 1979). The total leaf area of damaged plants usually is strongly reduced. Two reasons for this phenomenon can be given: a decreased rate of growth in leaf area (Avery, 1962; Avery and Briggs, 1968a; Summers and Stocking, 1972), and a decrease in the number of leaves per plant (Avery, 1962). One reason for a decrease in the number of leaves per plant can be defoliation. A reduction in the thickness of damaged leaves has also been reported, resulting from the lower number of cells in the leaf (Tanigoshi and Davis, 1978; Mothes and Seitz, 1982).

Roots of damaged plum plants were strongly reduced; they had a lower number of branches and after a prolonged period of feeding they turned brown. Accumulation of dry matter in these was also reduced (Avery and Briggs, 1968a).

Growth reduction is not always characteristic for plants damaged by spider mites. No influence on the growth of plants attacked by spider mites was observed by Zwick et al. (1976) and Hoyt and Tanigoshi (1982). In some cases, growth is stimulated. This phenomenon occurs most frequently when population density is low or when the feeding period is short (Avery and Briggs, 1968a; Storms, 1971; Papaioannou-Souliotis, 1979). This phenomenon of growth stimulation indicates the possibility of alteration in the level of growth regulators and in the possible compensatory capacity by the plants. Flowers of damaged ornamental plants often had a smaller diameter and shorter stems, while the number of flowers usually was reduced (Jesiotr, 1978a, 1978b).

The influence on flowering of fruit trees was sometimes very conspicuous in the year following the attack; reduction occasionally amounts to 75% (Golik, 1975).

Chapter 1.4.7. references, p. 327

The influence of spider mite damage on cropping in fruit trees may occur in the early stage of fruit development; in many cases early dropping of immature fruits and later on the ripening of the fruits were reported (Van de Vrie et al., 1972; Golik, 1975; Papaioannou-Souliotis, 1979; Hoyt and Tanigoshi, 1983). Sometimes fruits from damaged trees are more intensely coloured (Golik, 1975). These changes in fruit performance do not always occur in the year of attack; sometimes this effect was detectable even after 2 or 3 years. The intensity of reaction depends on the plant species. Pear trees have a very low tolerance to mite feeding and low population densities of *P. ulmi* may cause serious leaf burning and yield losses (Hoyt and Tanigoshi, 1983). There are, however, also reports of stimulative effects of feeding by low populations; Oatman et al. (1982) observed that the number and size of strawberries increased under these circumstances. Some information is available on the influence of mite attack on the quality of fruits. Zwick et al. (1976) found that in the apple variety 'Golden Delicious' the storage quality was lowered. Laing et al. (1972) did not find any differences in sugar content in grapes from infested and non-infested plants. Mite feeding caused decreases in the size and number of seeds in cotton (Papaioannou-Souliotis, 1979).

CONCLUDING REMARKS

The evidence of mite—host plant relations provides the possibility of explaining the probable interactions between mites and plants. The first consequence of mite attack is mechanical damage caused by piercing the tissues followed by injection of saliva. Mostly, in leaves inhabited by spider mites, 3 categories of cells can be observed: empty cells, cells with degenerated organelles, and normal healthy cells. Organs damaged in this way do not function normally. We can expect even healthy, undamaged cells to experience changes in functioning, owing to loss of contact with other cells. Mechanical injury to epidermal and mesophyll tissues can be a direct reason for water losses. Simultaneously, cell chloroplasts are damaged and the loss of a considerable amount of chlorophyll can be observed. Because of water losses, water stress may occur, and chlorophyll losses are important reasons for a reduction in intensity of photosynthesis. Plants under water stress have reduced growth, which has an influence on their cropping. In these plants, the intensity of hydrolytic processes is higher and consequently a higher concentration of these components increases osmotic tension; increased turgor allows for easier food uptake by spider mites. Changes in the concentration of soluble sugars and amino acids not only create better physical conditions for feeding but also provide a higher nutritive value of the food for spider mites.

Decrease in the water content of the plant and high concentrations of the primary products of photosynthesis (sugars and amino acids) can lower photosynthetic activity.

It is also possible that, apart from water stress and chlorophyll content, the injected saliva exerts some influence on photosynthesis. Components of saliva may exert an influence on the activity of photosynthetic enzymes and cause changes in the pathways of C in this process. As a result, qualitative and quantitative changes in photosynthetic products can occur. On the basis of our data we are of the opinion that the most important change in photosynthetic reactions after spider mite attack is the blocking of starch synthesis. This restriction of starch synthesis can result in higher sugar and PGA (phosphoglycine) levels. PGA levels which are too high can be reduced to lower levels through transformation of PGA to serine. Consequently, an increased concentration of amino acids will be observed (Kolodziej, 1976). Thus, accumulation of photosynthetic products can take place in 2 different ways: as a result of hydrolytic processes, or as a result of changed pathways of C in photosynthesis.

All the above-mentioned metabolic processes in injured plants may have some influence on their growth, flowering and cropping. Changes in plant hormone levels and in mineral content can also influence these processes.

Differences between host plants are responsible for the fact that not all host plants show identical reactions or intensity of reaction to mite feeding. We can expect that in some plants, 1 or more of the above-mentioned reactions may occur, or other more or less complicated relations exist. For instance, after mite feeding which governs defoliation, the reduction in the intensity of photosynthesis will be due mainly to a decrease in the total active photosynthetic leaf area. It is necessary to realize that mite—host plant relationships are mutual in effect. There are not only mechanisms favouring mite development; other reactions may be considered as defence mechanisms. The interpretations of mechanisms discussed above are an attempt to understand the very complicated relations between mites and their host plants. For a full understanding of the processes involved, more sophisticated techniques for fundamental studies are needed. These studies should not only clarify the problems connected with the susceptibility of plants to mite feeding, they also should concentrate on defence reactions and resistance mechanisms in the host plants. Knowledge of the fundamental processes governing spider mite—host plant relations will allow a better understanding of the effects of spider mites on the crop level. Economic injury levels could thus be given a sound basis, and plant growth analysis in the presence and absence of spider mites should accompany studies of the influence of spider mites on the crop level. This ultimately could facilitate a faster introduction of biological or integrated control of phytophagous mites.

REFERENCES

Andrews, K.L. and LaPré, L.F., 1979. Effects of Pacific spider mite on physiological processes of almond foliage. J. Econ. Entomol., 72: 651—654.

Atanasov, M., 1973. Physiological functions of plants as affected by damage caused by *Tetranychus atlanticus* McGregor. In: M. Daniel and B. Rosický (Editors), Proc. 3rd Int. Congr. Acarol., Prague, 1971, Publishing House of the Czechoslovakian Acad. of Sciences, Prague. pp. 183—186.

Avery, D.J., 1962. The Vegetative Growth of Young Plants of Brompton Plum Infested with Fruit Tree Red Spider Mite. Annu. Rep. East Mallin Res. Stn., 1961, pp. 77—80.

Avery, D.J. and Briggs, J.B., 1968a. Damage to leaves caused by fruit tree red spider mite, *Panonychus ulmi* (Koch). J. Hortic. Sci., 43: 463—473.

Avery, D.J. and Briggs, J.B., 1968b. The aetiology and development of damage in young fruit trees infested with tree red spider mite, *Panonychus ulmi* (Koch). Ann. Appl. Biol., 61: 277—288.

Avery, D.J. and Lacey, H.J., 1968. Changes in the growth-regulator content of plum infested with fruit tree red spider mite, *Panonychus ulmi* (Koch). J. Exp. Bot., 19, 61: 760—769.

Barnes, M.M. and Andrews, K.L., 1978. Effects of spider mites on almond tree growth and productivity. J. Econ. Entomol., 71: 555—558.

Borichenko, N. and Manolov, P., 1982. Changes in some physiological indicators and biochemical processes of apple leaves infested to various extents by *Panonychus ulmi* Koch. Hortic. Viticultural Sci., 19:44—50 (in Bulgarian, with English summary).

Boulanger, L.W., 1958. The effect of European red mite feeding injury on certain metabolic activities on Red Delicious apple leaves. Maine Agric. Exp. Stn. Bull., 570: 1—34.

De Angelis, J.D., Berry, R.E. and Krantz, G.W., 1983a. Photosynthesis, leaf conductance and leaf chlorophyll content in spider mite (Acari: Tetranychidae) injures peppermint leaves. Environ. Entomol., 12: 345—349.

De Angelis, J.D., Berry, R.E. and Krantz, G.W., 1983b. Evidence for spider mite (Acari: Tetranychidae) injury — induced leaf water deficits and osmotic adjustments in peppermint. Environ. Entomol., 12: 336—339.

De Angelis, J.D., Larson, K.C., Berry, R.E. and Krantz, G.S., 1982. Effects of spider mite injury on transpiration and leaf water status in peppermint. Environ. Entomol., 11: 975—978.

De Angelis, J.D., Marin, A.B., Berry, R.E. and Krantz, G.W., 1983c. Effects of spider mite (Acari: Tetranychidae) injury on essential oil metabolism in peppermint. Environ. Entomol., 12: 522—527.

Golik, Z., 1975. A study of the destructiveness of the fruit tree red spider mite, *Panonychus ulmi* (Koch) on apple. Zesz. Probl. Postepow Nauk Roln., 171: 15—34.

Hall, F.R. and Ferree, D.C., 1975. Influence of two-spotted spider mite populations on photosynthesis of apple leaves. J. Econ. Entomol., 68: 517—520.

Herbert, H.J. and Butler, K.P., 1973. The effect of European red mite, *Panonychus ulmi* (Acarina: Tetranychidae) infestations on N, P and K concentrations in apple foliage throughout the season. Can. Entomol., 105: 263—269.

Hoyt, S.C. and Tanigoshi, L.K., 1983. Economic injury levels for apple mite pests. In: B.A. Croft and S.C. Hoyt (Editors), Integrated Management of Insect Pests of Pome and Stone Fruit. John Wiley and Sons, New York, NY, pp. 203—218.

Jesiotr, L.J., 1978a. Further study on the injurious effects of the two-spotted spider mite (*Tetranychus urticae* Koch) on greenhouse carnations. Ekol. Polsk., 26: 303—310.

Jesiotr, L.J., 1978b. The injurious effects of the two-spotted spider mite (*Tetranychus urticae* Koch) on greenhouse roses. Ekol. Polsk., 26: 311—318.

Kielkiewicz, M., 1981. Physiological, Anatomical and Cytological Changes in Leaves of Two Strawberry Varieties (*Fragaria grandiflora* Duch.) Resulting from Feeding by the Two-spotted Spider Mite (*Tetranychus urticae* Koch). Dissertation, Agricultural University of Warsaw, 95 pp (in Polish).

Kolodziej, A., 1976. Exchange of CO_2 and Metabolism of C in Photosynthesis of Chrysanthemum (*Chrysanthemum morifolium* L.) Damaged by the Two-spotted Spider Mite (*Tetranychus urticae* Koch). Dissertation, Agricultural University of Warsaw, 93 pp (in Polish).

Kolodziej, A., Kropczyńska, D. and Poskuta, J., 1975. The effect of two-spotted spider mite (*Tetranychus urticae* Koch) injury on carbon metabolism in *Chrysanthemum morifolium* L. In: Proc. 7th Int. Congr. Plant Protect., Moscow, 1975, Part II, pp. 217—229.

Kolodziej, A., Kropczyńska, D. and Poskuta, J., 1979. Comparative studies on carbon dioxide exchange rates of strawberry and chrysanthemum plants infested with *Tetranychus urticae* Koch. In: E. Pifft (Editor), Proc. 4th Int. Congr. Acarol., Saalfelden, 1974. Akadémiai Kiadó, Budapest, pp. 209—214.

Lal, L. and Mukharji, S.P., 1979. Observations of the injury symptoms caused by phytophagous mites. Zool. Beitr., 25: 13—17.

Laing, J.E., Calvert, D.L. and Huffaker, C.B., 1972. Preliminary studies of the effects of *Tetranychus pacificus* on yield and quality of grapes in the San Joaquin Valley, California. Environ. Entomol., 1: 658—663.

Mothes, U. and Seitz, K.A., 1981. Fine structure and function of the propodosomal glands of *Tetranychus urticae* K. (Acari: Tetranychidae). Cell Tissue Res., 221: 339—349.

Mothes, U. and Seitz, K.A., 1982. Fine structural alternations of bean plant leaves by feeding injury of *Tetranychus urticae* Koch (Acari: Tetranychidae). Acarologia, 23: 149—157.

Oatman, E.R., Sances, F.V., LaPré, L.F., Toscano, N.C. and Voth, V., 1982. Effect of different infestation levels of the two-spotted spider mite on strawberry yield in winter plantings in southern California. J. Econ. Entomol., 75: 94—96.

Papaioannou-Souliotis, P., 1979. Effects of the population of *Tetranychus urticae* (Koch) on bean plants (*Phaseolus vulgaris* L.). Ann. Inst. Phytopathol., Benaki, 12: 138—143.

Plourde, D.F., Goonewardene, H.F., Kwolek, W.F. and Nielsen, N.C., 1983. The effect of European red mite, *Panonychus ulmi*, on chlorophyll a/b ratios of apple, *Malus domestica* Borkh., leaves in a growth chamber study (Acarina: Tetranychidae). Int. J. Acarol., 9: 11—18.

Poskuta, J. Kolodziej, A. and Kropczyńska, D., 1975. Photosynthesis, photorespiration and respiration of strawberry plants as influenced by infestation with *Tetranychus urticae* (Koch). Fruit Sci. Rep., 2: 1—17.

Sances, F.V., Wyman, J.A. and Ting, J.P., 1979a. Morphological responses of strawberry leaves to infestations of the two-spotted spider mite. J. Econ. Entomol., 72: 710—713.

Sances, F.V., Wyman, J.A. and Ting, J.P., 1979b. Physiological responses to spider mite infestation on strawberry. Environ. Entomol., 8: 711—714.

Sances, F.V., Wyman, J.A., Ting, J.P., Steenwijk, R. van and Oatman, E.R., 1981. Spider mite interactions with photosynthesis, transpiration and productivity of strawberry. Environ. Entomol., 10: 442—448.

Sances, F.V., Toscano, N.C., Hoffman, M.P., LaPré, L.F., Johnson, M.W. and Bailey, J.B., 1982a. Physiological responses of avocado leaves to Avocado brown mite feeding injury. Environ. Entomol., 11: 516—518.

Sances, F.V., Toscano, N.C., Oatman, E.R., LaPré, L.F., Johnson, M.W. and Voth, V., 1982b. Reductions in plant processes by *Tetranychus urticae* (Acari: Tetranychidae) feeding on strawberry. Environ. Entomol., 11: 733—737.

Simons, J.N., 1964. Tetranychid mites as defoliators of cotton cotyledons. J. Econ. Entomol., 57: 145—148.

Storms, J.J.H., 1971. Some physiological effects of spider mite infestations on bean plants. Neth. J. Plant Pathol., 77: 154—167.

Summers, F.M. and Stocking, C.R., 1972. Some immediate effects on almond leaves of feeding by *Bryobia rubrioculus* (Scheuten). Acarologia, 14: 170—178.

Tanigoshi, L.K. and Davis, R.W., 1978. An ultrastructural study of *Tetranychus mcdanieli* feeding injury to the leaves of 'Red Delicious' apple (Acari: Tetranychidae). Int. J. Acarol., 4: 47—56.

Tomczyk, A. and Vrie, M. van de, 1982. Physiological and biochemical changes in three cultivars of chrysanthemum after feeding by *Tetranychus urticae*. In: J.H. Visser and A.K. Minks (Editors), Proc. 5th Int. Symp. Insect—Plant Relationships, Wageningen, 1982. PUDOC, Wageningen, The Netherlands, pp. 391—392.

Van de Vrie, M., McMurtry, J.A. and Huffaker, C.B., 1972. Ecology of Tetranychid mites and their natural enemies. A review. III. Biology, ecology and pest status, and host plant relations of Tetranychids. Hilgardia, 41: 343—432.

Zukova, V.P., 1963. Feeding mechanisms of the spider mite *Tetranychus telarius*. Tr. Nauchno-Issled. Inst. Zashch. Rast. Uzb. SSR, 6: 13—18 (in Russian).

Zwick, R.W., Fields, G.J. and Mellenthin, W.M., 1976. Effects of mite population density on 'Newton' and 'Golden Delicious' apple tree performance. J. Am. Soc. Hortic. Sci., 101: 123—125.

Chapter 1.5 Techniques

1.5.1 Rearing Techniques

W. HELLE and W.P.J. OVERMEER

INTRODUCTION

It is possible to rear spider mites for experimental purposes in various ways, and different systems have been adopted by different workers. This paper will not give a survey. We briefly present only 3 systems, as applied in the Laboratory of Experimental Entomology in Amsterdam during the past 20 years, and we will bring the merits and the failings of each system to the notice of the reader.

DETACHED LEAF AND LEAF DISK CULTURES

Detached bean leaf cultures are prepared as follows. A young but full-grown primary bean leaf with a short piece of the petiole is pressed firmly, with the upper surface uppermost, on a wad of wet cotton wool contained in a small dish, such as a petri dish or a tin foil dish (Fig. 1.5.1.1). Species such as *Phaseolus vulgaris* L., *Phaseolus limensis* MacFadyen and *Ricinus communis* L. can be used. Any holes occurring between the margin of the leaf and the cotton wool are filled up by pulling the cotton wool up with the aid of a pair of needles and pressing it against the leaf. This is to prevent mites moving to the undersurface of the leaf, and thereby escaping observation. It is important that the end of the petiole touches the wet cotton wool. After some days the leaf will take root and then the leaf culture may last for several weeks. A detached bean leaf culture will last for much longer if the cotton wool is soaked in a nutrient solution instead of tap water. A recommended solution contains the following: 213 mg KNO_3, 127 mg $MgSO_4 \cdot 7H_2O$, 141 mg KH_2PO_4, 5 mg $(NH_4)_2SO_4$, and 186 mg NH_4NO_3 per litre of tap water (Helle, 1962). The nutrient solution should be used once only, namely when the cotton wool wad is moistened for the first time before the leaf is placed onto the wad. Thereafter the cotton wool can be kept wet by adding tap water at regular intervals, depending on the rate of evaporation. When the leaves become too densely populated, specimens should be transferred to fresh detached leaf cultures.

Several species of *Tetranychus* have been reared on this kind of leaf culture. They include several members of the *T. urticae* complex, *T. lombardinii* Baker & Pritchard, *T. pacificus* McGregor, *T. neocaledonicus* André, *T. turkestani* Ugarov & Nikolski, *T. ludeni* Zacher, *T. tchadi* Gutierrez & Bolland, *T. tumidus* Banks, *T. marianae* McGregor, *T. piercei* McGregor, *T. fijiensis* Hirst, *T. kanzawai* Kishida, *T viennensis* Zacher, *T. lambi* Pritchard & Baker and *T. yusti* McGregor. Furthermore, several oligophagous species

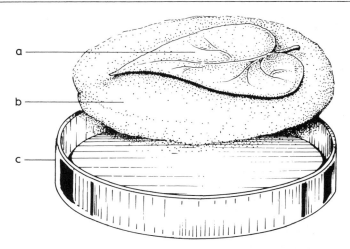

Fig. 1.5.1.1. Detached bean leaf culture. (a) leaf, (b) cotton wool, (c) petri dish.

belonging to *Eutetranychus*, *Oligonychus* and *Eotetranychus* are readily maintained on bean leaf cultures over longer periods, such as *Oligonychus bessardi* Gutierrez, *O. gossypii* (Zacher), *O. licinus* Baker & Pritchard and *O. gramineus* (McGregor); *Eutetranychus banksi* (McGregor), *Eu. orientalis* (Klein), *Eu. eliei* Gutierrez & Helle and *Eu. africanus* (Tucker); *Eotetranychus imerinae* Gutierrez, *Eo. paracybelus* Gutierrez and *Eo. rubiphilus* (Reck).

If required, the bean leaf can also be placed on the cotton wool with the undersurface uppermost. As a consequence, different webbing patterns may arise, because the threads can be attached to the leaf veins. Certain species can be reared only on the undersurface of *Phaseolus vulgaris* leaves, e.g. *Tetranychus fijiensis*, species inhabiting palm trees, and *Eo. rubiphilus* (H.R. Bolland, personal communication).

Leaf disks or parts of leaves can be used instead of whole leaves for similar rearing procedures. In experiments with *Panonychus ulmi* (Koch) and *Panonychus citri* (McGregor), leaf disk cultures of apple or plum and of different agrumes, respectively, are used. Many oligophagous and monophagous species can be reared on leaf disks or detached leaves from their host plants, provided that the excised leaves remain turgid and are of suitable form and size. Most *Eotetranychus* species from trees are conveniently reared in this way. Detached leaf cultures also appear feasible for the rearing of *Schizotetranychus* and *Oligonychus* species living on grasses. Small pieces of grass on a wad of wet cotton wool proved suitable. Observational arenas for the study of *Oligonychus pratensis* (Banks) on the grass *Bouteloua gracilis* (H.B.K.) Griffith have been described by Congdon and Logan (1983).

There are many advantages of detached leaf and leaf disk rearing methods. The cultures can easily be observed under the stereomicroscope, and the handling of the mites offers no difficulties. Furthermore, mass production of uniform material is feasible. By transferring mated adult females to the detached leaf cultures and leaving them for a period of 24 h, batches of eggs of a defined age interval are obtained. These 'egg-waves' will hatch within a similar time interval and will yield material of nearly the same age which will develop synchronously. This is convenient if one wishes to collect large numbers of a particular stage, for instance teleiochrysalids for crossing experiments, or uniform adult material for toxicological tests.

There are usually no serious problems with the control of the environment. Detached leaf cultures evaporate a limited amount of water as compared to plants, and can be maintained in a controlled environment

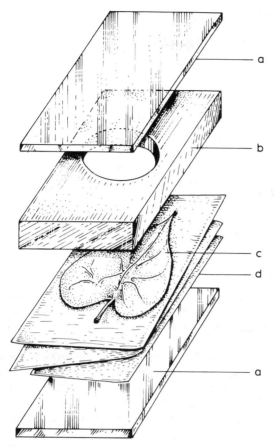

Fig. 1.5.1.2. Plexiglas cell with a single hole. (a) glass, (b) Plexiglas with hole, (c) leaf, (d) filter paper.

without special devices to prevent an excessively high humidity. Several stocks can be maintained in the same room, with their isolation reasonably well assured. The wet cotton wool is an efficient barrier for most species. Large species, however, such as the tumid spider mite (*Tetranychus tumidus*) are able to walk over the cotton wool unless it is very wet. It is advisable to place the detached leaf cultures in trays with water, thus providing an extra barrier. When tin foil dishes are used, the bottoms can be perforated so that the cotton wool will remain wet for a longer period by contact with water in the tray (Overmeer, 1967). Contamination by alien mite material can confidently be prevented by inspection of the leaves under a dissection microscope before use.

REARING IN CLOSED CELLS

Rearing in closed plexiglas cells is recommended for species which tend to migrate from the leaf, such as species of the genus *Bryobia*. It is also recommended where the leaf of a host plant cannot be properly adjusted to a wad of cotton wool, as for instance with leathery leaves, so that the detached leaf culture method is not feasible.

In a piece of plexiglas (of thickness 5—10 mm), one or more circular holes of diameter 5—15 mm are made, in such a configuration that these holes fit within the outline of the leaf lamina. A plate of plexiglas of the same size forms the base of the construction (see Fig. 1.5.1.2). On this base plate 3

Chapter 1.5.1. references p. 335

layers of thick, moistened filter paper are laid, on which the leaf is placed. The plexiglas with the holes is then placed carefully on the leaf. The whole construction is held together with rubber bands. Cover slips are used for closing the cells, and can be stuck over the plexiglas holes with a slip of transparent sticky tape, so that a hole can easily be opened for manipulation of the mites inside. Using small holes, the method is appropriate for rearing mites on grasses. It has the advantage that the mite colony in the holes is easily observed under a dissection microscope. Manipulation of the mites does not offer serious problems. There is no way to control the relative humidity in the holes, unless the cover slip is removed and isolation is achieved by a glue barrier round the hole in the plexiglas. The method is rather laborious and is not suited to rearing large quantities of mites. Clearly, however, this method guarantees perfect isolation.

REARING ON PLANTS

The simplest way of rearing is on potted host plants. In the case of the polyphagous *Tetranychus* species, beans are most appropriate (French beans, lima beans and Castor beans). Young plants with 2 primary leaves are most suitable for this purpose. The plants are lightly infested with mites, which will multiply on the foliage. As soon as the plants become densely populated, replacement by fresh plants is essential; these can be infested by leaves taken from the old culture. This method allows the continuous availability of large numbers of mites with a minimum of labour.

Adult mites can be collected when required by tapping the leaves over a white piece of paper, and thereafter picking them from the paper with a fine artist's brush. This can be done with the naked eye. The easily obtained yield of masses of adult females makes the rearing method on plants useful for students dealing with acaricide screening and similar investigations. However, it has certain disadvantages. Observation and manipulation under the dissection microscope is more difficult because mites are present on both sides of the leaf, often in a dense webbing. Isolation of mite strains when reared on whole plants meets with considerable difficulties. Even if the plants are placed in a cabinet, contamination with other material may occur, especially from infestations born on host plants raised in a greenhouse. Introduction of this alien material into the strain is difficult to prevent, since adequate examination of whole plants is not feasible. The necessary preventative action to be taken is a methylbromide fumigation of the potted bean plants, before they are used in the rearing cabinets; this requires special equipment. With a closed cabinet it is also difficult to prevent excessive relative humidity. The most appropriate solution is a device by which dried air is injected into the cabinet, so that the air in the cabinet is replaced continuously.

For mass rearing of *P. ulmi* on apple or plum, seedlings or potted rootstocks are used. The plum cultivar Myrobalan is very often used, because it is less liable to infection by mildew.

For mites that have to be reared on such host plants as *Asparagus*, *Spergula*, *Buxus* or *Pinus* there is no alternative to rearing on whole plants or on branches of such plants put in a glass beaker. For rearing of *Oligonychus ununguis* (Jacobi) on conifers, see Reeves (1963); this species can also be reared on chestnut instead of on coniferous plants, according to Wanibuchi and Saitô (1983).

REFERENCES

Congdon, B.D. and Logan, J.A., 1983. Temperature effects on development and fecundity of *Oligonychus pratensis* (Acari: Tetranychidae). Environ. Entomol., 12: 359—362.

Helle, W., 1962. Genetics of resistance to organophosphorus compounds and its relation to diapause in *Tetranychus urticae* Koch (Acari). Tijdschr. Plantenziekten, 68: 155—195.

Overmeer, W.P.J., 1967. Genetics of resistance to tedion in *Tetranychus urticae* C.L. Koch. Arch. Neerl. Zool., 17: 296—349.

Reeves, R.M., 1963. Tetranychidae infesting woody plants in New York State, and a life history study of the elm spider mite, *Eotetranychus matthyssei* n. sp., Mem. Cornell Univ. Agric. Exp. Stn., Ithaca, NY, 380: 1—99.

Wanibuchi, K. and Saitô, Y., 1983. The process of population increase and patterns of resource utilization of two spider mites, *Oligonychus ununguis* (Jacobi) and *Panonychus citri* (McGregor), under experimental conditions. (Acari: Tetranychidae). Res. Popul. Ecol., 25: 116—129.

1.5.2 Sampling Techniques

M.W. SABELIS

INTRODUCTION

In the context of spider-mite control, information on mite abundance is required to evaluate (the necessity of) control measures. For research purposes this information should concern the population size per se, but in agricultural practice it is sufficient to know whether the population size is below or above a predetermined critical threshold size, i.e. the economic damage threshold. Absolute counts are usually not feasible, so sampling should provide the desired information. Only by having a clear understanding of the objectives is it possible to develop or select adequate sampling methods. To this end the following questions should be answered: (1) what is the definition of the population to be sampled? What are the population elements?; (2) what is the sampling unit?; (3) which sampling technique should be used?; (4) where and when should sampling take place?; (5) what is the optimal sample size given the level of precision? How is this precision defined? and (6) what is the risk that the sampler is willing to incur that the actual precision exceeds the precision required?

A statistical treatise of the sampling methodology has been presented by Cochran (1963) and Seber (1973). Morris (1960), Southwood (1978), Karandinos (1976) and Kogan and Herzog (1980) have also discussed statistical aspects, but with emphasis on the practical aspects of sampling arthropods. The material presented in this chapter is largely based on these references, but the emphasis will be on the literature on spider-mite sampling, which has recently yielded some important advances of general interest to sampling theory (Nachman, 1981a, b, 1984; Wilson et al., 1983).

POPULATION ELEMENTS

Although the term 'population' is well defined in sampling statistics, it has been subject to much debate in population ecology. However, despite the fact that current definitions lack generality, it is usually not difficult to define populations in the context of pest control. They consist of a number of elements in a specified space, such as the number of European red mites in an apple orchard. Because mite counts are laborious and time consuming, population elements are often chosen so as to minimize the counting effort. In some studies only the mobile stages of spider mites are counted, whereas in others only one particular stage is taken into consideration, such as adult females or winter eggs. Since the work of Pielou (1960) several authors have advocated scoring only the absence or presence of mites in sampling units

---

Chapter 1.5.2. references, p. 348

because of savings in time and because it is still possible to estimate population size from this type of data (Nachman, 1981b, 1984; Wilson et al., 1983). Indices of population abundance could provide another way of minimizing counting effort. For example, Sabelis et al. (1983) and Rauwerdink et al. (in preparation) measured webbed leaf area in greenhouse experiments on the biological control of two-spotted spider mites and compared the measured area with that calculated by a simulation model. Because the model success-fully predicted the expansion of the webbed leaf area in the course of time, it may be worthwhile to use it as an indicator of population size in forth-coming sampling studies. However, it is important to realize that the webbed leaf area is a cumulative index of population abundance; once a leaf is covered by web, it continues to be so until leaf abcission or leaf picking by the grower (as is usual in the greenhouse culture of cucumbers). Hence, one needs data of at least 2 successive samples to estimate population size and, moreover, it may be that a variable degree of leaf abcission due to environ-mental influences, or a variable degree of leaf picking causes too much variation in the webbed area for it to be a population index of practical value.

SAMPLING UNIT

To estimate population size it is necessary to quantify the part of the habitat that has been sampled. Therefore, the habitat should be subdivided into parts of equal size, called the sampling units. These units must cover the whole of the population and they must not overlap. Every element in the population belongs to 1 and only 1 unit. For obvious reasons leaves are usually selected as sampling units in censuses of spider mite populations (e.g. Van de Vrie, 1966). Putman and Herne (1964) considered all leaves on a certain number of twigs of peach trees as one sampling unit, a method which necessi-tates correction for the increasing number of leaves per twig in the course of the season.

In some cases counts of mites on leaves are not sufficient because the mites also reside on other parts of the plant. For example, Summers and Baker (1952) and Herbert (1965) found that a significant proportion of *Bryobia* mites could be found on the wood; thus, the woody parts, such as branchlets, must be included in sampling. In other cases leaves are inadequate as sampl-ing units; clearly, winter eggs of *Panonychus ulmi* (Koch) should be sampled on the wood only. Fauvel et al. (1978) compared 4 sampling methods of winter eggs using a standard OILB method (OILB, 1969) developed by Vogel and Bachmann (1956). These methods differed mainly in the sampling unit selected and were proposed to simplify the work involved in the assessment of infestation levels. These sampling units were: branches of old wood (2 years or older) (Vogel and Bachmann, 1956); pairs of spurs on 2-year-old wood (Baillod and Fiaux, 1975); basal pairs of buds on 1-year-old wood (Touzeau, 1973); nodes on both sides of bud scale scars (Goonewardene and Kwolek, 1975); and the first 10 buds at the base of a 1-year-old twig (Fauvel et al., 1978). As winter eggs are deposited on parts of the tree other than the sampling units proposed, the methods can only provide relative estimates of winter egg abundance. They should therefore be calibrated; for example, to the infestation level of European red mites that have emerged in the spring.

SAMPLING METHODS

Several techniques have been devised for the sampling of field populations of spider mites, each of which has its advantages and disadvantages. These methods have been reviewed by Van de Vrie (1966), Jeppson et al. (1975) and Poe (1980).

Direct counting

This method is usually assumed to be the most accurate. Leaf samples are collected in the field and then inspected under a binocular microscope to make actual counts of the mites present per leaf. It is a time-consuming procedure, and may be less accurate than one might assume, especially when counting walking mites. Handling the leaf may stimulate the mites to crawl around and, when the area under examination is larger than the field of vision of the microscope, moving mites may be counted more than once. This particular disadvantage may be reduced by cold storage of the leaves (ca. 5°C) prior to counting or by spraying the leaves with 'Super Hold' hairspray immediately after sampling (Jones and Parrella, 1984).

A simplified, but much less accurate version of the direct counting method is leaf inspection by means of a hand lens (Van de Vrie, 1966).

Imprint counting

The imprint, or paper impression method, was developed by Venables and Dennys (1941) in Canada and improved by Austin and Massee (1947) in England. The equipment is a simple clothes wringer in which the leaves are pressed between a folded sheet of mimeograph paper or other paper of suitable absorbance. The mites are crushed and leave an imprint which, after some experience, can be distinguished from imprints of other material. This method has the advantage of providing a semi-permanent record of the mite infestation. It inactivates the mites at the most opportune time and counting is much easier. It allows a more accurate evaluation of all stages of the particular species present on the collected leaves. The imprint method, however, relies on distinctive colours or other features to allow the worker to distinguish between different species. If there are 2 different tetranychid species of the same colour present on the leaves, it may not be possible to distinguish between them after crushing. Poe (1980) reports that under high mite densities the imprints frequently coalesce, thus making the counting procedure very inaccurate.

Counting mites dislodged with the aid of a brushing machine

A brushing machine was developed by Henderson and McBurnie (1943). To use this machine infested leaves are passed between rotating brushes; and as the mites are brushed off, they fall onto a revolving disc. The disc has a sticky coating and is marked into aliquot sections (Klostermeyer and Rasmussen, 1956). This method has most of the advantages of the imprint method and allows a better identification of mite species and stages removed from the leaves. A recording of the results is possible by taking photographs of portions of the plate, by cold storage of the plates or by smearing the plate with polyurethane before and after brushing (Muir and Easterbrook, personal communication, 1983). Van de Vrie (1966) found that the effectiveness of the brushing machine depends on the smoothness of the leaves. Lower proportions of mites are brushed off when the leaves have pronounced ribs and

Chapter 1.5.2. references, p. 348

a dense field of hairs. Poe (1980) considers it to be a great disadvantage that tender or large leaves are either folded, crumpled, or shredded as they pass between the opposing roller brushes, resulting in a messy if not inaccurate sample.

Counting mites washed off their host plant leaves

Henderson (1960) has described another method to remove mites from leaves before counting. The mites are washed from the leaves and, after stirring, an aliquot of the solution is taken. Leigh, Maggi and Wilson (see Wilson et al., 1983) developed a 'mite rinse' machine in order to separate the mites from the silk and soil particles on the leaves and to get them out of the leaf folds. The rinse machine circulates the infested leaves in a hypochlorite solution, dissolves the silk, separates the mites from the leaves and silt, and enables them to be counted rapidly (all age classes) on a filter paper, using a binocular microscope.

Counting mites dislodged by beating foliage

Summers and Baker (1952) described a beating method for collecting *Bryobia* spp. from leaves and twigs. The branches are beaten over a funnel and the mites are collected in a jar below the funnel. It is a useful method to obtain estimates of the relative abundance of tetranychid species that do not produce a complex web on the leaf. Dislodgement of web-spinning mites, such as *Tetranychus* spp., is much more difficult to achieve as the web ensures their attachment to the leaf. In the latter case the beating method is only useful for taxonomic purposes. Alternatively, shaking the plant over a ground cloth or white paper may be used to detect the presence of mites and even relative estimates of their number may be obtained. Movement of mites on a white background is essential if they are to be counted with the unaided eye (Poe, 1980). Both the shaking and beating methods should be utilized with caution, realizing that an undetermined portion of the population will not be dislodged from the plants and that the age distribution of the sampled population may vary considerably from that measured (Poe, 1980).

The value of the above methods, except for the washing method, has been discussed by Morgan et al. (1955), Mathys and Van de Vrie (1965), Van de Vrie (1966) and Poe (1980). The brushing method was rated to be the most efficient. Putman (1966), however, found that great care is needed to obtain reliable results. Perhaps the washing method is a useful way of overcoming some of the problems associated with handling the brushing machine. Evaluation of these methods is desirable, the more so if attention is paid to the contribution of each method in reducing errors due to differences between individual samplers. The latter aspect is often overlooked in the development of sampling plans.

SAMPLING FREQUENCY

Sampling of phytophagous mites in the growing season usually starts within a month of the new leaves unfolding. The time interval between successive sample dates is proportional to the time constant of population growth, which equals the inverse of the intrinsic rate of increase (r_m^{-1} day). Potentially, spider mite populations double in 2—4 days. However, because of low night temperatures, the influence of predators and unstable age distributions time intervals of 1 week are generally sufficient to record 2-fold increases in spider-mite populations. Population models of spider mites (in- or ex-cluding interactions with predators) that take into account abiotic factors and are free of assumptions regarding (stable) age distributions, could be extremely useful in determining time intervals between successive sample dates (Croft et al., 1976; Rabbinge, 1976). There are, however, few or no reports demonstrating their usefulness.

STRATIFICATION OF THE HABITAT

If population elements were to be homogeneously distributed over their habitat, variation in the number of elements per sampling unit will be low. Clearly, under these conditions the sample size can be smaller than when variation is large as a consequence of a clumped spatial distribution of the mites. In the latter case it is useful to divide the habitat into strata so as to reduce within-stratum variation at the expense of between-stratum variation. Stratified random sampling can be a powerful tool in reducing the sampling effort without loss of precision (Cochran, 1963).

Between-plant distribution

Numerous authors have stressed that within-tree (or within-plant) variation in the number of mites on leaves is much smaller than between-tree variation (Daum and Dewey, 1960; Herbert and Butler, 1973; Croft et al., 1976; Mowery et al., 1980; Nachman, 1981a; Zahner and Baumgartner, 1984; Jones and Parrella, 1984). Croft et al. (1976) suggest that within-tree populations of European red mites are rendered less contiguous by the presence of an effective predator population that readily disperses through a tree. They are of the opinion that between-tree variation is large as a consequence of the limited ability of both tetranychid and phytoseiid mites to disperse from one tree to another. In some trees, predators are effective and in others they are less so; this variability in predator effectiveness tends to increase variation of either species and hence also the number of trees to be sampled. A recent study by Karg (1983) showed that *P. ulmi* tends to form infestation focusses extending over 5—25 neighbouring apple trees. This leads me to suggest that it is worthwhile to consider within- and between-focus variations in future sampling plans provided the focusses are easily detectable. Such an approach may further reduce the number of units to be sampled from the orchard.

Within-plant distribution

In trees mite abundance does not seem to vary with compass directions (Westigard and Calvin, 1971; Herbert and Butler, 1973; Wilson et al., 1984). However, exposure to the sun may not be homogeneously distributed over the plant, thereby affecting temperature-dependent growth rates of the population. For this reason, Flaherty and Huffaker (1970) kept separate records for the tops and the morning-sun and afternoon-sun sides of grapevines.

Several studies of orchards have confirmed the view that spider-mite infestations usually start in the lower portions and centres of trees (Van de Vrie, 1964; Tanigoshi et al., 1975; Zalom et al., 1984). This may be explained by the position of overwintering sites in a tree or may reflect the survival chances of the hibernating population, but several other explanations are also possible (e.g. allocation of defensive plant compounds to the young

Chapter 1.5.2. references, p.348

leaves). Watson (1964) noted that mites tend to be more fecund on young leaves. This may explain why Tanigoshi et al. (1975) found fewer, but larger mite colonies on the upper (youngest) leaves than on the lower (oldest) leaves of apple trees.

During the summer season the within-tree distribution may become more homogeneous, but this tendency is certainly not general. Gilstrap et al. (1980) found that Banks grass mites, *Oligonychus pratensis* (Banks), aggregated on the bottom portions of corn plants. Chandler and Corcoran (1981) found that most mites were present on the upper third of an ornamental foliage plant of commercial importance. In cotton plants most of the *Tetranychus* mites colonize midway mainstem node leaves during much of the season and the number of mites on these leaves can be correlated with mite densities on the whole plant (Carey, 1982; Wilson et al., 1983; Mollet et al., 1984).

ESTIMATION OF POPULATION SIZE

As argued above, sample size will depend on the distribution of the mites in the area to be sampled. However, knowledge of this distribution can only be obtained from prior experience of sampling the population. The time invested will only be justified by subsequent savings at subsequent population censuses. For this reason it is instructive first to have a quantitative notion of the effects of distribution on sample size. This will be done below in an imaginary sampling experiment using a population whose size and distribution are known a priori.

Numerical sampling from a population of known size and distribution

The objective is to estimate the population size X by random sampling from a population that is assumed to be randomly distributed (i.e. the probability of finding an individual mite at any particular point in the area is the same for all points, and the presence of one mite does not influence the position occupied by another). The number of mites found in the sampled area, x, is binomially distributed with parameters p (the probability of sampling a mite) and q ($= 1 - p$). An estimate of X is then

$$\hat{X} = x/p$$

which is unbiased with variance

$$\text{var}(\hat{X}) = Xq/p$$

and coefficient of variation

$$C = (\text{var}(\hat{X}))^{1/2}/X = (q/(Xp))^{1/2} \tag{1}$$

Rearranging this equation gives

$$p = (1 + XC^2)^{-1} \tag{2}$$

C is usually referred to as the level of precision and p can be re-interpreted as the fraction of the leaf area to be sampled given total population size X and the level of precision C. The lower the value of C, the higher p and thus the larger the sample size required.

In Dutch spindle bush orchards the leaf area per tree is about $9\,\text{m}^2$ and the number of leaves per tree is 3000. The range of spider-mite densities of practical interest is between 0.5 and 5 European red mites per leaf. Thus, X is between 1500 and 15 000 and by demanding that C be less than or equal to

0.25 the fraction of the leaf area per tree to be sampled can be calculated from eqn. (2). The result is that p varies between 0.01 and 0.001 in the density range of practical interest, therefore the sample size must range between 30 and 3 leaves per tree. Usually, orchards are sampled rather than a single tree. Therefore, the limits of sample size will be set by the time and costs involved in sampling. The value of p will consequently be small, so that $q^{1/2}$ in eqn. (1) approximates unity. As $Xp = x$, the precision C is approximated by $x^{-1/2}$, i.e. it depends solely on the total number of mites in the sample. This shows that C can be kept small by sampling a sufficiently large leaf area so as to make x sufficiently large.

When the population is patchily distributed rather than randomly, then $\text{var}(\hat{X})$ will exceed Xq/p and p will be underestimated by eqn. (2). For example, if the population of European red mites is distributed over apple leaves according to the negative binomial distribution with parameters X/L and k, then the fraction p can be shown to be equal to:

$$p = \frac{1}{C^2}\left(\frac{1}{X} + \frac{1}{kL}\right) \tag{3}$$

where L is the total number of sampling units (leaves) available in the total (leaf) area. If k tends to infinity, then the negative binomial approaches the Poisson distribution assumed in deriving eqn. (2). In that case eqns. (2) and (3) are very similar indeed. If, however, for given X and L the value of k decreases, then the population distribution becomes more and more clustered. Hence, p and the sample size calculated from eqn. (3) are larger than when calculated from eqn. (2). For example, Sabelis and Van de Vrie (unpublished data) found k to be linearly related to the mean density of European red mites per apple leaf, with slope 0.536 and intercept 0.09. Hence, when the mean density of mites is 0.5, $X = 1500$, $L = 3000$ (leaves per tree) and $C = 0.25$, then p of eqn. (3) equals 0.0256, which corresponds to a sample of ca. 75 leaves. When the mean density is 5 mites per leaf and $X = 15\,000$, then the required sample size is 9 leaves. Clearly, these sample sizes are 2 to 3 times larger than calculated by eqn. (2) under the assumption of a Poisson distribution.

Numerical sampling from a population of unknown size and distribution

As a rule, population size is not known a priori, as it is the quantity to be estimated. Knowledge of the distribution would require intensive sampling in preceding surveys. However, despite this lack of knowledge it is possible to determine the sample size by considering the distribution of the averages of random samples. According to the Central Limit Theorem this distribution tends to normality independent of the population distribution in the area, but this will only be true if the sample size is sufficiently large. Cochran (1963, p. 38) presents an example of a positively skewed frequency distribution of the averages of 200 random samples each containing 49 sampling units, which resembles a normal distribution, but still displays some positive skewness. It is, therefore, not easy to indicate how large the sample size should be for the normal approximation to be sufficient. Alternatively, data may be normalized by applying an appropriate transformation.

If we may assume normality, then d, half the length of the confidence interval around the expectation, is estimated by $Z_{1/2\,\alpha}(s^2/n)^{1/2}$, which upon solving for n gives:

$$n = (Z_{(1/2)\alpha}s/d)^2 \tag{4}$$

Chapter 1.5.2. references. p. 348

where $Z_{(1/2)\alpha}$ is the upper $\frac{1}{2}\alpha$ point of the cumulative standard normal distribution ($Z \approx 2$ for $\alpha = 0.05$) and s^2 is an estimate of the variance of y_i (the number of mites in sampling unit i). The value of s^2 is obtained from:

$$s^2 = \sum_{i=1}^{n} (y_i - \bar{y})^2 /(n-1)$$

After taking a first sample of size n_0 the value of s^2 can be calculated, and then n from eqn. (4) after setting d to the value desired by the sampler. If n exceeds n_0, then additional $n-n_0$ sampling units should be drawn at random. Subsequently, first s^2 and then n should be calculated repeatedly until no additional samples are required. Such sequential sampling procedures are further elaborated by Kuno (1969, 1972) and Green (1970).

The above method yields the sample size satisfying the precision set by d, half the length of the confidence interval. In population dynamic studies it is often not required to set d to a fixed value because, for example, one may tolerate larger d values when $\bar{y}$ is large. Precision may then be defined by $D = d/\bar{y}$, for example, so that sample size can be calculated from:

$$n = \left(\frac{Z_{(1/2)\alpha}s}{D\bar{y}}\right)^2 \tag{5}$$

Several other definitions of precision are possible (see, for example, Karandinos, 1976; Ruesink, 1980). The selection of appropriate definitions depends on the objectives of sampling.

According to Karandinos and Ruesink another way of expressing precision is that the standard error should be smaller than a certain value (SE = $s/n^{1/2} = d'$) or within a certain fraction of the mean (SE = $\bar{y}D'$). Ruesink states that this approach leaves out probablistic statements. However, it is in fact equivalent to eqns. (4) and (5) when α is chosen so as to make $Z_{(1/2)\alpha}$ equal to unity, i.e. $\alpha = 0.1587$. Hence, it will be clear that sample size based on precision defined in terms of SE will always be smaller than that based on precision defined in terms of confidence intervals with α usually taken to be 0.05.

Numerical sampling and the variance—mean relationship

There is good evidence that the sample variance tends to increase with the sample mean. Four different relationships (Taylor, 1984) have been derived:

(1) from the negative binomial distribution with common k (Bliss and Owen, 1958); $s^2 = \bar{y} + y^2/k_c$; examples have been reported by Croft et al. (1976), Mowery et al. (1980), and Zahner and Baumgartner (1984);

(2) from the negative binomial distribution with density dependent k; $k = f(\bar{y})$; (see Taylor et al., 1979); examples of this density dependence have been found by Nachman (1981a, b) for the two-spotted spider mite and Sabelis and Van de Vrie (unpublished data) for the European red mite ($k = 0.536\,\bar{y} + 0.09$);

(3) from linear regression of mean crowding (see Lloyd, 1967) and the mean (Iwao and Kuno, 1968); $s^2 = (a + 1)\,\bar{y} + (b-1)\,\bar{y}^2$; examples have been reported by Tanigoshi et al. (1975), Zahner and Baumgartner (1984) and Jones an Parrella (1984);

(4) from Taylor's power law (Taylor, 1961, 1971); $s^2 = a\,\bar{y}^b$; examples have been reported by Daum and Dewey (1960), Herbert and Butler (1973), Croft et al. (1976), Nachman (1981a, b), Wilson et al. (1983),

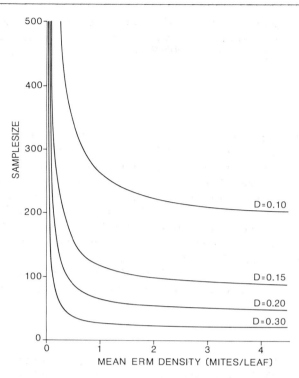

Fig. 1.5.2.1. The sample size required to estimate the mean number of European red mites (all stages) per leaf given that precision is defined as an upper limit to the ratio of the standard error to the mean and that the mean—variance relationship is derived from the negative binomial distribution with density-dependent k ($k = 0.536\bar{y} + 0.09$; Bartlett's method of fitting a straight line when both variables are subject to error is used; a note of caution should be made because the scatter of $k - \bar{y}$ points around the fitted line is quite large, i.e. k ranges between 40 and 160% of the value predicted by the fitted line). Data from Sabelis and Van de Vrie (unpublished).

Zalom et al. (1984), Wilson et al. (1984), Mollet et al. (1984), Margolies et al. (1984), Jones and Parrella (1984) and Zahner and Baumgartner (1984).

Because all of these relationships are no more than empirical descriptions, inferences as to the mechanisms that generate them are doomed to be speculative.

In Fig. 1.5.2.1 an example is given of the relation between sample size and the mean number of European red mites per leaf. It is based on a variance—mean relationship derived from the negative binomial distribution with density dependent k. Precision is defined in terms of the standard error which should not exceed a certain fraction of the mean ($= D'$). Two general features ensue from the calculations presented in the figure. Firstly, the relation between sample size and the mean number of mites per leaf is hyperbolic, and below 1 mite per leaf the sample size becomes impracticable. Secondly, increasing the sample size is associated with progressively smaller gains in precision (the law of diminishing returns). For population censuses of European red mites in Dutch apple orchards (Golden Delicious) it is recommended that ca. 100 leaves are sampled from a row of 20 adjacent trees. A similar rule of thumb has been proposed by Van de Vrie (1966).

Presence—absence sampling

With regard to spider mites that produce a conspicuous web or cause visible damage to the leaves, it is rather easy to distinguish between colonized

Chapter 1.5.2. references, p. 348

and uncolonized leaves. If there is a relation between p_0, the fraction of uncolonized leaves in a sample and the mean number of mites per sampling unit, then a great deal of time can be saved at subsequent population censuses. Ever since the first successful attempt by Pielou (1960) numerous authors have reported such relations (Baillod et al., 1979; Baillod and Bassino, 1979; Mowery et al., 1980; Nachman, 1981b, 1984; Wilson et al., 1983; Wilson et al., 1984). According to Gerrard and Chiang (1970) these relationships are linearized by regression of $\ln(\bar{y})$ on $\ln(-\ln(\hat{p}_0))$. Assuming $n\hat{p}_0$ is normally distributed, half the length of the confidence interval of p_0 equals

$$d = Z_{(1/2)\alpha}\sqrt{n\hat{p}_0\hat{q}_0}/n \tag{6}$$

Solving for n leads to the sample size formulae given by Wilson and Room (1983) and Wilson et al. (1983). However, as pointed out by Nachman (1984) the error associated with linear regression should be taken into account as well as the sampling error due to spatial heterogeneity. Nachman (1984) provided formulae for the variance and confidence limits about $\ln(\bar{y})$. Moreover, he investigated the relative precision of the direct counting method and the presence/absence sampling method (see also Chapter 2.1.4.3).

SAMPLING AND PEST MANAGEMENT DECISIONS

In the context of pest management, sampling should provide information on population size relative to a control decision threshold. The aim is to collect just enough sampling units to justify the decision whether the pest should be treated by acaricides or not. Sequential sampling is then appropriate, the rationale of which is discussed by Wald (1945) and Onsager (1972). The discussion presented in this section is largely drawn from Wilson et al. (1983).

Figure 1.5.2.2 illustrates a hypothetical population of mites through a season and is provided with a horizontal dashed line to indicate the control decision threshold. Four cases are highlighted: (A) population is far below the threshold; (B) it is close to but below the threshold; (C) it is far above the threshold; and (D) it is close to but above the threshold. The 4 'bell-shaped' vertical curves are probability distributions of a sample estimate of

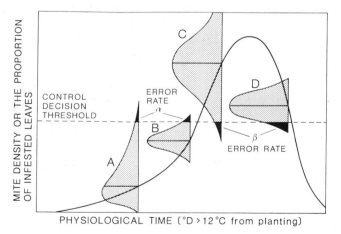

Fig. 1.5.2.2. The hypothetical relationship between spider mite density $(\bar{y})$ or the fraction of colonized leaves and the physiological time (degree-days above 12°C starting from the date of planting) (———). (— — — —) indicates the control decision threshold. For explanation of curves A—D and the error rates see text. Redrawn after Wilson et al. (1983).

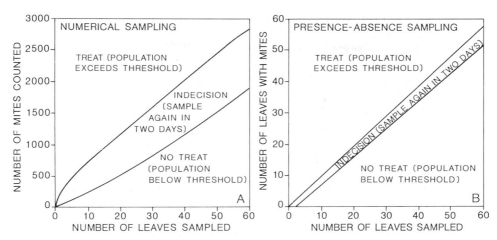

Fig. 1.5.2.3. Sequential sampling decision rules for spider mites on cotton using (A) numerical and (B) presence—absence sampling. Redrawn after Wilson et al. (1983).

population size. This estimate differs from the real density by varying amounts. For A and B, the dark tail end of the curve which rises above the threshold is the α error: the risk of treating when the true population density is below the control decision threshold. For C and D, the dark tail end of the curve, which drops below the threshold, is called the β error: the risk of not treating when the true population density is above the threshold. For the α error rates of both A and B to be equal requires that a greater number of samples be taken for B which is closer to the threshold. The difference in shape of curves A and B (A is less peaked) is due to: (1) a difference between the population densities at the 2 times and therefore a different variance— mean ratio; and (2) the larger number of sample units per sample for B. A similar relationship exists for C and D.

Using eqn. (6) for numerical or presence/absence sampling it is possible to calculate the sample size required to estimate with a given α-error rate that the population density (or the proportion of infested leaves) is below the control action threshold (lower control decision line), and the sample size required to estimate with a given β-error rate that the population density is above the threshold (upper control decision line). Tentative control decision lines for numerical and presence/absence sequential sampling are shown in Fig. 1.5.2.3A and B (Wilson et al., 1983) for spider mite infestation of cotton using 5000 mites per m^2 as a threshold (which is equivalent to 40 mites per leaf). For conventional sampling, control decision lines are usually parallel (see Wald, 1945 and Onsager, 1972), whereas the lines in Fig. 1.5.2.3 diverge and are curved. This difference is due to the use of 1 threshold instead of 2 and to the use of a dynamic instead of a static variance—mean relationship (see also Iwao, 1975). With respect to field implementation the utility of presence—absence sequential sampling is amply illustrated by comparing the number of mites which must be counted to reach a decision with the number of infested leaves which must be counted (Fig. 1.5.2.3, Wilson et al., 1983).

FUTURE PROSPECTS

When reconsidering what has been achieved in sampling of spider-mite populations, 2 points emerge that should receive more attention in future studies. Firstly, the sample size determinations ensuing from eqns. (4), (5) and

Chapter 1.5.2. references, p. 348

(6) are all based on the normal distribution. This assumption is only valid for sufficiently large samples and this conflicts with the goal of minimizing sampling effort. It would be interesting to evaluate existing sampling plans by use of Monte Carlo simulations of the sampling procedure in an area with realistically distributed populations of mites. The second point concerns the errors caused by the samplers. Wilson et al. (1983) discussed the case of a fast sampler who did not examine the leaves sufficiently. He rated damaged leaves as colonized by mites even when these leaves had been cleared of spider mites by the action of predators. Standardization of samplers is certainly not an easy task to accomplish. Sabelis et al. (1983) reported that despite extensive instructions before the start of population experiments the samplers evolved to an individually based error in estimating the webbed area per leaf. The selection of methods of assessment that reduce the effect of the sampler's bias, should therefore receive more attention.

ACKNOWLEDGEMENTS

I thank Gösta Nachman for comments and Martin Brittijn for (re)drawing the figures.

REFERENCES

Austin, M.D. and Massee, A.M., 1947. Investigations on the control of the fruit-tree red spider mite (*Metatetranychus ulmi* Koch) during the dormant season. J. Pomol. Hortic. Sci., 23: 227—253.

Baillod, M. and Fiaux, G., 1975. Description d'une méthode simplifiée pour le côntrole des oeufs hivernants de l'acarien rouge des arbres fruitiers (*P. ulmi* Koch) en verger de pommier. Station fédérale de Recherches agronomiques de Changins, Rapport interne, non-published, 5 pp.

Baillod, M. and Bassino, J.P., 1979, L'estimation du risque provoqué par l'acarien rouge (*Panonychus ulmi* Koch) en viticulture et perspectives ouvertes par la lutte biologique à l'aide de Typhlodromes. Proceedings of the International Symposium on IOBC/-WPRS on Integrated Control in Agriculture and Forestry, Vienna.

Baillod, M., Bassino, J.P. and Piganeau, P., 1979. L'estimation du risque provoqué par l'acarien rouge (*Panonychus ulmi* Koch) et l'acarien des charmilles (*Eotetranychus carpini* Oudemans) en viticulture. Rev. Suisse Vitic., Arboric. Hortic., 11(3): 123—130.

Bliss, C.I. and Owen, A.R.G., 1958. Negative binomial distributions with a common k. Biometrika, 45: 37—58.

Carey, J.R., 1982. Within-plant distribution of tetranychid mites on cotton. Environ. Entomol., 11: 796—800.

Chandler, L.D. and Corcoran, S.M., 1981. Distribution densities of *Tetranychus cinnabarinus* on greenhouse-grown *Codiaeum variegatum*. Environ. Entomol., 10: 721—723.

Cochran, W.G., 1963. Sampling Techniques. Wiley, New York, 413 pp.

Croft, B.A., Welch, S.M. and Dover, M.J., 1976. Dispersion statistics and sample size estimates for populations of the mite species *Panonychus ulmi* and *Amblyseius fallacis* on apple. Environ. Entomol., 5(2): 227—234.

Daum, R.J. and Dewey, J.E., 1960. Designing orchard experiments for European red mite control. J. Econ. Entomol., 53: 892—898.

Fauvel, G., Audemard, H., Rodolphe, F. and Rambier, A., 1978. Le récensement des oeufs d'hiver de l'acarien rouge *Panonychus ulmi* (Koch) sur le pommier. I. Ann. Zool.—Ecol. Anim., 10(3): 461—482.

Flaherty, D.L. and Huffaker, C.B., 1970. Biological control of Pacific mites and Willamette mites in San Joaquin Valley vineyards. II. Influence of dispersion patterns of *Metaseiulus occidentalis*. Hilgardia, 40(10): 309—330.

Gerrard, D.J. and Chiang, H.C., 1970. Density estimation of corn rootworm egg populations based upon frequency of occurrence. Ecology, 51: 237—245.

Gilstrap, F.E., Summy, K.R., Chandler, L.D., Archer, T.L. and Ward, C.R., 1980. Within-plant distribution of Banks grass mite on corn in West-Texas. Environ. Entomol., 9: 546—548.

Goonewardene, H.F. and Kwolek, W.F., 1975. Sampling dormant apple twigs to estimate the density of eggs of the European red mite. Environ. Entomol., 4(6): 923—928.

Green, R.H., 1970. On fixed precision level sequential sampling. Res. Popul. Ecol., 12: 249—251.

Henderson, C.F., 1960. A sampling technique for estimating populations of small arthropods in soil and vegetation. J. Econ. Entomol., 53: 115—121.

Henderson, C.F. and McBurnie, H.V., 1943. Sampling techniques for determining populations of the citrus red mite and its predators. U.S. Dep. Agric., Circ. No. 671:1—11, Washington DC.

Herbert, H.J., 1965. The brown mite, *Bryobia arborea* M. and A. (Acarina: Tetranychidae) on apple in Nova Scotia. Can. Entomol., 97: 1303—1318.

Herbert, H.J. and Butler, K.P., 1973. Sampling systems for European red mite, *Panonychus ulmi* (Acarina: Tetranychidae), eggs on apple in Nova Scotia. Can. Entomol., 105: 1519—1523.

Iwao, S., 1975. A new method of sequential sampling to classify populations relative to a critical density. Res. Popul. Ecol., 16: 281—288.

Iwao, S. and Kuno, E., 1968. Use of the regression of mean crowding on mean density for estimating sample size and the transformation of data for the analysis of variance. Res. Popul. Ecol., 10: 210—214.

Jeppson, L.R., Keifer, H. and Baker, E.W., 1975. Mites Injurious to Economic Plants. University of California Press, Berkeley.

Jones, V.P. and Parrella, M.P., 1984. Dispersion indices and sequential sampling plans for the citrus red mite (Acari: Tetranychidae). J. Econ. Entomol., 77: 75—79.

Karandinos, M.G., 1976. Optimum sample size and comments on some published formulae. Bull. Entomol. Soc. Am., 22: 417—421.

Karg, W., 1983. Untersuchungen zur Flächendispersion und Befallsentwicklung der Obstbaumspinnmilbe *Panonychus ulmi* Koch in Sortenblocken von Apfelintensivanlagen als Grundlage für eine rationelle Überwachung. Z. Angew. Entomol., 96: 433—442.

Klostermeyer, E.C. and Rasmussen, W.B., 1956. A counting plate for sampling mite populations. J. Econ. Entomol., 49: 705—706.

Kogan, M. and Herzog, D.C., 1980. Sampling Methods in Soybean Entomology. Springer, New York.

Kuno, E., 1969. A new method of sequential sampling to obtain the population estimates with a fixed level of precision. Res. Popul. Ecol., 11(2): 127—136.

Kuno, E., 1972. Some notes on population estimation by sequential sampling. Res. Popul. Ecol., 14(1): 58—73.

Lloyd, M., 1967. Mean crowding. J. Anim. Ecol., 36: 1—30.

Margolies, D.C., Lampert, E.P. and Kennedy, G.G., 1984. Sampling program for the two-spotted spider mite (Acari: Tetranychidae) in peanut. J. Econ. Entomol., 77: 1024—1028.

Mathys, G. and Van de Vrie, M., 1965. Etude comparative des méthodes de récensement de l'acarien rouge *Panonychus ulmi* Koch. Entomophaga, 10: 265—271.

Mollet, J.A., Trumble, J.T., Walker, G.P. and Sevacherian, V., 1984. Sampling scheme for determining population intensity of *Tetranychus cinnabarinus* (Boisduval) (Acarina: Tetranychidae) in cotton. Environ. Entomol., 13: 1015—1017.

Morgan, C.V.G., Chant, D.A., Anderson, N.H. and Ayre, G.L., 1955. Methods for estimating orchard mite populations, especially with the mite brushing machine. Can. Entomol., 87: 207—214.

Morris, R.F., 1960. Sampling insect populations. Annu. Rev. Entomol., 5: 243—264.

Mowery, P.D., Hull, L.A. and Asquith, D., 1980. Two new sampling plans for European red mite surveys on apple utilizing the negative binomial distribution. Environ. Entomol., 9: 159—163.

Nachman, G., 1981a. Temporal and spatial dynamics of an acarine predator—prey system. J. Anim. Ecol., 50: 435—451.

Nachman, G., 1981b. A mathematical model of the functional relationship between density and spatial distribution of a population. J. Anim. Ecol., 50: 453—460.

Nachman, G., 1984. Estimates of mean population density and spatial distribution of *Tetranychus urticae* (Acarina: Tetranychidae) and *Phytoseiulus persimilis* (Acarina: Phytoseiidae) based upon the proportion of empty sampling units. J. Anim. Ecol., 21: 903—913.

O.I.L.B., 1969. Introduction à la lutte intégrée en vergers de pommiers. Brochure No. 1, 50 pp.

Onsager, J.A., 1972. The rationale of sequential sampling, with emphasis on its use in pest management. U.S. Dep. Agric., Tech. Bull. No. 1526, Washington DC.

Pielou, D.P., 1960. Contagious distribution in the European red mite, *Panonychus ulmi* (Koch), and a method of grading population densities from a count of mite-free leaves. Can. J. Zool., 38: 645—653.

Poe, S.L., 1980. Sampling mites on soybean. In: M.Kogan and D.C. Herzog (Editors), Sampling Methods in Soybean Entomology. Springer Verlag, New York, pp. 313—323.

Putman, W.L., 1966. Sampling mites on peach leaves with the Henderson—McBurnie machine. J. Econ. Entomol., 59: 224—225.

Putman, W.L. and Herne, D.H.C., 1964. Relations between *Typhlodromus caudiglans* and phytophagous mites in Ontario peach orchards. Can. Entomol., 96: 925—943.

Rabbinge, R., 1976. Biological control of the fruit-tree red spider mite. Simulation Monographs, Pudoc, Wageningen, 228 pp.

Ruesink, W.G., 1980. Introduction to sampling theory. In: M. Kogan and D.C. Herzog (Editors), Sampling Methods in Soybean Entomology. Springer Verlag, New York.

Sabelis, M.W., Van Alebeek, F., Bal, A., Van Bilsen, J., Van Heijningen, T., Kaizer, P., Kramer, G., Snellen, H., Veenebos, R. and Vogelezang, J., 1983. Experimental validation of a simulation model of the interaction between *Phytoseiulus persimilis* and *Tetranychus urticae* on cucumber. OILB Bull. SROP/WPRS 6(3): 207—229.

Seber, G.A.F., 1973. The Estimation of Animal Abundance. Griffin, London, 506 pp.

Southwood, T.R.E., 1978. Ecological Methods With Particular Reference to the Study of Insect Populations. Chapman and Hall, London (first edn. 1966).

Summers, F.M. and Baker, G.A., 1952. A procedure for determining relative densities of brown almond mite populations on almond trees. Hilgardia, 21(13): 369—382.

Tanigoshi, L.K., Browne, R.W. and Hoyt, S.C., 1975. A study on the dispersion pattern and foliage injury by *Tetranychus mcdanieli* (Acarina: Tetranychidae) in simple apple ecosystems. Can. Entomol., 107: 439—446.

Taylor, L.R., 1961. Aggregation, variance and the mean. Nature (London), 189: 732—735.

Taylor, L.R., 1971. Aggregation as a species characteristic. Stat. Ecol., 1: 357—377.

Taylor, L.R., 1984. Assessing and interpreting the spatial distributions of insect populations Annu. Rev. Entomol., 29: 321—357.

Taylor, L.R., Woiwod, I.P. and Perry, J.N., 1979. The negative binomial as a dynamic ecological model for aggregation, and the density dependence of k. J. Anim. Ecol., 48: 289—304.

Touzeau, J., 1973. La répartition des oeufs d'hiver de *Panonychus ulmi* sur arbres fruitiers et propositions concernant la prognose hivernale. ACTA Note d'Information, 11: 5—9.

Van de Vrie, M., 1964. The distribution of phytophagous and predacious mites on leaves and shoots of apples trees. Entomophaga, 9: 233—238.

Van de Vrie, M., 1966. Population sampling for integrated control. Proceedings of the FAO Symposium on integrated control, 2: 57—75.

Venables, E.P. and Dennys, A.A., 1941. A new method of counting orchard mites. J. Econ. Entomol., 34: 324.

Vogel, W. and Bachmann, F., 1956. referred to in Fauvel et al., 1978.

Wald, A., 1945. Sequential Analysis in Inspection and Experimentation. Columbia University Press, New York.

Watson, T.F., 1964. Influence of host plant condition on population increase of *Tetranychus telarius* (Linnaeus) (Acarina: Tetranychidae). Hilgardia, 35: 273—322.

Westigard, P.H. and Calvin, L.D., 1971. Estimating mite populations in Southern Oregon pear orchards. Can. Entomol., 103: 67—71.

Wilson, L.T. and Room, P.M., 1983. Clumping patterns of fruit and arthropods in cotton with implications for binomial sampling. Environ. Entomol., 12: 50—54.

Wilson, L.T., Hoy, M.A., Zalom, F.G. and Smilanick, J.M., 1984. Sampling mites in almonds: I. Within-tree distribution and clumping pattern of mites with comments on predator—prey interactions. Hilgardia, 52(7): 1—13.

Wilson, L.T., Gonzalez, D., Leigh, T.F., Maggi, V., Foristiere, C. and Goodell, P., 1983. Within-plant distribution of spider mites (Acari: Tetranychidae) on cotton: A developing implementable monitoring program. Environ. Entomol., 12: 128—134.

Zahner, Ph. and Baumgartner, J., 1984. Sampling statistics for *Panonychus ulmi* (Koch) (Acarina: Tetranychidae) and *Tetranychus urticae* Koch (Acarina: Tetranychidae) feeding on apple trees. Res. Popul. Ecol., 26: 97—112.

Zalom, F.G., Hoy, M.A., Wilson, L.T. and Barnett, W.W., 1984. Presence—absence sequential sampling for *Tetranychus* mite species. Hilgardia, 52(7): 14—24.

1.5.3 Mounting Techniques

J. GUTIERREZ

PRESERVATION

Tetranychidae removed from the leaves may be killed in 70% ethyl alcohol. They can be maintained for several years in this liquid, in small tubes about 45 mm long with a diameter of 10 mm which are hermetically sealed with three-winged polyethylene caps. This is also the simplest way of sending specimens to specialists for identification. Labels written in Indian ink are fixed onto the outside and placed inside each tube.

According to some authors, alcohol hardens the mites after a long period of preservation. Evans et al. (1961) recommend 70—80% industrial methyl alcohol with up to 5% glycerol.

Tuttle (cited in Jeppson et al., 1975) prefers the AGA solution (alcohol, glycerol and acetic acid), which is a modification of Oudemans' fluid and has the following composition:

8 parts 70% isopropyl or ethyl alcohol;
1 part glacial acetic acid;
1 part glycerol;
1 part sorbitol (or sugar) may be added to AGA.

CLEARING

Tetranychidae kept in alcohol should be removed and dipped in a clearing agent, in order to examine morphological features under a microscope. Pure lactic acid coloured with lignin pink, as indicated by Evans and Browning (1955), seems the best agent. Lactic acid inflates the body and extends the legs, it attacks soft tissues while chitinized cuticles remain. Lignin pink, while keeping the bodies translucid, concentrates slightly in the mites and enables them to be clearly distinguished under the stereomicroscope.

In practice, the preliminary period in alcohol should last a minimum of 1 day (some authors recommend 2 weeks), to avoid quick disintegration of mites dipped in lactic acid. The contents of the small alcohol vial ($2\,cm^3$) are poured into a thick glass cupel ($4\,cm^3$) with a flat bottom. About 8—10 drops of stained lactic acid (cherry-coloured) are then added. The cupel is deposited on an electric slide warmer maintained at $40-60°C$. Alcohol quickly evaporates and the clearing generally lasts 1—2 h. As bodies may burst, it is necessary to be most attentive and to monitor the operation with frequent controls under a stereomicroscope.

In certain cases, for very dark specimens, it may be useful to dip the mites for several minutes in André's fluid, which has the following composition:

Chapter 1.5.3. references, p. 353

water, 30 cm^3;

chloral hydrate, 40 g;

glacial acetic acid, 30 cm^3.

This operation weakens the integument and should be carried out under a stereomicroscope.

MOUNTING

Temporary mounting

All specimens are best studied in temporary preparations according to the principle described by Grandjean (1949) and modified by Evans et al. (1961). One uses a cavity ground in glass slides of thickness 1.0—1.2 mm. The cavity is made with a small tungsten carbide grindstone mounted on a drill used by dental technicians and is shaped like a segment of a sphere with a diameter of about 3 mm. One may adjust the depth of the cavity to the size of the mite under study; for instance males need smaller cavities than the females of the same species.

A drop of coloured lactic acid is placed in the cavity, the mite is deposited and pushed into the liquid with a curl-tipped minutien pin fastened to a glass rod. A square cover glass (22 × 22 mm, extra thin), with a minute droplet in its centre is set on its edge and lowered slowly by means of a needle. If there is not enough lactic acid, a few more droplets should be added at the edges of the cover glass, if there is too much, it should be absorbed with a piece of filter paper. If the specimen is not clear enough, the slide may be heated gently on the electric warmer. The use of ordinary cover glasses would stop the objective of the microscope and would not allow focusing on every part of the mite.

Under the stereomicroscope, or even under a microscope at low magnification, the specimen may be orientated in any direction inside the cavity, by sliding the cover glass back and forth. The mite is not distorted and retains the appearance of a living tetranychid.

A series of such temporary mounts allows the examination of several characteristics for each specimen. With such large cover glasses, slides may be studied with caution under the oil-immersion objective. This technique is very useful for quick identification of samples, even with few specimens.

Temporary mounts sealed with Glyceel or Euparal may be transformed into semi-permanent preparations and kept for 2 or 3 years.

Permanent mounting

None of the mounting media are fully satisfactory. Pritchard and Baker (1955) have attempted the use of different media and concluded that the best one seems to be Hoyer's solution, consisting of

distilled water, 50 cm^3;

gum arabic, 30 g;

chloral hydrate, 200 g;

glycerol, 16 cm^3.

The ingredients are mixed in the above sequence with a stirring rod. It is necessary to heat the liquid slowly at about 50°C to accelerate the dissolution of the gum arabic. Use of purified gum in fine powder form simplies the process and makes filtration superfluous.

Jeppson et al. (1975) proposed a reduction in the amount of distilled water to 40 cm^3. Females are orientated dorsoventrally or in profile, males preferably are mounted in profile to see the aedeagus. It is best to put only one specimen in each slide.

Permanent mounting may be done with cavity slides or with simple slides. Mitchell and Cook (1952) also recommended the double cover-slip technique but the present author has never used it.

With cavity slides, one should use square cover glasses as for temporary mounting. The technique requires time to remove the air bubbles from the medium. Slides should be heated on the electric warmer at 50°C for at least 2 days. The slides are fragile but the specimens keep their live appearance and may be easily reprocessed if necessary (using water as solvent). They are suitable for ordinary microscopic examination only.

With simple slides, it is preferable to use small, round cover glasses (12 mm in diameter). Slides are easier to prepare and are dried in 1 day, but the specimens are squeezed, often crushed and deformed, and very difficult if not impossible to remount. They are suitable for ordinary and phase-contrast microscopes.

When dried, all permanent slides should be ringed with Glyceel or Euparal because the medium is hygroscopic and particularly unstable in a humid climate.

Two labels should be affixed to each slide. One should indicate the host plant, locality, date of collection and collector's name. The other should give the scientific name of the species, the sex of the specimen and its position (dorsal, ventral or profile), and the name of the scientist responsible for the determination.

Slides should be laid flat and stored in a dark dry place.

REFERENCES

Evans, G.O. and Browning, E., 1955. Techniques for the preparation of mites for study. Ann. Mag. Nat. Hist., 8(12): 631—635.

Evans, G.O., Sheals, J.G. and Macfarlane, D., 1961. The Terrestrial Acari of the British Isles. An Introduction to their Morphology, Biology and Classification. Trustees of the British Museum, London, 219 pp.

Grandjean, F., 1949. Observation et conservation des très petits Arthropodes. Bull. Mus. Hist. Nat. Zool., Ser. 2, 21(3): 363—370.

Jeppson, L.R., Keifer, H.H. and Baker, E.W., 1975. Mites injurious to economic plants. University of California Press, Berkeley, CA, 614 pp.

Mitchell, R.D. and Cook, D.R., 1952. The preservation and mounting of water-mites. Turtox News, 30(9): pages not numbered.

Pritchard, A.E. and Baker, E.W., 1955. A revision of the spider mite family Tetranychidae. Mem. Pac. Coast Entomol. Soc., 2: 1—472.

1.5.4 Karyotyping

L.P. PIJNACKER and W. HELLE

INTRODUCTION

Though the chromosome numbers of tetranychids are low, identification of the individual chromosomes is rather difficult. The chromosomes are small, similarly sized, and devoid of primary and secondary constrictions (Chapter 1.2.4). A diagrammatic representation of the karyotypes in idiograms has not yet been carried out, except for *Tetranychus urticae* (Pijnacker and Ferwerda, 1976). Of course, the technique of handling the chromosomes plays an important role. During the last decade, new methods for identifying eukaryotic chromosomes have been developed. These concern, among others, banding techniques and techniques for the detection of repetitive DNA (Darlington and La Cour, 1969; Sharma and Sharma, 1972; Bostock and Sumner, 1978).

Until now, most research on tetranychid chromosomes has been carried out with standard chromosome techniques, two of which are described below (Schedules I and II). Modern methods still have to be elaborated for mite chromosomes. The only modern example is given in Schedule III. A survey of the chromosome numbers in Tetranychidae is given in Chapter 1.2.3.

MATERIALS

The longest chromosomes occur during the cleavage divisions in the egg; this phase of embryonic development is thus most suited for chromosome identification. Before development of the chitinous skeleton, embryos have many mitotic cells; since the chromosomes are much smaller than during cleavage divisions, they can best be used for chromosome counting. Mitoses also take place in developing organs of older embryos and pre-adult life stages and thus these tissues can also be used for chromosome counting; however, when the particular organ cannot be dissected, debris must be removed before or during staining.

Eggs and individuals are put in the fixative at the first convenient moment, even during field collections. If slide preparation does not take place soon, the fixed specimens can be stored in a deep freeze for several months. Care must be taken that the material sinks in the fixative, which it generally does when alcohol—acetic acid mixtures are used. The specimens can best be sampled in small glass vials and handled further with a Pasteur pipette, forceps, needles, etc. The methods do not require an excessive amount of equipment.

Chapter 1.5.4. references, p. 357

METHODS

Schedule I: Acetic—orcein squash method

(1) Fix in absolute ethyl alcohol—glacial acetic acid (3:1) mixture for at least 24 h.
(2) Put material on a slide and add a drop of acetic—orcein (2% orcein in 70% acetic acid).
(3) Stain for 3—15 min; be aware of the evaporation of the stain.
(4) Place cover slip; this can also be done after step (2).
(5) Heat over a flame; take care the solution does not boil.
(6) Spread material by gently tapping with a needle holder on the cover slip above the material.
(7) Squash by applying firm pressure to the cover slip; this can best be done between blotting paper in order to prevent movement of the cover slip.

Notes:
A. Step (1) can be omitted if the material is processed immediately.
B. Instead of step (1), the material can be pretreated in 1% sodium citrate on a slide for 1—2 min (Helle et al., 1980).
C. Squash may be made semi-permanent by sealing with nail polish.
D. Squash may be made permanent as follows:
   (a) Freeze slide below − 20°C.
   (b) Separate slide and cover slip with a razor blade.
   (c) Transfer slide and cover slip to absolute ethyl alcohol at − 20°C for 1 min.
   (d) Mount in Euparal.
The following method is simpler but can cause shrivelling of the tissue:
   (a′) Lift cover slip.
   (b′) Air-dry.
   (c′) Mount in Euparal.

Results: Quick method for counting chromosomes.

Schedule II: Aceto-iron haematoxylin—chloral hydrate squash method (Wittman, 1965)

(1) Fix in absolute ethyl alcohol—glacial acetic acid (3:1) mixture for at least 4 days.
(2) Put material on a slide.
(3) Add a drop of aceto-iron haematoxylin—chloral hydrate solution when the fixative has almost evaporated.
(4) Place cover slip and stain for about 2 min.
(5) Heat over a flame until colour of stain changes.
(6) Spread, squash, and make semi-permanent as in Schedule I.

Notes:
A. Staining solution is prepared as follows:
   Stock solution: dissolve 4 g haematoxylin and 1 g ferric ammonium sulphate (iron alum) in 100 ml 45% acetic acid; ripe for at least 24 h.
   Staining solution: dissolved 2 g chloral hydrate in 5 ml of the stock solution.
B. Slides cannot be made permanent.
Results: Quick method for counting chromosomes; more details of chromosomes than in Schedule I; low background staining.

Schedule III: Smear method for differential Giemsa staining

(1) Put unfixed cleavage egg on a slide; if more than one egg, they can be placed in a row.

(2) Prick and at the same time smear the egg with a needle; if more eggs are present smear one by one in the same direction.

(3) Air-dry for 15 min.

(4) Fix in absolute ethyl alcohol—glacial acetic acid 3:1 mixture for 30 min.

(5) Air-dry.

(6) Incubate in 5 M HCl at room temperature for 10 min.

(7) Rinse with tap water, then with distilled water.

(8) Stain with 2% Giemsa solution in Sörensen buffer pH 6.9 at room temperature for 15 min.

(9) Rinse in Sörensen buffer pH 6.9, then in distilled water.

(10) Air-dry.

(11) Place in xylene and mount in DePeX under a cover slip.

Notes:

A. Slides are handled in slide jars.

B. Slides can be stored in boxes after step (5).

C. After step (5) chromosomes can be treated/stained according to all kinds of banding methods (Pijnacker and Ferwerda, 1976). Moreover, the slides can be stained directly, for instance with Giemsa (omit steps (6) and (7)) or even with acetic—orcein (Wysoki, 1968) for simple chromosome counting.

Results: Low but reproducible yield of G-banded chromosomes; smear method demonstrates morphology of chromosomes better than squashes.

REFERENCES

Bostock, C.J. and Sumner, A.T., 1978. The Eukaryotic Chromosome. North-Holland Publishing Company, Amsterdam, 525 pp.

Darlington, C.D. and La Cour, L.F., 1969. The Handling of Chromosomes. George Allen and Unwin, London, 272 pp.

Helle, W., Bolland, H.R. and Heitmans, W.R.B., 1980. Chromosomes and types of parthenogenesis in the false spider mites (Acari: Tenuipalpidae). Genetica, 54: 45—50.

Pijnacker, L.P. and Ferwerda, M.A., 1976. Differential Giemsa staining of the Holokinetic Chromosomes of the Two-spotted Spider Mite, *Tetranychus urticae* Koch (Acari, Tetranychidae). Experientia, 32: 158—159.

Sharma, A.K. and Sharma, A., 1972. Chromosome Techniques: Theory and Practice. University Park Press, Baltimore, MD, 575 pp.

Wittmann, W., 1965. Aceto-iron haematoxylin—chloral hydrate for chromosome staining. Stain Technol., 40: 161—164.

Wysoki, M., 1968. A smear method for making permanent mounts of the metaphase chromosomes in eggs of phytoseiid mites (Acarina: Mesostigmata). Isr. J. Entomol., 3: 119—122.

1.5.5 Histological Techniques

A.R. CROOKER, L.J. DRENTH-DIEPHUIS, M.A. FERWERDA and F. WEYDA

LIGHT MICROSCOPY

L.J. Drenth-Diephuis and M.A. Ferwerda

Microtechnique for light microscopic (LM) studies of mites is derived from standard methods found in basic histotechnique handbooks. Useful references for these are Gray (1964), Gurr (1965), Romeis (1968) and Pearse (1972). In this contribution, the successive steps of fixation, dehydration, embedding, sectioning, staining and mounting of the two-spotted spider mite, *Tetranychus urticae* Koch, are discussed, with special attention given to the handling of the mites. This information is applicable to spider mites in general, and to other small arthropods as well.

Fixation

The choice of fixative depends on the organ studied and the stain used. Alcoholic fixatives like Carnoy (absolute ethyl alcohol:glacial acetic acid = 3:1 or absolute alcohol:chloroform:glacial acetic acid = 6:3:1) have the advantage that the mites sink immediately, contrary to watery fixatives, on which the mites often float. Considerable shaking is then needed to immerse the mites in such a fixative. It is advisable to use small test tubes (6 × 30 mm), stoppered with a cork. Those mites which remain floating are not handled further. Of the watery fixatives, Sanfelice (1% chromic acid:40% formaldehyde:glacial acetic acid = 16:8:1) gives an excellent fixation of all tissues with few artefacts and minimal shrinkage. The fixation time in Sanfelice is 12—24 h, that in Carnoy is 1 h or longer. (The mites can best be transferred into test tubes by means of a fine brush.)

Dehydration and clearing

When Carnoy is used, 2 changes in absolute alcohol (2 × 30 min) and 2 changes in toluene (2 × 1 h) are sufficient. Most other fixatives must first be washed out by water (for Sanfelice, 8 changes of 1 h each); mites are then dehydrated via 70% alcohol (12—24 h), 96% alcohol (2 × 30 min), absolute alcohol (2 × 30 min) and brought to toluene (2 × 1 h).

Eosin, which stains mites red, is often added to one of the alcohol steps to facilitate recognition of the mites during further processing. (Replacement of reagents can best be carried out with a Pasteur pipette.)

Embedding

Paraffin (or paraplast) is used as the embedding medium. Melted paraffin (60°C) is introduced into a porcelain embedding dish (50 × 35 × 10 mm), the inside of which is covered with a thin film of glycerin. To prevent scattering of the mites throughout the paraffin block, a small hole is made in the paraffin after 10 min hardening, and the mites, with as little toluene as possible, are introduced into the hole. The paraffin is melted again and left to infiltrate the mites for 4—24 h at 60°C. Then one blows on the surface until the paraffin is sufficiently solidified and the dish can be transferred to luke-warm water in order to harden the paraffin further. The paraffin block separates easily from the dish because of the glycerin film. The mites are found together on the bottom of the block. (Care should be taken not to move too much with the dish, because this may move the mites out of position. It is possible to rearrange the mites with a warm needle when the paraffin is still liquid.)

Sectioning

Four to 8 μm sections are made with a standard microtome. Several mites embedded together always gives some sections in the desired direction (frontal, sagittal or transverse). The paraffin ribbon is handled with brushes and the sections are mounted on the slide with albumen –glycerin, stretched on a warm plate and left to dry at 40°C for 24 h.

When semi-thin sections (1—4 μm) are required for high-resolution LM or for survey prior to electron microscopy (EM), EM embedding techniques are necessary (see the section below on electron microscopy). The plastic sections are cut on a Pyramitone or ultramicrotome with a glass knife and are transferred, in order, with the aid of a wire loop or a pair of forceps to a drop of water on a slide. The sections are dried on a hot plate at approximately 90°C for between 30 min and 2 h.

Staining and mounting

Paraffin sections

The paraffin is melted over a flame and the slides are introduced into distilled water in jars via 2 short rinses in xylene, 2 in absolute alcohol and 2 in 96% alcohol. The appropriate stain has to be chosen for the subject to be investigated. Some staining techniques frequently used at the Groningen Institute are the following:
(1) Heidenhain's haematoxylin
    (a) mordant in 3% iron alum (12—24 h);
    (b) rinse in distilled water;
    (c) stain in haematoxylin (2—24 h);
    (d) rinse with water;
    (e) differentiate in 1.5% iron alum (monitor this process under the microscope);
    (f) wash in tap water;
    (g) dehydrate via 2 short rinses of 96% alcohol, absolute alcohol and xylene and mount under a cover slip in DePex, Caedax or Canada balsam.
The haematoxylin staining solution is prepared as follows: ripen a stock solution of 1 g haematoxylin in 10 ml 96% alcohol for 1 month. Before use, this solution is diluted 10 times with distilled water.
Result: tissues are stained in various shades of brown-grey-purple to black.

(2) Toluidine blue—Kernechtrot
    (a) 0.1% Kernechtrot in 5% aluminium sulphate for 10 min;
    (b) rinse in distilled water;
    (c) 1% toluidine blue in distilled water for 10 min;
    (d) rinse in distilled water;
    (e) dehydrate and mount in balsam (see 1(g)).
Result: tissues are stained in various shades of blue-purple-red.
(3) Feulgen reaction
    (a) hydrolyse in $5N$ HCl for 30—60 min at room temperature;
    (b) rinse in distilled water;
    (c) stain in Schiff's reagent for 24 h at $4°C$ in the dark;
    (d) wash in running tap water for 10 min;
    (e) rinse in distilled water;
    (f) dehydrate and mount in balsam (see 1(g)).
Schiff's reagent is prepared as follows:
    (1) dissolve 1 g pararosaniline in 30 ml $1N$ HCl;
    (2) mix with 1 g $K_2S_2O_5$ dissolved in 170 ml distilled water;
    (3) leave the solution overnight in the refrigerator;
    (4) to decolour the solution, shake with 0.6 g charcoal and filter.
Result: nuclei and chromosomes are stained red.

Plastic sections (see also electron microscopy section below)

Plastic sections are brought directly to the staining solution. A methylene blue—azur II (without basic fuchsin) staining (Humphrey and Pittman, 1974) is used as follows:
    (a) stain in methylene blue—azur II for 30 s to 1 h or more at $65°C$;
    (b) rinse in distilled water;
    (c) dry the sections (mounting not necessary).
The methylene blue—azur II solution consists of 0.130 g methylene blue, 0.020 g azur II, 10 ml glycerol, 10 ml methanol, 30 ml phosphate buffer pH 6.9 and 50 ml distilled water.
Result: tissues are stained in various shades of blue.

Storage

Store slides in a dark and cool place to prevent decolouring.

ELECTRON MICROSCOPY
A.R. Crooker

Chemical fixation and subsequent processing is the most widely used method of preparing spider mites for both the transmission electron microscope (TEM) and the scanning electron microscope (SEM). Cellular components and surface structures are well preserved by this technique. In general, the preparation of spider mites follows the standard techniques of biological electron microscopy, with special consideration given to specimen handling and fixation. Proper specimen preparation is a prerequisite for obtaining satisfactory electron micrographs; it is often the factor limiting the quality of a micrograph.

The purpose of this section is not to reiterate portions of the extensive literature on biological specimen preparation, but rather to emphasize certain practical approaches which will better enable the researcher to use electron microscopy as a tool for the study of spider mites. Much of this

---

Chapter 1.5.5. references, p. 380

information applies to the preparation of other mites, as well as to other arachnids and insects. A brief Appendix citing the formulation of some of the more common fixatives, buffers, embedding media, and stains used with spider mites is included. More complete information and additional formulations can be found in books dealing exclusively or almost exclusively with specimen preparation (Glauert, 1974; Millonig, 1976; Hayat, 1978, 1981a,b).

Transmission electron microscopy

The basic steps involved in the preparation of ultra-thin sections of spider mites for the TEM are chemical fixation, dehydration, and embedding, followed by sectioning and staining. The first step in this process, fixation, is of critical importance; lack of attention to detail during fixation is the most common reason for failure to attain good ultrastructural preservation.

Fixation

The goals of chemical fixation are halt postmortem change and to preserve the tissues in a condition which resembles the structure of the living animal as closely as possible. Chemical fixation, as the name implies, relies upon chemical means to kill and 'fix' the tissue, rather than physical means such as rapid freezing. The fixative consists of a fixing agent such as glutaraldehyde or osmium tetroxide in a suitable buffer to maintain a constant pH. Salts may be added to balance the ionic concentration and osmolarity of the fixative so that the tissue components neither swell nor shrink during fixation.

The most popular method of fixation for most biological specimens, including spider mites, is the two-stage fixation procedure introduced by Sabatini et al. (1963), in which primary or pre-fixation using glutaraldehyde buffered with phosphate or cacodylate is followed by secondary or post-fixation in osmium tetroxide. The glutaraldehyde preserves protein structure and carbohydrate, whereas the osmium preserves many lipids and imparts electron density to cell components. Acrolein or mixtures of formaldehyde—glutaraldehyde have also been used as primary fixatives, and occasionally osmium tetroxide has been used as the only fixative. Fixation is commonly undertaken at room temperature, but may also be conducted in ice at around $4°C$.

Typically, spider mites are fixed by immersion in 2.5—4% glutaraldehyde in a 0.05 to $0.1 M$ buffer at pH 7.2—7.6 (see, e.g., Mills, 1973b; Alberti and Storch, 1976; Mothes and Seitz, 1981). The optimal duration of fixation is not known, but an arbitrary time of 1—4 h is common. In the laboratory, fixation in glutaraldehyde may extend from 4 h to overnight. Mites can be left in glutaraldehyde for longer periods, such as when collecting in the field, without adverse effect. Two or three 10—15 min $0.1 M$ buffer washes (rinses) follow the primary fixation prior to post-fixation in osmium tetroxide. Occasionally, mites are rinsed in the buffer for an hour, and sometimes 2 h to overnight. The glutaraldehyde must be washed out of the tissue very thoroughly since it combines with and reduces osmium tetroxide to cause unwanted precipitation in the tissue.

Phosphate, cacodylate, or collidine are the buffers most commonly used to control the pH of fixatives for spider mites. Most formulations of phosphate buffers are based on that of Sorensen (Dawson et al., 1969). Phosphate buffers are non-toxic and produce excellent results, but they do become contaminated with microorganisms and are prone to formation of precipitates during fixation. Cacodylate buffer is very stable and is resistant to bacterial contamination during storage, but it is toxic and incompatible with uranyl acetate. Collidine buffer is stable but toxic, and causes more tissue extraction than phosphate or cacodylate buffers (Luft and Wood, 1963). Recently, PIPES (1,4-piperazine diethane sulphonic acid) buffer has seen increased usage in biological electron microscopy because it is not excessively toxic, does not precipitate common ions, and is an effective buffer. However, it has been little used with spider mites.

Secondary or post-fixation is carried out equally well with either 2% osmium tetroxide in distilled water or osmium tetroxide prepared in the same buffer used for the primary fixative. The use of 2 different buffers may result in the formation of artefactual electron-dense granules, as when cacodylate buffer is used during primary fixation and phosphate buffer during post-fixation (Kuthy and Csapo, 1976). Post-fixation usually lasts for 1 h at room temperature. Two 10—15 min 0.1 M buffer washes follow post-fixation if buffered osmium tetroxide is used; 2 distilled water washes follow post-fixation in osmium tetroxide in distilled water. When osmium tetroxide has been used alone as a primary fixative, collidine was the buffer (see, e.g., Penman and Cone, 1974). The main disadvantages of using osmium tetroxide as the primary fixative are its slow rate of penetration and its inability to cross-link most proteins such as microtubules. As a result, fine structure may be changed prior to completion of fixation. Osmium tetroxide is very volatile and toxic. Like many substances used in specimen preparation, it should be handled with rubber gloves in a fume hood.

It is obvious that for fixation to occur, the fixative must penetrate the tissue; however, this is one aspect of fixation commonly ignored by investigators who are beginners in the field. Surface tension of the fixative combined with the small size of the mites and their hydrofuge cuticle with projecting setae permit mites to float on the fixative. The integument itself is a significant potential barrier to the penetration of fixatives. Mites may remain alive for several hours on the fixative surface; many subcellular changes then take place before fixation occurs. Some mites may eventually die with little fixation having occurred.

Mites may be fixed whole, but for critical work this is not recommended. Fixation of whole mites is not uniform from specimen to specimen. The relatively impermeable integument prevents rapid penetration of the fixative, and agonal and postmortem changes take place before fixation occurs. Good fixative penetration can be attained by puncturing or cutting the integument and by keeping the mite submerged in the fixative. Puncturing or cutting is best performed on the stage of a dissecting microscope placed in a fume hood. The mite is first immobilized or slowed down by placing it in a petri dish over ice; with care, the mite can also be immobilized with sticky tape or melted wax attached to the appendages. The use of low temperature has the advantage of avoiding the potentially harmful effects of reagents such as anaesthetics, solvent fumes, or hot wax. The mite is then flooded with drops of chilled fixative (glutaraldehyde). Next, the specimen is punctured with sharpened or etched minuten pins (Penman and Cone, 1974), sharp needles (Mills, 1973a), tungsten or other metallic microneedles, glass slivers (Mills, 1973b) or glass microneedles. The integument may also be cut with a micro-scalpel or razor blade piece held in forceps or a special holder. Even removal of legs by cutting is sufficient to allow fixative penetration. Naturally, the microcuts or punctures must be made with as little damage to the mite as possible, and away from areas to be examined by electron microscopy.

After cutting or puncturing, the mites are transferred into fresh fixative in short, wide-mouthed snap-cap vials of 5—10 ml volume. Total immersion of the mites in the fixative is a prerequisite for satisfactory fixation. Surfactants

Chapter 1.5.5. references, p. 380

are not necessary to cause immersion of mites in the fixative; surfactants may also affect mite ultrastructure. Specimens which do not sink can be submerged by gently poking them into the fixative with a blunt probe or by lowering a bell-shaped, fine-mesh stainless steel or nylon screen over the mites. Even with these precautions, air bubbles may surround portions of the mite, requiring the specimen vial to be tapped against the fume hood to dislodge the bubbles. Alternatively, small capsules with finely screened ends can contain and keep the mites submerged during fixation and subsequent processing up to embedding. The plastic BEEM capsule used in transmission electron microscopy for embedding is convenient. The capsule can be modified by removing the conical end and capping both ends of the cylinder with BEEM capsule lids which have been punched out to leave large openings, then covered with a fine screen or filter to trap the mites. An example of such a device is that of Day (1974). Tolerances of the capsule must be kept very close or mites will become lost or trapped and destroyed under unevenly cut surfaces.

During fixation and subsequent processing, solutions must be replaced with fresh or different solutions. Typically, mites are left in the same covered vial in which they were fixed until they are ready to be transferred to gelatin capsules or moulds for embedding. The appropriate solutions are pipetted into and out of this vial. Exchange of solutions should be rapid, and specimens should not be allowed to become dry. Because of the small size of the mites, it is easy to pipette the mites along with the processing fluid. This is especially true during glutaraldehyde fixation; once the mites have been post-fixed in osmium tetroxide, they are darkened and more readily seen. Care must also be taken that the pipette tip does not damage the specimen. Pipetting problems can be minimized by affixing a small stainless steel screen (such as a stainless steel EM grid) or nylon screen to the pipette tip to prevent mites from being drawn into the pipette. If a small screen capsule device is used to contain the mites during processing, then the pipetting problem is avoided.

Dehydration and embedding

The epoxy resins commonly used as embedding media for spider mites are immiscible with water. Therefore, before infiltration with embedding medium, water must be removed from the fixed and washed specimens. Since epoxy resins are soluble in ethanol and acetone, dehydration is commonly carried out with one of these solvents. A graded series of 30%, 50%, 70%, 90% or 95%, and 100% solvent is often used. The duration of dehydration should be short to minimize extraction of cell contents by the solvent, but long enough to allow complete removal of the water. If dehydration is incomplete, resin infiltration is poor and sectioning problems result. If dehydration is too rapid or a graded series of dehydrant is not used, violent osmotic changes and surface tension forces may cause structural distortion. Ten minutes in each solvent concentration, with a second and third bath of 10—15 min each in the 100% solvent, is usually sufficient.

Propylene oxide (1,2-epoxypropane) is used as a transitional fluid at the last stage of dehydration when ethanol is used as the main dehydrating agent. The reason is that epoxy resins are more readily soluble in propylene oxide than in ethanol. If propylene oxide is not used following ethanol, then 2 changes of a 1:1 mixture of ethanol:resin are required, followed by 2 or 3 changes of fresh 100% embedding medium. Resins such as Epon can be mixed with concentrations of alcohol of 70% or higher. These additional changes of medium are time-consuming and unnecessary unless special procedures such as minimization of lipid loss during dehydration are being undertaken. There is also the risk that the ethanol will not be completely removed. Normally, two 10—15 min changes of propylene oxide are used. When acetone is employed as a dehydrating agent, the use of propylene oxide as a transitional solvent is unnecessary. ERL 4206, commonly known as Spurr's embedding medium (Spurr, 1969), is miscible with alcohol or acetone in all proportions, and thus does not require a transitional solvent. Propylene oxide is volatile and carcinogenic, and should be handled with care.

Epoxy resins such as Epon, Araldite, Epon—Araldite mixtures, and ERL 4206 are the commonly used embedding resins for spider mites. These materials, when infiltrated into the mite and polymerized, lend sufficient strength to the tissue to allow an ultra-thin section to be cut. During infiltration, if a transitional fluid such as propylene oxide is used, it is replaced with a 1:1 mixture of embedding medium:propylene oxide. After about 2 h, this mixture is replaced with fresh, unpolymerized embedding medium for an additional 2 h; the mites are then transferred to a small plastic vessel containing additional unpolymerized embedding medium. The change can be made without damage to the specimens by transferring the mites individually in a drop of embedding medium on a teflon rod or wooden applicator stick.

Mites must be left in unpolymerized embedding medium long enough for the medium to completely penetrate the tissues. Two hours to overnight is adequate, the time depending upon the type of embedding medium. Spurr's medium, for example, can be left for a shorter time since it polymerizes more rapidly than Epon. When infiltration is complete and before polymerization makes the embedding medium very viscous, mites are transferred to fresh embedding medium in the final embedding mould or capsule; an identifying label is placed in the mould. This final embedment can be placed in the oven for polymerization (hardening) or left at room temperature for a few hours to overnight to allow further infiltration of the new medium before placement in the oven. Time and temperature schedules for polymerization vary with the resin. Spurr's epoxy might be polymerized for 8 h at 75°C, whereas Epon might be polymerized for 2—3 days at 60°C or 1 day at 50°C, followed by overnight at 75°C and 1 h at 100°C.

Complete and uniform penetration of tissues by the embedding medium is required for satisfactory sectioning. Infiltration should be gradual and continuous, but not unnecessarily prolonged because cell contents can still be extracted. Mechanical shakers, especially rotary shakers, aid infiltration of embedding medium. Excellent penetration can also be obtained in a stationary specimen-infiltration mixture at slightly longer times; however, there may be greater extraction of cellular materials. The duration of infiltration can be decreased by increasing the temperature to approximately 40°C for a short time. This temperature reduces viscosity without causing significant polymerization. Brief treatment in a vacuum oven may be beneficial, but this is not necessary. It might be expected that spider mites, because of their relatively impermeable integument, would be troublesome to infiltrate with embedding medium, especially the more viscous Epon and Araldite. In practice, however, few problems are experienced.

The presence of water in ingredients of the embedding mixture, in the atmosphere, or in dehydration solvents can cause poor embedding, which in turn causes specimens to be difficult or even impossible to section. Solvents and embedding materials must be kept dry. If atmospheric humidity is high, embedding procedures should be carried out in a desiccator. Spurr's epoxy is the most sensitive to moisture, and routine use of a desiccator jar is recommended. The ingredients of any of the embedding media must be thoroughly mixed.

Chapter 1.5.5. references, p. 380

Epoxy resins will cause severe irritation on prolonged or repeated contact (epoxy resin dermatitis); some individuals have a strong allergy to epoxies and their components. Work with any component should be conducted under a fume hood, with cleanliness and careful handling strictly observed. ERL 4206 (vinyl cyclohexene dioxide) is a carcinogen.

Specimen orientation

A knowledge of the plane of sectioning is often essential for correct interpretation of an electron micrograph. If specimen orientation is known, much sectioning time can be saved, and fewer sections will have to be examined to find the desired structures. The simplest method of orientating the specimen so that the plane of sectioning can be deduced is to orientate the specimen in the embedding mould prior to polymerization. For mites placed in flat embedding moulds, this approach is often successful, although slight shifts in specimen position sometimes occur during polymerization, owing to the initial decrease in viscosity as the embedding medium is heated. Success rates are higher if orientation is attempted after the embedding medium has been heated to raise the viscosity and thus minimize specimen displacement. Orientation in capsules is more difficult. A simple approach is to place the mite in a drop of embedding medium at the bottom of a capsule. The tissue is orientated after the drop has been heated to raise the viscosity; then the rest of the capsule is filled with resin.

The usual approach employed by the present authors is to orientate the mite in a flat embedding mould after the embedding medium has become viscous, then fully polymerize the resin. If a specimen shifts during polymerization, the resin containing the specimen is cut from the block with a jeweller's saw and trimmed with a razor blade to the correct angle, so that when the specimen is glued to a lucite rod or aluminium stub which fits the ultramicrotome chuck, the desired orientation is achieved. Glueing may be done with epoxy left over from embedding or with a quick-setting epxoy. Surfaces to be glued should be sanded or cut flat, and bubbles should be eliminated from the glue.

More specialized methods of achieving the desired specimen orientation which do not deal with spider mites but are applicable to them are discussed in the work by Hayat (1981b).

An example of a 'standard' fixing and embedding schedule for the preparation of spider mites for TEM is presented below:

Spider mite fixation—embedment procedure

Primary fixation (3% glutaraldehyde in $0.1\,M$ cacodylate or phosphate buffer)	1 h
3 buffer rinses ($0.1\,M$ cacodylate or phosphate buffer)	10 min each
Post-fixation (2% osmium tetroxide in distilled water)	1 h
2 distilled water rinses	10 min each
30% ethanol	10 min
50% ethanol	10 min
70% ethanol	10 min
95% ethanol	10 min
3 changes of 100% ethanol	10 min each
2 changes of propylene oxide	10 min each
Propylene oxide—Epon (1:1)	2 h to overnight
Epon	2 h
Epon	2 h

Room temperature in moulds	8—12 h
50°C	8—12 h
75°C	8—12 h
100°C	1 h

Sectioning

After the resin is polymerized, trimming is necessary to expose the mite to the microtome knife and to provide a small surface for good cutting. While viewing the embedment through a dissecting microscope, the block is trimmed manually with a razor blade or mechanically by a Pyramitome or similar device to form a four-sided pyramid with a flat top. The flat top is the cutting or block face; the specimen is located at the surface of this face. The shape of the block face, as seen from above, is typically trapezoidal, with parallel upper and lower faces and 2 sloped sides. During ultramicrotomy, the base of the trapezoid contacts the knife edge first; the base length should be 0.2 mm or less if glass knives are to be used. This small cutting surface means that only portions of large spider mites can be sectioned at a time, especially if longitudinal sections are being examined. If a diamond knife is used, ultra-thin sections can be taken from entire mites, since block faces measuring 1 mm or more on the longest side can be utilized.

Silver-grey (500—600 Å), silver (600—900 Å), or silver-pale gold (900—1200 Å) ribbons of ultra-thin sections are cut with an ultramicrotome. The thickness used depends on the resolution desired. The ribbon is then picked up on a copper grid. Excellent clarity of micrographs is obtained by using 200-mesh grids with no support film; however, grid bars may obscure desirable areas. If structures are obscured too often, grids with larger openings such as 50- or 75-mesh should be used. Slot grids may also be used, especially for serial sectioning. Support films such as Formvar, Parlodion, or Pioloform are necessary for the coarser mesh grids. These films are usually given a light carbon coating for further stabilization and support, either before picking up the sections or after the sections have been picked up and stained.

Staining

Staining increases the differential electron-scattering power of tissue components, thereby making structures more visible. The staining method most commonly used with spider mites and biological specimens in general is uranyl acetate followed by Reynolds' lead citrate (Reynolds, 1963). To stain, a grid bearing ultra-thin sections is floated, sections downward, on the surface of a drop of aqueous, methanolic, or ethanolic uranyl acetate solution for 5—15 min, then rinsed in distilled water before being floated on a drop of lead citrate for 5—10 min. The grids are then rinsed again in distilled water and dried. Staining times will vary with the type and concentration of stain used, type and hardness of embedding medium, section thickness, and the investigator's preference for stain intensity. Typically, drops of stain are placed on the surface of dental wax in the bottom of a covered petri dish. Some workers surround lead stain drops with a few pellets of NaOH to absorb atmospheric CO_2 and help prevent the deposition of lead carbonate contamination on the section. Many home-made and commercial devices are available for staining large numbers of grids simultaneously.

Ultra-thin sections of spider mites have also been stained with potassium permanganate and lead citrate (Mothes and Seitz, 1981).

Chapter 1.5.5. references, p. 380

Scanning electron microscopy

The time and labour involved in preparing biological specimens for the SEM is generally less than that necessary before a biological specimen can be examined in the TEM. The reason for this is the time spent during embedding and ultramicrotomy for the TEM. Preparative techniques such as osmium tetroxide—thiocarbohydrazide—osmium tetroxide (OTO), cryo-fixation, and others have been applied to the scanning electron microscopy of spider mites, but the usual procedure involves fixation, dehydration, drying, mounting, and coating.

Fixation

Fixation of spider mites for the SEM can be achieved by following the same schedule employed for the TEM. This is convenient, since specimens for both the TEM and SEM can be fixed at the same time. Some workers, such as Mothes and Seitz (1981), have used a different procedure by fixing first in 1.4% glutaraldehyde for 2 h, then 4% glutaraldehyde for 2—3 days before placing in osmium tetroxide. The present author has found this procedure to give results similar to the schedule for the TEM. Puncture of the integument away from the structure of interest can be helpful since fixation of internal tissues renders rigidity to the flexible exoskeleton in subsequent steps. Care must be exercised, however, not to distort the integument during puncture. If mites are being fixed in situ on a plant surface, or if micrographs of entire mites are desired, then this step may be omitted.

Fixatives for SEM usually have a lower osmolarity (often isotonic) than fixatives for TEM, which are hypertonic. Isotonic fixatives are used for SEM because the surface layers of the specimen require this to avoid osmotic shock. Cells and tissues revealed by sectioning require hypertonic fixatives because they become diluted with tissue fluids at the time of actual fixation of the deeper cell layers. Since the chitinous integument of the mite is being examined rather than an external layer of cells, osmolarity has not been of major concern in the preparation of spider mites for the SEM.

Ethanol, a coagulant type of fixative, can be used to fix mites for the SEM. Concentrations of 70—100% are used. Preparation is rapid because the ethanol dehydrates the mites at the same time as fixation occurs. However, distortion of the flexible integument is frequent using this technique, so that even though micrographs of selected hard parts such as setae, ambulacral apparatus, or mouthparts may be excellent, micrographs of entire mites are often unsatisfactory. Ethanol fixation—dehydration does work well on hard-bodied mites.

Because the SEM is used to view surface structure at high spatial resolution, the surface of the specimen must be clean. There are many techniques for the surface preparation of biological specimens which can be applied both before and after fixation. Surface cleanliness, however, is not usually a problem for spider mites. If a group of mites should happen to possess surface debris, the debris is usually loosely attached and comes off during processing, or it can be removed easily by a brief period of sonication during fixation or dehydration. After sonication, the debris containing fluid is discarded and replaced by the appropriate fresh processing fluid. For mites to be examined on a plant surface, sonication is usually avoided because it may disrupt the position of mites and webbing.

Dehydration

Dehydration follows the schedule employed for the TEM. Ethanol is commonly used as the dehydrating agent, although acetone may also be used.

Drying

Drying is the final step of dehydration; specimens must be dry before they are introduced into the vacuum chamber of the SEM. Air-drying is the easiest method and works well for hard-bodied mites dried after treatment with 100% ethanol, but when applied to spider mites, it usually results in some collapse of the integument owing to the high surface tension associated with the drying. Freeze drying has been used by the present author with favourable results; however, this technique will not be examined here because it is used infrequently for spider mites. The most commonly used drying technique is critical-point drying. Results have been very favourable because this method dries tissues without subjecting the mite to damaging surface tension. Specimens in a suitable liquid are placed in the chamber of a critical-point dryer and the temperature and pressure are gradually raised. At a certain temperature and pressure, called the critical point, the vapour and liquid phases of the volatile liquid are in equilibrium; the phase boundary disappears and there is no surface tension exerted on the sample. After the critical point has been exceeded, gas in the specimen chamber is gradually exhausted. The samples are then removed from the chamber completely dry without having been subjected to surface tension effects.

In practice, mites to be critical-point dried are transferred from the dehydrating solvent, usually ethanol, to a final solvent (transitional fluid) in the critical-point apparatus. The transitional fluid, often CO_2, is then driven beyond its critical point. For the transfer, the specimen chamber of the critical-point apparatus is partially filled with ethanol to prevent drying of the specimen. Next, the ethanol is purged from the chamber by the CO_2, since the drying procedure tolerates a low proportion of ethanol. An additional, intermediate fluid is necessary if the dehydrating agent is not soluble in the transitional fluid. In most instances, dehydrating solvents such as ethanol are used as the intermediate fluid.

Mounting

After drying, the specimen is mounted on a specimen stub with a conductive cement, paste, or tape. Pastes or cements of silver or carbon are in common use, but have a tendency to be drawn up around a small specimen, obscuring features. This can be a problem for mites mounted singly, but there should be no difficulty with mites examined on plant surfaces. Individual mites, as well as mites on plant surfaces, can be mounted easily on adhesive copper tape affixed to the stub. The tape adhesive has less tendency to be drawn around the specimen.

Coating

Coating with a thin metallic film prevents or reduces the build-up of electric charge on the specimen surface, and increases the emission of low-energy secondary electrons responsible for image formation. A 60:40 gold—palladium mixture is commonly used as a fine-grained conductive coating; pure gold or pure palladium are also often used as coating metals.

Chapter 1.5.5. references, p. 380

Coating may be done by either vacuum evaporation or sputtering. In vacuum evaporation, the metal vapour atoms travel in a straight line to the specimen, where they condense and adhere to form a thin film. Because of the complex surface structure of spider mites, a rotary specimen holder should be used to present as much of the specimen surface as possible to the evaporation source. On the other hand, sputtered atoms have a short mean free path, undergo multiple collisions, and travel in many directions before finally condensing on the specimen. The metal will thus be deposited on parts of the specimen out of direct line of sight of the source. Spider mites can be well-coated by this procedure, but specimens may at times require additional coating to provide a continuous conductive surface. This can be accomplished by placing the specimen stub at an angle or on its side in the sputter coating device to expose previously uncoated areas of the mite. The coating thickness with either technique is normally 100–150 Å, as judged by a thickness monitor or calculated from mathematical expression.

Specimen storage at ambient humidity after coating may be satisfactory, but processed mites should preferably be stored in a desiccator jar.

Correlative microscopy

Correlation techniques for microscopy are numerous. Specimens prepared for the light microscope (LM) can be examined in the TEM or SEM and vice versa, and specimens prepared for the SEM can be processed for the TEM and vice versa. Rather than reiterate the literature, some of the more useful techniques for investigators working with spider mites are presented. This information pertains mainly to the correlation between light microscopy and transmission electron microscopy.

Determining the location of a specific structure in a specimen block is a well-known problem in sectioning for TEM. The problem results from the difficulty of locating the desired cells or tissue in 3 dimensions. First, the mite must be sectioned or trimmed to the depth at which the structure occurs, then the block face must be trimmed to contain the structure. It is not practical to take ultra-thin sections of an entire mite to locate a specific tissue because of the large number of sections involved.

The simplest method of locating specific tissues is careful observation of ultra-thin sections as they come off the specimen block during ultra-microtomy. The investigator should see some internal detail and the outline of the mite (integument) where it has not been cut away during block trimming. Examination of internal detail may be sufficient to locate a specific structure, but if not, the integument can be used as a guide to locate internal detail. Structural detail can also be seen in the block face and in thick sections as they come off the specimen block during microtomy.

Sampling problems can be partially alleviated by cutting thick (= semi-thin, 0.5–2.0 μm) sections from the specimen and examining them in the light microscope. Once the desired structure is located in a section, the specimen block is trimmed to contain the structure. This correlation is achieved when the orientation of the block face is maintained between the thick and ultra-thin sections. Some information will be lost because a portion of the desired area is contained in the thick section. Alternatively, thick sections can be taken to a depth believed to be slightly above the desired structure before ultra-thin sectioning is begun, or alternating thick and ultra-thin sections can be taken until the desired structure is located.

A more precise technique for locating specific structures within a volume of material involves the ultra-thin sectioning of thick sections. Thick sections are cut from the mite and examined by light microscopy. When sections

containing the desired structures are found, the thick sections are mounted
on an epoxy block or aluminium stub which fits the ultramicrotome chuck,
then are trimmed and sectioned for the TEM. There are several methods for
doing this, 3 of which will be mentioned here. These techniques have been
found by the present author to work well.

In the present author's adaptation of a technique developed by Robbins
and Gonatas (1964), sections 1—2 μm thick are picked up from the water-
filled trough of a microtome knife with a wire loop and transferred to a
carbon-coated glass microscope slide. After the water has dried and the
sections are adherent to the slide, a thin (1—2 mm) layer of epoxy is poured
onto the slide to cover the thick sections. The epoxy is then polymerized as
in specimen embedment. The epoxy with embedded sections is separated
from the glass slide (separation is possible because of the carbon coat),
resulting in a thin layer of epoxy and embedded thick sections with the
dimensions of a glass microscope slide. The epoxy 'slide' is placed on a light
microscope stage and examined by phase optics for structures of interest.
Sections containing the desired structures are cut from the slide with a razor
blade, scalpel, or jeweller's saw and are mounted with epoxy on a specimen
stub or epoxy block. The side of the section previously contacting the
carbon is mounted facing the knife to offer a very flat surface for block
alignment and sectioning. The section is trimmed and ultra-thin sections are
taken from the thick section. The original article should be consulted for
details critical to the success of this technique.

A related approach involves the use of a release agent, MS-122 fluoro-
carbon, to allow separation of the embedded thick sections from a glass
microscope slide (Kloetzel, 1973). Thick sections can also be mounted in
Falcon plastic petri dishes, then re-embedded in epoxy (Springer, 1973).
After separation from the dishes, selected areas can be resectioned for
electron microscopy. The polystyrene of the dishes is attacked by the epoxy
mixture of Spurr (1969).

Correlative techniques for resectioning thick sections require the
investigator to be a skilled ultramicrotomist. The specimen block must be
precisely aligned with the knife edge, since few (approximately 15) ultra-
thin sections are contained in a section 1 μm thick. For this reason, sections
of thickness 2 μm or more may be desirable.

Thick (semi-thin) sections of mites embedded in epoxy are useful not only
in correlative microscopy, but also in the study of specimen structure by
light microscopy. Sections nearly as wide as the glass or diamond knife edge
can be cut. As in the case of ultra-thin sections, thick sections are floated on
a trough fluid, usually distilled water. The investigator can examine these
plastic sections by light microscopy to become familiar with specimen
anatomy prior to examining structure at higher magnification in the electron
microscope. Unknown tissues visualized in the electron microscope can be
located and identified by comparing ultra-thin sections with thick sections.
Plastic sections can often be used in place of paraffin sections for light
microscopy. Plastic embedments offer the advantage that sections in the
thickness range 0.5—2.0 μm are easily obtained, whereas traditional paraffin
sections are usually 4—8 μm thick (see previous section on light microscopy).
The thinner plastic sections improve the resolution of tissue components
attained in the microscope; there is also greater freedom from shrinkage
distortion and other artefacts. Much time is saved by being able to use the
same specimen preparation for both light microscopy and electron
microscopy.

Semi-thin epoxy sections for light microscope histology may be stained
intact or after the plastic has been removed; they may also be examined

unstained by phase contrast microscopy, as is done during correlative microscopy. Although both overall and differential staining of tissue is possible, staining techniques for epoxy sections are not as numerous and versatile as staining techniques for paraffin sections. The staining of semi-thin sections of epoxies and other plastics has been discussed by Hayat (1970, 1975, 1981a) and Lewis and Knight (1977).

Conclusion

Guidelines for the preparation of spider mites for electron microscopy have been outlined above. The methods presented can be adapted to an investigator's particular bias, or modified for special techniques. In general, the processing follows standard procedure for biological electron microscopy, with special attention given to fixation and handling.

Conventional chemical fixation is the most widely used method of preparing spider mites for scanning and transmission electron microscopy. Variables to be considered during processing are numerous, and there are differences of opinion about such matters as time and temperature of fixation, importance of pH and tonicity of the fixative, and the best routines for dehydration, embedding, staining, and other aspects of processing. These variables have not been adequately or faithfully examined in spider mites.

In future studies of spider mite ultrastructure and functional anatomy, the continued use of chemical fixation methods is to be expected because of their ease of application and excellent preservation of ultrastructure. Increased use of other common electron microscopy techniques such as cytochemistry, autoradiography, freeze-fracture/etch, freeze-substitution, and cryo-ultramicrotomy and x-ray microanalysis seem likely. These techniques have not been commonly applied to spider mites, but hold the promise of a wealth of complementary and new information in the future.

APPENDIX

Fixatives

Glutaraldehyde

Glutaraldehyde fixative may be prepared by diluting an 8% EM-grade glutaraldehyde solution with an equal volume of $0.2\,M$ buffer.

Osmium tetroxide

A thoroughly cleaned 1 g ampoule of osmium tetroxide is broken open under 50 ml distilled water in a glass-stoppered dark glass bottle. The solid tetroxide will slowly dissolve to yield a 2% solution in distilled water. The bottle of osmium tetroxide solution should be kept inside a closed container, e.g. a tin or jar, in a refrigerator. If buffered fixative is desired, a 2% or stronger solution of osmium tetroxide in distilled water is mixed with the appropriate amount of the desired buffer.

Buffers

Sorensen's phosphate

A $0.2\,M$ solution of disodium hydrogen phosphate is prepared by dissolving 28.39 g Na_2HPO_4 (anhydrous), 35.61 g $Na_2HPO_4 \cdot 2H_2O$, 53.65 g

$Na_2 HPO_4 \cdot 7H_2 O$, or $71.64\,g$ $Na_2 HPO_4 \cdot 12H_2 O$, in distilled water to make $1000\,ml$. A $0.2\,M$ solution of sodium dihydrogen phosphate is then prepared by dissolving $27.6\,g$ $NaH_2 PO_4 \cdot H_2 O$ or $31.21\,g$ $NaH_2 PO_4 \cdot 2H_2 O$ in distilled water to make $1000\,ml$. The $0.2\,M$ buffer is prepared by mixing x ml of $0.2\,M$ disodium hydrogen phosphate with y ml $0.2\,M$ sodium dihydrogen phosphate, according to the desired pH, thus:

pH at $25\,^{\circ}$C	x (ml)	y (ml)
6.8	24.5	25.5
7.0	30.5	19.5
7.2	36.0	14.0
7.4	40.5	9.5
7.6	43.5	6.5

Cacodylate

A $0.2\,M$ buffer is prepared by dissolving $21.4\,g$ of sodium cacodylate in $400\,ml$ of distilled water. The pH of this solution is adjusted to the desired value with $0.2\,M$ hydrochloric acid and then made up to $500\,ml$ with additional distilled water.

Collidine

The $0.2\,M$ s-collidine (2, 4, 6-trimethylpyridine) buffer is made by mixing $2.67\,ml$ of pure s-collidine with $50\,ml$ distilled water and $9\,ml$ $1\,N$ HCl, then made up to $100\,ml$ with additional distilled water. The desired pH is obtained by adding additional $1\,N$ HCl to the solution. For pH 7.4, approximately $9\,ml$ HCl must be added to the $100\,ml$ solution.

Embedding media

Spurr's

ERL 4206 (vinyl cyclohexene dioxide)	$5\,g$
DER 736	$3\,g$
NSA	$13\,g$
DMAE (accelerator)	$0.2\,ml$

This mixture produces a firm block which will polymerize in 8 h at $70\,^{\circ}$C.

Araldite

Araldite 502	$45\,g$
DDSA	$40\,g$
DBP (dibutylphthalate)	$4\,g$
DMP-30 (accelerator)	$2\,ml$

Epon

Epon 812	5 parts by volume
DDSA	3 parts by volume
NMA	2 parts by volume
DMP-30 (accelerator)	1% to $1\frac{1}{2}$%

Chapter 1.5.5. references, p. 380

Epon 812 has been used in many studies of spider mites with excellent results. It has been largely replaced by similar embedding media under names such as EMbed (Electron Microscopy Sciences, Fort Washington, PA, U.S.A.), Poly/Bed 812 (Polysciences, Inc., Warrington, PA, U.S.A.), and Medcast (Ted Pella, Tustin, CA, U.S.A.). These resins can be substituted for Epon 812 without change in procedure.

Epon—Araldite standard mixture

Epon	12.5 ml
Araldite 6005	10 ml
DDSA	30 ml
DMP-30 (accelerator)	1.0 ml

Place into polyethylene vials or syringes and store below $-20\,°C$ (in bag containing silica gel).

Stains

Uranyl acetate (aqueous)

A 2% solution is prepared by dissolving 1 g uranyl acetate salt in 50 ml distilled water.

Reynolds' lead citrate

30 ml distilled water, 1.33 g lead nitrate, and 1.76 g sodium citrate are shaken at intervals, for 30 min, in a 50 ml Erlenmeyer flask. An 8 ml portion of $1N$ NaOH is added and mixed to completely dissolve the lead. Then a 12 ml portion of distilled water is added to the solution to make 50 ml of stain.

ARTEFACTS IN ELECTRON MICROSCOPY
F. Weyda

A living organism that is being prepared for conventional electron microscopy undergoes various kinds of handling and processing (cf. the previous parts of this chapter). Many minor or major changes in the original state of the organism may and do occur at all stages of preparation. Such changes, which usually are incorporated in their characteristic way in the final information about the organism, i.e. in electron micrographs, are called artefacts. It is of utmost importance to distinguish and correctly interpret them if one wishes to obtain objective information on the organism being examined. Some of the artefacts characteristic of the preparation of spider mites for SEM and TEM are mentioned below.

It is usually easier to prepare specimens for SEM than for TEM, so that the occurrence of artefacts in them is less frequent than with TEM. The most common are artefacts caused by dehydration and drying. Figures 1.5.5.1 and 1.5.5.3 show surface artefacts produced by air drying (Fig. 1.5.5.1) or insufficient freeze drying (Fig. 1.5.5.3). The same parts of spider mites, well desiccated by freeze drying, are shown in Figs. 1.5.5.2 and 1.5.5.4. Artefacts of another group often occur in connection with the conductive metallic coating of specimens. Well known are complex-image artefacts, commonly referred to as 'charging', which arise from variations in the surface potential of specimens. Figure 1.5.5.5 shows a local charging effect on the body

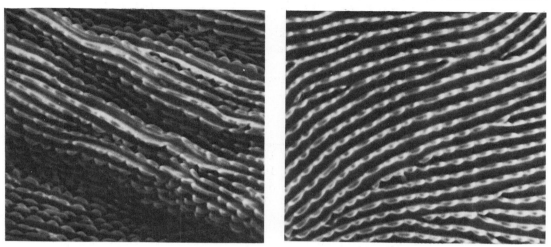

Figs. 1.5.5.1 and 1.5.5.2. Details of the dorsal cuticle of spider mite females. Air-dried (1) and freeze-dried (2) specimens. Magnification: 2300×. (Photographs by P. Berkovský and F. Weyda.)

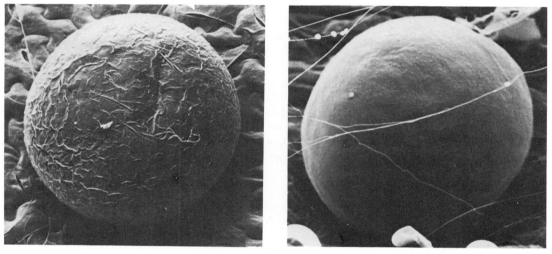

Figs. 1.5.5.3 and 1.5.5.4. Eggs of spider mites. Insufficiently (3) and sufficiently (4) freeze- dried specimens. Magnification: 600×. (Photographs by P. Berkovský and F. Weyda.)

surface of a spider mite, and a charging effect manifested as a sparkling line on an image is documented in Fig. 1.5.5.6. This defect can be corrected by additional coating. Figure 1.5.5.7 shows ramified fissures on the surface of a part of the body of a spider mite. These may be due to:

(a) damage by heat radiation during thermal evaporation coating (this kind of damage can be reduced either by placing a cold plate with an aperture over the specimen, or by using sputter coating);

(b) cracking of the coating layer in the electron microscope owing to insufficient desiccation of the object;

(c) beam damage during observation in the SEM (this type of damage can be reduced by working at lower beam currents). Characteristic artefacts appear after faulty cryofracturing, i.e. when a frozen object has been allowed to defrost during or after fracturing and to freeze again in a lyophilizer (Fig. 1.5.5.8).

An unpleasant artefact, which is not directly connected with the object, is a structured background (or cracking of the adhesive layer) on which the object is mounted. Another kind of adhesive should be used in such cases.

Chapter 1.5.5. references, p. 380

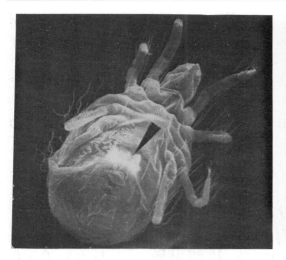
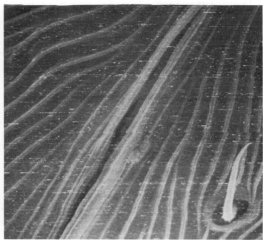

Fig. 1.5.5.5. Local charging effect (arrow) on the body surface of spider mite female. Magnification: 150×. (Photograph by P. Berkovský and F. Weyda.)

Fig. 1.5.5.6. Sparkling lines on image of spider mite anus caused by charging. Magnification: 3000×. (Photograph by P. Berkovský and F. Weyda.)

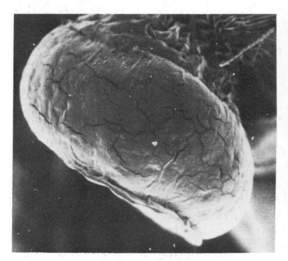
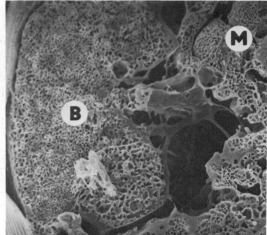

Fig. 1.5.5.7. Cracking of the metallic coating layer on the surface of a newly laid egg. Magnification: 600×. (Photograph by P. Berkovský and F. Weyda.)

Fig. 1.5.5.8. Brain (B) and part of mesenteron (M) of cryofractured spider mite female. Note the bubble-like artefacts induced by insufficient freeze drying. Magnification: 1000×. (Photograph by P. Berkovský and F. Weyda.)

Various artefacts occurring in SEM have been discussed in detail by Goldstein et al. (1981).

The problem of artefacts is more urgent and complex in TEM because the preparation of objects is more complicated. The final picture of tissues and cytological details is particularly affected by fixation, dehydration, and embedding in resin. It depends on the type of fixation, dehydrating agents and resins, also on the quality of chemicals, temperature at individual procedures, timing of individual operations, and on many other factors. Their combination represents a vast and highly variable complex whose final effect includes a certain size and shape of cells and tissues, fine structure of organelles, stability of macromolecules and other substances etc., which more or less (but always only to a certain extent) approximate the original

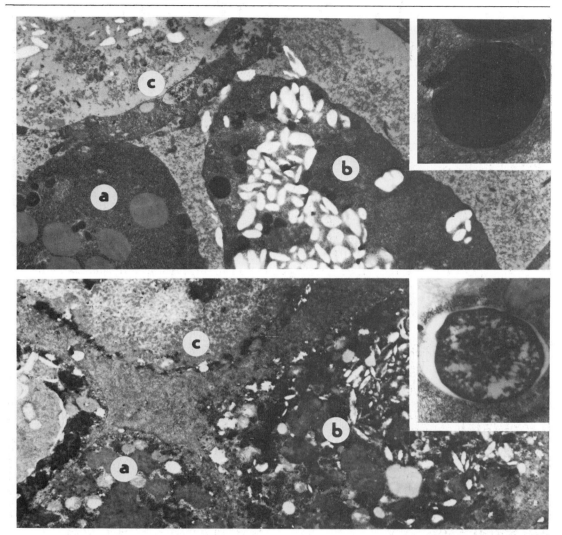

Figs. 1.5.5.9 and 1.5.5.10. Phagocytes in various stages of development (a, b, c) from the mesenteron of a female (inset shows osmiophilic granula). Cells fixed with osmium tetroxide vapour only (9) or with 2% glutaraldehyde following 1% osmium tetroxide solution (10). Magnification: 6600× (insets: 20000×). (Photographs by F. Weyda.)

living tissue. Pictures of the same tissue processed differently may vary substantially. Figures 1.5.5.9 and 1.5.5.10, which show the results of applying different fixatives to the same tissue (the mesenteron of a female spider mite), serve as good examples: the tissue in Fig. 1.5.5.9 was fixed in osmium tetroxide vapour, whereas the tissue in Fig. 1.5.5.10 was fixed in 2% gluataraldehyde and post-fixed in 1% osmium tetroxide (dehydration, embedding and staining were the same for both tissues). The ensuing pictures of the 2 tissues are different, e.g. the appearance of osmiophilic inclusions (inset) of the mesenteron cells.

As mentioned above, the hydrophobic cuticle of spider mites prevents rapid penetration of glutaraldehyde, and agonal and postmortem changes take place before fixation is completed. Artefacts produced in this way appear; e.g. in Fig. 1.5.5.11, disrupted mitochondria and swollen cisternae of rough endoplasmic reticulum are produced, which can fuse artificially with mitochondria. It is commonly known that substances of different kinds are more or less well stabilized by different fixatives. For example, proteins are very well stabilized by glutaraldehyde, whereas osmium tetroxide is more

Chapter 1.5.5. references, p. 380

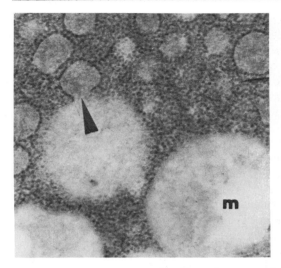 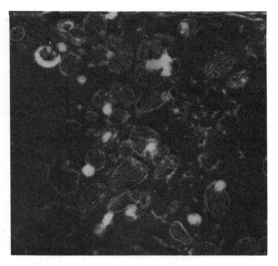

Fig. 1.5.5.11. Swollen cisternae of the rough endoplasmic reticulum artificially fused (arrow) with disrupted mitochondria (m) in the basal cells of the spider mite mesenteron. Magnification: 60000×. (Photograph by F. Weyda.)

Fig. 1.5.5.12. Membrane artefacts looking like negative contrast caused by dimethoxy-propane containing higher amount of HCl during dehydration. Nutritive cells of diapausing female. Magnification: 10000×. (Photograph by F. Weyda.)

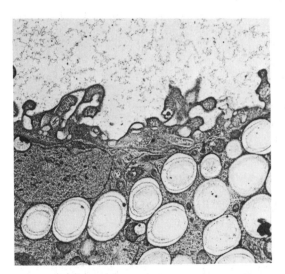 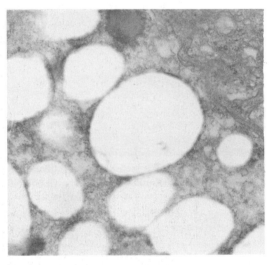

Figs 1.5.5.13 and 1.5.5.14. Specialized vacuolar cells in spider mite mesenteron. Presence of secretion (appearing as concentric layers) after glutaraldehyde following osmium tetroxide fixation (13) or absence of secretion after osmium tetroxide vapour fixation (14). Magnification: Fig. 13, 9000×; Fig. 14, 24000×. (Fig. 13: photograph by G. Šuťáková; Fig. 14: photograph by F. Weyda.)

suitable for lipids. This means that some insufficiently stabilized substances may be selectively washed out by different fixatives. Figure 1.5.5.13 shows vacuoles containing inclusions of concentric layers in specialized midgut cells (described by Mothes and Seitz, 1981 and by Weyda, 1981) fixed in gluta-raldehyde and post-fixed in osmium tetroxide. Figure 1.5.5.14 shows the same cells fixed by osmium tetroxide vapour; the contents of vacuoles had not been sufficiently stabilized and were washed out during dehydration and embedding. Various kinds of dehydration can also affect the final picture of tissues. The nutritive cells of the ovary of a diapausing female spider mite shown in Figure 1.5.5.12 have been fixed in osmium tetroxide vapour and dehydrated by dimethoxypropane according to the original method of Muller and Jacks (1975). The absence of contrast in the membranes, resembling negative contrast, corresponds to the damage caused to membranes by the originally recommended high concentration of HCl in dimethoxypropane (the concentration of HCl used at the present time is 20 times lower and such artefacts no longer appear).

Various artefacts may be induced by ultra-thin sectioning: occasional vibrations of the resin block are reflected in the ultra-thin sections as characteristic stripes (so-called chatter) perpendicular to the direction of movement of the block (Fig. 1.5.5.15). Similar artefacts may be formed by an imperfect edge of the knife, and stripes in this case are parallel with the direction of movement of the block. Multilayered objects whose layers differ in hardness also constitute certain problems; this particularly concerns arthropods in which cuticle sometimes separates from resin (Fig. 1.5.5.16), or cuticle from epidermis. For details see Allizard and Zylberberg (1982).

The staining of ultra-thin sections is another critical point in the preparation of objects for TEM. Different methods produce somewhat different results. Precipitation of heavy metals on sections is problematic, as the making of quality micrographs suitable for publication is prevented by precipitates. Fortunately, there are methods of removing the precipitates (see, e.g. Kuo et al., 1981). Artefacts can also be induced by examining ultra-thin sections in an electron microscope.

The problem of artefacts in electron microscopy is serious. Avoidance of artefacts is a prerequisite for successful study of ultrastructure (certain intrinsic artefacts resulting from the method of electron microscopy cannot be prevented). Attention to detail is important in all stages of the preparation of specimens, such as the cleanness of glass, purity of chemicals, and correct preparation of solutions. It is also advisable not to rely on only 1 type of preparation but to use various kinds of fixatives or to combine a standard method with a gentler one, e.g. cryo-ultramicrotomy. Electron microscopy should not be done out of context with the other methods of

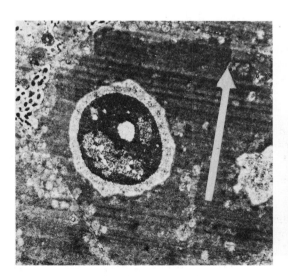

Fig. 1.5.5.15. Lines (chatter) perpendicular to the direction (arrow) of movement of the specimen block caused by vibrations during ultra-thin sectioning of nutritive cells of the spider mite ovary. Magnification: 5000×. (Photograph by F. Weyda.)

Fig. 1.5.5.16. Artificial empty spaces (asterisk) caused by separation of spider mite cuticle from the resin in the course of ultra-thin sectioning and handling of sections. Magnification: 4500×. (Photograph by F. Weyda.)

Chapter 1.5.5. references, p. 380

research; a combination of results obtained by several methods is optimal, such as correlative microscopy (combining the advantages of light microscopy with those of TEM and SEM) supplemented by histochemical and cytochemical data (at the level of optical and electron microscopy), and possibly biochemical ones.

REFERENCES

Alberti, G. and Storch, V., 1976. Ultrastruktur-Untersuchungen am männlichen Genitaltrakt und an Spermien von *Tetranychus urticae* (Tetranychidae, Acari). Zoomorphologie, 83: 283—296.

Allizard, F. and Zylberberg, L., 1982. A technical improvement for sectioning hard laminated fibrous tissues for electron microscopic studies. Stain Technol., 57: 335—339.

Dawson, R.M.C., Elliott, D.C., Elliott, W.H. and Jones, K.M., 1969. Data for Biochemical Research, 2nd edn. Clarendon Press, Oxford.

Day, J.W., 1974. A BEEM capsule chamber-pipette for handling small specimens for electron microscopy. Stain Technol., 49: 408—410.

Glauert, A.M., 1974. Fixation, dehydration and embedding of biological specimens. In: A.M. Glauert (Editor), Practical Methods in Electron Microscopy, Vol. 3, Part I. North-Holland, Amsterdam, 208 pp.

Goldstein, J.I., Newbury, D.E., Echlin, P., Joy, D.C., Fiori, Ch. and Lifshin, E., 1981. Scanning Electron Microscopy and X-Ray Microanalysis. A Text for Biologists, Materials Scientists, and Geologists. Plenum Press, New York, NY, 673 pp.

Gray, P., 1964. Handbook of Basic Microtechnique. McGraw-Hill, London, 301 pp.

Gurr, E., 1965. The Rational Use of Dyes in Biology and General Staining Methods. Leonard Hill, London, 422 pp.

Hayat, M.A., 1970. Principles and Techniques of Electron Microscopy. Biological Applications, Vol. 1, 1st edn. Van Nostrand-Reinhold, New York, NY, xvi, 412 pp.

Hayat, M.A., 1975. Positive Staining for Electron Microscopy. Van Nostrand-Reinhold, New York, NY, xxi, 361 pp.

Hayat, M.A., 1978. Introduction to Biological Scanning Electron Microscopy. University Park Press, Baltimore, MD, xviii, 323 pp.

Hayat, M.A., 1981a. Principles and Techniques of Electron Microscopy. Biological Applications, Vol. 1, 2nd edn. University Park Press, Baltimore, MD, xv, 522 pp.

Hayat, M.A., 1981b. Fixation for Electron Microscopy. Academic Press, New York, NY, xix, 501 pp.

Humphrey, C.D. and Pittman, F.E., 1974. A simple methylene blue—azur II—basic fuchsin stain for epoxy-embedded tissue sections. Stain Technol., 49: 9—14.

Kloetzel, J.A., 1973. A simplified method of preparing optically clear flat embedments with epoxy resin. Stain Technol., 48: 349—351.

Kuo, J., Husca, G.L. and Lucas, L.N.D., 1981. Forming and removing stain precipitates on ultrathin sections. Stain Technol., 56: 199—204.

Kuthy, E. and Csapo, Z., 1976. Peculiar artifacts after fixation with glutaraldehyde and osmium tetroxide. J. Microsc., 107: 177—182.

Lewis, P.R. and Knight, D.P., 1977. Staining methods for sectioned material. In: A.M. Glauert (Editor), Practical Methods in Electron Microscopy, Vol. 5, Part I. North-Holland, New York, NY, xvi, 312 pp.

Luft, J.H. and Wood, R.L., 1963. The extraction of tissue protein during and after fixation with osmium tetroxide in various buffer systems. J. Cell Biol., 19: 46A.

Millonig, G., 1976. In: Mario Saviolo (Editor), Laboratory Manual of Biological Electron Microscopy. Vercelli, viii, 67 pp.

Mills, L.R., 1973a. Morphology of glands and ducts in the two-spotted spider-mite, *Tetranychus urticae* Koch, 1836. Acarologia, 15: 218—236.

Mills, L.R., 1973b. On the Detailed Morphology of the Visual System, Dorsal Setae, and Glands of the Two-Spotted Spider-Mite, *Tetranychus urticae* Koch. Ph.D. Dissertation, Stanford University, xiv, 141 pp.

Mothes, U. and Seitz, K.A., 1981. Functional microscopic anatomy of the digestive system of *Tetranychus urticae* (Acari, Tetranychidae). Acarologia, 22: 257—270.

Muller, L.L. and Jacks, T.J., 1975. Rapid chemical dehydration of samples for electron microscopic examinations. J. Histochem. Cytochem., 23: 107—110.

Pearse, A.G.E., 1972. Histochemistry, Theoretical and Applied, Vols. I and II. Churchill Livingstone, Edinburgh, London, 1518 pp.

Penman, D.R. and Cone, W.W., 1974. Structure of cuticular lyrifissures in *Tetranychus urticae*. Ann. Entomol. Soc. Am., 67: 1—4.

Reynolds, E.S., 1963. The use of lead citrate at high pH as an electron opaque stain in electron microscopy. J. Cell Biol., 17: 208—212.

Robbins, E. and Gonatas, N.K., 1964. In vitro selection of the mitotic cell for subsequent electron microscopy. J. Cell Biol., 20: 356—358.

Romeis, B., 1968. Mikroskopische Technik. R. Oldenbourg Verlag, München, Wien, 757 pp.

Sabatini, D.D., Bensch, K. and Barrnett, R.J., 1963. Cytochemistry and electron microscopy. The preservation of cellular structures and enzymatic activity by aldehyde fixation. J. Cell Biol., 17: 19—58.

Springer, M., 1973. Casting Epon slides containing serial 3—5 μm sections for scanning prior to electron microscopy. Stain Technol., 48: 45—46.

Spurr, A.R., 1969. A low-viscosity epoxy resin embedding medium for electron microscopy. J. Ultrastruct. Res., 26: 31—43.

Weyda, F., 1981. Biotypy svilušky chmelové, Tetranychus urticae ve vztahu k rezistenci — Biotypes of the Two-spotted Spider Mite, Tetranychus urticae in Relation to Resistance, Vol. 1. Ph.D. Thesis, Czechoslovak Academy of Sciences, Prague, 320 pp. (in Czech).

1.5.6 Studies on Artificial Diets for Spider Mites

L.P.S. VAN DER GEEST

INTRODUCTION

Early work on the nutrition of spider mites was aimed at a better understanding of the effect of host plant physiology on the dynamics of spider mites and of host plant resistance. A great deal of this work was done by studying the effects of mineral nutrition of the host plant on spider mite reproduction and longevity (see, e.g., Hussey and Huffaker, 1976). In this chapter, only those nutritional studies on spider mites will be reviewed that make use of artificial feeding systems for the mites. Until now, all artificial diet studies in spider mites have concerned the species *Tetranychus urticae* Koch.

FEEDING SYSTEMS

Artificial diets for spider mites are offered to the mites under a membrane as an imitation of feeding through the epidermis of plant leaves. The first attempt to maintain spider mites on an artificial diet was made by Fritzsche (1960). He offered the diet to the mites as a hanging drop under a collodion membrane. The mites were placed on the membrane inside a glass ring which was smeared with a sticky substance in order to confine the mites to the membrane. Storms (1965) tried membranes of collodion and of Parafilm M®, but had better results with dried inner epidermis of onion bulbs, although in later studies Parafilm was used (Storms and Noordink, 1972). Rodriguez and Hampton (1966) also used a flexible collodion membrane inside a plastic Gelman filter holder that contained the diet. They usually positioned their containers in such a way that the membrane was facing downwards. The mites were then placed on the underside of the feeding containers, as in the natural situation where *Tetranychus* normally feeds on the underside of the leaves. Walling et al. (1968) prepared membranes of a resin of Butvar B-76, consisting of polyvinyl butyral. Their diet containers were polypropylene tube closures with a pad of polyurethane foam inside which absorbed the diet and supported the membrane. In later studies, use was made of membranes of stretched Parafilm M. Ekka et al. (1971), Dabrowski and Rodriguez (1972), Storms and Noordink (1972) and Kantaratanakul and Rodriguez (1979a, b) all used 2 layers of Parafilm, between which the diet was contained. These sachets were pulled over the open end of a tube or similar object for support. Mites were confined to the membrane by means of a sticky barrier of Tanglefoot. Bosse et al. (1981) and Van der Geest et al. (1983) used Parafilm stretched over a 50 mm petri dish lid, on top of which the diet was placed. They used a barrier of talcum powder for the confinement of the mites to the membrane. This barrier, a steep circular wall, cannot be traversed by the mites. It has the advantage that the mites do not become trapped, as is the case with a sticky barrier. The form of the wall, however, is important: the inner edge should be steep and the height of the wall should be at least 3 mm.

Conditions under which the mites are grown on the membrane are also of importance. Rodriguez et al. (1967) observed a higher ingestion by teneral adult mites under green light (535 nm) than under light of another colour. The wavelength of the light was found to have no effect on mites which were still in the protonymphal stage. The intensity of the light did not affect ingestion of either adult mites or protonymphs. Van der Geest et al. (1983) also observed a favourable effect of green light on survival and development of mites when grown from the egg stage to adulthood on a meridic diet. However, in their experiments, the intensity of the light also appeared to be important, as the performance of the mites under diffuse light was considerably better than when they were placed directly under a fluorescent lamp.

PREPARATION OF DIETS

Spider mites puncture the wall of plant cells with their stylets and with the aid of their pharyngeal pump suck the cell contents into their oesophagus and thence into the ventriculus. The food they take is liquid, although it also contains small particles such as chloroplasts. Research on artificial diets for spider mites has, therefore, always been aimed at the development of a fluid medium which may be ingested easily by the mites through a membrane. When formulating such liquid media one encounters special difficulties with respect to the 'solubilization' of lipoids. Rodriguez and Hampton (1966) used lecithin as an emulsifying agent for the lipoid fraction: the emulsion was stabilized after autoclaving by sonication. The disadvantage of using lecithin is that it is not a chemically well-defined product. It is a mixture of several lecithins and is often contaminated with other chemicals and therefore not well-suited for incorporation into a holidic diet. For that reason, Storms and Noordink (1972), Ekka et al. (1971) and Kantaratanakul and Rodriguez (1979a, b) used Tween 80 as an emulsifier, often in combination with sonication or homogenization. Bosse et al. (1981) and Van der Geest et al. (1983) used no emulsifier, as the necessary lipids were still contained in the wheat germ fraction, but they also added a water-soluble derivative of cholesterol to their (meridic) diet.

Preparation of diets is usually done under sterile conditions in order to avoid contamination by micro-organisms. Soluble, heat labile components are sterilized by filtration through a micropore filter, while non-soluble components (lipids, wheat germ etc.) are sterilized by heat treatment (see, e.g., Kantaratanakul and Rodriguez, 1979a and Bosse et al., 1981). The latter authors autoclaved their germ and casein fraction twice in order to kill all spore-forming bacteria present. Preparation of the diet containers is usually done in a laminar flow cabinet or similar sterile box and membranes of Parafilm are sterilized by immersion in 70% ethanol. Addition of preservatives to the media to reduce growth of micro-organisms has only been done by Rodriguez (1969) and Bosse et al. (1981). Most authors find the addition of such compounds undesirable, because of any deleterious effect they may have on the mites. Van der Geest et al. (1983) proved that addition of preservatives is not necessary, as they were able to maintain mites on a diet for over a week without any infection of the medium occurring.

NUTRITIONAL STUDIES

The qualitative nutritional requirements of organisms, irrespective of their systematic position, differ only slightly although a few exceptions should be made (House, 1974). It is therefore surprising that the first artificial diet on which spider mites can complete their life cycle was published only recently (Bosse et al., 1981), despite research on this subject for over 20 years. Not the absence of certain essential nutrients, but the manner in which the diet is formulated — the art of dietetics — is probably the factor which has retarded the development of artificial diets for spider mites for so long.

As stated above, the first attempt to grow spider mites on an artificial medium was made by Fritzsche (1960). He composed a medium consisting of sucrose, asparagine, aspartic acid, glutamine, glutamic acid, peptone, a vitamin mixture and mineral salts, as a watery solution. Adult females transferred from bean leaves to the diet could be kept alive for about 20 days, while eggs deposited on the membrane hatched. Larvae which hatched from these eggs failed to moult towards the protonymphal stage and eventually died. His studies were concerned especially with the effects of sugar concentration and of glutamine, glutamic acid, asparagine and aspartic acid on the reproduction of *T. urticae*. It was found that 0.7% sucrose in a medium without amino acids resulted in a higher reproduction of the mites than a medium with 0.3% sucrose. Addition of asparagine, aspartic acid, glutamine or glutamic acid to these sugar-containing media increased reproduction even more, although it remained lower than when mites were grown on bean leaves.

An important contribution to nutritional studies in spider mites has been made by Rodriguez and co-workers. Rodriguez and Hampton (1966) presented a holidic diet on which the mites could be grown from the protonymphal stage to adulthood. In contrast to Fritzsche's diet, their medium contained a large number of different amino acids, several lipid compounds, and furthermore a vitamin mixture, mineral salts, RNA and sucrose. The composition of the amino acid mixture was based on an analysis of bean leaves, although the concentration in the diet was considerably higher, as a result of a preliminary investigation by Rodriguez (1964). Using this diet, the essentiality of amino acids was determined, following the method described by Kasting and McGinnis (1962). When the mites were fed on this diet with glucose-U-C^{14} added to it and subsequently analysed, high radioactivity was found in alanine, aspartic acid, cysteic acid, cystine, glutamic acid, glycine, proline, serine and threonine. These amino acids apparently can be synthesized by the mites from glucose and need not be offered in their food, if synthesized by the mites in sufficient quantities. Low or no radioactivity was found in arginine, histidine, isoleucine, leucine, lysine, methionine, phenylalanine, tyrosine and valine, an indication that these amino acids are essential. In addition, the following non-protein amino acids were found: ornithine, citrulline and α-amino butyric acid, which cannot be synthesized from glucose, as they appeared to be unlabelled. In a later study, using a slightly modified diet, Walling et al. (1968) determined the distribution, catabolism and synthesis of fatty acids in *T. urticae*. Fatty acids in spider mites consist of normal saturated and monoenoic acids of 12—20 carbon atoms, but polyunsaturated linoleic and linolenic acids were also found. Over 80% of the total fatty acids belong to the C_{18} series. During development from the protonymphal stage to adulthood, an increase was found in linolenic acid and a decrease in palmitic and palmitoleic acids. Starvation of the mites resulted in a rapid decrease in linolenic acid. Adult mites which were kept on a liquid diet with tritiated acetate incorporated the label into saturated and monoenoic and polyenoic fatty acids.

Chapter 1.5.6. references, p. 389

Rodriguez (1969) presented a holidic diet which differed from that of Rodriguez and Hampton (1966) in having a much lower concentration of the lipid fraction and a higher concentration of sugars. The low level of lipids was chosen because of the observation that egg production by females on a diet with a low content of lipids was much higher. Ekka et al. (1971) used this diet to study oviposition and egg viability of *T. urticae* when reared on the diet. They made slight alterations in the composition, such as the omission of cholesterol and the addition of vitamin E (α-tocopherol). A lower concentration of fatty acids, in combination with a higher content of total sterols (plant sterols only) resulted in a higher egg production (0.64 egg per female per day) and an egg viability of about 20%. On Rodriguez's original diet, these figures were 0.34 and 3.6%, respectively. The preoviposition period was also shortened from 3.5 to 1.5 days.

Another attempt to develop an artificial diet for spider mites was made by Storms and Noordink (1972), using the diet of Rodriguez (1966). The concentration of amino acids in their medium, however, was much lower (0.14% instead of 2%) and was more a reflection of the actual concentration in bean leaves (0.06%). Ingestion of the diet by the mites was measured by the uptake of ^{32}P added to the medium, while egg production, monitored over the 48 h after mites were transferred from leaves to the diet, was used as a second criterion. Ingestion could be increased considerably by the addition of 167 mg ATP, 100 mg stearic and 100 mg palmitic acids to 100 ml of diet. Also, pH appeared to be of importance: at pH 12, ingestion was highest, but a less extreme pH, i.e. pH 10, was preferred in further experiments. Egg production was higher than reported in previous studies and amounted to about 2.25 eggs per female per day. Although mites could be kept alive for about three weeks on the diet, no development from egg to adult was obtained.

Kantaratanakul and Rodriguez (1979a) studied the effect of concentration of amino acids and sugars on egg production and egg viability. Reproduction of the mites was improved, compared to that on the diet of Ekka et al. (1971), by a lower concentration of amino acids (0.88%) and total sugars (1%), and by the formulation of an amino acid mixture based on the analysis of sodium caseinate. Oviposition was no better on their final, chemically defined medium, but reproduction was enhanced by a higher egg viability. In a subsequent paper Kantaratanakul and Rodriguez (1979b) described the effect of adding a number of natural products to the diet on the reproduction of the mites. A pronounced positive effect on oviposition was obtained by the addition of casein and yeast hydrolyzate to the medium. Corn starch and pectin showed no effect, while the addition of soluble starch to the diet even lowered egg production of the mites. Oviposition rates, as reported by these authors, are in general lower than those given by Storms and Noordink (1972). Comparison, however, is not very easy, as these latter authors followed egg production for only 48 h, while Kantaratanakul and Rodriguez (1979b) monitored oviposition over a period of 14 days. Any effect of nutrients which the mites had obtained from the leaves will therefore be smaller in the latter case.

On all previously described diets, it was not possible for spider mites to complete their entire life cycle. Larvae hatched from eggs deposited by females feeding on an artificial diet never progressed towards the protonymphal stage, although feeding occurred. The first successful culture of spider mites on an artificial diet for more than one generation was reported by Bosse et al. (1981). Their diet contained crude materials, such as casein and wheat germ, and is therefore a meridic diet. Detailed data are given by Van der Geest et al. (1983) with respect to development and survival for two

TABLE 1.5.6.1

Composition of a holidic diet for the spider mite *Tetranychus urticae* Koch (Kantaratanakul and Rodriguez, 1979a)

Constituent	mg/100 ml diet	Constituent	mg/100 ml diet
L-Amino acids		Pyridoxine-HCl	0.10
L-Alanine	37.5	Riboflavin	0.20
L-Arginine	50	Thiamin-HCl	0.20
L-Aspartic acid	50	Nicotinic acid	0.30
L-Cysteine	5	Biotin	0.0001
L-Glutamic acid	45	Ca-pantothenate	0.50
L-Glycine	50	Choline chloride	10.00
L-Histidine	25	Vitamin B_{12}	0.002
L-Isoleucine	78	Ascorbic acid	100.00
L-Leucine	119		
L-Lysine	95	Lipids	
L-Methionine	37	β-Sitosterol	1.00
L-Phenylalanine	64	Stigmasterol	0.88
L-Proline	15	Stearic acid	0.50
L-Serine	20	Palmitic acid	0.25
L-Threonine	51	Oleic acid	0.25
L-Tryptophan	16	Linoleic acid	0.25
L-Tyrosine	12.5	Linolenic acid	1.50
L-Valine	109	Soybean lecithin	5.00
Sugars		Major salts	
Sucrose	750	K_2HPO_4	18.75
Glucose	125	$Na_2HPO_4 \cdot 12H_2O$	5.30
Levulose	125	$MgSO_4 \cdot 7H_2O$	7.80
Fat-soluble vitamins		$CaCl_2 \cdot 2H_2O$	1.81
Vitamin A	3	$CoCl_2 \cdot 6H_2O$	0.04
Vitamin E		Minor salts (chelates)	
Water-soluble vitamins		NaFe	2.00
Folic acid	0.34	Na_2Mn	1.00
Inositol	6.00	Na_2Cu	0.16
p-Aminobenzoic acid	0.50	Na_2Zn	0.24

generations of *T. urticae* maintained on this diet. The time of development was approximately twice as long as on bean leaves, while egg production of the first generation was only about 1.5 eggs per female per day, in contrast to *Phaseolus* leaves, on which egg production may be as high as 7—8 eggs per female per day. In a few instances, mites could be grown for more than two successive generations on the diet, but usually a third generation was not possible as hardly any adult females were produced. This could be due to poor insemination power of the males, resulting in only haploid eggs being deposited, which give rise to males.

Autoclaving of the wheat germ casein fraction was necessary to allow completion of the development of the mites on this diet, as the life cycle was not completed on diets to which non-autoclaved wheat germ was added. It was not necessary that the mites fed on the solid wheat germ particles: the supernatant of the autoclaved wheat germ suspension proved sufficient. It seems likely that lipids present in wheat germ become available to the mites by the treatment, and serve as precursors for hormones which regulate ecdysis.

In Table 1.5.6.1 is given the composition of a holidic diet, developed by Kantaratanakul and Rodriguez (1979a). This is at present the best holidic diet available, although spider mites cannot complete their development on it. Table 1.5.6.2 gives the composition of the meridic diet of Bosse et al. (1981), on which *T. urticae* can be successfully reared for at least two successive generations.

Chapter 1.5.6. references, p. 389

TABLE 1.5.6.2

Composition of a meridic diet for the spider mite *Tetranychus urticae* Koch (Bosse et al., 1981)

Constituent	mg/100 ml diet	Constituent	mg/100 ml diet
Amino acids		Vitamins	
L-Alanine	4	*para*-Aminobenzoic acid	0.1
β-Alanine	4	Aneurine-HCl	0.1
L-Arginine-HCl	12	L-Ascorbic acid	50
L-Asparagine $\cdot$ H_2O	26	D-Biotin	0.005
L-Aspartic acid	8	Choline chloride	3
L-α-Amino-n-butyric acid	2	Cyanocobalamine	0.003
γ-Amino-n-butyric acid	3	Folic acid	0.005
DL-Carnithine-HCl	3	*meso*-Inositol	3
L-Citrulline	3	Nicotinamide	0.5
L-Cysteine-HCl	4	(+)-Pantothenic acid	
L-Cystine	3	(calcium salt)	0.3
L-Glutamic acid	4	Pyridoxine-HCl	0.3
L-Glutamine	12	Riboflavin	0.05
L-Glycine	3	Minerals	
L-Histidine	12	$CaCl_2 \cdot 2H_2O$	2
L-Homoserine	3	$CoCl_2 \cdot 6H_2O$	0.05
L-Hydroxyproline	3	$CuCl_2 \cdot 2H_2O$	0.08
L-Isoleucine	12	$FeCl_3 \cdot 6H_2O$	0.6
L-Leucine	8	$K_2HPO_4 \cdot 3H_2O$	15
L-Lysine-HCl	8	$MgSO_4 \cdot 7H_2O$	8
L-Methionine	12	$MnCl_2 \cdot 4H_2O$	0.2
L-Ornithine-HCl	8	$ZnCl_2$	0.4
L-Phenylalanine	8		
L-Proline	7	Other components	
L-Serine	3	Cholesterol (water-	
L-Threonine	7	soluble)	10
L-Tryptophan	4	D-Glucose	1000
L-Tyrosine	8	Casein (sodium salt)	2000
L-Valine	8	Wheat germ (powdered)	2000

STUDIES ON FEEDING RESPONSES

In a series of experiments, Dabrowski studied the gustatory effect of homogenates of leaves of different plant species and of solutions of chemicals from different groups on the two-spotted spider mite *T. urticae*. Labelled homogenates and solutions were offered to the mites in Parafilm sachets and ingestion was determined by measuring the radioactivity of the mites after a certain period of feeding. Distinct differences were found in the gustatory response to 4 different plant extracts. Ingestion of extracts of *Fragaria grandiflora* Ehrh. and of *Phaseolus vulgaris* L. was considerably higher than that of extracts of *Nicotiana tabacum* L. and of *Gingko biloba* L. (Dabrowski, 1973a). It was assumed that the last 2 extracts possessed a feeding suppressant effect. In other studies, the effect of amino acids, carbohydrates and water-soluble vitamins was investigated on the feeding response of the spider mites (Dabrowski, 1973b, c, 1974). Phagostimulatory effects were found for a number of amino acids and sugars, but the concentration in which they were tested was often higher than that present in leaves. The stimulatory effect of vitamins was only found to a slight degree for folic acid, vitamin B_{12} and inositol. It is very questionable whether one can use these data for the analysis of the feeding response of spider mites. Polyphagous insects and mites do not depend for their feeding on the presence of one or a few chemicals in their food, as oligophagous organisms do. Acceptance of food, however, is a result of all nutrients present in the diet and of their mutual ratios.

Dabrowski and Rodriguez (1972) also investigated the effect of 14 different phenolic compounds, known to occur in strawberry leaves, on the feeding response of *T. urticae*, as these compounds were thought to play a role in the resistance of certain varieties of strawberry to spider mites. Nearly all compounds showed a repellent or deterrent effect on the mites. The observations that *T. urticae* fecundity decreases with increasing concentration of phenols (as is the case in older leaves of *Fragaria*) were in agreement with the results of this study.

CONCLUDING REMARKS

Little information is available on the nutritional requirements of spider mites, despite research on this subject for over > 20 years. The main reason is that no holidic diet is yet available on which spider mites can complete their life cycle, in contrast to aphids, which have been grown on holidic diets for many years. This is ascribed to the presence of symbionts in aphids, which do not seem to play a role in spider mites (Sologic and Rodriguez, 1971). Data on the nutritional essentiality of nutrients can, therefore, only be obtained by indirect means, as has been done by Rodriguez and Hampton (1966) for amino acids.

The bottleneck in the development of a satisfactory holidic diet for spider mites is the first moult from larva to protonymph. Deficiency in hormone balance of the larva seems a logical explanation of this phenomenon, and may be caused by the absence of an essential, probably lipoid precursor of the moulting hormone(s) in the diet or by a form of such a precursor which is inaccessible for the mites.

A meridic diet has recently become available, on which *T. urticae* cultures can be grown for more than one generation. Such a diet cannot be used to establish the nutritional requirements of spider mites, but it may prove to be of great value for the study of the physiological effects of substances on spider mites.

REFERENCES

Bosse, Th.C., Van der Geest, L.P.S. and Veerman, A., 1981. A meridic diet for the two-spotted spider mite *Tetranychus urticae* Koch (Acarina: Tetranychidae). Meded. Fac. Landbouwwet. Rijksuniv. Gent, 46: 499—502.

Dabrowski, Z.T., 1973a. Studies on the relationships of *Tetranychus urticae* Koch and host plants. II. Gustatory effect of some plant extracts. Pol. Pismo Entomol., 43: 127—138.

Dabrowski, Z.T., 1973b. Studies on the relationships of *Tetranychus urticae* Koch and host plants. III. Gustatory effect of some amino acids. Pol. Pismo Entomol., 43: 309—330.

Dabrowski, Z.T., 1973c. Studies on the relationships of *Tetranychus urticae* Koch and host plants. IV. Gustatory effect of some carbohydrates. Pol. Pismo Entomol., 43: 521—533.

Dabrowski, Z.T., 1974. Studies on the relationships of *Tetranychus urticae* Koch and host plants. V. Gustatory effect of water-soluble vitamins. Pol. Pismo Entomol., 44: 359—372.

Dabrowski, Z.T. and Rodriguez, J.G., 1972. Gustatory responses of *Tetranychus urticae* Koch to phenolic compounds of strawberry foliage. Zesz. Probl. Postepow Nauk Roln., 129: 69—78.

Ekka, I., Rodriguez, J.G. and Davis, D.L., 1971. Influence of dietary improvement and egg viability of the mite *Tetranychus urticae*. J. Insect Physiol., 17: 1393—1399.

Fritzsche, R., 1960. Morphologische, biologische und physiologische Variabilität und ihre Bedeutung für die Epidemiologie und Bekämptung von *Tetranychus urticae* Koch. Biol. Zentralbl., 79: 521—576.

House, H.L., 1974. Nutrition. In: M. Rockstein (Editor), The Physiology of Insects, Vol. 5. Academic Press, New York, NY, pp. 1—62.

Hussey, N.W. and Huffaker, C.B., 1976. Spider mites. In: V.L. Delucchi (Editor), Studies in Biological Control. Cambridge University Press, Cambridge, pp. 179—228.

Kantaratanakul, S. and Rodriguez, J.G., 1979b. Nutritional studies in *Tetranychus urticae*. (Koch) (Acarina, Tetranychidae). I. Development of a chemically defined diet. Int. J. Acarol., 5: 83—92.

Kantaratanakul, S. and Rodriguez, J.G., 1979b. Nutritional studies in *Tetranuchus urticae*. II. Development of a meridic diet. In: J.G. Rodriguez (Editor), Recent Advances in Acarology, Vol. 1. Academic Press, New York, NY, pp. 405—411.

Kasting, R. and McGinnis, A.J., 1962. Nutrition of the pale western cutworm, *Agrotis orthogonia* Morr. (Lepidoptera, Noctuidae). Amino acid requirements determined with glucose-U-C^{14}. J. Insect Physiol., 8: 589—596.

Rodriguez, J.G., 1964. Nutritional Studies in the Acarina. Acarologia, Facsimile Hors Ser., 6: 324—337.

Rodriguez, J.G., 1969. Dietetics and nutrition of *Tetranychus urticae* Koch. Proc. 2nd Int. Congr. Acarol., 1967, pp. 469—475.

Rodriguez, J.G. and Hampton, R.E., 1966. Essential amino acids determined in the two-spotted spider mite *Tetranychus urticae* Koch (Acarina, Tetranychidae) with glucose-U-C^{14}. J. Insect Physiol., 12: 1209—1216.

Rodriguez, J.G., Singh, P., Seay, T.N. and Walling, M.V., 1967. Ingestion in the two-spotted spider mite, *Tetranychus urticae* Koch, as influenced by wavelength of light. J. Insect Physiol., 13: 925—932.

Sologic, D.S. and Rodriguez, J.G., 1971. Micro-organisms associated with the two-spotted spider mite *Tetranychus urticae*. J. Invertebr. Pathol., 17: 48—52.

Storms, J.J.H., 1965. Rearing methods for studying the effect of the physiological condition of the host plant on the population development of *Panonychus ulmi* (Koch). Boll. Zool. Agrar. Bachic., Ser. 2, 7: 79—85.

Storms, J.J.H. and Noordink, P.Ph.W., 1972. Nutritional requirement of the two-spotted spider mite *Tetranychus urticae* Koch (Acarina, Tetranychidae). Zesz. Probl. Postepow Nauk Roln., 129: 59—67.

Van der Geest, L.P.S., Bosse, Th.C. and Veerman, A., 1983. Development of a meridic diet for the two-spotted spider mite *Tetranychus urticae* Koch. Entomol. Exp. Appl., 33: 297—302.

Walling, M.V., White, D.C. and Rodriguez, J.G., 1968. Characterization, distribution, catabolism and synthesis of the fatty acids of the two-spotted spider mite, *Tetranychus urticae*. J. Insect Physiol., 14: 1445—1458.

1.5.7 Toxicological Test Methods

W. HELLE and W.P.J. OVERMEER

GENERAL REMARKS

Several test methods are in use for evaluation of the toxicity of pesticides to spider mites. It is impossible to recommend just one of these methods, since the methods very often serve different aims. In analytical studies on acaricide resistance, the precision and repeatability of the test results deserve high priority. Students may in that case decide in favour of methods in which the toxicant is applied directly to the mite, rather than to the host plant, and may use the topical application method (Harrison, 1961) or the slide-dip method (Voss, 1961; Dittrich, 1962). For screening purposes, these methods may be too sophisticated and laborious, and simple plant-spray, plant-dip or leaf-dip methods, such as those described by Ebeling (1960), Jeppson (1966) and Saba (1971) are most satisfactory. If special aspects, such as the residual effect, or the effect of the toxicant on developmental stages, are to be investigated, the leaf- or plant-dip method can be used.

With respect to the choice of test species, the two-spotted spider mite (*Tetranychus urticae* Koch) is by far the most convenient test subject, because it can be reared easily and tolerates a wide range of temperature and humidity. It accepts host plants of various kinds. Many other pest species of the genus *Tetranychus* are also appropriate test subjects. Species of the genera *Eutetranychus*, *Panonychus* and *Oligonychus* offer more difficulties as test animals. Rearing *Panonychus ulmi* (Koch) on potted Myrobalan root-stocks offers no difficulties in itself but contamination with *T. urticae* can be a serious nuisance since the latter species tends to take over from *P. ulmi*.

It should be taken into account that species may exhibit quite different susceptibilities to acaricides. The comparative responses of different spider mite species to various pesticides have been studied by Abo-El-Ghar and Boudreaux (1958).

Before describing some of the test methods most commonly used, it is important to take account of a number of rules which are essential for the reproducibility of the test methods. One should standardize the rearing of the host plant (the use of *Phaseolus vulgaris* L. is highly recommended) and be able to control the environmental conditions under which the plants are grown, in order to obtain uniform plant material. The same applies to the rearing of the mite material. One should avoid excessive temperatures (e.g. higher than $30°C$), since this may temporarily affect the intrinsic susceptibility of the test material. Long-day illumination is essential to prevent induction of diapause. The susceptibility of the summer females and the diapausing females of *T. urticae* are widely different, the latter being less susceptible to many acaricides (van Houten, 1976; Hurkova and Weyda,

1980). For the same reason, the mites (whether on plants or not) should be stored during the testing period in a controlled environment at a temperature of 20—27°C and a relative humidity of 60—90%.

Dip methods do not offer an accurate way of applying the pesticide; no more accurate, say, than spraying by hand. Also, the subsequent drying process is an important variable factor. It is therefore strongly recommended that the user of a dip method standardizes the different actions. Adding a detergent (wetter) to the solution is advisable under certain circumstances, especially when the pesticide is applied at concentrations well below field strength, in order to improve the spread of the pesticide on the substrate. If dipping can be replaced by spraying with a spray tower, the latter method is definitely preferred, since it allows more accurate dosing.

If a reference strain is maintained for toxicological studies over longer periods, it is strongly recommended that the reference strain is provided with a visible marker gene. This guarantees that contamination with other material is readily recognized. Busvine (1980) reported that in order to avoid contamination, susceptible colonies should always be kept in a separate room. J.E. Cranham (personal communication, 1984) suggested that if one has to mass-rear mites on plants to obtain required numbers, it is important to keep a 'nuclear' stock of the strain on detached leaf culture and material of this stock must be used to initiate plant cultures, rather than renewing plant cultures by simply putting across a few infested leaves from the old culture plants.

For the storage of treated leaves or leaf disks, several possibilities are appropriate. Disks can be floated on water or on a nutrient solution in a tray. Disks and leaves can be laid on wet filter paper, or on cotton wool. Cotton wool does not dry out as rapidly as filter paper, and gaps between the leaf and the substrate can be filled up by tearing the cotton threads. Free water may prove the most adequate barrier to prevent escape of mites, but it is easier to remove mites from a stationary leaf or leaf disk than from a floating one (Foott and Boyce, 1966). In order to keep the leaves in a good condition, the use of a nutrient solution is recommended. The formula for a nutrient solution is given in Chapter 1.5.1.

It is important to use homogeneous test material. Adult mites should preferably be of approximately the same age, within the range 1—8 days old; for ovicidal tests, eggs which are 0—2 days old are recommended.

TOPICAL APPLICATION

This method was designed by Harrison (1961). Each adult female mite is treated by expelling the insecticidal contents of a capillary over its ventral or dorsal surface. The mite is held in position on the fine end of a glass tube by suction. The capillary is self-filling with kerosene in which the toxicant is dissolved. The capillary is of a minute calibrated volume, approximately 0.0005 μl, being about 1/40 of the body weight of an adult female mite. Batches of 10 treated mites are collected under CO_2 anaesthesia in small perspex cages. Each perspex cage with treated mites is held to a bean leaf in such a way that the mites can feed on the leaf. Mortality is determined after 24 h.

By this method, the dosage of toxicant per mite is known. It has a high accuracy and repeatability, since the toxicant is not applied to a leaf surface, or to another variable substrate, but directly to the mite itself. Contrary to the slide-dip method, the method permits selection of survivors. The disadvantage is that the method requires considerable skill and is rather

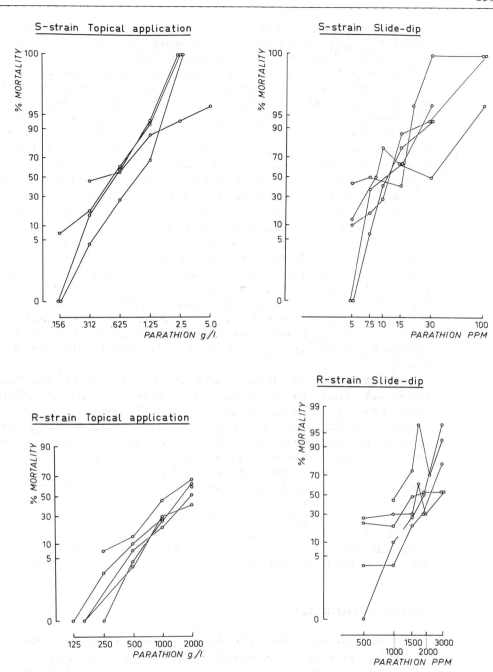

Figs. 1.5.7.1—4. Dose—mortality data obtained from experiments with parathion on a susceptible (S) and a resistant (R) strain. Each dot represents the mortality in 25—30 female adults. The dots belonging to the same experiment are connected by a line. Figs. 1.5.7.1 and 1.5.7.3 refer to the topical application technique; Figs. 1.5.7.2 and 1.5.7.4 refer to the slide-dip technique. Unpublished data of the Laboratory of Experimental Entomology, Amsterdam, 1967.

laborious. It is applicable to several acarine species and, according to Ballantyne and Harrison (1967), can be used satisfactorily for testing male spider mites. A comparison between the topical application technique and the slide-dip method or T. urticae is given in Figs. 1.5.7.1—4.

SLIDE-DIP METHOD

A small piece of double-sided Scotch tape is pressed onto a slide. Adult females are affixed to the tape on their dorsums. The slides are dipped in a toxicant solution and gently agitated for 5 s. When withdrawn they should be placed on edge on absorbant material and allowed to drain for 15 min. To ensure a uniform residue, excess of toxicant is blotted off the slide with filter paper. Slides with mites are stored in an incubator at 20—25°C and a high relative humidity, preferably higher than 90%. Mortality is checked 24 h later. Mites responding to prodding with a brush are considered to be alive. Cranham (1974) sprayed the toxicant solution on the slides, using a Potter tower; his modification of the slide-dip to a slide-spray method certainly improves precision.

The method is simple to operate and gives repeatable results (see Figs. 1.5.7.3 and 1.5.7.4). A disadvantage is that surviving mites are lost. The method is very suitable for testing adult *Tetranychus* females but is not so good for males. It is not very suitable for long-legged species (such as *Eutetranychus*) or species with long and tough dorsocentral setae (such as *Panonychus*).

In the recommended methods for the detection and measurement of resistance of spider mites by the FAO (1974), the slide-dip method is advised for both *Tetranychus* spp. and *P. ulmi* but it is the present authors' experience that *P. ulmi* cannot be properly affixed to the slides. Furthermore, the FAO recommended that the slides should be kept after treatment at a temperature of 27°C. This is a rather high temperature and may result in a high check mortality. Therefore, the present authors suggest working at a lower temperature. In a later FAO paper (Busvine, 1980), a leaf-residue test for adults of *P. ulmi* has been described.

ADULT LEAF-DISK TEST

This method is advised for spider mites not amenable to the slide-dip. Leaves are either dipped or sprayed in a spray tower to produce leaf residues. Disks are cut from the leaves and placed on wet cotton wool. Ten adult female mites are transferred to each disk, using three replicates for each treatment and the controls. Mortality is recorded after an appropriate period, depending on the speed of action of the acaricide. Mites that have walked off the disks are disregarded. In addition, the numbers of eggs produced may be recorded.

Clearly, this method can also be used for mites that are suitable for the slide-dip. The present method makes it possible to carry out selection experiments. Survivors can be isolated to propagate on fresh leaves.

OVICIDAL/OVOLARVICIDAL TESTS

Ovicidal tests in current use mostly involve leaves or leaf disks bearing eggs, which are dipped or sprayed. A method developed by Harrison and Smith (1961), involving the use of glass slides with eggs, is mentioned here but is not recommended. The following test, as described by Overmeer and Van Zon (1973), is advised. Adult females are placed on detached bean leaf cultures (or leaf disks) on cotton wool in small dishes (petri dishes or tin foil dishes) for 24 or 48 h to deposit eggs. Subsequently, the leaf cultures with eggs are sprayed with the pesticide, at the chosen concentrations, using a Potter tower, if possible. The detached leaf cultures are placed in trays with water and stored in cabinets at 25°C and a relative humidity of 85%, under long-day photoperiods. Mortality (eggs and larvae) is checked after 5 days.

Cranham (1968) used a similar method for *P. ulmi* with leaf disks of Myrobalan leaves held on wet filter paper. This method is recommended by the FAO (Busvine, 1980).

If no spraying tower is available, the leaves or leaf disks, with eggs attached, are dipped in the test liquid.

REFERENCES

Abo-El-Ghar, M.R. and Boudreaux, H.B., 1958. Comparative responses of five species of spider mites to four acaricides. J. Econ. Entomol., 51: 518—522.

Ballantyne, G.H. and Harrison, R.A., 1967. Genetic and biochemical comparisons of organophosphate resistance between strains of spider mites (*Tetranychus* species: Acari). Entomol. Exp. Appl., 10: 231—239.

Busvine, J.R., 1980. Revised method for spider mites and their eggs (e.g. *Tetranychus* spp. and *Panonychus ulmi* Koch). FAO method No. 10a. In: Recommended methods for measurement of pest resistance to pesticides. FAO Plant Prod. Prot. Pap., 21: 49—53.

Cranham, J.E., 1968. Laboratory determination of resistance to tetradifon in the fruit tree red spider mite, *Panonychus ulmi* (Koch). Rep. East Malling Research Station for 1967, pp. 165—168.

Cranham, J.E., 1974. Resistance to organophosphates in the red spider mite, *Tetranychus urticae*, from English hop gardens. Ann. Appl. Biol., 78: 99—111.

Dittrich, V., 1962. A comparative study of toxicological test methods on a population of the two-spotted spider mite (*Tetranychus telarius*). J. Econ. Entomol., 55: 644—648.

Ebeling, W., 1960. In: H.H. Shepard (Editor), Methods of Testing Chemicals on Insects. Burgess Publishing Company, Minneapolis, Ch. 11.

FAO, 1974. Recommended methods for the detection and measurement of resistance of agricultural pests to pesticides. Tentative methods for spider mites and their eggs, *Tetranychus* spp. and *Panonychus ulmi* Koch. FAO method No. 10. FAO Plant Prot. Bull., 22(5/6): 103—106.

Foott, W.H. and Boyce, H.R., 1966. A modification of the leaf-disk technique for acaricide tests. Proc. Entomol. Soc. Ont., 96: 117—119.

Harrison, R.A., 1961. Topical application of insecticide solutions to mites and small insects. N. Z. J. Sci., 4: 534—539.

Harrison, R.A. and Smith, A.G., 1961. The influence of temperature and relative humidity on the development of ovicides against *Tetranychus telarius* (L.) (Acarina: Tetranychidae). N. Z. J. Sci., 4: 540—549.

Hurkova, J. and Weyda, F., 1982. Response to pesticides of diapausing females *Tetranychus urticae* (Acarina, Tetranychidae). Vestn. Cesk. Spol. Zool., 46: 92—99.

Jeppson, L.R., 1966. Evaluating toxicity of chemicals to mites. Down Earth, 22: 19—20.

Overmeer, W.P.J. and Van Zon, A.Q., 1973. Genetics of dicofol resistance in *Tetranychus urticae* Koch (Acarina: Tetranychidae). Z. Angew. Entomol., 73: 225—230.

Saba, F., 1971. A simple test method for evaluating response to toxicants in mite populations. J. Econ. Entomol., 64: 321.

Van Houten, J.S., 1976. The effect of acaricides on active and diapausing females of *Tetranychus urticae* Koch, the two-spotted spider mite. Z. Angew. Entomol., 81: 248—252.

Voss, G., 1961. Ein neues Akarizid-Austestverfahren für Spinnmilben. Anz. Schaedlingskd., 34: 76—77.

General Index

Index to the Spider Mites

Index to the Predators